Biotechnological Production of Bioactive Compounds

Biotechnological Production of Bioactive Compounds

Edited by

Madan L. Verma
Department of Biotechnology, Dr Y. S. Parmar University of Horticulture and Forestry, Himachal Pradesh, India; Centre for Chemistry and Biotechnology, Deakin University, VIC, Australia

Anuj K. Chandel
Department of Biotechnology, Engineering School of Lorena (EEL), University of São Paulo, Estrada Municipal do Campinho, Lorena, São Paulo, Brazil

ELSEVIER

Elsevier
Radarweg 29, PO Box 211, 1000 AE Amsterdam, Netherlands
The Boulevard, Langford Lane, Kidlington, Oxford OX5 1GB, United Kingdom
50 Hampshire Street, 5th Floor, Cambridge, MA 02139, United States

Notices

Knowledge and best practice in this field are constantly changing. As new research and experience broaden our understanding, changes in research methods, professional practices, or medical treatment may become necessary.

Practitioners and researchers must always rely on their own experience and knowledge in evaluating and using any information, methods, compounds, or experiments described herein. In using such information or methods they should be mindful of their own safety and the safety of others, including parties for whom they have a professional responsibility.

To the fullest extent of the law, neither the Publisher nor the authors, contributors, or editors, assume any liability for any injury and/or damage to persons or property as a matter of products liability, negligence or otherwise, or from any use or operation of any methods, products, instructions, or ideas contained in the material herein.

Library of Congress Cataloging-in-Publication Data

A catalog record for this book is available from the Library of Congress

British Library Cataloguing-in-Publication Data

A catalogue record for this book is available from the British Library

ISBN: 978-0-444-64323-0

For information on all Elsevier publications visit our website at
https://www.elsevier.com/books-and-journals

Publisher: Susan Dennis
Acquisition Editor: Kostas KI Marinakis
Editorial Project Manager: Devlin Person
Production Project Manager: Vignesh Tamil
Cover Designer: Mark Rogers

Typeset by TNQ Technologies

Contents

Contributors

Ajay Bansal
Department of Chemical Engineering, Dr B R Ambedkar National Institute of Technology, Jalandhar, India

Anuj K. Chandel
Department of Biotechnology, Engineering School of Lorena (EEL), University of São Paulo, Estrada Municipal do Campinho, Lorena, São Paulo, Brazil

Vivek Chauhan
Department of Biotechnology, Himachal Pradesh University, Summer Hill, Shimla, India

Silvio S. da Silva
Department of Biotechnology, Engineering School of Lorena – University of São Paulo, Lorena, São Paulo, Brazil

Sarah de Souza Queiroz
Departamento de Biotecnologia, Escola de Engenharia de Lorena, Universidade de São Paulo, Lorena, SP, Brasil

Maria das Graças de Almeida Felipe
Departamento de Biotecnologia, Escola de Engenharia de Lorena, Universidade de São Paulo, Lorena, SP, Brasil

Balraj Singh Gill
Department of Higher Education, Shimla, Himachal Pradesh, India

Praveen Guleria
Department of Biotechnology, Faculty of Life Sciences, DAV University, Jalandhar, Punjab, India

Indarchand Gupta
Department of Biotechnology, Institute of Science, Aurangabad, Maharashtra, India

Andrés Felipe Hernández-Pérez
Departamento de Biotecnologia, Escola de Engenharia de Lorena, Universidade de São Paulo, Lorena, SP, Brasil

Avinash P. Ingle
Department of Biotechnology, Engineering School of Lorena, University of Sao Paulo, Lorena, SP, Brazil

Fanny Machado Jofre
Departamento de Biotecnologia, Escola de Engenharia de Lorena, Universidade de São Paulo, Lorena, SP, Brasil

Shamsher Singh Kanwar
Department of Biotechnology, Himachal Pradesh University, Summer Hill, Shimla, India

Suman Kapur
Birla Institute of Technology & Science, Pilani, Hyderabad Campus, Telangana, India

Rupali Kaur
Centre of Biotechnology, University of Allahabad, Prayagraj, India

Kaushal Kishor
Technology Research and Advisory, Aranca Pvt Ltd, Mumbai, India

Pankaj Kumar
Department of Biotechnology, Dr Y. S. Parmar University of Horticulture and Forestry, Neri Campus, Himachal Pradesh, India

Rakesh Kumar
Department of Biotechnology, Himachal Pradesh University, Summer Hill, Shimla, India

Santosh Kumar
Department of Biochemistry, University of Missouri, Columbia, MO, USA

Vineet Kumar
Department of Biotechnology, Faculty of Technology and Sciences, Lovely Professional University (LPU), Phagwara, Punjab, India

Sanjeev Kumar
Department of Basic Sciences, Dr Y. S. Parmar University of Horticulture and Forestry, Hamirpur, Himachal Pradesh, India

Rekha Kushwaha
Department of Biochemistry, University of Missouri, Columbia, MO, USA

Marcela O. Leite
Department of Biotechnology, Engineering School of Lorena – University of São Paulo, Lorena, São Paulo, Brazil

Moumita Majumdar
Department of Chemistry, National Institute of Technology, Agartala, Tripura, India

Gilda Mariano-Silva
Department of Biotechnology, Engineering School of Lorena – University of São Paulo, Lorena, São Paulo, Brazil

Fabiana B. Mura
Department of Biotechnology, Engineering School of Lorena — University of São Paulo, Lorena, São Paulo, Brazil

Navgeet
Department of Biotechnology, KMV College, Jalandhar, Punjab, India

Ayantika Pal
Department of Human Physiology, Tripura University, Agartala, Tripura, India

Feng Qiu
University of Missouri, Christopher S. Bond Life Sciences Center, Columbia, MO, USA

Mahendra Rai
Department of Biotechnology, Sant Gadge Baba Amravati University, Amravati, Maharashtra, India

Varsha Rani
Department of Biotechnology, Shoolini University Solan, Himachal Pradesh, India

Dijendra Nath Roy
Department of Bio Engineering, National Institute of Technology, Agartala, Tripura, India

Amaresh Kumar Sahoo
Department of Applied Sciences, Indian Institute of Information Technology, Allahabad, Allahabad, India

Raj Saini
Department of Basic Sciences, Dr Y. S. Parmar University of Horticulture and Forestry, Himachal Pradesh, India

Priya Sharma
Department of Biotechnology, Faculty of Life Sciences, DAV University, Jalandhar, Punjab, India

Sneh Sharma
Department of Biotechnology, Dr Y. S. Parmar University of Horticulture and Forestry, Neri Campus, Himachal Pradesh, India

Deepka Sharma
Department of Biotechnology, Dr Y. S. Parmar University of Horticulture and Forestry, Himachal Pradesh, India

Krishan D. Sharma
Department of Food Science and Technology, Dr Y. S. Parmar University of Horticulture and Forestry, Solan, Himachal Pradesh, India

Ishani Shaunak
Department of Biotechnology, Dr Y.S. Parmar University of Horticulture and Forestry, Solan, Himachal Pradesh, India

Sudhir Shende
Department of Biotechnology, Sant Gadge Baba Amravati University, Amravati, Maharashtra, India

Saurabh Shivalkar
Department of Applied Sciences, Indian Institute of Information Technology, Allahabad, Allahabad, India

Shailendra Kumar Singh
Centre of Biotechnology, University of Allahabad, Prayagraj, India

Salvador Sánchez-Muñoz
Department of Biotechnology, Engineering School of Lorena – University of São Paulo, Lorena, São Paulo, Brazil

Shanthy Sundaram
Centre of Biotechnology, University of Allahabad, Prayagraj, India

Meenu Thakur
Department of Biotechnology, Shoolini Institute of Life Sciences and Business Management, Solan, Himachal Pradesh, India

Priscila Vaz de Arruda
Universidade Tecnológica Federal do Paraná, Câmpus Toledo, PR, Brasil

Madan L. Verma
Department of Biotechnology, Dr Y. S. Parmar University of Horticulture and Forestry, Neri Campus, Himachal Pradesh, India; Centre for Chemistry and Biotechnology, Deakin University, Geelong Campus, VIC, Australia

Aruna Verma
Department of Biosciences, Himachal Pradesh University Shimla, India

Biographies

Dr. Madan Verma is a visiting researcher and professor in the Department of Biotechnology at Dr. Y. S. Parmar University of Horticulture and Forestry, Himachal Pradesh, India. Dr. Verma is an established senior researcher who has developed sustainable processes through his expertise in bioprocessing and nanobiotechnology. He employs nanotechnology approaches for enhancing the efficiency of various bioprocesses that have application in food biotechnology, pharmaceutical, and bioenergy sector. He has many international and national awards to his credits. Dr. Verma has published 46 research articles in peer-reviewed journals and 35 book chapters and and has edited 7 books on industrial biotechnology.

Dr. Anuj Chandel is a visiting researcher and professor in the Department of Biotechnology, Engineering School of Lorena, University of São Paulo, Brazil. He has over 18 years' experience working on process optimization and large-scale production of industrial enzymes and on the production of vaccine particles, biofuels, and membrane-based separation of fats, proteins, and viruses. His primary research interest is developing sustainable processes for bioconversion of lignocellulosics into renewable fuels and biochemicals by bridging the gap between research laboratories and industries. He has published 65 articles in peer-reviewed journals and 34 book chapters and has edited 10 books on various aspects of industrial biotechnology.

Preface

Presently, the prime focus of biotechnology research is to ameliorate the health benefits and basic nutrition values from the original source. With the recent advances in nanobiotechnology, enzyme biotechnology and purification technology, it is feasible to harness the natural extracts having bioactive compounds from the original source with the desired purity and yield. Human society is becoming health and calorie conscious, therefore preference for consumption for bioactive compounds has increased immensely in last two decades. For example, according to World Health Organization around 1.6 million people worldwide died due to diabetes in 2016 and number will rise to 629 million diabetics globally by 2045.

Natural extracts from plants, metabolites from microorganisms or marine based sources are a rich source of bioactive compounds and are reported to have healthier effects on prevention of cardiovascular diseases, mental disorders, inflammation, obesity and several other health issues.

Recently, researchers have employed novel methods for the purification and extraction of bioactive compounds with desired purity and recovery. Now extraction methodologies for bioactives production with improved efficiency has revamped with the advancement in nanobiotechnology and analytical molecular chemistry based techniques. At the outset of new nano-biotechnology applications, there is a pressing need to explore this area and thus seeks special attention to publish the advancement in this area at this high time. The present book is an attempt to bring together leading scientists to contribute review articles that cover the focal theme of all aspects of Bioactive compounds production and extraction, Enzyme technology, and Nano-technological interventions of bioactive compounds production and recovery.

The first chapter summarizes various techniques and their applications in the extraction and production of bioactive compounds from natural sources. The second chapter emphasis on the recovery and utilization of bioactives from food processing waste. The third chapter discusses the physical and chemical methods employed for flavonoid extraction and various applications in the biotechnology sector. The fourth chapter highlights the bioactive peptides, their sources, production techniques, major uses and biotechnological applications. The fifth chapter is focused on the recent progress made in the field of enhancing the production of stevioside through biotechnological interventions. The sixth chapter addresses the biotechnological applications of health promising bioactive molecules present in vegetable crops and medicinal herbs/plants. The seventh chapter summarizes the various recent studies employed to enhance the yield and productivity of two high valued algal compounds including astaxanthin and lutein under various growth conditions. The eighth chapter provides an overview of the biologically active compounds, extraction of algal biomass and their potential role for the human welfare, in a structured way. The ninth chapter presents the classification of sweeteners, microbial production of sweeteners, commercial outlook and demand, health effects and regulations for consumption. The 10th chapter discusses the recent advances made in the bioprocessing of oleaginous microalgae to omega-3 fatty acid production. The 11th chapter describes the potential of filamentous fungi, yeast and bacteria for the production of biopigments, and their potential applications. The 12th chapter highlights the mechanism(s) of cytotoxicity of RNases toward the cancerous cells, which make(s) RNases prominent chemotherapeutic or anticancer agents. The thirteen chapter discusses the nanoparticles synthesized using greener methods that significantly enhances the pharmacological properties of biomolecules. The 14th chapter presents the significance of nanotechnology in bioactive compounds. The chapter focuses on the health benefits of bioactive compounds, their bio-accessibility and bioavailability. The last 15th chapter describes the procedure of enhancement of bioactivity of compound by using nanotechnology.

The editors sincerely thank all the authors for their outstanding efforts to provide state of-the-art information on the subject matter of their respective chapters. We thank several reviewers who evaluated the manuscripts and provided critical suggestions to improve these further. We hope that this book will provide information about the latest research and advances, especially the innovations in biotechnological production of bioactive compounds. We wish to thank everyone involved in making possible publication of the book in a timely manner, especially Dr. Kostas Marinakis, Senior Book Acquisition Editor, Devlin Person, Editorial Project Manager and Vignesh Tamil, Project Manager, Reference Content Production, Elsevier Inc. It is our hope that this book will be useful to the students and researchers from both academia and industry.

Editors
Dr. Madan L. Verma
Dr. Anuj K. Chandel

CHAPTER

Technologies for extraction and production of bioactive compounds

1

Balraj Singh Gill[a,1], Navgeet[b,1], Feng Qiu[c]

Department of Higher Education, Shimla, Himachal Pradesh, India[a]; Department of Biotechnology, KMV College, Jalandhar, Punjab, India[b]; University of Missouri, Christopher S. Bond Life Sciences Center, Columbia, MO, USA[c]

Chapter outline

[1] Contributed equally.

Biotechnological Production of Bioactive Compounds. https://doi.org/10.1016/B978-0-444-64323-0.00001-1

1. Introduction

Continuous alteration in the behavior of numerous genes consequences in the modification of genetic material. These genetic changes may be beneficial or harmful depending upon modification of the particular gene targeting signaling pathway (Gill et al., 2016a,b). Positive or beneficial changes result in progressive evolution, whereas harmful alteration may cause some dreadful diseases and even causes death. These diseases are treated with surgery, chemotherapy, radiotherapy, and other modern techniques using natural or synthetic drugs (Joshi et al., 2015). Synthetic drugs potentially curb these dreadful diseases but cause some side effects. These side effects can be minimized by introducing new and better bioactive compounds which fit best to the moiety. The food crisis is another common problem in underdeveloping and developing countries. Natural bioactive compounds can not only provide a platform for drug discovery but also remediate the food crisis. Natural bioactive compounds interact with biological molecules such as proteins, DNA and others to yield the desired outcome. Therefore, both food and pharmaceutical industry have strong interest to characterize new bioactive compounds for developing nutraceuticals, functional foods, and therapeutic agents. Natural compounds are biosynthesized in conjugation or in combination with another form in small quantity (Altemimi et al., 2017). Therefore, concentration and purification of these conjugated products requires repeated separation of complex extracts into individual bioactive compounds and this process is time-consuming and tedious which makes it unprofitable for industries. Various modern techniques have been developed for extraction and production of bioactive compounds from natural sources such as plants and marine algae. For example, natural compounds can be efficiently purified with much lower cost using metabolic engineering and other molecular biological techniques (Sweetlove et al., 2017).

2. History of bioactive compounds

Plants have been used since the ancient time for nutritional value, and with the passage of time their medicinal value has been glorified. Natural products have been used to prevent and treat various diseases for a long history (Gill et al., 2017). In the Greek and Roman era, renowned scholars described the use of various herbal plants in routine (Paulsen, 2010). The historians discovered the use of coriander and castor oil in various recipes, medicinal applications, and remedies by Egyptian (Vinatoru, 2001). Plants produce two types of metabolites namely primary and secondary metabolites. Primary metabolites mainly include sugars, amino acids, fatty acids, and nucleic acids primarily used for growth and development. On the other hand, secondary metabolites are synthesized in a particular phase of development at a particular time. Secondary metabolites are important for the interaction between plants and environment and enhance the overall potential of plants for better

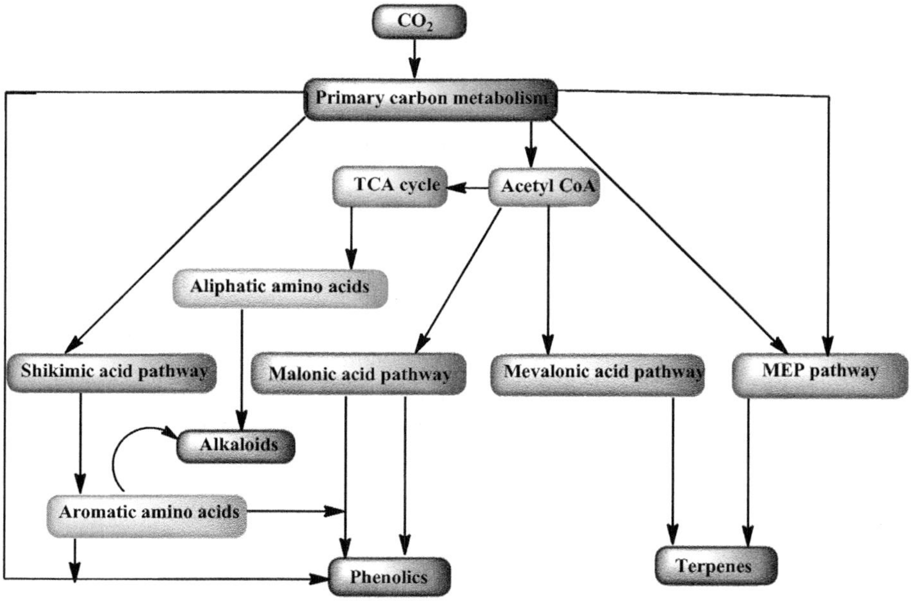

FIG. 1.1

The biosynthetic pathways of secondary metabolites.

survivability. For example, plants synthesize aroma to attract insects and other pollinators during pollination (Gill et al., 2016a,b). In allelopathy, plants secrete chemicals to inhibit the growth of surrounding plants. Secondary metabolites are classified into three main classes represented in (Fig. 1.1). Many secondary metabolites are bioactive compounds exhibiting pharmaceutical effects.

3. Synthesis of bioactive compounds

Secondary metabolites are classified into three main classes namely (a) terpenes/terpenoids, (b) alkaloids and (c) phenolic compounds. The four major pathways for the biosynthesis of secondary metabolites are shikimic acid pathway, malonic acid pathway, mevalonic acid pathway, and non-mevalonate (MEP) pathway (Fig. 1.1).

The majority of bioactive compounds belong to one of the abovementioned families, each of which has particular structural characteristics arising from the way in which they are built up in nature (biosynthesis). There are four major biosynthetic pathways for the secondary metabolites: (1) shikimic acid pathway, (2) malonic acid pathway, (3) mevalonic acid pathway and (4) non-mevalonate (MEP) pathway (Taiz and Zeiger, 2006). Terpenes are synthesized through mevalonic acid and MEP pathways, whereas phenolics are synthesized through malonic acid or shikimic acid pathway. Aromatic amino and alkaloids are synthesized through the shikimic acid pathway.

4. Source of bioactive compounds

Bioactive compound have been extracted from various sources including plant tissues, marine organisms, algae, and microorganisms. The isolated compounds include terpenes, phenolics, alkaloids, lipids, carbohydrates, peptides, and proteins.

4.1 Plant tissues

Plants possess various medicinal properties such as antimicrobial, antioxidative, antifungal, anti-inflammatory, and antiparasitic effects. These medicinal properties are due to plant secondary metabolites which are thought to enhance plant adaptability to prevailing conditions. Specific compounds are synthesized in plants at a particular time for better growth and development (Gill, 2013). Various compounds such as carotenoids, terpenoids, alkaloids, phenylpropanoids, vinblastines, and vincristines play important roles in plant growth and defense system (Nobili et al., 2009). Phenolics are another important class of natural products and widely distributed in plants particularly in leaves and stems. The pivotal role of phenolics ranges from natural defense to better adaptation and survivability. These functions result from physiological and genetic variations which are critical for the evolution (Figueiredo et al., 2008). Secondary metabolites possess a wide range of therapeutic effects and form a large candidate pool for drug discovery. In the last 25 years, around 60–70% of approved drugs were derived from the scaffold of plant secondary metabolites (Newman and Cragg, 2012).

4.2 Marine system

The marine system comprises of unique bioactive compounds which are important for pharmaceutical and food industry. The marine inhabitants serve as excellent models to study the process of nerve transmission. The extraction and characterization of marine bioactive compounds is critical to understand the marine life adaptation and predator-prey relationship. Marine organisms and their byproducts are important sources for fertilizer, fish oil, fish meat, pet food, and fish silage (Agatonovic-Kustrin et al., 2018). In addition, other marine products such as fish muscle proteins, collagen, and gelatin, fish bone, shellfish and crustacean shells are valuable sources of bioactive compounds. For example, peptides from fish protein hydrolysates have shown immunomodulation, antioxidative, and antihypertensive properties. The bioactivity and separation of biomolecules depend upon the molecular mass of peptide fragments (Rajapakse et al., 2005). Fish skin waste acts as a good source of collagen and gelatin and used extensively in food and pharmaceutical industry. The products derived from bovine and gelatin (bovine-derived) are useful to treat mad cow disease and bovine spongiform encephalopathy. Collagen is usually extracted with acid treatment and commonly used in pharmaceutical industry, especially microfibrous collagen in cancer treatment (Lee et al., 2001). In addition, fish oil is rich in omega-3 fatty acids which can potentially enhance immunity and cardiovascular health. These acids are found to reduce the risk of thrombosis (Von Schacky, 2000), decrease serum triglyceride levels, improve the functioning of vascular endothelial, and decrease the level of blood pressure (Kris-Etherton et al., 2003). Rajaganapathi et al. isolated an anti-HIV protein called Bursatellanin-P from the purple fluid of sea hare which exhibited the property of resistance to digestion under the influence of proteinase K and mercaptoethanol (Rajaganapathi et al., 2002). Furthermore, high phenolic content in alga *Corallina pilulifera* was effective to inhibit free radicals by reducing the expression of UV-induced MMP-2 and -9 in human dermal fibroblast (Ryu et al., 2009).

4.3 Microorganisms

More than 2300 bioactive compounds have been isolated from microorganisms (Olano et al., 2008). These compounds exhibit various biological functions. For example, bioactive compounds found in fungi such as *Penicillium, Aspergillus* and *Streptomyces* provide protection against lethal photooxidation in environmental stress and act as cofactors in many enzyme reactions (Mapari et al., 2005). These compounds have been used for developing antibiotics, enzymes, and organic acids (Liu, 2013). Bioactive compounds from microorganisms have been also used as flavor enhancers, preservatives, emulsifiers, and food supplement. The extraction of bioactive compounds from microorganisms often has high yields and microbial genes are more readily manipulated. However, only 1% of microorganisms have been exploited for searching bioactive compounds (Demain, 2000).

4.4 Algae and microalgae

Algae are a heterogeneous group of photosynthetic organisms with simple reproductive organs. There are more than 30000 microalga species on earth, from which more than 15000 bioactive compounds under extreme environmental conditions have been characterized (Mehra et al., 2018). These environmental variations such as light, salinity, and temperature increase the survivability of the microalga resulting in the production of new secondary metabolites for defense system (Rodríguez-Meizoso et al., 2010). Most of the algae are easy to cultivate on large scale for industrial use purposes. The compound bioactivities are dependent of various parameters such as molecular weight, solubility, heat resistance and the process of extraction and identification. Importantly, green chemistry technologies allow the extraction of target compounds without using organic toxic solvents (Ibañez et al., 2012).

4.5 Metabolic engineering (ME)

Metabolic engineering is the process of alteration of metabolic pathways to produce the metabolites of interest for the use in chemical, energy, food, and pharmaceutical industry. Genetic materials control and modulate the production of bioactive compounds (Negi et al., 2014). The modulation results in overexpression or downregulation of genes which are responsible for the synthesis of target metabolites. The approaches for genetic engineering mainly include gene insertion or deletion, heterogonous expression of gene clusters, and alteration in gene expression (Weber et al., 2015). Metabolic engineering has been used to modulate the synthesis of volatile organic compounds in plants which protect plants from herbivore and enhances their defense system. Metabolic engineering also assists in pollination system and seed dispersals to enhance the production of volatile products in plants (Dudareva et al., 2013). For example, metabolic engineering has been used to modulate several genes which are responsible for the synthesis phenylalanine-derived volatile compounds (Peled-Zehavi et al., 2015).

5. Methods for extracting bioactive compounds

The extraction and bioactivity screening of natural compounds comprises of several steps. First, raw materials are selected based on their nutritional or medicinal effects. The selected materials are checked for toxicity with standard protocols. Then, elemental analysis is carried out to determine the chemical composition and potential bioactivities of the materials (Gill et al., 2016). The compounds are isolated from the crude extracts and their potential activities are tested *in vitro* and *in vivo* (Gill et al., 2017). Finally, the bioactive compounds are commercialized into medicinal products and proved to be fruitful in curbing various diseases (Fig. 1.2).

5.1 Solvent extraction

Solvents extraction is one of conventional methods for extracting compounds from bacteria, algae, fungi, and plants. Raw material is often ground into powder form to increase the extraction efficiency. Both polar and nonpolar solvents have been used for the extraction, such as ethanol, ether, chloroform, hexane, benzene, and water as well as their combinations in different ratios (Negi and Gill, 2013). This technique has been widely used due to its easy availability and low cost. However, some of organic solvents are highly toxic and/or flammable and a large amount of solvents are often used during processing and extraction. Thus, users should follow appropriate handling procedures to ensure safety and environmental compliance. It is important to note that the organic solvents may cause thermal degradation of bioactive compounds (Teo and Idris, 2014). Furthermore, the extraction process is time-consuming and labor-intensive. To overcome these issues, other advanced methods have been developed such as soxhlet, ultrasound, and microwave extraction which are mentioned below.

5.2 Nonconventional methods

The conventional extraction methods have certain limitations which have prompted researchers to develop more advanced methods such as supercritical fluid extraction,

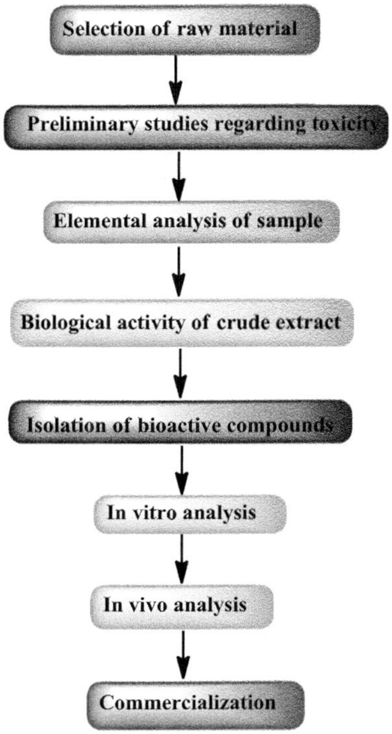

FIG. 1.2

Schematic representation of extraction and commercialization of bioactive compounds.

microwave-assisted extraction, ultrasound-assisted extraction, pressurized liquid extraction, and pulsed-electric field extraction (Gill & Kumar, 2015). These new methods have reduced the extraction time and prevented thermal decomposition of compounds (De Castro and Garcıa-Ayuso, 1998).

5.3 Green extraction techniques

Conventional extraction methods often use large amount of organic solvents which pose the risk of chemical exposure to the environment (Gill et al., 2016). The concept of green chemistry has been introduced to reduce the chemical hazard and limit their use and exposure to the environment. Green chemistry also promotes the sustainability and allows us to exploit the nature without depleting the environment. This concept has been incorporated in various chemical processes including synthesis, catalysis, separation, and monitoring. The main aspects behind green chemistry are waste, energy, and hazard which dominate among the twelve principles laid by Anastas and Warner (Anastas and Warner, 1998).

5.4 Supercritical fluid extraction (SFE)

Hannay and Hogarth first introduced supercritical fluid (SFE) in 1879, as an alternative extraction method, whereas detailed study was conducted until 1960s (Hosikian et al., 2010) Elaborative study was done by Zosel in which he used SFE to decaffeinate coffee beans (Zosel, 1964). Supercritical state of a solvent is achieved when temperature and pressure is beyond its critical point at which the solvent possess both gas and liquid-like properties such as diffusion, surface tension, viscosity, density and salvation (Sihvonen et al., 1999). This principle of extraction is based on various parameters such as fluid, density, the viscosity of the solvent above their critical points. SFE is an eco-friendly and highly selective method. Carbon dioxide (CO_2) is the most common solvent for SFE and other solvents are also used, such as ethylene, methane, nitrogen, xenon, or fluorocarbons (Daintree et al., 2008). The CO_2 is cost-effective and "Generally Recognized As Safe" (GRAS) in food industry. The supercritical CO_2 exhibits high diffusivity and ease with respect to temperature and pressure. During SFE, raw materials are kept in an extractor under controlled temperature and pressure. The dissolved material is transferred to a separator. The extracts are then collected from the separator and the regenerated fluid is released to the outer environment (Sihvonen et al., 1999). After the extraction, the system is depressurized to convert CO_2 from liquid into gas form. This technique has been used to extract bioactive compounds from macroalgae, microalgae, cyanobacteria and marine invertebrates (such as crustacean, crawfish, crab, or shrimp, squid, urchin, or starfish) (Wang and Weller, 2006). As mentioned earlier, plants release volatile and semivolatile compounds to assist their survival and food gathering. Many of these chemicals are terpenes which are used by plants to attack herbivores chemically (Quintero and Bowers, 2018). For example, *Dictyopteris membranacea*, which is a brown algae, releases terpenoids and sulfur-containing compounds (El Hattab et al., 2007). Volatile compounds released by microalgae assist in defense and have antifungal, antibacterial, and antiprotozoal properties (Mehra et al., 2018). For example, a bioactive compound isolated from *Dunaliella salina* through SFE exhibits antimicrobial activity against *Escherichia coli, Staphylococcus aureus, Candida albicans,* and *Aspergillus niger* (Macías-Sánchez et al., 2009). In addition, saponifiable (essential fatty acids) and unsaponifiable compounds were also extracted from *D. salina* using SFE. The supercritical CO_2 is commonly used to extract nonpolar components in SFE due to its low polarity. For example, omega-3 (w-3) fatty acids was efficiently extracted from an algae, *Hypnea charoide*, using SFE and CO_2 as solvent (Xu et al., 2015). Nonpolar compounds were also isolated from *Botrycoccus braunii*, such as long-chain hydrocarbons which can be used as a substitute of paraffin and natural waxes (Uquiche et al., 2016). However, SFE was less frequently used for the extraction of other secondary metabolites such as phenolics and isoflavones. To expand the applicability of SFE, Klejdus et al. developed a new hyphenated technique for the extraction of isoflavones from macroalgae by supercritical CO_2 extraction followed by fast chromatography using 3% (v/v) of MeOH/H_2O as solvent (Klejdus et al., 2010).

The waste products from a marine organism such as fish head, tail, blood, skin, and other parts contain useful components such as fish oil or PUFA from microalgae. These waste or by-products have been used for the extraction of omega-3 fatty acids using SFE at pressure 300 bar and temperature 75 °C. The extracts contained 10.95% EPA and 13.01% DHA (Rubio-Rodríguez et al., 2012). There are many other applications of SFE such as coffee decaffeination, phenol and flavonoid extraction, fatty acid refining, nutraceutical and functional food preparation. The yield of SFE depends upon various operational conditions and parameters, such as raw materials, solvents, pressure, and temperature, which need be optimized for the components to be extracted (Tables 1.1–1.4).

5.5 Microwave-assisted extraction (MAE)

Microwave-assisted extraction (MAE) was first explained by Ganzler and his team in 1986 (Ganzler et al., 1986). MAE is a method for extracting products in solvents using microwave energy. MAE uses electromagnetic radiations with a wavelength from 0.001 m to 1 m. It is composed of two oscillating field which is perpendicular to the magnetic and electric field. Microwave is converted into thermal energy which causes direct impacts on polar materials by exerting pressure on cell wall (Letellier and Budzinski, 1999). This conversion is the result of ionic conduction and dipole rotation causing resistance to flow. The pressure developed inside the cells modulates the expression of genes responsible for the physical and biological properties of cells. The extraction process comprises of different phases such as separation of solute under high temperature and high pressure from raw materials, diffusion of solvent across matrix and discharge of solute into solvent from matrix (Alupului et al., 2012). MAE heats the matrix both externally and internally and thus, reduces thermal gradient. MAE is more advantageous than the conventional Soxhlet method because MAE reduces equipment size, uses less solvent, and increases the productivity. Compared to other methods, the product yield was much higher while extracting caffeine and polyphenols at room temperature (Belwal et al., 2017). Many bioactive compounds were isolated from different pant species using MAE. For example, the phenolics such as kaempferol, quercetin, and their glycosides derivatives were efficiently extracted from the leaves of *Moringa olerifera* (Rodríguez-Pérez et al., 2016). MAE also enabled rapid extraction of pigments such as carotenoids from *Dunaliella* species due to non-availability of frustules in cells of microalgae (Akyil et al., 2018). Twelve phenolics were extracted from *Berberis asiatica* using MAE for the exploration of their potential to prevent the damage of erythrocytes and deoxyribonucleic acid. This study also discovered the higher concentration of chlorogenic and catechin than other phenolics present in the plant (Belwal et al., 2017). In addition, MAE increased the yield of an anti-inflammatory agent acetyl-11-keto-β-boswellic acid (AKBA) from *Boswellia serrata* while reducing the extraction time and solvent use (Niphadkar and Rathod, 2018).

Table 1.1 Examples of supercritical extraction of bioactive compounds.

S. No	Raw material	Extraction method	Bioactive compound	Use	References
1	Olive leaves	Subcritical water 3–11 MPa, 323–423 K 0–3.3 m^3/s	Mannitol	Pharmaceutical industry, Diabetic food products.	Ghoreishi and Shahrestani (2009)
2	Black tea, celery, ginseng leaf	Subcritical (110–200 °C), extraction time (5–15 min), and pressure (about 10 MPa)	Myricetin, quercetin, and kaempferol	Antioxidant and food industry	Cheigh et al. (2015)
3	*Kaempferia galangal* L.	120 °C, extraction time 20 min, extraction pressure 10 Mpa, ultrasonic power density 250 W/L	Essential oils	Antioxidants, food and pharmaceutical industry	Ma et al. (2015)
4	Basil and oregano	Subcritical water 100, 150, 200, and 250 °C 30 and 300 min	α-pinene, limonene, camphor, citronellol, carvacrol	Anti-inflammatory, antioxidant activities.	Yang et al. (2007)
5	Onion skin	Subcritical water 100–190 °C 5–30 min 90–131 bar	Quercetin	Anticancer, antivirus, and anti-inflammatory	Ko et al. (2011)
6	Rosemary	Subcritical water 25–200 °C	Carnosol, rosmanol, methyl carnosate, carnosic acid, cirsimaritin, genkwanin	Antioxidant	Ibanez et al. (2003)
7	*Coffea arabica* L.	(160–180 °C), time (35–55 min) and the solid-to-liquid ratio (14.1–26.3 g/L)	3-*O*-caffeoylquinic acid, 4-*O*-caffeoylquinic acid, and 5-*O*-caffeoylquinic acid	Antioxidant, anticancer and food industry	Xu et al. (2015)
8	Wet algae	220 °C and microwave heating	Proteins, omega-3 fatty acids, sugars	Biodiesel production	Reddy et al. (2014)
9	Sunflower seeds	130 °C, 30 min of extraction	Oil	Antioxidants and food industry	Ravber et al. (2015)
10	*Salvia officinalis* L.	201.5 °C, extraction time of 15.8 min	Total phenols and total flavonoids	Food industry and anticancer property	Pavlić et al. (2016)
11	Winery grape seeds	Subcritical water 50, 100, and 150 °C 1500 psi	Catechins and Proanthocyanidins	Pharmaceutical and food industry	Duba & Fiori (2015)
12	*Zingiber officinale*	130 °C/25 min, and 190 °C/15 min	6-gingerol and 6-shogaol	Antioxidant and food industry	Ko et al. (2019)

13	Apple and peach pomaces (Plant fruit)	Subcritical carbon dioxide + ethanol 20–60 MPa, 40–60 °C 14–20% EtOH and 10–40 min	Polyphenols	Health promoting effects	Adil et al. (2007)
14	Berberis Aristata roots	110–170 °C, the particle size of 0.65 mm, time of 70 min, using Microwave assisted subcritical water extraction	Isoquinoline Berberine	Food and Pharmaceuticals industry	Manikyam et al. (2017)
15	Sorghum bran	144.5 °C of temperature, 21 min of time, and 35 mL/g of the solid-liquid ratio	Taxifolin, taxifolin hexoside, oligomeric procyanidins, epicatechin	Antioxidant and medicinal use	Luo et al. (2018)
16	*Pistacia vera* L.	6.9 MPa Pressure, 110 and 190 °C, and a flow rate of 4 mL/min	Gallotannin, gallic acid, penta-*O*-galloyl-*β*-D-glucose, anacardic acid,	Antioxidants	Erşan et al. (2018)
17	*Lycium ruthenicum* Murr.	170 °C, 55 min, the flow rate was 3 mL/min	Anthocyanin	Antioxidants, food industry, the pharmaceutical industry	Wang et al. (2018)
18	*Panax ginseng* C.A. meyer	(150–200 °C) and extraction time (5–30 min)	Total phenolic content, maltol content, ginsenoside	Antioxidant and medicinal properties	Lee et al. (2018)
19	*Saccharina japonica*	(100 °C-150 °C), pressure (10–50 bar), water content (50%–70%), and liquid/solid (L/S) ratio (30–50 mL/g)	Alginate, fucoidan	Antioxidant	Saravana et al. (2018)
20	Chamomile	Isothermal conditions (100 °C) at five different pressures (10, 30, 45, 60 and 90 bar)	Polyphenols, apigenin	Antioxidant, food industry, medicinal use	Cvetanović et al. (2018)
21	*Citrus grandis* (L.) Osbeck	120 °C and 30 bar	Pectin	Food industry and antioxidant properties	Liew et al. (2018)
22	*Teucrium montanum*	160 °C and pressure of 10 bar.	Naringin and gallic acid	Pharmaceutical industry	Nastić et al. (2018)
23	Mangosteen pericarps	120–160 °C and 1–10 MPa	Xanthone and phenolic compounds	Antioxidants and medicinal use	Machmudah et al. (2018)

Continued

Table 1.1 Examples of supercritical extraction of bioactive compounds. *Continued*

S. No	Raw material	Extraction method	Bioactive compound	Use	References
24	Orange peel	Temperature (110–150 °C) and water flow rate (10–30 mL/min)	Hesperidin, narirutin	Antioxidants and food industry	Lachos-Perez et al. (2018)
25	*Sinopodophyllum hexandrum*	180 °C, 4 MPa, flow rate 2.5 mL/min	Podophyllotoxin	Pharmaceutical industry	Wang et al. (2018)
26	*Satureja montana*	Temperature of 200 °C and extraction time of 20.8 min.	Total phenols, flavonoids, thymoquinone	Antioxidants,	Vladić et al. (2017)
27	*Quercus ilex*	(130–170 °C) and reaction time (5–220 min)	Hemicelluloses	Medicinal use	Yedro et al. (2017)
28	*Allium ursinum*	(120–200 °C), extraction time (10–30 min) and acidifier, HCl (0–1.5%)	5hydroximethylfurfural (5-HMF) and furfural (F), kaempferol	Antioxidants	Tomšik et al. (2017)
29	*Morus nigra* L.	120 °C, 60 min, 2 mL/min	Phenolics, flavonoids, and anthocyanins	Pharmaceutical industry, medical use	Koyu et al. (2017)
30	*Phlomis umbrosa* Turcz	100 °C–200 °C, time (from 5 to 25 min)	Polyphenols, flavonoids	Antioxidants, industrial use	Ko et al. (2017)
31	*Camellia oleifera* Abel.	Temperature (110–150 °C), extraction time (20–40 min) and solvent to material ratios (5:1–15:1 mL/g).	Oil, tea saponin	Food industrial and pharmaceutical industry	Wu et al. (2018)
32	Carrot leaves	(110–230 °C), time (0–114 min), and solid-liquid ratio (15 and 35 g/L)	Phenolics, luteolin	Agriculture purpose and food industry	Song et al. (2018)

Table 1.2 Examples of microwave and ultrasonic-assisted extraction of bioactive compounds.

S. No	Method and conditions	Materials	Bioactive compounds	Use	References
1	UAE, MetOH 20%, 60 °C, 60 min	*Forsythia suspense*	Phillyrin	Anti-inflammatory, antioxidant, antiviral and vasorelaxant activities.	Sheng et al. (2012)
2	UAE, EtOH(30%–70%), 80 °C (40–80) min	*Allium ursinum* L.	Phenols, flavonoids,	Antioxidants, Anti-inflammatory	Tomšik et al. (2016)
3	UAE, EtOH 41%, 79 °C, 30.5 min	*Prunella vulgaris* L.	Flavonoids	Reduces sore throat, fever and wound healing.	Zhang et al. (2011)
4	UAE, MAE, 25% Water	Grapes	Phenolics	Antioxidant	Bubalo et al. (2016)
5	MAE, EtOH, water, 7 min in cycles of 30 s at 250 W	Rosemary leaves	Phenolics, rosmarinic carnosic Acids	Antioxidants and in food industry	Rodríguez-Rojo et al. (2012)
6	MAE, EtOH 42%, 500 W microwave power, 62 s	*Myrtus communis*	Phenolics	Antioxidants	Dahmoune et al. (2015)
7	UAE, Water, 45 min, 222 W	Litchi seeds	Polysaccharides	Antitumoral, antioxidant, hypoglycemic properties	Chen et al. (2011)
8	MAE, Acetone, 56 °C, 5 min at 50 W	*Dunaliella tertiolecta, Cylindrotheca closterium*	Chlorophyll *a*, *b* carotene	Health and food industry	Pasquet et al. (2011)
9	UAE, 20 min, 50 °C, 20 W	*Nephelium lappaceum* L.	Anthocyanin, phenolics, flavonoid	Antioxidants, healing power	Maran et al. (2017)
10	MAE, EtOH 80%, 65 °C, 300 W	Pigeon pea leaves	Cajaninstilbene acid, pinostrobin	Postmenopausal, osteoporosis	Kong et al. (2010)

Continued

Table 1.2 Examples of microwave and ultrasonic-assisted extraction of bioactive compounds. *Continued*

S. No	Method and conditions	Materials	Bioactive compounds	Use	References
11	MAE, EtOH 96%, 6 min, 450 W	*Citrus aurantifolia*	Pectin	Food industry	Megawati et al. (2017)
12	MAE, MetOH 60%, 500 W	*Berberis asiatica*	Phenolics, Chlorogenic acid, catechin	Antioxidant and food industry	Belwal et al. (2017)
13	MAE 120 W for 25 min, UAE 200 W for 80 min	*Momordica cochinchinensis* Spreng.	Carotenoid and antioxidant	Antioxidant, food industry	Chuyen et al. (2018)
14	MAE, 272 W, 4 min, EtOH 90%	*Boswellia serrata (gum)*	Acetyl 11 keto β boswellic acid	Anti-inflammatory agent	Niphadkar & Rathod (2018)
15	MAE, 90 °C, 10 min,	*Solieria chordalis*	Carrageenan	Antiviral activity	Boulho et al. (2017)
16	MAE, 250–350 W, 2–3 min	*Olea europaea*	Total phenolic content, oleuropein	Antibacterial activity	Şahin et al. (2017)
17	MAE, 60 °C, Hexane as a solvent, 15 min	Coconut copra residue	Oil, fatty acids, free fatty acids	Oil and food industry, Medicinal use	Hakimi et al. (2018)
18	MAE, 517 W, 2.15 min	*Opuntia ficus indica*	Galacturonic acid, Pectin-related proteins	Food industry and medicinal purpose	Lefsih et al. (2017)
19	MAE, 360–720 W EtOH 60–100%, 10 s	*Momordica cochinchinensis* Spreng	Saponins	Nutraceuticals and pharmaceutical industries	Le et al. (2018)
20	MAE, 70 °C, 4min, 1.3 mL EtOH, chlorobenzene as solvent	Litchi fruit	Pyrethroids	Pesticides and food industry	Wang et al. (2018)
21	Natural deep eutectic solvents (NADESs), MAE 63.68 °C, 17.08 min, water 32.19%	*Lippia citriodora*	Iridoids, phenylpropanoids, flavonoids	Antioxidants and food industry	Ivanović et al. (2018)
21	MAE, 4 min, EtOH 75%, 160 W	*Stevia rebaudiana*	Stevioside	Food industry and antidiabetic.	Ameer et al. (2017)
22	MAE, 40 °C, 5 min, EtOH	*Corydalis decumbens*	Alkaloids	Medicinal use and research work	Mao et al. (2017)

23	MAE, (50–150 °C), (15–75% EtOH), extraction time (5–20 min)	*Hibiscus sabdariffa*	Glycoside flavonoids, phenolics	Antioxidants, industrial use	Pimentel-Moral et al. (2018)
24	UAE,37.11 °C, vessel diameter 6 cm, solvent ratio 1:28.42 g/mL	*Rheum moorcroftianum*	Polyphenolic	Industrial use	Pandey et al. (2018)
25	UAE, 56 °C, 3min, 0.6s, solvent ratio 3.6%	Olive cake	Protocatechuic acid, cinnamic acid	Food industry	Mojerlou & Elhamirad (2018)
26	UAE, 5 min, 90 °C and 80% of ethanol	*Arbutus unedo* L.	Cyanidin-3-glucoside	Industrial application	López et al. (2018)
27	UAE, 76 °C, 10 min, amplitude 100%	*Laminaria digitata*	Fucose, Glucans	Food and industrial use	Garcia-Vaquero et al. (2018)
28	UAE, 60 °C, 15 (*v/w*),	*Olea europaea* L.	Oleuropein, verbascoside, luteolin-4′-*O*-glucoside	Industrial use	Giacometti et al. (2018)

Table 1.3 Examples of pressurized liquid extraction of bioactive compounds.

S. No	Method and conditions	Materials	Bioactive compounds	Use	References
1	PLE, water, ethanol and temperature 60, 80 and 100 °C	*Rubus fruticosus* L.	Total phenolics, monomeric anthocyanins, antioxidant activity	Antioxidants and food industry	Machado et al. (2015)
2	SE, methanol, acetone, and water	Bunga kantan inflorescence	Phenols, tannins, flavonoids, and anthocyanins	Antioxidant/natural antioxidants for food and nutraceuticals	Wijekoon et al. (2011)
3	PLE, hexane, acetonitrile UAE acetonitrile, hexane	Liver, muscle, kidney samples	Sulfonamides	Medicinal use	Hoff et al. (2015)
4	PLE, EtOAc in MeOH	*Oryza sativa*	Guaiacol, ellagic acid, vanillic acid and protocatechuic acid	Food industry	Setyaningsih et al. (2016)
5	MAE, 60 °C for 3 min of 0.3 g PLE 75 °C for 26.7 min	*Cynara scolymus* L.	Inositols and inulin	Industrial use	Ruiz-Aceituno et al. (2016)
6	PLE, Ethanol	*Schinus terebinthifolius* Raddi	Anthocyanins, biflavonoids	Antioxidants and food industry	Feuereisen et al. (2017)
7	PLE, 190 °C for 3 consecutive cycles	Olive leaves	Oleuropein content, antioxidant activity	Antioxidant, medicinal and pharmaceutical industry	Xynos et al. (2014)
8	PLE with effect of nitrogen, light intensity or carbon supplied	*Neochloris oleoabundans*	Carotenoids, lutein	Food industry, anticancerous	Castro-Puyana et al. (2017)

9	PLE, 192.4 °C and time of 15 min UAE 7 min, ultrasonic intensity of 301 W/cm^2	*Phyllanthus amarus*	Total phenolics, gallic acid, hydrolyzable tannins, flavonoids, and lignans	Antioxidants and food industry	Sousa et al. (2016)
10	PLE, Gas chromatography-tandem mass spectrometry (GC–MS/MS)	Cosmetics products	Methoxycinnamates, benzophenones, salicylates, *p*-aminobenzoic acid derivatives	Cosmetic industry	Vila et al. (2015)
11	Soxhlet extraction (LPSE–SOX), percolation (LPSE–PE) and pressurized liquid extraction (PLE) Ethanol, (35–55 °C), pressure (0.1–8 MPa) and flow rate (1.6, 2.4 g/min)	Pressed palm fiber	Carotenoids	Food purpose and medicinal use	Cardenas-Toro et al. (2015)
12	MAE Extraction time of 20 min, 42% ethanol, 158 °C PLE 128 °C, 35% of ethanol, and 20 min	*Moringa olerifera*	Phenolics, flavonoid, kaempferol diglycoside	Food and pharmaceutical industry	Rodríguez-Pérez et al. (2016)
13	PLE, MAE, capillary electrophoresis	*Padina pavonica*	Hyaluronidase	Cosmetic industry, artificial skin	Fayad et al. (2017)
14	PLE, 195 °C, water, ethanol	Coffee	Caffeine	Food and cosmetics industry	Shang et al. (2017)
15	Supercritical carbon dioxide and pressurized liquid (ethanol and water)	*Salvia*	Tocopherols, rosmarinic acid, Apigenin-7-*O*-β-D-glucuronide, caffeic, and carnosic acids	Functional foods and nutraceuticals	Šulniūtė et al. (2017)

Continued

Table 1.3 Examples of pressurized liquid extraction of bioactive compounds. *Continued*

S. No	Method and conditions	Materials	Bioactive compounds	Use	References
16	Supercritical carbon dioxide extraction, PLE ($P = 55–57$ MPa, $t = 120–131$ min, $T = 50$ °C and $v = 2.5$ L min^{-1})	*Viburnum opulus pomace* and berries	Oleic and linoleic fatty acids, tocopherols	Foods, nutraceuticals	Kraujalis et al. (2017)
17	Supercritical carbon dioxide, PLE at 140 °C	*Solidago virgaurea*	Tocopherol	Antioxidant and industrial purpose	Kraujalienė et al. (2017)
18	PLE, 180 °C and 86% ethanol in water	*Lycium barbarum* L.	Total phenolic content, flavonoid, antioxidant activity	Pharmaceutical and food industry	Tripodo et al. (2018)
19	PLE, 47.2 in methanol, 75.5 °C, 200 atm, 90 s, pH 3.01, and 50.2% for flushing	*Morus nigra* L	Anthocyanins, phenolic compounds	Pharmaceutical and antioxidant property	Espada-Bellido et al. (2018)
20	PLE, 50 °C, water-ethanol mixture 25:75 (% v/v)	*Ocimum basilicum* L.	Rosmarinic acid and caffeic acid	Nutraceutical, cosmetic, food applications.	Pagano et al. (2018)
21	PLE	Cocoa bean	Catechin, epicatechin, procyanidin B2, caffeine, theobromine	Fertilizers and animal feed	Okiyama et al. (2018)
22	PLE, 200 °C, 1:1 v/v as ethanol/water ratio,	Avocado peel	Phenolic compounds	Food, cosmetic, pharmaceutical	Figueroa et al. (2018)

23	PLE, hexane, ethyl acetate, acetone, ethanol and ethanol: water 50:50, 80 °C, 120 °C and 160 °C	Galician Algae	Antioxidant, antibacterial activity	Nutraceutical agents	Otero et al. (2018)
24	UAE, PLE, water, ethanol + water 30, 50 and 70% v:v, 50–100 °C, ultrasound power (0–800 W at the generator, or 0–38.5 W	*Punica granatum* L.	Phenolics	Food industry	Sumere et al. (2018)
25	Pressurized hot water extraction, (100, 130, 160 °C), (5 and 10 min), 10.34 MPa	*Stevia rebaudiana* Bertoni	Antioxidants and steviol glycosides	Industrial use	Kovačević et al. (2018)
26	Pressurized hot water extraction, 140 °C, 10 min,	*Mentha spicata* L.	phenolics and essential oil	Industrial use	Çam et al. (2018)

Table 1.4 Examples of pulse-electric field extraction of bioactive compounds.

S. No	Method and conditions	Materials	Bioactive compounds	Use	References
1	PEF, 75 pulses, a pulse duration of 5.70 μs, the electric field strength of 2.964 kV cm^{-1}, 5 min, the force of 45 daN cm^{-2}	*Ulva*	Proteins	Food and fuel	Polikovsky et al. (2016)
2	PEF, UAE, High voltage electrical discharges (0–109 kJ/kg), pH (2.5–12), and EtOH (0–50 %)	*Olea europaea*	Total phenolic compounds, antioxidant, proteins, pigments	Antioxidant, pharmaceutical and food industry	Roselló-Soto et al. (2015)
3	PEF, sonication (pH 8.5, 11)	*Nannochloropsis*	Pigments, proteins, carbohydrates, total phenolic compounds, antioxidant capacity	Pharmaceutical, food industry	Parniakov et al. (2015)
4	PEF, electric field intensity 12.4–38.4 kV/cm, 2 μs pulse duration,	*Agaricus bisporus*	Polysaccharide, proteins, polyphenols	Food industry and medicinal use	Xue & Farid (2015)
5	PEF, (2 kV/cm; 11.25 kJ/kg)	Olive oil	Oil extraction	Health benefits	Puértolas & de Marañón (2015)
6	PEF, (Field strength of 1, 3, and 5 kV/cm), energy input of 10 kJ/kg, cell disintegration index 0.70, 0.80, and 0.87	*Vaccinium myrtillus* L.	Total phenolic content, anthocyanin content, antioxidant activity	Food industry and medicinal use	Bobinaitė et al. (2015)

7	PFE, (25–55 °C), total specific energy (0.55–1.11 kWh $kg_{DW^{-1}}$)	***Chlorella vulgaris***	Carbohydrates, proteins	Medicinal use and food industry	Postma et al. (2016)
8	PEF, (10–30 kV/cm), (20–100 kJ/kg), High-pressure homogenization (HPH)	*Chlorella* vulgaris	Carbohydrates, proteins	Food and pharmaceutical industry	Carullo et al. (2018)
9	PEF field strength 0.75 kV/cm and 600 μs of treatment time, Solid-liquid extraction (SLE) with methanol	Potato peel	Glycoalkaloids, aglycone alkaloids	Industrial use and food industry	Hossain et al. (2015)
10	PEF, 15 kV/cm, 150 μs	*Artrosphira platensis*	C-phycocyanin	Food and industrial use	Martínez et al. (2017)
11	PEF, 3 kV/cm and 10 kV/cm	Citrus fruits	Polyphenols	Industrial use	El Kantar et al. (2018)
12	PEF, 10–25 kV/cm, 100 μs	*Chlorella vulgaris*	Lutein	Industrial use	Luengo et al. (2015)
13	PFE 13.3 kV/cm	Mango peels	Bioactive compounds	Food and industrial use	Parniakov et al. (2016)
14	PEF and Ultrasound 1 h	*Opuntia stricta* Haw	Betanin and isobetanin	Industrial use	Koubaa et al. (2016)
15	PEF, 15 kV/cm, 150 μs, electroporation with propidium iodide	*Rhodotorula glutinis*	Carotenoid	Food industry	Martínez et al. (2018)
16	PEF	Grapes	Anthocyanins, phenolic acids, flavanols, and flavonols	Industrial use	López-Giral et al. (2015)

Continued

Table 1.4 Examples of pulse-electric field extraction of bioactive compounds. *Continued*

S. No	Method and conditions	Materials	Bioactive compounds	Use	References
17	PEF, 1.5 MJ per kilogram ethanol-hexane blends	*Auxenochlorella protothecoides*	Lipids	Industrial use	Silve et al. (2018)
18	PEF, (40, 120 and 200 kV m^{-1}) (5, 18 and 30 pulses), 4 °C for 24 h	Tomato	Carotenoids	Food industry	González-Casado et al. (2018)
19	PEF with 80% methanol	Peach	Phenols, flavonoids and antioxidant compounds	Food purposes and antioxidants	Redondo et al. (2018)
20	PEF	Cocoa bean shell, coffee silver skin	Polyphenols and methylxanthines	Industrial purpose	Barbosa-Pereira et al. (2018)
21	PEF, 30 pulses of 30 μs, 7 kV/cm	Lemon	Total phenols, flavanones hesperidin and eriocitrin	Food and industrial use	Peiró et al. (2017)
22	PEF	*Ulva*	Superoxide dismutase, troponin C, thioredoxin h, aldolase A	Food and pharmaceutical industry	Polikovsky et al. (2018)
23	PEF and microwave 3.25 kV/cm electric field intensity and 30 pulse number, In Microwave, 540 W for 180 s	*Nigella sativa* seeds	Oil extraction	Industrial and medicinal use	Bakhshabadi et al. (2018)
24	PEF and high hydrostatic pressure (HHP)	*Opuntia dillenii* cactus juice	Ascorbic acid, flavonols, betacyanins, and antioxidants	Antioxidants, food and pharmaceutical industry	Moussa-Ayoub et al. (2017)

5.6 Ultrasound-assisted extraction (UAE)

Ultrasound-assisted extraction uses sound wave at 20 kHz–100 MHz for creating compression and expansion of the cells to extract the chemical components. UAE is based on the mechanism of diffusion across cells and breaking of the cells due to mass transfer (Mason et al., 2011). Specifically, ultrasound creates cavities or microbubbles which absorb energy and expand in size in the expansion cycle and then, recompress in the compression cycle. These cycles cause a sequence of production, growth, and collapse of bubbles in cells. The collapsing process generates sound waves of high pressure (1000 atm) and high temperature (about 5000 K) at a heating and cooling rate above 1010 K/s (Herrera and De Castro, 2005). The explosion of bubbles results in disintegration of the cells and release of bioactive compounds. Only liquid or liquid with solid exhibits this effect. UAE has been used for the extraction of sugars, polysaccharides, proteins, oils, phenolics, and carotenoids (Shi et al., 2014). Hossain et al. developed an UAE-based solid-liquid extraction method using methanol as solvent to extract multiple steroidal alkaloids from potato peel waste, including α-solanine, α-chaconine, solanidine and demissidine, which have various immunological functions (Hossain et al., 2014). UAE also assisted in the isolation of isoflavones from *Pueraria lobata* (Wild.) Ohwi stem and the extraction efficiency was much higher than conventional methods (Huaneng et al., 2007). Furthermore, phenols, carboxylic acids, carnosic acid and rosmarinic acid were isolated from *Rosmarinus officinalis* using UAE and the yield was higher than using conventional methods (Zu et al., 2012). UAE was also used for the isolation of polysaccharides from *Gynura divaricata* which have antioxidant and anti-glycation potential and can serve as drug candidates for the pharmaceutical industry (Deetae et al., 2017). Compared to conventional methods, UAE consumes less solvent and is more cost-effective and efficient to extract polyphenols and other compounds.

5.7 Pressurized liquid extraction (PLE)

Pressurized liquid extraction (PLE) was firstly described by Richter and also known as pressurized fluid extraction (PFE) and enhanced solvent extraction. PLE works under high pressure (1450–2175 psi) and high temperature (50–200 °C) to keep solvent liquid above the normal boiling point (Dunford et al., 2010). High temperature enhances the dielectric constant and thus, the polarity of the solvent decreases while the solubility and mass transfer rate increase. At high pressure, the liquid is forced to enter into the extracting cells and thus, fewer amounts of solvent are needed and the extraction yield is higher. In addition, automated techniques decrease the extraction time and does not require any solvents (Plaza et al., 2010). PLE has been used to extract bioactive compounds from marine sponges, organic pollutants, and plant materials. For example, 17 phenolic compounds were extracted from *Oryza sativa* and guaiacol, ellagic acid, vanillic acid, and protocatechuic acid were the most prominent among them (Setyaningsih et al., 2016). Anthocyanins and flavonoids were extracted from Brazilian pepper *Schinus terebinthifolius* Raddi (Feuereisen et al., 2017).

5.8 Pulsed-electric field extraction (PEF)

Pulsed-electric field extraction (PEF) is a nonthermal technique which uses short electric field pulse to extract the bioactive compounds and improve the quality of food materials. During PEF, cell membrane is distorted or destroyed by the electric field and electric potential is transferred to the cells (Bryant and Wolfe, 1987). This electric potential generates charge in the cell membrane and bioactive compounds are separated from the cells when the electric potential exceeds 1V. The effectiveness of the extraction is dependent of various operational parameters such as field strength, energy input, temperature, the material used and pulse number (Heinz et al., 2003). For example, Ganeva et al. carried out PEF extraction using an electric field of 2.75 kV/cm to treat beer yeast for 5 h. Technological advancements have enabled the use of high electric field intensity (up to 80 kV/cm) in PEF and significantly increased the extraction efficiency. PEF has been used for various processes varying from pressing, extraction, drying, to diffusion. This technology decreases the time of extraction and increases mass transfer by distorting the membrane of the raw material. The distortion or the damage of cell membrane is important to the enhancement of permeability and prove to be advantageous over conventional methods of extraction.

6. Bioactive compounds as a source of functional foods

The concept of functional food was introduced in Japan and highlighted the need for foods with nutrition, sensory and physiological functions (Hilliam, 2003). Functional foods provide health benefits beyond basic nutritions (Bech-Larsen and Grunert, 2003). It was estimated that more than 1700 functional foods were developed in period of 1988–98 and the market value was around US$14 billion in 1999 (Menrad, 2003). The market of functional foods reached US$15 billion in 2006 (Kotilainen, 2006). Currently, about 17% of foods in Spain market are functional foods and it is predicted to increase to 40% in 2020 (Granato et al., 2010). Bioactive compounds in functional food cannot only fulfill the need for basic nutritional values but also have certain health benefits, such as decreasing the level of cholesterol, reducing lactose intolerance, enhancing immunity enhancement, and reducing the risks of cardiovascular disease (McClements, 2018). In addition, they have many other biological activities such as antioxidant, antiviral, and antihypertensive. The demand of functional foods continues to grow as the result of the increasing cost of health care as well as increasing consumer awareness of a good diet for healthy living and longer life expectancy (Norton et al., 2015).

7. Conclusion

Numerous diseases prevailing around the world have caused high mortality. Pharmaceutical industry has been searching novel bioactive compounds for developing more effective treatments. In addition, food and nutraceutical industry has also used more natural products to develop functional foods that assist improve functions of the body. The extraction of bioactive compounds using conventional methods was tedious and time-consuming. Advanced extraction methods such as supercritical fluid extraction, ultrasound-assisted extractions, pressurized liquid extraction, and microwave extraction have improved the extraction efficiency and yield. The increasing demand of natural products in food and pharmaceutical industry requires more efficient, productive and environmental-friendly extraction techniques.

Acknowledgments

We thank Kanya Maha Vidyalaya (Jalandhar, Punjab, India) and Department of Higher Education (Shimla, Himachal Pradesh, India) for providing the necessary resources to carry out the present work.

Conflict of Interest

The authors declare no conflict of interest.

References

Adil, I.H., Cetin, H., Yener, M., Bayındırlı, A., 2007. Subcritical (carbon dioxide + ethanol) extraction of polyphenols from apple and peach pomaces, and determination of the antioxidant activities of the extracts. J. Supercrit. Fluids 43 (1), 55–63.

Agatonovic-Kustrin, S., Kustrin, E., Angove, M., Morton, D., 2018. A screening method for cardiovascular active compounds in marine algae. J. Chromatogr. A 1550, 57–62.

Akyil, S., İlter, I., Koç, M., Ertekin, F., 2018. Recent trends in extraction techniques for high value compounds from algae as food additives. Turkish J. Agric. Food Sci. Technol. 6 (8), 1008–1014.

Altemimi, A., Lakhssassi, N., Baharlouei, A., Watson, D.G., Lightfoot, D.A., 2017. Phytochemicals: extraction, isolation, and identification of bioactive compounds from plant extracts. Plants 6 (4), 42.

Alupului, A., Calinescu, I., Lavric, V., 2012. Microwave extraction of active principles from medicinal plants. UPB Sci. Bull. Seri. B. 74 (2), 1454–2331.

Ameer, K., Bae, S.-W., Jo, Y., Lee, H.-G., Ameer, A., Kwon, J.-H., 2017. Optimization of microwave-assisted extraction of total extract, stevioside and rebaudioside-A from Stevia rebaudiana (Bertoni) leaves, using response surface methodology (RSM) and artificial neural network (ANN) modelling. Food Chem. 229, 198–207.

Anastas, P.T., Warner, J.C., 1998. Principles of green chemistry. Green Chem. Theor. Pract. 29–56.

Bakhshabadi, H., Mirzaei, H., Ghodsvali, A., Jafari, S.M., Ziaiifar, A.M., 2018. The influence of pulsed electric fields and microwave pretreatments on some selected physicochemical properties of oil extracted from black cumin seed. Food Sci. Nutri. 6 (1), 111–118.

Barbosa-Pereira, L., Guglielmetti, A., Zeppa, G., 2018. Pulsed electric field assisted extraction of bioactive compounds from cocoa bean shell and coffee silverskin. Food Bioprocess Technol. 11 (4), 818–835.

Bech-Larsen, T., Grunert, K.G., 2003. The perceived healthiness of functional foods: a conjoint study of Danish, Finnish and American consumers' perception of functional foods. Appetite 40 (1), 9–14.

Belwal, T., Bhatt, I.D., Rawal, R.S., Pande, V., 2017. Microwave-assisted extraction (MAE) conditions using polynomial design for improving antioxidant phytochemicals in *Berberis asiatica* Roxb. ex DC. leaves. Ind. Crops Prod. 95, 393–403.

Bobinaitė, R., Pataro, G., Lamanauskas, N., Šatkauskas, S., Viškelis, P., Ferrari, G., 2015. Application of pulsed electric field in the production of juice and extraction of bioactive compounds from blueberry fruits and their by-products. J. Food Sci. Technol. 52 (9), 5898–5905.

Boulho, R., Marty, C., Freile-Pelegrín, Y., Robledo, D., Bourgougnon, N., Bedoux, G., 2017. Antiherpetic (HSV-1) activity of carrageenans from the red seaweed *Solieria chordalis* (Rhodophyta, Gigartinales) extracted by microwave-assisted extraction (MAE). J. Appl. Phycol. 29 (5), 2219–2228.

Bryant, G., Wolfe, J., 1987. Electromechanical stresses produced in the plasma membranes of suspended cells by applied electric fields. J. Membr. Biol. 96 (2), 129–139.

Bubalo, M.C., Ćurko, N., Tomašević, M., Ganić, K.K., Redovniković, I.R., 2016. Green extraction of grape skin phenolics by using deep eutectic solvents. Food Chem. 200, 159–166.

Çam, M., Işıklı, M.D., Yüksel, E., Alaşalvar, H., Başyiğit, B., 2018. Application of pressurized water extraction and spray drying techniques to produce soluble spearmint tea. J. Food Measure. Character. 1–8.

Cardenas-Toro, F.P., Alcázar-Alay, S.C., Coutinho, J.P., Godoy, H.T., Forster-Carneiro, T., Meireles, M.A.A., 2015. Pressurized liquid extraction and low-pressure solvent extraction of carotenoids from pressed palm fiber: experimental and economical evaluation. Food Bioprod. Process. 94, 90–100.

Carullo, D., Abera, B.D., Casazza, A.A., Donsì, F., Perego, P., Ferrari, G., et al., 2018. Effect of pulsed electric fields and high pressure homogenization on the aqueous extraction of intracellular compounds from the microalgae *Chlorella vulgaris*. Algal Res. 31, 60–69.

Castro-Puyana, M., Pérez-Sánchez, A., Valdés, A., Ibrahim, O., Suarez-Alvarez, S., Ferragut, J., et al., 2017. Pressurized liquid extraction of *Neochloris oleoabundans* for the recovery of bioactive carotenoids with anti-proliferative activity against human colon cancer cells. Food Res. Int. 99, 1048–1055.

Cheigh, C.-I., Yoo, S.-Y., Ko, M.-J., Chang, P.-S., Chung, M.-S., 2015. Extraction characteristics of subcritical water depending on the number of hydroxyl group in flavonols. Food Chem. 168, 21–26.

Chen, Y., Luo, H., Gao, A., Zhu, M., 2011. Ultrasound-assisted extraction of polysaccharides from litchi (*Litchi chinensis* Sonn.) seed by response surface methodology and their structural characteristics. Innov. Food Sci. Emerg. Technol. 12 (3), 305–309.

Chuyen, H.V., Nguyen, M.H., Roach, P.D., Golding, J.B., Parks, S.E., 2018. Microwave-assisted extraction and ultrasound-assisted extraction for recovering carotenoids from Gac peel and their effects on antioxidant capacity of the extracts. Food Sci. Nutri. 6 (1), 189–196.

Cvetanović, A., Švarc-Gajić, J., Zeković, Z., Gašić, U., Tešić, Ž., Zengin, G., et al., 2018. Subcritical water extraction as a cutting edge technology for the extraction of bioactive compounds from chamomile: influence of pressure on chemical composition and bioactivity of extracts. Food Chem. 266, 389–396.

Dahmoune, F., Nayak, B., Moussi, K., Remini, H., Madani, K., 2015. Optimization of microwave-assisted extraction of polyphenols from *Myrtus communis* L. leaves. Food Chem. 166, 585–595.

Daintree, L., Kordikowski, A., York, P., 2008. Separation processes for organic molecules using SCF technologies. Adv. Drug Deliv. Rev. 60 (3), 351–372.

De Castro, M.L., Garcıa-Ayuso, L., 1998. Soxhlet extraction of solid materials: an outdated technique with a promising innovative future. Anal. Chim. Acta 369 (1–2), 1–10.

Deetae, P., Tseng, K., Pasuphan, C., Panya, A., Methacanon, P., Temthawee, W., et al., 2017. Ultrasonic extraction, antioxidant and anti-glycation activities of polysaccharides from *Gynura divaricata* leaves. Chiang Mai J. Sci. 44 (1), 144–156.

Demain, A.L., 2000. Microbial biotechnology. Trends Biotechnol. 18 (1), 26–31.

Duba, K.S., Fiori, L., 2015. Supercritical CO_2 extraction of grape seed oil: effect of process parameters on the extraction kinetics. J. Supercrit. Fluids 98, 33–43.

Dudareva, N., Klempien, A., Muhlemann, J.K., Kaplan, I., 2013. Biosynthesis, function and metabolic engineering of plant volatile organic compounds. New Phytologist 198 (1), 16–32.

Dunford, N.T., Irmak, S., Jonnala, R., 2010. Pressurised solvent extraction of policosanol from wheat straw, germ and bran. Food Chem. 119 (3), 1246–1249.

El Hattab, M., Culioli, G., Piovetti, L., Chitour, S.E., Valls, R., 2007. Comparison of various extraction methods for identification and determination of volatile metabolites from the brown alga *Dictyopteris membranacea*. J. Chromatogr. A 1143 (1–2), 1–7.

El Kantar, S., Boussetta, N., Lebovka, N., Foucart, F., Rajha, H.N., Maroun, R.G., et al., 2018. Pulsed electric field treatment of citrus fruits: improvement of juice and polyphenols extraction. Innov. Food Sci. Emerg. Technol. 46, 153–161.

Erşan, S., Üstündağ, Ö.G., Carle, R., Schweiggert, R.M., 2018. Subcritical water extraction of phenolic and antioxidant constituents from pistachio (*Pistacia vera* L.) hulls. Food Chemistry 253, 46–54.

Espada-Bellido, E., Ferreiro-González, M., Barbero, G.F., Carrera, C., Palma, M., Barroso, C.G., 2018. Alternative extraction method of bioactive compounds from mulberry (*Morus nigra* L.) pulp using pressurized-liquid extraction. Food Anal. Methods 1–12.

Fayad, S., Nehmé, R., Tannoury, M., Lesellier, E., Pichon, C., Morin, P., 2017. Macroalga Padina pavonica water extracts obtained by pressurized liquid extraction and microwave-assisted extraction inhibit hyaluronidase activity as shown by capillary electrophoresis. J. Chromatogr. A 1497, 19–27.

Feuereisen, M.M., Barraza, M.G., Zimmermann, B.F., Schieber, A., Schulze-Kaysers, N., 2017. Pressurized liquid extraction of anthocyanins and biflavonoids from *Schinus terebinthifolius* Raddi: a multivariate optimization. Food Chem. 214, 564–571.

Figueiredo, A.C., Barroso, J.G., Pedro, L.G., Scheffer, J.J., 2008. Factors affecting secondary metabolite production in plants: volatile components and essential oils. Flavour Fragrance J. 23 (4), 213–226.

Figueroa, J.G., Borrás-Linares, I., Lozano-Sánchez, J., Quirantes-Piné, R., Segura-Carretero, A., 2018. Optimization of drying process and pressurized liquid extraction for recovery of bioactive compounds from avocado peel by-product. Electrophoresis 39, 1908–1916.

Ganzler, K., Salgó, A., Valkó, K., 1986. Microwave extraction: a novel sample preparation method for chromatography. J. Chromatogr. A 371, 299–306.

Garcia-Vaquero, M., Rajauria, G., Tiwari, B., Sweeney, T., O'Doherty, J., 2018. Extraction and yield optimisation of fucose, glucans and associated antioxidant activities from *Laminaria digitata* by applying response surface methodology to high intensity ultrasound-assisted extraction. Mar. Drugs 16 (8), 257.

Ghoreishi, S., Shahrestani, R.G., 2009. Subcritical water extraction of mannitol from olive leaves. J. Food Eng. 93 (4), 474–481.

Giacometti, J., Žauhar, G., Žuvić, M., 2018. Optimization of ultrasonic-assisted extraction of major phenolic compounds from olive leaves (*Olea europaea* L.) using response surface methodology. Foods 7 (9), 149.

Gill, B.S., Kumar, S., 2015. Differential algorithms-assisted molecular modeling-based identification of mechanistic binding of ganoderic acids. Med. Chem. Res. 24 (9), 3483–3493.

Gill, B.S., Kumar, S., 2016a. Ganoderic acid targeting multiple receptors in cancer: in silico and in vitro study. Tumor Biol. 37 (10), 14271–14290.

Gill, B.S., Kumar, S., 2016b. Triterpenes in cancer: significance and their influence. Mol. Biol. Rep. 43 (9), 881–896.

Gill, B.S., Kumar, S., 2017. Ganoderic acid modulating TNF and its receptors: in silico and in vitro study. Med. Chem. Res. 26 (6), 1336–1348.

Gill, B.S., Alex, J.M., Kumar, S., 2016a. Missing link between microRNA and prostate cancer. Tumor Biol. 37 (5), 5683–5704.

Gill, B.S., Sharma, P., Kumar, S., 2016b. Chemical composition and antiproliferative, antioxidant, and proapoptotic effects of fruiting body extracts of the Lingzhi or Reishi medicinal mushroom, *Ganoderma lucidum* (Agaricomycetes), from India. Int. J. Med. Mushrooms 18 (7).

Gill, B.S., Mehra, R., Kumar, V., Kumar, S., 2017. Ganoderic acid, lanostanoid triterpene: a key player in apoptosis. Investig. New Drugs 1–8.

Gill, B.S., 2013. Chemical Profiling of Ganoderma Lucidum of Bathinda Region. Central University of Punjab.

González-Casado, S., Martín-Belloso, O., Elez-Martínez, P., Soliva-Fortuny, R., 2018. Enhancing the carotenoid content of tomato fruit with pulsed electric field treatments: effects on respiratory activity and quality attributes. Postharvest Biol. Technol. 137, 113–118.

Granato, D., Branco, G.F., Nazzaro, F., Cruz, A.G., Faria, J.A., 2010. Functional foods and nondairy probiotic food development: trends, concepts, and products. Compr. Rev. Food Sci. Food Saf. 9 (3), 292–302.

Hakimi, M.I., Ilham, Z., Kohar, R.A.A., 2018. Enhancement of agro-industrial copra residue oil yield using microwave-assisted extraction. Waste and Biomass Valorization 1–8.

Heinz, V., Toepfl, S., Knorr, D., 2003. Impact of temperature on lethality and energy efficiency of apple juice pasteurization by pulsed electric fields treatment. Innov. Food Sci. Emerg. Technol. 4 (2), 167–175.

Herrera, M., De Castro, M.L., 2005. Ultrasound-assisted extraction of phenolic compounds from strawberries prior to liquid chromatographic separation and photodiode array ultraviolet detection. J. Chromatogr. A 1100 (1), 1–7.

Hilliam, M., 2003. Future for dairy products and ingredients in the functional foods market. Aust. J. Dairy Technol. 58 (2), 98.

Hoff, R.B., Pizzolato, T.M., Peralba, M.C.R., Díaz-Cruz, M.S., Barceló, D., 2015. Determination of sulfonamide antibiotics and metabolites in liver, muscle and kidney samples by pressurized liquid extraction or ultrasound-assisted extraction followed by liquid chromatography–quadrupole linear ion trap-tandem mass spectrometry (HPLC–QqLIT-MS/MS). Talanta 134, 768–778.

Hosikian, A., Lim, S., Halim, R., Danquah, M.K., 2010. Chlorophyll extraction from microalgae: a review on the process engineering aspects. Int. J. Chem. Eng. 2010.

Hossain, M.B., Tiwari, B.K., Gangopadhyay, N., O'Donnell, C.P., Brunton, N.P., Rai, D.K., 2014. Ultrasonic extraction of steroidal alkaloids from potato peel waste. Ultrason. Sonochem. 21 (4), 1470–1476.

Hossain, M.B., Aguiló-Aguayo, I., Lyng, J.G., Brunton, N.P., Rai, D.K., 2015. Effect of pulsed electric field and pulsed light pre-treatment on the extraction of steroidal alkaloids from potato peels. Innov. Food Sci. Emerg. Technol. 29, 9–14.

Huaneng, X., Zhang, Y., Chaohong, H., 2007. Ultrasonically assisted extraction of isoflavones from stem of *Pueraria lobata* (Willd.) Ohwi and its mathematical model. Chin. J. Chem. Eng. 15 (6), 861–867.

Ibanez, E., Kubátová, A., Señoráns, F.J., Cavero, S., Reglero, G., Hawthorne, S.B., 2003. Subcritical water extraction of antioxidant compounds from rosemary plants. J. Agric. Food Chem. 51 (2), 375–382.

Ibañez, E., Herrero, M., Mendiola, J.A., Castro-Puyana, M., 2012. Extraction and characterization of bioactive compounds with health benefits from marine resources: macro and micro algae, cyanobacteria, and invertebrates. In: Marine Bioactive Compounds. Springer, pp. 55–98.

Ivanović, M., Alañón, M., Arráez-Román, D., Segura-Carretero, A., 2018. Enhanced and green extraction of bioactive compounds from *Lippia citriodora* by tailor-made natural deep eutectic solvents. Food Res. Int. 111, 67–76.

Joshi, G., Singh, P.K., Negi, A., Rana, A., Singh, S., Kumar, R., 2015. Growth factors mediated cell signalling in prostate cancer progression: implications in discovery of anti-prostate cancer agents. Chem. Biol. Interact. 240, 120–133.

Klejdus, B., Lojková, L., Plaza, M., Šnóblová, M., Štěrbová, D., 2010. Hyphenated technique for the extraction and determination of isoflavones in algae: ultrasound-assisted supercritical fluid extraction followed by fast chromatography with tandem mass spectrometry. J. Chromatogr. A 1217 (51), 7956–7965.

Ko, M.-J., Cheigh, C.-I., Cho, S.-W., Chung, M.-S., 2011. Subcritical water extraction of flavonol quercetin from onion skin. J. Food Eng. 102 (4), 327–333.

Ko, M.-J., Lee, J.-H., Nam, H.-H., Chung, M.-S., 2017. Subcritical water extraction of phytochemicals from *Phlomis umbrosa* Turcz. Innov. Food Sci. Emerg. Technol. 42, 1–7.

Ko, M.-J., Nam, H.-H., Chung, M.-S., 2019. Conversion of 6-gingerol to 6-shogaol in ginger (*Zingiber officinale*) pulp and peel during subcritical water extraction. Food Chem. 270, 149–155.

Kong, Y., Zu, Y.-G., Fu, Y.-J., Liu, W., Chang, F.-R., Li, J., et al., 2010. Optimization of microwave-assisted extraction of cajaninstilbene acid and pinostrobin from pigeonpea leaves followed by RP-HPLC-DAD determination. J. Food Compos. Anal. 23 (4), 382–388.

Kotilainen, L., 2006. Health Enhancing Foods: Opportunities for Strengthening the Sector in Developing Countries: International Bank for Reconstruction and Development. The World Bank.

Koubaa, M., Barba, F.J., Grimi, N., Mhemdi, H., Koubaa, W., Boussetta, N., et al., 2016. Recovery of colorants from red prickly pear peels and pulps enhanced by pulsed electric field and ultrasound. Innov. Food Sci. Emerg. Technol. 37, 336–344.

Kovačević, D.B., Barba, F.J., Granato, D., Galanakis, C.M., Herceg, Z., Dragović-Uzelac, V., et al., 2018. Pressurized hot water extraction (PHWE) for the green recovery of bioactive compounds and steviol glycosides from Stevia rebaudiana Bertoni leaves. Food Chem. 254, 150–157.

Koyu, H., Kazan, A., Ozturk, T.K., Yesil-Celiktas, O., Haznedaroglu, M.Z., 2017. Optimizing subcritical water extraction of *Morus nigra* L. fruits for maximization of tyrosinase inhibitory activity. J. Supercrit. Fluids 127, 15–22.

Kraujalienė, V., Pukalskas, A., Venskutonis, P., 2017. Biorefining of goldenrod (*Solidago virgaurea* L.) leaves by supercritical fluid and pressurized liquid extraction and evaluation of antioxidant properties and main phytochemicals in the fractions and plant material. J. Function. Foods 37, 200–208.

Kraujalis, P., Kraujalienė, V., Kazernavičiūtė, R., Venskutonis, P.R., 2017. Supercritical carbon dioxide and pressurized liquid extraction of valuable ingredients from *Viburnum opulus* pomace and berries and evaluation of product characteristics. J. Supercrit. Fluids 122, 99–108.

Kris-Etherton, P.M., Harris, W.S., Appel, L.J., 2003. Omega-3 fatty acids and cardiovascular disease. Arterioscler. Thromb. Vasc. Biol. 23 (2), 151–152.

Lachos-Perez, D., Baseggio, A.M., Mayanga-Torres, P., Junior, M.R.M., Rostagno, M., Martínez, J., et al., 2018. Subcritical water extraction of flavanones from defatted orange peel. J. Supercrit. Fluids 138, 7–16.

Le, A., Parks, S., Nguyen, M., Roach, P., 2018. Optimisation of the microwave-assisted ethanol extraction of saponins from Gac (*Momordica cochinchinensis* Spreng.) seeds. Medicines 5 (3), 70.

Lee, C.H., Singla, A., Lee, Y., 2001. Biomedical applications of collagen. Int. J. Pharm. 221 (1–2), 1–22.

Lee, J.-H., Ko, M.-J., Chung, M.-S., 2018. Subcritical water extraction of bioactive components from red ginseng (*Panax ginseng* CA Meyer). J. Supercrit. Fluids 133, 177–183.

Lefsih, K., Giacomazza, D., Dahmoune, F., Mangione, M.R., Bulone, D., San Biagio, P.L., et al., 2017. Pectin from *Opuntia ficus* indica: optimization of microwave-assisted extraction and preliminary characterization. Food Chem. 221, 91–99.

Letellier, M., Budzinski, H., 1999. Microwave assisted extraction of organic compounds. Analusis 27 (3), 259–270.

Liew, S.Q., Teoh, W.H., Tan, C.K., Yusoff, R., Ngoh, G.C., 2018. Subcritical water extraction of low methoxyl pectin from pomelo (*Citrus grandis* (L.) Osbeck) peels. Int. J. Biol. Macromol. 116, 128–135.

Liu, R.H., 2013. Dietary bioactive compounds and their health implications. J. Food Sci. 78 (s1), A18–A25.

López, C.J., Caleja, C., Prieto, M., Barreiro, M.F., Barros, L., Ferreira, I.C., 2018. Optimization and comparison of heat and ultrasound assisted extraction techniques to obtain anthocyanin compounds from *Arbutus unedo* L. Fruits. Food Chem. 264, 81–91.

López-Giral, N., González-Arenzana, L., González-Ferrero, C., López, R., Santamaría, P., López-Alfaro, I., et al., 2015. Pulsed electric field treatment to improve the phenolic compound extraction from Graciano, Tempranillo and Grenache grape varieties during two vintages. Innov. Food Sci. Emerg. Technol. 28, 31–39.

Luengo, E., Martínez, J.M., Bordetas, A., Álvarez, I., Raso, J., 2015. Influence of the treatment medium temperature on lutein extraction assisted by pulsed electric fields from *Chlorella vulgaris*. Innov. Food Sci. Emerg. Technol. 29, 15–22.

Luo, X., Cui, J., Zhang, H., Duan, Y., 2018. Subcritical water extraction of polyphenolic compounds from sorghum (*Sorghum bicolor* L.) bran and their biological activities. Food Chem. 262, 14–20.

Ma, Q., Fan, X.-D., Liu, X.-C., Qiu, T.-Q., Jiang, J.-G., 2015. Ultrasound-enhanced subcritical water extraction of essential oils from *Kaempferia galangal* L. and their comparative antioxidant activities. Separ. Purif. Technol. 150, 73–79.

Machado, A.P.D.F., Pasquel-Reátegui, J.L., Barbero, G.F., Martínez, J., 2015. Pressurized liquid extraction of bioactive compounds from blackberry (*Rubus fruticosus* L.) residues: a comparison with conventional methods. Food Res. Int. 77, 675–683.

Machmudah, S., Lestari, S.D., Kanda, H., Winardi, S., Goto, M., 2018. Subcritical water extraction enhancement by adding deep eutectic solvent for extracting xanthone from mangosteen pericarps. J. Supercrit. Fluids 133, 615–624.

Macías-Sánchez, M., Mantell, C., Rodriguez, M., De La Ossa, E.M., Lubián, L., Montero, O., 2009. Comparison of supercritical fluid and ultrasound-assisted extraction of carotenoids and chlorophyll a from *Dunaliella salina*. Talanta 77 (3), 948–952.

Manikyam, H.K., Ramesh, C., Poluri, K.M., Assad, A., 2017. Microwave assisted Subcritical water extraction of Berberine hydrochloride from the roots of Berberis Aristata using Harmony search algorithm. J. Herbal Med. Res. 2.

Mao, Z., Di, X., Zhang, J., Wang, X., Liu, Y., 2017. Rapid and cost-effective method for the simultaneous quantification of seven alkaloids in Corydalis decumbens by microwave-assisted extraction and capillary electrophoresis. J. Sep. Sci. 40 (14), 3008–3014.

Mapari, S.A., Nielsen, K.F., Larsen, T.O., Frisvad, J.C., Meyer, A.S., Thrane, U., 2005. Exploring fungal biodiversity for the production of water-soluble pigments as potential natural food colorants. Curr. Opin. Biotechnol. 16 (2), 231–238.

Maran, J.P., Manikandan, S., Nivetha, C.V., Dinesh, R., 2017. Ultrasound assisted extraction of bioactive compounds from *Nephelium lappaceum* L. fruit peel using central composite face centered response surface design. Arab. J. Chem. 10, S1145–S1157.

Martínez, J.M., Luengo, E., Saldaña, G., Álvarez, I., Raso, J., 2017. C-phycocyanin extraction assisted by pulsed electric field from *Artrosphira platensis*. Food Res. Int. 99, 1042–1047.

Martínez, J.M., Delso, C., Angulo, J., Álvarez, I., Raso, J., 2018. Pulsed electric field-assisted extraction of carotenoids from fresh biomass of Rhodotorula glutinis. Innov. Food Sci. Emerg. Technol. 47, 421–427.

Mason, J., Chemat, F., Vinatoru, M., 2011. The extraction of natural products using ultrasound or microwaves. Curr. Org. Chem. 15 (2), 237–247.

McClements, D.J., 2018. Recent developments in encapsulation and release of functional food ingredients: delivery by design. Curr. Opin. Food Sci. 47, 80–84.

Megawati, M., Pardede, R.S., Nawangsari, D.P. (Eds.), 2017. Pectin Extraction of Citrus aurantifolia's Peel Using Microwave Assisted Extraction. Seminar Nasional Teknik Kimia Kejuangan.

Mehra, R., Bhushan, S., Gill, B.S., 2018. Algae-Based Composites and Their Applications. Biocomposites. Pan Stanford, pp. 183–200.

Menrad, K., 2003. Market and marketing of functional food in Europe. J. Food Eng. 56 (2–3), 181–188.

Mojerlou, Z., Elhamirad, A., 2018. Optimization of ultrasound-assisted extraction (UAE) of phenolic compounds from olive cake. J. Food Sci. Technol. 55 (3), 977–984.

Moussa-Ayoub, T.E., Jäger, H., Knorr, D., El-Samahy, S.K., Kroh, L.W., Rohn, S., 2017. Impact of pulsed electric fields, high hydrostatic pressure, and thermal pasteurization on selected characteristics of Opuntia dillenii cactus juice. LWT-Food Sci. Technol. 79, 534–542.

Nastić, N., Švarc-Gajić, J., Delerue-Matos, C., Morais, S., Barroso, M.F., Moreira, M.M., 2018. Subcritical water extraction of antioxidants from mountain germander (Teucrium montanum L.). J. Supercrit. Fluids 138, 200–206.

Negi, A., Gill, B.S., 2013. Success stories of enolate form of drugs. PharmaTutor 1 (2), 45–53.

Negi, A., Gill, B.S., Navgeet, N., Anand, S.S., 2014. Tilling: versatile reverse genetic tool. PharmaTutor 2 (1), 26–32.

Newman, D.J., Cragg, G.M., 2012. Natural products as sources of new drugs over the 30 years from 1981 to 2010. J. Nat. Prod. 75 (3), 311–335.

Niphadkar, S., Rathod, V., 2018. Optimization of microwave-assisted extraction of acetyl 11 keto β boswellic acid (AKBA) from Boswellia serrata by using orthogonal array design (OAD). J. Biol. Active Prod. Nat. 8 (2), 90–103.

Nobili, S., Lippi, D., Witort, E., Donnini, M., Bausi, L., Mini, E., et al., 2009. Natural compounds for cancer treatment and prevention. Pharmacol. Res. 59 (6), 365–378.

Norton, J., Espinosa, Y.G., Watson, R., Spyropoulos, F., Norton, I., 2015. Functional food microstructures for macronutrient release and delivery. Food Funct. 6 (3), 663–678.

Okiyama, D.C., Soares, I.D., Cuevas, M.S., Crevelin, E.J., Moraes, L.A., Melo, M.P., et al., 2018. Pressurized liquid extraction of flavanols and alkaloids from cocoa bean shell using ethanol as solvent. Food Res. Int. 114, 20–29.

Olano, C., Lombo, F., Mendez, C., Salas, J.A., 2008. Improving production of bioactive secondary metabolites in actinomycetes by metabolic engineering. Metab. Eng. 10 (5), 281–292.

Otero, P., Quintana, S.E., Reglero, G., Fornari, T., García-Risco, M.R., 2018. Pressurized liquid extraction (PLE) as an innovative green technology for the effective enrichment of Galician algae extracts with high quality fatty acids and antimicrobial and antioxidant properties. Mar. Drugs 16 (5), 156.

Pagano, I., Sánchez-Camargo, A.P., Mendiola, J.A., Campone, L., Cifuentes, A., Rastrelli, L., et al., 2018. Selective extraction of high-value phenolic compounds from distillation wastewater of basil (Ocimum basilicum L.) by pressurized liquid extraction. Electrophoresis 39, 1884–1891.

Pandey, A., Belwal, T., Sekar, K.C., Bhatt, I.D., Rawal, R.S., 2018. Optimization of ultrasonic-assisted extraction (UAE) of phenolics and antioxidant compounds from rhizomes of Rheum moorcroftianum using response surface methodology (RSM). Ind. Crops Prod. 119, 218–225.

Parniakov, O., Barba, F.J., Grimi, N., Marchal, L., Jubeau, S., Lebovka, N., et al., 2015. Pulsed electric field and pH assisted selective extraction of intracellular components from microalgae Nannochloropsis. Algal Research 8, 128–134.

Parniakov, O., Barba, F.J., Grimi, N., Lebovka, N., Vorobiev, E., 2016. Extraction assisted by pulsed electric energy as a potential tool for green and sustainable recovery of nutritionally valuable compounds from mango peels. Food Chem. 192, 842–848.

Pasquet, V., Chérouvrier, J.-R., Farhat, F., Thiéry, V., Piot, J.-M., Bérard, J.-B., et al., 2011. Study on the microalgal pigments extraction process: performance of microwave assisted extraction. Process Biochem. 46 (1), 59–67.

Paulsen, B.S., 2010. Highlights through the history of plant medicine. In: Bioactive Compounds in Plants-Benefits and Risks for Man and Animals, pp. 18–29.

Pavlić, B., Vidović, S., Vladić, J., Radosavljević, R., Cindrić, M., Zeković, Z., 2016. Subcritical water extraction of sage (Salvia officinalis L.) by-products—process optimization by response surface methodology. J. Supercrit. Fluids 116, 36–45.

Peiró, S., Luengo, E., Segovia, F., Raso, J., Almajano, M.P., 2017. Improving polyphenol extraction from Lemon residues by pulsed electric fields. Waste Biomass Valorization 1–9.

Peled-Zehavi, H., Oliva, M., Xie, Q., Tzin, V., Oren-Shamir, M., Aharoni, A., et al., 2015. Metabolic engineering of the phenylpropanoid and its primary, precursor pathway to enhance the flavor of fruits and the aroma of flowers. Bioengineering 2 (4), 204–212.

Pimentel-Moral, S., Borrás-Linares, I., Lozano-Sánchez, J., Arráez-Román, D., Martínez-Férez, A., Segura-Carretero, A., 2018. Microwave-assisted extraction for Hibiscus sabdariffa bioactive compounds. J. Pharm. Biomed. Anal. 156, 313–322.

Plaza, M., Santoyo, S., Jaime, L., Reina, G.G.-B., Herrero, M., Señoráns, F.J., et al., 2010. Screening for bioactive compounds from algae. J. Pharm. Biomed. Anal. 51 (2), 450–455.

Polikovsky, M., Fernand, F., Sack, M., Frey, W., Müller, G., Golberg, A., 2016. Towards marine biorefineries: selective proteins extractions from marine macroalgae Ulva with pulsed electric fields. Innov. Food Sci. Emerg. Technol. 37, 194–200.

Polikovsky, M., Fernand, F., Sack, M., Frey, W., Müller, G., Golberg, A., 2018. In silico food allergenic risk evaluation of proteins extracted from macroalgae Ulva sp. with pulsed electric fields. Food Chem. 276, 735–744.

Postma, P., Pataro, G., Capitoli, M., Barbosa, M., Wijffels, R.H., Eppink, M., et al., 2016. Selective extraction of intracellular components from the microalga Chlorella vulgaris by combined pulsed electric field–temperature treatment. Bioresour. Technol. 203, 80–88.

Puértolas, E., de Marañón, I.M., 2015. Olive oil pilot-production assisted by pulsed electric field: impact on extraction yield, chemical parameters and sensory properties. Food Chem. 167, 497–502.

Quintero, C., Bowers, M.D., 2018. Plant and herbivore ontogeny interact to shape the preference, performance and chemical defense of a specialist herbivore. Oecologia 1–12.

Rajaganapathi, J., Kathiresan, K., Singh, T., 2002. Purification of anti-HIV protein from purple fluid of the sea hare Bursatella leachii de Blainville. Mar. Biotechnol. 4 (5), 447–453.

Rajapakse, N., Mendis, E., Byun, H.-G., Kim, S.-K., 2005. Purification and in vitro antioxidative effects of giant squid muscle peptides on free radical-mediated oxidative systems. J. Nutr. Biochem. 16 (9), 562–569.

Ravber, M., Knez, Ž., Škerget, M., 2015. Simultaneous extraction of oil-and water-soluble phase from sunflower seeds with subcritical water. Food Chem. 166, 316–323.

Reddy, H.K., Muppaneni, T., Sun, Y., Li, Y., Ponnusamy, S., Patil, P.D., et al., 2014. Subcritical water extraction of lipids from wet algae for biodiesel production. Fuel 133, 73–81.

Redondo, D., Venturini, M.E., Luengo, E., Raso, J., Arias, E., 2018. Pulsed electric fields as a green technology for the extraction of bioactive compounds from thinned peach by-products. Innov. Food Sci. Emerg. Technol. 45, 335–343.

Rodríguez-Meizoso, I., Jaime, L., Santoyo, S., Señoráns, F.J., Cifuentes, A., Ibáñez, E., 2010. Subcritical water extraction and characterization of bioactive compounds from Haematococcus pluvialis microalga. J. Pharm. Biomed. Anal. 51 (2), 456–463.

Rodríguez-Pérez, C., Gilbert-López, B., Mendiola, J.A., Quirantes-Piné, R., Segura-Carretero, A., Ibáñez, E., 2016. Optimization of microwave-assisted extraction and pressurized liquid extraction of phenolic compounds from Moringa oleifera leaves by multiresponse surface methodology. Electrophoresis 37 (13), 1938–1946.

Rodríguez-Rojo, S., Visentin, A., Maestri, D., Cocero, M.J., 2012. Assisted extraction of rosemary antioxidants with green solvents. J. Food Eng. 109 (1), 98–103.
Roselló-Soto, E., Barba, F.J., Parniakov, O., Galanakis, C.M., Lebovka, N., Grimi, N., et al., 2015. High voltage electrical discharges, pulsed electric field, and ultrasound assisted extraction of protein and phenolic compounds from olive kernel. Food Bioprocess Technol. 8 (4), 885–894.
Rubio-Rodríguez, N., Sara, M., Beltrán, S., Jaime, I., Sanz, M.T., Rovira, J., 2012. Supercritical fluid extraction of fish oil from fish by-products: a comparison with other extraction methods. J. Food Eng. 109 (2), 238–248.
Ruiz-Aceituno, L., García-Sarrió, M.J., Alonso-Rodriguez, B., Ramos, L., Sanz, M.L., 2016. Extraction of bioactive carbohydrates from artichoke (Cynara scolymus L.) external bracts using microwave assisted extraction and pressurized liquid extraction. Food Chem. 196, 1156–1162.
Ryu, B., Qian, Z.-J., Kim, M.-M., Nam, K.W., Kim, S.-K., 2009. Anti-photoaging activity and inhibition of matrix metalloproteinase (MMP) by marine red alga, Corallina pilulifera methanol extract. Radiat. Phys. Chem. 78 (2), 98–105.
Şahin, S., Samli, R., Tan, A.S.B., Barba, F.J., Chemat, F., Cravotto, G., et al., 2017. Solvent-free microwave-assisted extraction of polyphenols from olive tree leaves: antioxidant and antimicrobial properties. Molecules 22 (7), 1056.
Saravana, P.S., Cho, Y.-N., Woo, H.-C., Chun, B.-S., 2018. Green and efficient extraction of polysaccharides from brown seaweed by adding deep eutectic solvent in subcritical water hydrolysis. J. Clean. Prod. 198, 1474–1484.
Setyaningsih, W., Saputro, I., Palma, M., Barroso, C., 2016. Pressurized liquid extraction of phenolic compounds from rice (Oryza sativa) grains. Food Chem. 192, 452–459.
Shang, Y.-F., Xu, J.-L., Lee, W.-J., Um, B.-H., 2017. Antioxidative polyphenolics obtained from spent coffee grounds by pressurized liquid extraction. South Afr. J. Bot. 109, 75–80.
Sheng, Z., Li, J., Li, Y., 2012. Optimization of ultrasonic-assisted extraction of phillyrin from Forsythia suspensa using response surface methodology. J. Med. Plants Res. 6 (9), 1633–1644.
Shi, M., Yang, Y., Hu, X., Zhang, Z., 2014. Effect of ultrasonic extraction conditions on antioxidative and immunomodulatory activities of a Ganoderma lucidum polysaccharide originated from fermented soybean curd residue. Food Chem. 155, 50–56.
Sihvonen, M., Järvenpää, E., Hietaniemi, V., Huopalahti, R., 1999. Advances in supercritical carbon dioxide technologies. Trends Food Sci. Technol. 10 (6–7), 217–222.
Silve, A., Papachristou, I., Wüstner, R., Sträßner, R., Schirmer, M., Leber, K., et al., 2018. Extraction of lipids from wet microalga Auxenochlorella protothecoides using pulsed electric field treatment and ethanol-hexane blends. Algal Res. 29, 212–222.
Song, R., Ismail, M., Baroutian, S., Farid, M., 2018. Effect of subcritical water on the extraction of bioactive compounds from carrot leaves. Food Bioprocess Technol. 11 (10), 1895–1903.
Sousa, A.D., Maia, A.I.V., Rodrigues, T.H.S., Canuto, K.M., Ribeiro, P.R.V., Pereira, R.C.A., et al., 2016. Ultrasound-assisted and pressurized liquid extraction of phenolic compounds from Phyllanthus amarus and its composition evaluation by UPLC-QTOF. Ind. Crops Prod. 79, 91–103.
Šulniūtė, V., Pukalskas, A., Venskutonis, P.R., 2017. Phytochemical composition of fractions isolated from ten Salvia species by supercritical carbon dioxide and pressurized liquid extraction methods. Food Chem. 224, 37–47.

Sumere, B.R., de Souza, M.C., dos Santos, M.P., Bezerra, R.M.N., da Cunha, D.T., Martinez, J., et al., 2018. Combining pressurized liquids with ultrasound to improve the extraction of phenolic compounds from pomegranate peel (Punica granatum L.). Ultrason. Sonochem. 48, 151–162.

Sweetlove, L.J., Nielsen, J., Fernie, A.R., 2017. Engineering central metabolism–a grand challenge for plant biologists. Plant J. 90 (4), 749–763.

Taiz, L., Zeiger, E., 2006. Plant Physiology, 4th. Sinauer Associate, Sunderland, Mass. EUA.

Teo, C.L., Idris, A., 2014. Enhancing the various solvent extraction method via microwave irradiation for extraction of lipids from marine microalgae in biodiesel production. Bioresour. Technol. 171, 477–481.

Tomšik, A., Pavlić, B., Vladić, J., Ramić, M., Brindza, J., Vidović, S., 2016. Optimization of ultrasound-assisted extraction of bioactive compounds from wild garlic (Allium ursinum L.). Ultrason. Sonochem. 29, 502–511.

Tomšik, A., Pavlić, B., Vladić, J., Cindrić, M., Jovanov, P., Sakač, M., et al., 2017. Subcritical water extraction of wild garlic (Allium ursinum L.) and process optimization by response surface methodology. J. Supercrit. Fluids 128, 79–88.

Tripodo, G., Ibáñez, E., Cifuentes, A., Gilbert-López, B., Fanali, C., 2018. Optimization of pressurized liquid extraction by response surface methodology of Goji berry (Lycium barbarum L.) phenolic bioactive compounds. Electrophoresis 39, 1673–1682.

Uquiche, E., Antilaf, I., Millao, S., 2016. Enhancement of pigment extraction from B. braunii pretreated using CO_2 rapid depressurization. Braz. J. Microbiol. 47 (2), 497–505.

Vila, M., Lamas, J.P., Garcia-Jares, C., Dagnac, T., Llompart, M., 2015. Optimization of an analytical methodology for the simultaneous determination of different classes of ultraviolet filters in cosmetics by pressurized liquid extraction–gas chromatography tandem mass spectrometry. J. Chromatogr. A 1405, 12–22.

Vinatoru, M., 2001. An overview of the ultrasonically assisted extraction of bioactive principles from herbs. Ultrason. Sonochem. 8 (3), 303–313.

Vladić, J., Canli, O., Pavlić, B., Zeković, Z., Vidović, S., Kaplan, M., 2017. Optimization of Satureja Montana subcritical water extraction process and chemical characterization of volatile fraction of extracts. J. Supercrit. Fluids 120, 86–94.

Von Schacky, C., 2000. n–3 fatty acids and the prevention of coronary atherosclerosis. Am. J. Clin. Nutr. 71 (1), 224s–227s.

Wang, L., Weller, C.L., 2006. Recent advances in extraction of nutraceuticals from plants. Trends Food Sci. Technol. 17 (6), 300–312.

Wang, K., Xie, X., Zhang, Y., Huang, Y., Zhou, S., Zhang, W., et al., 2018. Combination of microwave-assisted extraction and ultrasonic-assisted dispersive liquid-liquid microextraction for separation and enrichment of pyrethroids residues in Litchi fruit prior to HPLC determination. Food Chem. 240, 1233–1242.

Wang, Y., Luan, G., Zhou, W., Meng, J., Wang, H., Hu, N., et al., 2018. Subcritical water extraction, UPLC-Triple-TOF/MS analysis and antioxidant activity of anthocyanins from Lycium ruthenicum Murr. Food Chem. 249, 119–126.

Wang, Y., Zhang, G., Chi, X., Chen, S., 2018. Green and efficient extraction of podophyllotoxin from Sinopodophyllum hexandrum by optimized subcritical water extraction combined with macroporous resin enrichment. Ind. Crops Prod. 121, 267–276.

Weber, T., Charusanti, P., Musiol-Kroll, E.M., Jiang, X., Tong, Y., Kim, H.U., et al., 2015. Metabolic engineering of antibiotic factories: new tools for antibiotic production in actinomycetes. Trends Biotechnol. 33 (1), 15–26.

Wijekoon, M.J.O., Bhat, R., Karim, A.A., 2011. Effect of extraction solvents on the phenolic compounds and antioxidant activities of bunga kantan (Etlingera elatior Jack.) inflorescence. J. Food Compos. Anal. 24 (4–5), 615–619.

Wu, H., Li, C., Li, Z., Liu, R., Zhang, A., Xiao, Z., et al., 2018. Simultaneous extraction of oil and tea saponin from Camellia oleifera Abel. seeds under subcritical water conditions. Fuel Process. Technol. 174, 88–94.

Xu, H., Wang, W., Liu, X., Yuan, F., Gao, Y., 2015. Antioxidative phenolics obtained from spent coffee grounds (Coffea arabica L.) by subcritical water extraction. Ind. Crops Prod. 76, 946–954.

Xu, T., Sutour, S., Casabianca, H., Tomi, F., Paoli, M., Garrido, M., et al., 2015. Rapid screening of chemical compositions of Gracilaria dura and Hypnea musciformis (Rhodophyta) from Corsican lagoon. Int. J. Phytocosmetic. Nat. Ingredients 2 (1), 8.

Xue, D., Farid, M.M., 2015. Pulsed electric field extraction of valuable compounds from white button mushroom (Agaricus bisporus). Innov. Food Sci. Emerg. Technol. 29, 178–186.

Xynos, N., Papaefstathiou, G., Gikas, E., Argyropoulou, A., Aligiannis, N., Skaltsounis, A.-L., 2014. Design optimization study of the extraction of olive leaves performed with pressurized liquid extraction using response surface methodology. Separ. Purif. Technol. 122, 323–330.

Yang, Y., Kayan, B., Bozer, N., Pate, B., Baker, C., Gizir, A.M., 2007. Terpene degradation and extraction from basil and oregano leaves using subcritical water. J. Chromatogr. A 1152 (1–2), 262–267.

Yedro, F.M., Grénman, H., Rissanen, J.V., Salmi, T., Garcia-Serna, J., Cocero, M.J., 2017. Chemical composition and extraction kinetics of Holm oak (Quercus ilex) hemicelluloses using subcritical water. J. Supercrit. Fluids 129, 56–62.

Zhang, G., He, L., Hu, M., 2011. Optimized ultrasonic-assisted extraction of flavonoids from Prunella vulgaris L. and evaluation of antioxidant activities in vitro. Innov. Food Sci. Emerg. Technol. 12 (1), 18–25.

Zosel, K., 1964. Process for the Decaffeination of Coffee. US Patent.

Zu, G., Zhang, R., Yang, L., Ma, C., Zu, Y., Wang, W., et al., 2012. Ultrasound-assisted extraction of carnosic acid and rosmarinic acid using ionic liquid solution from Rosmarinus officinalis. Int. J. Mol. Sci. 13 (9), 11027–11043.

CHAPTER

Recovery and utilization of bioactives from food processing waste

2

Santosh Kumar[a,]*, Rekha Kushwaha[a,]*, Madan L. Verma[b,c]

Department of Biochemistry, University of Missouri, Columbia, MO, USA[a]; Department of Biotechnology, Dr Y. S. Parmar University of Horticulture and Forestry, Neri Campus, Himachal Pradesh, India[b]; Centre for Chemistry and Biotechnology, Deakin University, Geelong Campus, VIC, Australia[c]

Chapter outline

1. Introduction

According to the United Nation's Save Food initiative, Food and Agriculture Organization (FAO), United Nations Environment Programme' (UNEP), and

* Contributed equally.

Biotechnological Production of Bioactive Compounds. https://doi.org/10.1016/B978-0-444-64323-0.00002-3

stakeholders: ReFED (Rethink Food Waste) food waste can be defined as "any removal of food from the food supply chain which is or was at some point fit for human consumption, or which has spoiled or expired, mainly caused by economic behavior, poor stock management or neglect" (http://www.thinkeatsave.org/index.php/be-informed/definition-of-food-loss-and-waste). European project FUSIONS defined food waste as "any food, and inedible parts of food, removed from (lost to or diverted from) the food supply chain to be recovered or disposed including composted, crops ploughed in/not harvested, anaerobic digestion, bio-energy production, co-generation, incineration, disposal to sewer, landfill or discarded to sea" (Östergren et al., 2014).

According to FAO, annual 1.4 billion tons of food waste is roughly equal to one-third of the global food supply, can feed approx. 2 billion people each year. (http://www.fao.org/food-loss-and-food-waste/en/). As per USDA (2014), these loss or wastes are around 133, billion pounds/per year. In terms of money, this food loss was worth about $161.6 billion in 2010 (https://www.usda.gov/foodlossandwaste). By 2020 this food waste will be expected to rise to about 126 Mt (Mirabella et al., 2014) (Fig. 2.1A and B).

Food losses and waste per year are roughly 30% for cereals, 40%–50% for root crops, fruits and vegetables, 20% for oilseeds, meat, and dairy, plus 35% for fish globally (http://www.fao.org/save-food/resources/keyfindings/en/ data accessed on Nov. 2018) (Fig. 2.2). Food waste production is higher in developed countries followed by developing and undeveloped countries (Sagar et al., 2018). In Europe and North America, food waste production ranges from 95 to 115 kg/year/per person, while in sub-Saharan Africa and South/Southeast Asia is only 6–11 kg/year. This high extent of food waste in medium- and high-income countries can be explained simply by throwing away the food, which is still suitable for human consumption. In low-income countries, this loss is mostly during the production-to-processing stages.

The food waste occurs at various stages of food production, post-harvest handling, storage, processing, distribution and consumption (Gustavsson et al., 2011; Parfitt et al., 2010). The highest amount of food loss occurs at the farm level (Berkenkamp and Nennich, 2015). According to Rethink Food Waste through Economics and Data 9.2 billion kilograms of food lost at the farm level annually in the US (ReFED Collaborative, 2016). During food processing in food industries food losses and food waste occurs at all stages of processing such as damage during transport or non-appropriate transport systems, problems during storage, losses during processing or contamination, inappropriate packaging. An erratic way of handling or conservation and lack of cooling/cold storage are the main cause of food loss and waste at the market level (Parfitt et al., 2010). At consumer level food is wasted due to over- or non-appropriate purchasing, bad storage conditions, over-preparation, portioning and cooking, etc (Papargyropoulou et al., 2014; Gunders, 2012). Food waste is a complex problem rather than the socioeconomic problem. Billions of dollars spent on the treatment of agricultural and food waste in order to lower the risks for humans, animals, and the environment. Food waste is mostly disposed of as a landfill which becomes a big source of harmful methane gas

(A)

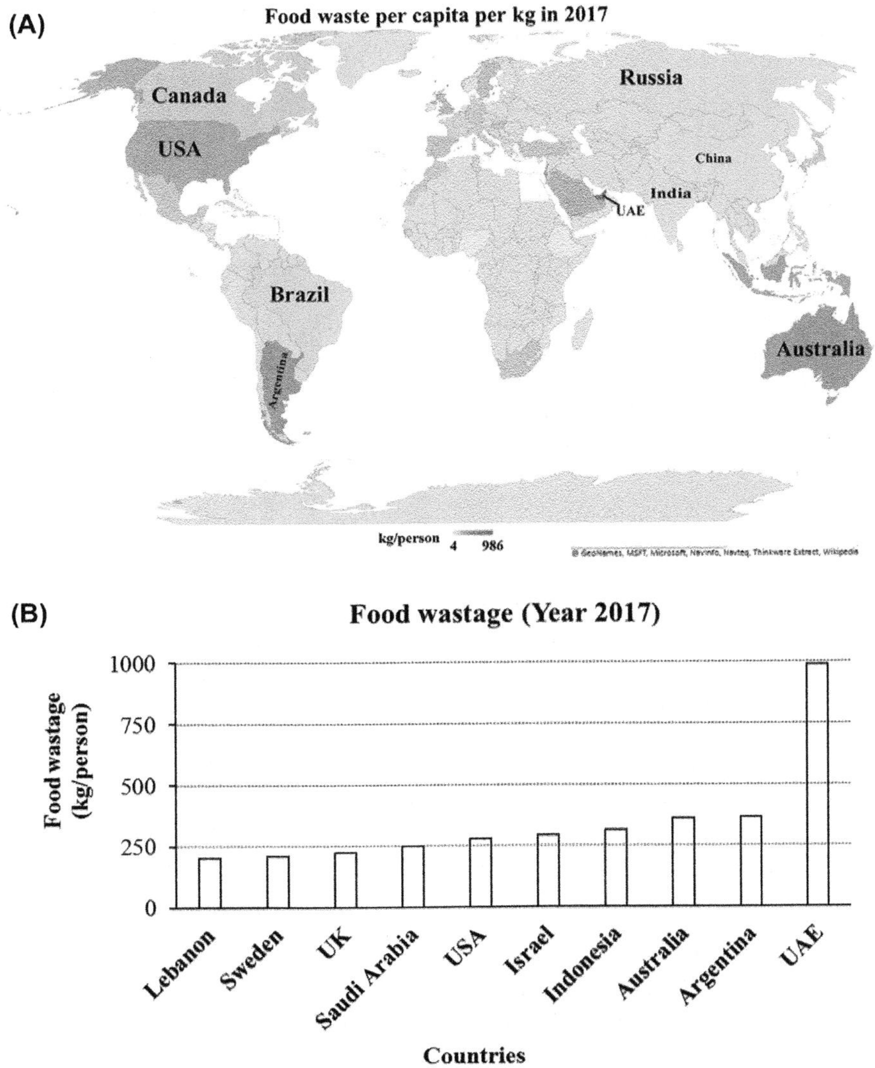

FIG. 2.1

(A) Distribution of food waste known around the world per capita per year (in kg/person/year). The map was prepared from the data retrieved from Food sustainability index 2017 on Nov 19–, 2018. (B) Top 10 countries with the highest food waste known around the world per capita per year (in kg/person/year) during the year 2017.

The graph was prepared from the data retrieved from Food sustainability index 2017 on Nov 19–, 2018.

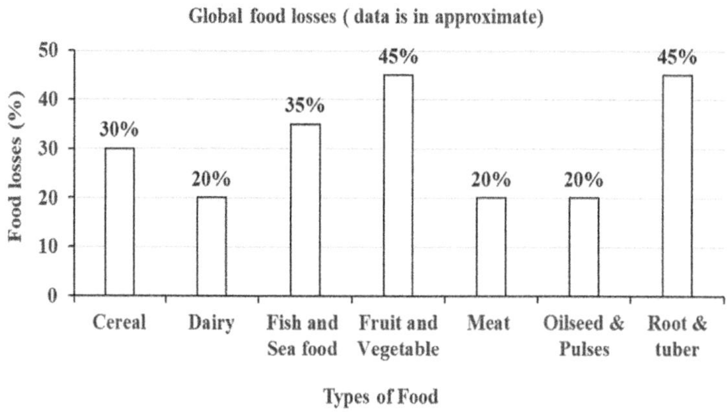

FIG. 2.2

Graph showing percentage loss of food in different food industries globally. The data are presented as rough estimates of food losses, which occurs every year.

Graph is generated from the data downloaded from http://www.fao.org/save-food/resources/keyfindings/en/.

production (Venkat, 2011). When incinerated, it causes harmful air pollution and loss of chemical values. Interactions among food waste, water & energy resources, environmental quality, and social justice make it a more serious problem and require immediate attention from individual to global level (Kibler et al., 2018) as well as appropriate management (Ma et al., 2009). Though not impossible but it is hard to eliminate food waste. However, there are significant ways to reduce food waste and find a good and beneficial use for it.

Food wastes are rich in carbohydrate polymers (starch, cellulose, and hemicelluloses), lignin, proteins, lipids, organic acids, smaller inorganic parts minerals, phenolic compounds antioxidants and vitamins (Ferrentino et al., 2018). These bioactive compounds possess antibacterial, antitumor, antiviral, antimutagenic, and cardioprotective activities (Sagar et al., 2018; Djilas et al., 2009; Yahia et al., 2017; Lemes et al., 2016). Thus, food wastes can be used to extract and isolate potential bioactives that can be used in the food, pharmaceutical, cosmetics, and textile industries. This utilization of bioactives from food wastes not only could reduce the risks and the costs for treatment of waste, but also could potentially add more value for agricultural and food production. Recovery and utilization of these bioactives is a very challenging process. The biggest challenge for the recovery of these compounds is to find the most appropriate and environment-friendly extraction techniques able to achieve the maximum extraction yield without compromising the stability of the extracted products. The extraction of the bioactives must be economically feasible to perform. This target can be accomplished by isolating the interested compounds through individual and additionally consolidated physical and biochemical methodologies with the end goal to give a scope of segments, all of which would add to accomplishing whole-waste exploitation (Baiano, 2014; Socaci et al., 2017; Fărcaş et al., 2017). Therefore, in a present chapter, an attempt has been made to review the recent literature for various methods of recovery of these bioactives and re-utilized them in the food, cosmetic and pharmaceutical industries. Our main emphasis is on plant-based food wastes.

2. Recovery of bioactives from various plant-based food waste

Food waste related to cereals comprise of rice bran, wheat bran and brewers spent grain. Root and tubers wastes include potato peel, sugar beet, and molasses. Wastes from oil crops and pulses include sunflower seeds, soybean seed, and olive pomace etc. (Fig. 2.3). Fruits, vegetables, roots, and tubers have the highest wastage rates (https://www.usda.gov/foodlossandwaste). Food wastes from these plant-based

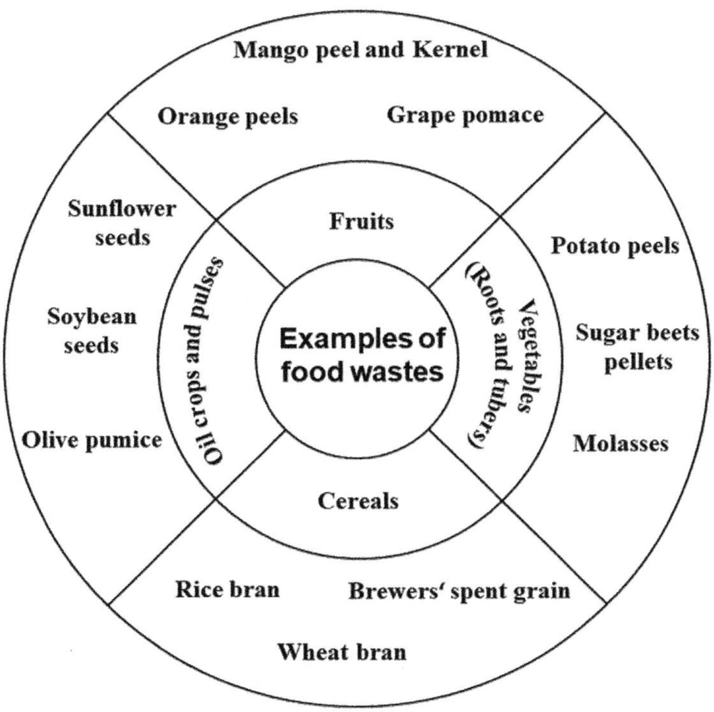

FIG. 2.3

A few examples of food waste classified based on their origin.

food industries comprised of proteins, polysaccharides, phenolic compounds, carotenoids, and other compounds.

2.1 Proteins

Proteins are large biomolecules consisting of one or more long chains of amino acid residues known to perform many functions within organisms. Proteins serve as macronutrients in our diet and known for their building blocks activity in the body. They are not a good source of energy because each gram of protein contains only 4 calories. The recent advancement in nutrition science leads to the identification of many nutritional benefits of plant proteins (FAO, 2011). In the edible oil industry, proteins are the main components of food waste, with a relatively high content of protein are present in the defatted meals obtained from oil sunflower, canola, rapeseed, palm, etc. These defatted by-products generated from oil refineries (oil cake, stem, and grain husk) are not only good sources of proteins but are also available in large quantities and at a low cost. Due to the rapidly growing plant protein market, there is great competition to evaluate oil-seeds, fruits and vegetables, cereals and soy alternative legumes and their processing wastes as a source of commercial proteins. There are few studies had been reported for the bioactive and functional properties of protein extracted from meals of major oilseeds such as for canola (Aydemir et al., 2014).

The recovery of protein from different biomass matrices has been intensively studied to obtain nutritionally important products. First, based on their solubility, plant proteins have typically been extracted with alkali. This technique involves alkaline solubilization of proteins, removal of insoluble material by centrifugation, precipitation of the protein at the pH corresponding to the isoelectric point, and collection of precipitated proteins by centrifugation. The second technique is based on micellization involving protein extraction using a salt solution, centrifugation, and precipitation. Here, the protein is recovered from a salt extract by ultrafiltration, diafiltration membranes, or dilution in cold water, followed by centrifugation. The isoelectric precipitation technique leads to higher yields of extracted proteins than the micellization methods (Hadnadjev et al., 2017). Below are the examples of how different food waste is utilized for protein extraction.

2.1.1 Soybean food industries

Soybean is a legume with no cholesterol and is low in saturated fat. Soya seeds are a rich source of plant proteins where dry seeds almost contain 42% of proteins relative to other components. Soybean seed protein is also called soy protein and it is often used to replace animal proteins in an individual's diet. Soybeans are the only vegetable food that contains all eight essential amino acids among nine essential amino acids required for human body. Besides this, they are also a good source of fiber, iron, calcium, zinc, and B vitamins (Montgomery, 2003; Imai, 2015; Hoffman and Falvo, 2004; Rizzo and Baroni, 2018). Food wastes from the soy oil and milk industries are rich in protein and are commercially available as protein supplements.

Soybean curd residue is the main waste from the soy-industries responsible for making soymilk, tofu, etc. In Japan, soybean curd residue is called as okara and is the main waste of soybean products. 1.1 (kg) of soybean curd residue generated from each kg of soybeans processed into soymilk or tofu (Li et al., 2013). Soybean curd residue is a relatively inexpensive source of protein that is known for its high nutritional and excellent functional properties. Soybean curd residue contains 27% protein with good nutritional quality. This can be a potential source of low-cost vegetable protein for human consumption (Wang and Cavins, 1989; Zee et al., 1988). Protein extraction from soybean curd residue is mainly affected by the low water solubility of the protein in soybean curd residue. Therefore, solubility improvement is a key process reported in the literature. One specific example is the recovery of up to 53% of proteins at pH 9.0 and 80 °C for 30 min (Ma et al., 1996). This method was further modified where mild acid treatments were done in order to improve the emulsifying and foaming properties (Chan and Ma, 1999). 93.4% of the protein was recovered from the okara flour by using a three-step-sequential extraction which includes, the introduction of primary and secondary grinding steps (Vishwanathan et al., 2011). This method also showed how emulsification, water, and fat binding, and foaming properties of protein from soybean curd residue were comparable to the commercial soy isolate (Vishwanathan et al., 2011). Free amino acid and soy peptide were produced after fermentation of soybean curd residue (Chan and Ma, 1999). Food waste from tofu (a kind of curd made from mashed soybean seeds) industry includes whey protein, which is rich in branched-chain amino acids. About 14 million pounds of soybean whey protein of high biological value is disposed of as waste (Rackis et al., 1971). Smith et al. (Smith et al., 1962) describes the recovery of proteins from whey by forming insoluble complexes with anionic materials. Whole whey protein was prepared by dialysis, and whey was fractionated by heating into heat-coagulable and supernatant proteins (Rackis et al., 1971).

The residue left after oil extraction from soybean is known as soybean meal. Defatted soybean meal, containing no hulls has an intermediate energy concentration. The main reason for the popularity of soybean meal is the unique composition of amino acids. It is particularly a good source of both lysine and tryptophan. The higher nutritional value of soybean meal is responsible for its excellence as poultry, livestock, and companion animals feed. Dehulled and defatted soybeans residue generated in the soy industry is used to produce soy flour, concentrates, and isolates. These are commercial products used as a protein supplement with varying protein concentration soybean flour (~40% protein), soy protein concentrate (SPC, ~70% protein), and soy protein isolate (SPI, >90% protein). Protein extraction for production of soy protein concentrates and isolates were done by wet extraction methods and dry fractionation methods (Xing et al., 2018). Preparing protein isolates through alkali extraction and isoelectric precipitation is commonly used for soybean (Campbell et al., 2016). Lee et al. (Lee et al., 2015) Extracted and purified the protein of soybean flakes and meals by lime treatment followed by ultrafiltration. Xing et al. (Xing et al., 2018) used a custom-built bench scale electrostatic separator for soy protein enrichment.

2.1.2 Canola oil industries

Canola is the second largest oilseed produced in the world next to soybean, which produces protein-rich meal during oil extraction (Campbell et al., 2016; Oecd/Fao, 2015). Canola meal protein is very rich in essential amino acids including sulfur-containing amino acids. Total protein content in the defatted canola meal is being around 32%. Cruciferin, napin and lipid transfer proteins are the main three proteins present in canola meal. Out of these, cruciferin and napin are the predominant storage proteins. Their Protein Digestibility Corrected Amino Acid Score was reported to be similar to the soy protein isolate i.e., 0.86 (Campbell et al., 2016; Bos et al., 2007; Wanasundara, 2011). Protein Digestibility Corrected Amino Acid Score is a method of evaluating the quality of a protein, based on digestibility and requirement of the amino acids in humans. Peptide mixtures and hydrolysates derived from canola proteins are beneficial for human health such as lowering of blood pressure by inhibiting angiotensin I-converting enzyme (Alashi et al., 2014). Besides this, these peptides and hydrolysate fractions are rich in antioxidant, antidiabetic, anorexigenic, anticancer, antiviral, hypercholesterolemic and bile acid binding properties (Wanasundara, 2011; Aachary and Thiyam, 2012). Amino acid profile of isolates indicated high nutritional quality for use in products for 10–12-year old kids. More than 99% of the protein was extracted from crude commercial hexane defatted canola meal when a 5% w/v suspension in 0.4% w/v NaOH was agitated for 60 min at room temperature in baffled flasks on an orbital shaker at 180–200 rpm. Protein recovery was 87.5% upon precipitation with acetic acid (Klockeman et al., 1997).

Tzeng et al. (Tzeng et al., 1990) designed a method for canola protein production by alkaline extraction followed by precipitation and membrane processing. The Edward Donald (Murray, 1998) patented a protein recovery method using salt for forming a canola protein isolate of high protein content in a gentle non-denaturing process in which fat is substantially removed. Canola protein isolates can also be prepared through ultrafiltration/diafiltration of aqueous protein extracts. However, this wet processing of canola protein isolates involves high water usage and energy costs, which is not feasible at a commercial level. However, the use of advanced centrifuges with reduced energy requirements and higher gravitational forces, not only reduces the cost but enhance the effective separation with higher purity, and quality of protein. Another method is enzyme-assisted, chemical-free processes that can extract protein in presence of oil (such as from expeller-pressed meal). As there is no use of solvents or chemicals in these methods, these methods are environmentally safe but the production cost is still challenging. On commercial scale such as "Burcon Nutra Science" has been using aqueous extraction, combined with membrane filtration were able to produce three different canola protein isolates with excellent functionality and a neutral flavor by using a clean and gentle extraction process. In order to avoid heat damage during these extraction process, cold pressing and low-temperature desolventizing have been used (Campbell et al., 2016).

2.1.3 Peanut oil industries

Wastes from peanut processing byproducts are a rich source of natural high-quality protein, the protein amount varies from 50% to 55% (Beuchat et al., 1975; Zhao et al., 2012). An average of 5.78 million metric tons of peanut meal was produced during 2000–2010 all over the world (Zhao et al., 2012). Depending on the oil content peanut meals are divided into the fresh or dry meal. Protein digestibility corrected amino acid score, of peanut proteins, is nutritionally equivalent to meat and eggs for human growth and health (Zhao et al., 2012). Thus, it can serve as a cost-effective source of protein for poor countries. Peanut proteins possess high emulsifying activity, emulsifying stability, foaming capacity, excellent water retention and high solubility (Wu et al., 2009). Different methods like isoelectric precipitation, alcohol precipitation, isoelectric precipitation with alcohol precipitation, hot water extraction and alkali solution with isoelectric precipitation are used for the separation of peanut proteins (Yu et al., 2007). However, these separation methods resulted in the production of wastewater that is responsible for serious environmental pollution as these methods consumed a large amount of acid and alkali. To improve the extraction efficiency of proteins, many new methods have been designed. These include the enzyme, superfine grinding, radiation, microwave and ultrasonic (US) processing (Quist et al., 2009). For the production of protein concentrate, protein recovery has been done by isoelectric precipitation; aqueous precipitation; alcohol precipitation; isoelectric precipitation and alcohol precipitation; hexane and aqueous alcohol precipitation and ultrafiltration (Zhao et al., 2012; Wu et al., 2009; Yu et al., 2007; Krishna Kumar et al., 2004). Alkali solution and isoelectric precipitation; ultrafiltration is the recovery methods used for protein isolates productions (Wu et al., 2009; Krishna Kumar et al., 2004; Dumay and Morançais, 2016).

2.1.4 Global protein waste from hazelnut oil industry

Hazelnut oil cake is a residue from the hazelnut oil industry and it is also a source of functional proteins. The hazelnut cake contains 54.4% protein, 3.1% lipid, 7.0% ash and 35.5% carbohydrate. While studying the bioactive, functional and edible film-forming properties of isolated hazelnut (*Corylus avellana* L.) isolated proteins from untreated, hot extracted, acetone washed and acetone washed and hot extracted. Aydemir et al. (Aydemir et al., 2014) found that isolated meal protein is rich in antioxidants, iron chelation, antiproliferative activities on colon cancer cells and good oil absorption properties. Further, they showed that bioactive, solubility and gelation properties of hazelnut proteins could be improved by simple processes like acetone washing and/or heat treatment.

2.1.5 Protein wastes from sunflower oil industries

Protein contents in sunflower seeds range between 10% and 27.1% dry weight basis. 85% of these proteins mainly belongs to storage proteins (Pérez and Vereijken, 2007). In sunflower two major classes of globular proteins are present, 11S globulin (or helianthinin) and 2S albumins (sunflower albumins (SFAs)) (Gonzalez-Perez et al., 2005). The quality of sunflower meal depends on the plant characteristics (seed composition, hulls/kernel ratio, dehulling potential, growth and storage conditions) and on the processing (dehulling, mechanical and/or solvent extraction (Golob et al., 2008)). The protein content in the defatted meal depends on the processing. It is 40% during mechanical extraction, 50% during solvent extraction and when prepared from dehulled seeds protein content varies from 53% to 66% (Bau et al., 1983; Robertson, 1972). Although a good source of protein but the presence of phenolic compounds, especially chlorogenic acid, and protein denaturation during processing for oil extraction hampered the recovery and utilization of the functional proteins and also protein solubility, emulsion, foam, and gelling properties.

2.1.6 Protein waste generation from palm oil industries

Unlike other oil cake, the palm kernel cake is not as popular as is the other oil cake explained in the previous section. Protein content in palm kernel cake is 16%−18% (Alimon, 2005). The chemical composition of palm kernel cake is like that of corn gluten or rice. There are only 85% of amino acids present, which is lower than the most oilseed meals (Alimon, 2005; Onwudike, 1986; Chee et al., 2012). The essential amino acids like lysine, methionine, histidine and threonine contents are also less in palm kernel cake (Ezieshi and Olomu, 2007). The crude protein composition of palm kernel cake depends on the processing method employed and the type of palm kernel used (Ezieshi and Olomu, 2007). Due to low protein contents and lack of a commercially feasible extraction method, the protein concentrates and isolates are not commercially produced, rather the extract is used for the production of wood adhesive formulations (Chee et al., 2012; Ong et al., 2018). Among the several methods used for protein extraction from palm kernel cake, the alkaline method was able to recover 11.91 g/100 g (Balogun, 1982). Arifin et al. (Arifin et al., 2009) also confirmed the efficacy of saline treatment over alkaline treatment by using saline and alkali treatment method for extraction of palm kernel protein. They also applied the Central Composite Designs of Response surface methodology and identified the optimized conditions for better yield. For saline condition, protein recovery was 28.39%−88.38% and optimized conditions were pH-9.0, NaCl concentration-0.02 M and Solvent/meal 60:1 ratio. The optimized conditions in alkaline treatment were 0.03 M NaOH, at a temperature of 35 °C with a liquid/solid ratio of 30:1. The recovery in alkaline treatment varies from 10.5% to 74.5%.

Chee et al. (Chee et al., 2012) used the response surface methodology to optimize the trypsin-assisted assay conditions for palm kernel cake protein extraction. Here the recovery of protein was 61.99% i.e. 0.74 g/100 g of the original protein content of palm kernel cake. Their recovery yield is five times higher than the alkaline method yield from trypsin-assisted extraction procedure was almost (10.21 ± 0.24 g/100 g). This method not only significantly improved the palm kernel cake protein recovery but also improved the solubility and emulsifying properties. Ng and Mohd Khan (Ng and Mohd Khan, 2012) successfully recovered 68.50 ± 3.08% crude protein by using an alkaline solution at pH 11, at a ratio of 1:10 (g/mL).) They also used the enzymatic hydrolysis to produce palm kernel extract protein hydrolysates or crude palm kernel extract peptide. According to this study, pepsin was found to be the least efficient protease to hydrolyze the palm kernel extract peptide. In another study, they optimized the enzymatic hydrolysis of palm kernel cake protein with trypsin to obtain palm kernel extract peptide hydrolysates by using response surface methodology (Ng et al., 2013).

2.1.7 Protein waste data from cereal waste

The abundance of protein content in cereal waste is next to fibers, which is a byproduct in the brewing process and present as Brewer's spent grain. Brewer's spent grain is a major insoluble solid residue obtained after beer wort production in the brewing industry (Ikram et al., 2017; Salihu and Muntari, 2011). This represents the ~85% of the residue obtained after the mashing process (Mussatto, 2013; Xiros and Christakopoulos, 2012). Annual global production of brewer's spent grain is estimated to be ~39 million tons (Mussatto, 2013). Brewer's spent grain mainly contains fiber (30%–50% w/w) and protein (19%–30% w/w) (Mussatto, 2013). Besides proteins and fibers, brewer's spent grain also contains a variety of minerals, phenolic compounds, and sugars (Mussatto and Roberto, 2006; Meneses et al., 2013). The most abundant proteins in the brewer's spent grain include hordeins, glutelins, globulins, and albumins (Celus et al., 2006). Among proteins, the essential amino acids represent ~30% of the total content, with lysine being the most abundant (14.3%) (Lynch et al., 2016).

There are many methods has been used for the recovery of the proteins from the brewer's spent grain these include extraction with alkali, salt solutions, detergents and aqueous enzyme-assisted methods (Murray, 1998; Ervin et al., 1989; Treimo et al., 2008; Ernster, 1986; Wahlström et al., 2017). Ernster (1986) patented a process for extracting protein solids from spent brewer's grains includes alkaline extractions of the grains followed by ultrafiltration of the product to yield a highly purified protein solid. Celus et al. (Celus et al., 2007) prepared the brewer's spent grain protein concentrate by alkaline extraction of brewer's spent grain (17% w/v) with 0.1 M NaOH at 60 °C. He further enzymatically hydrolyzed brewer's spent grain protein concentrate in a pH-stat setup by commercially available proteases (alcalase, flavourzyme, and pepsin) to obtain hydrolysates with different degrees of hydrolysis. During physical processes, such as pressing and sieving of wet brewer's spent grain, resulted in two fractions, protein fraction (rich in protein and fat and low in fiber) and a fiber fraction (low in protein and rich in arabinoxylans (Schwencke et al., 2005; Schwencke, 2006). Schwencke (Schwencke, 2006) described chemical extractions of brewer's spent grain protein fraction in alkaline medium, namely the alkaline extraction of brewer's spent grain at pH 11–12 and 104–121 °C. The protein yield in ultrasonic-assisted extraction using response surface methodology was 96.4 ± 3.5 mg/g of brewer's spent grain which was in agreement with the predicted value 104.2 mg/g brewer's spent grain (Zou et al., 2014). Wahlström et al. (Wahlström et al., 2017) described the preparation and characterization of new carboxylate salt-urea called as deep eutectic solvents for Brewer's spent grain protein extraction. With 10 wt% water addition this is an excellent solvent for protein extraction from the brewer's spent grain. The yield of extracted protein was 80%, due to the dissolving of insoluble protein during the mashing process.

2.2 Polysaccharides

Polysaccharides are long chains of monosaccharides linked by glycosidic linkages, which on hydrolysis give the monosaccharides or oligosaccharides. Their structure varies from linear to branch. Polysaccharides are commonly heterogeneous, containing slight modifications of the repeating unit. About 90% of total natural polysaccharides produced on earth belongs to plants (Di D et al., 2015). Starch, cellulose, hemicelluloses, pectin, inulin etc. are some examples of polysaccharides and are called as a dietary fiber. Dietary fiber/fiber is the non-digestible constituents making up the plant cell wall. These are resistant to enzymatic digestion (Dhingra et al., 2012; Hipsley, 1953). Based on their water solubility these fibers can be divided into the insoluble dietary fiber (cellulose, hemicelluloses, lignin etc.) and soluble dietary fibers (pectin, inulin, gums, and mucilage). Dietary fiber exerts a great impact on health-promoting food for mankind (Praznik et al., 2015). Dietary fiber intake reduces the risk of coronary heart disease, stroke, hypertension, diabetes, obesity, gastrointestinal disorders (Lairon et al., 2005). In addition, dietary fiber improves serum lipid concentrations, lowers blood pressure, improves blood glucose control in diabetes, promotes regularity, aids in weight loss, and appears to improve immune function (Liu et al., 1999; Steffen et al., 2003; Whelton et al., 2005; Montonen et al., 2003; Petruzziello et al., 2006; Brown et al., 1999; Keenan et al., 2002; Anderson et al., 2004).

Cereals, vegetables, fruits, and nuts are a natural source of dietary fiber (Rodríguez et al., 2006). Wastes from the cereals, fruits, and vegetable processing are the most widely investigated substrates for the extraction of several types of dietary fibers (Galanakis, 2012). A number of methods have been used for dietary fiber extraction from these wastes which includes, dry processing, wet processing, chemical, gravimetric, enzymatic, physical, and microbial or a combination of many methods (Yang et al., 2017). Although these methods are being used for a long time, these are responsible for modifying the structure and functionality of the extracted fiber. Therefore, with advances in technology the latest methods for extractions like ultrasound, microwave, and high-pressure processing etc. are now in use. These improved methods not only reduce processing times and temperatures but also enhance the yield and quality (Tejada-Ortigoza et al., 2016, 2017). Below are the few examples of polysaccharides extraction from various food wastes.

2.2.1 Cereals

Wheat, rice, corn, barley, sugarcane are the main agro-wastes and are potential sources of polysaccharides. The polysaccharides from these agro wastes are produced during cultivation, harvesting, and post-harvesting steps. These polysaccharides rich wastes are also called as 'lignocellulosic residues': i.e. the complex of cellulose, hemicellulose, and lignin (Di D et al., 2015). Isolation of hemicelluloses involves alkaline hydrolysis of ester linkages to liberate them from the lignocellulosic matrix followed by extraction into aqueous media (Brienzo et al., 2016). Wheat straw is extensively studied for hemicelluloses extractions. Earlier, hemicellulose was extracted from lignified wheat straw using aqueous alkali method, but it often results in brown colored hemicellulose. Therefore, with the time alkali method was modified for better yield and quality of the hemicellulose. Lawther et al. (Lawther et al., 1996) studied the effect of varying concentrations of KOH, H_3BO_3 under different temperature and time of extraction. They reported the varying nature of the alkali (calcium hydroxide, sodium hydroxide, lithium hydroxide, and liquid ammonia) for the optimal extraction and isolation of hemicellulose and cellulose from wheat straw. At an optimum concentration of 24% KOH/2%H_3BO_3 and 20 °C for 2 h the maximum yields for hemicellulose and cellulose of 34.23 and 35.96%, respectively was achieved.

Sun et al. (Sun et al., 2000) when treated the wheat straw with 2% H_2O_2 at 50 °C and pH 11.5 for 4 ± 30 h or with 2% H_2O_2±0.05% anthraquinone at 50 °C and pH 11.5 for 4.5 h resulted in 79%–86% of the original lignin and 77%–91% of the original hemicelluloses solubilization, respectively. Ultrasonically assisted extraction produced a higher yield of wheat straw hemicelluloses and lignin than those of the classical alkali procedure (Sun and Tomkinson, 2002). When surface methodology was optimized with extrusion parameter it showed that extrusion cooking had a positive effect on total and soluble dietary fiber and negative on the insoluble dietary fiber (Rashid et al., 2015). For extraction of wheat bran dietary fiber, alkali (2% NaOH) in a combination of proteinase resulting in relatively pure dietary fiber (Tejada-Ortigoza et al., 2016).

Total polysaccharides yield from rice waste depends on the kind of plant and raw materials (husk or straw) The yield varies from 8.2% to 26.1% (Arefieva et al., 2017) Glucans are the main water-soluble polysaccharides in rice, Polysaccharides after alkaline extraction of rice husks contain arabinose, xylose, glucose, and galactose. Hemicelluloses are one of the most abundant natural polysaccharides and comprise over 30% of the dry matter of rice straw. A comparative study of the extraction of rice straw hemicellulose by alkaline and hydrogen peroxide treatments resulted in the development of a fractionated treatment procedure for rice straw hemicellulose with maximum yield but minimal degradation and light color (Sun et al., 2000).

Bagasse is a dry pulpy residue left after extraction of sugar cane juice. It has 32%–34% cellulose, 19%–24% hemicellulose, 25%–32% lignin, 6%–12% extractives and 2%–6% ash (Haghdan et al., 2015). Earlier, alkaline extraction was the best and most efficient way to remove lignin from bagasse since a higher concentration of the solution. Chemical extraction of sugarcane bagasse showed that high alkaline concentration is responsible for the production of thin fibers with a lack of tenacity, of bending rigidity, and of bending hysteresis (Michel et al., 2013). Ultrasounds and mechanical action after the chemical extraction lead to a further improvement (Michel et al., 2013). The treatment with alkaline hydrogen peroxide has also affected the physical and chemical properties of sugarcane bagasse where brightness, water-holding capacity, and oil binding capacity have increased by 34%, 96% and 55% (Sangnark and Noomhorm, 2003). It also leads to removal of lignin by 53% thereby color of Solka Floc® 900, a commercial dietary fiber, was pure white (L = 93.51) (Sangnark and Noomhorm, 2003). The reduction of loaf volume and softness of bread was found At 5% of alkaline hydrogen peroxide - sugarcane bagasse and Solka Floc® 900 substitutions (Sangnark and Noomhorm, 2003).

Corn hulls are the inexpensive byproducts of the dry milling and wet milling processes and it contains hemicelluloses (30%–50%), cellulose (approximately 20%), phenolic acids (approximately 4%, mainly ferulic and ferulic acid), starch (9%–23%), proteins (10%–13%), lipids (2%–3%) and ash (2%) (Dhingra et al., 2012; Saulnier et al., 1995; Wang et al., 2018). In defatted corn hull, the maximal yield of dietary fiber A which reached 33% was obtained by hot-compressed water 150 °C for 60 min (Wang et al., 2018). The yield of dietary fiber B has also increased from 2.0% to 56.9% as the temperature increased from 110 to 180 °C, while the yield of solid residue decreased from 88.7% to 27.7% (Wang et al., 2018).

2.2.2 Vegetables

Fruit and vegetable wastes including seeds, peels, leaves, roots, tubers, skin, pulp, seeds, stones, pomace etc. are the rich sources of dietary fibers, and other bioactives (Sagar et al., 2018; Rudra et al., 2015; Panouillé et al., 2007; Iriondo-DeHond et al., 2018). Many fruits and vegetables generate at least up to 25%–30% of unusable waste materials (Ajila et al., 2007). The food waste varies from type of fruits and vegetable as well as the part used for eating. The waste amount in apples is 10.91% while in papaya 53% of the final product. In Mandarin, peeling is responsible for almost 16% of wastes and while in pineapple this waste amount goes to the even higher level of 48% of the final product (Ayala-Zavala et al., 2010; Joshi et al., 2012). Onion peel, potato dietary fibers, by-products (stems and florets) from cauliflower, carrot pomace, tomato pomace are mainly used for the production of dietary fibers. Dietary fibers study on onion showed that the dietary fiber content depends upon the variety and processing methods (Jaime et al., 2002; Benítez et al., 2013). Cellulose and pectic polysaccharides are the main components of the dietary fibers are (Kay, 1982). The dietary fibers solubility in water depends on the presence of galactan side chain (Jaime et al., 2002). Potato peel, pulp and potato solid wastes all are a rich source of dietary fibers (Gumul et al., 2011; Sharoba et al., 2013). Potato pulp is richer in rhamnogalacturonan I (Byg et al., 2012). In cauliflower, after comparing the stems and florets wastes for non-starch polysaccharides it was found that higher amounts of polysaccharides are present in stems than the florets (Femenia et al., 1997). Both stems and florets wastes are rich in insoluble fibers with pectic polysaccharides as the main component of non-starch polysaccharides. Carrot pomace and tomato pomace are also rich in dietary fibers. On the basis of dry matter, carrot pomace has 63.6% and tomato has 50% of dietary fibers. (Chantaro et al., 2008; Del Valle et al., 2006). When edible snail's enzymes were used on carrot pomace it leads to the production of soluble dietary fiber from carrot pomace (Yoon et al., 2005). It was observed that 77.3 g water-soluble fiber/100 g of carrot pomace was produced after 96 h of enzymatic hydrolysis (Yoon et al., 2005).

2.2.3 Fruits

Fruit waste has been utilized for pectin as a soluble dietary fiber. A number of methods have been tried to extract pectin from various sources so far (Pagán et al., 2001; Minjares-Fuentes et al., 2014; Kulkarni and Rathod, 2014; Pinheiro et al., 2008; Panchami and Gunasekaran, 2017). Industrially pectin is chemically obtained from apple pomace and citrus peels by using strong acids such as oxalic, hydrochloric, nitric, and sulfuric acids (Koubala et al., 2008; Hwang et al., 1998; Constenla et al., 2002; Garna et al., 2007; Yapo, 2009). Acid treatment is not environmentally friendly and thus extraction methods were further modified. These modifications include enzymatic extraction i.e. use of enzymes such polygalacturonase (hemi) cellulose, protease and microbial mixed enzymes (Min et al., 2011; Ptichkina et al., 2008), ultrasonic, autoclave, microwave, and extrusion-assisted (Min et al., 2011; Panchev et al., 1988; Oosterveld et al., 1996, 2000; Liu et al., 2006) treatments to extract pectins from apple pomace and orange peel etc.

Apple pomace is a waste material from apple juice processing and it contains significant amounts of dietary fiber. Total dietary fiber in vacuum-dried pomace varies from 442 to 495 g/kg and 480 g/kg in freeze-dried pomace (Min et al., 2011; Yan and Kerr, 2012). Li et al. (Li et al., 2014) compared the extraction of water-soluble dietary fiber (SDF) from apple pomace (AP) by cellulase, microwave and ultrasound-assisted methods with the conventional acid method. Microwave-assisted methods showed a drastic efficiency and the cellulase method provided the highest soluble dietary fiber yield.

The citrus industry generates a considerable number of by-products (or waste) with high amounts of valuable bioactive components. The annual world production of citrus fruits is over 100 million metric tons. Skin, pulp, seeds, and wastewater represent the byproducts obtained from the lemon processing. These wastes are rich in bioactive molecules. Peel from citrus industry is a very rich source of insoluble dietary fiber. The dietary fibers in peels were significantly higher than in peeled fruits. Soluble dietary fibers in different citrus fruits range in between 34.2% and 46.6% of the total dietary fiber amount (Gorinstein et al., 2001; Chinapongtitiwat et al., 2013; Russo et al., 2015).

Chau and Huang (Chau and Hauun, 2003) evaluated and compared the chemical composition and physicochemical properties of various fractions of dietary fibers (soluble and insoluble dietary fiber, alcohol-insoluble solid, and water-insoluble solid) prepared from Liucheng sweet orange peel by different methods. They found that peel was rich in insoluble fiber-rich fractions and mainly composed of pectic polysaccharides and cellulose. Kratchanova et al. (Kratchanova et al., 2004) showed that microwave pretreatment of fresh orange peels led to an increase in the capillary-porous characteristics and thus the water absorption capacity. Heating also inactivated the pectinesterase activity in the oranges, leads to a considerable increase in the yield and quality of extractable pectin. Zykwinska et al. (Zykwinska et al., 2008) used proteases and cellulases to isolate pectins and pectic oligosaccharides from the citrus peel at 50 °C for 4 h. They used three different enzyme mixtures for extraction of high methoxy pectins of high molar mass. Wang et al. (Wang et al., 2008a) studied the pectin composition in peels of eight varieties of citrus by sequential extraction using ethanol and water. A similar method was used by Mandalari et al. (Mandalari et al., 2006) for pectin extraction from *Citrus bergamia* Risso fruit peel.

Apart from oranges, grapes are the world's largest fruit crop (Kammerer et al., 2005). Therefore, a large amount of solid remains of grapes fruit called 'grape pomace is produced after juice extraction. It contains the skins, pulp, seeds, and stems of the fruit. Grape pomace is a good source for many bioactives including dietary fibers, namely, hemicelluloses, cellulose, and small proportions of pectin (Kammerer et al., 2005). Llobera and Cañellas (Llobera and Cañellas, 2007) studied the composition of pomace and stem of the Manto Negro red grape (*Vitis vinifera*) variety and found that pomace and stem showed high contents of total dietary fiber. Besides this, they also studied the soluble dietary fiber, insoluble dietary fiber, uronic acids and Klason lignin in both samples. In another study by Llobera and Cañellas (Llobera and Cañellas, 2008) on white grape (*V. vinifera*) variety, Prensal Blanc was analyzed for its dietary fiber components. A comparative analysis of red and white grapes varieties was done by González et al. (González-Centeno et al., 2010). They evaluated the grape pomaces and stems, from ten different grape (*V. vinifera* L.) varieties (six red and four white) as raw materials for the production of dietary fiber concentrates. They also analyzed the carbohydrate and functional properties of the dietary fiber. In their study, they found that pectin was the main components of the cell wall while in the stem, cellulose was higher than the other components. The red grape cultivar "Tempranillo" had the highest dietary fiber content in the pomace (36.90 g/100 g fresh weight), stem (34.80 g/100 g fresh weight), and fruit (5.10 g/100 g fresh weight). Deng et al. (Deng et al., 2011) analyzed the skins of two white wine and three red wine grape pomace from US Pacific Northwest for their dietary fiber (DF) by their dietary fiber gravimetric–enzymatic method with sugar profiling by high-performance liquid chromatography with an evaporative light-scattering detector and reported that red grape pomace has significantly higher DF than those of white grape pomace but reverse was true in case of but soluble sugar with low soluble sugar in red grape pomace. Zheng et al. (Zhang et al., 2017) used enzymatic-gravimetric method for recovery of dietary fiber from grape juice pomace and evaluated their functional properties of red grape variety 'Amur' grape.

Mango (*Mangifera indica* L. Anacardiacea) is the most popular tropical fruit ranked fifth in total world production among the major fruit crops. 20% of the fruits are processed for products such as puree, nectar, leather, pickles, canned slices etc. (Ajila et al., 2007). Peel and Kernel are most important byproduct with high amounts of dietary fibers (Ajila et al., 2007; Ajila and Prasada Rao, 2013; Vergara-Valencia et al., 2007; Elegbede et al., 1995; Nzikou et al., 2010; Dhingra and Kapoor, 1985). Another common and important tropical fruit crop is Banana. It is also one of the earliest crops cultivated in the history of human agriculture. Due to its cultivation and consumption in recent decades, it is now second largest fruit crop (Padam et al., 2014). Banana by-products have been used for wrappings foods and clothes (Kennedy, 2009). The fiber content in the peel is approximately 50% which include both the soluble fraction (pectins, gums) and the insoluble fraction (Cellulose, lignin, hemicelluloses, β-glucans) (Happi Emaga et al., 2008). Banana peel is a good source fiber of lignin (6%−12%), pectin (10%−21%), cellulose (7.6%−9.6%), hemicelluloses (6.4%−9.4%) and galactouroninc acid (Mohapatra et al., 2010). Beside this, it also contains sugars including glucose, galactose, Arabinose, rhamnose, and xylose (Mohapatra et al., 2010). Sequential H_2O/Chelating agent/acid extraction was mainly used for fiber extraction (Happi Emaga et al., 2008).

In recent years sour cherry byproducts have received much more attention (Yılmaz et al., 2018). The sour cherry has an astringent taste since their acid/sugar ratios are higher in comparison to the sweet cherry (*Prunus avium* L.). Uronic acid is the main sugar component of the alcohol-insoluble solid following with cellulose, arabinose, and galactose in pomace. Pectin had a low methylation degree and are rich in simple sugars, arabinose, and galactose (Kosmala et al., 2009).

2.3 Phenolic compounds

Phenolic compounds are the plant secondary metabolites responsible for the sensory characteristic and are also natural antioxidants present in plants, foods, and beverages (Kalpna et al., 2011; Tomás-Barberan et al., 2000; Lapornik et al., 2005). Phenolic compounds are responsible for their chemopreventive properties such as antioxidant, anticarcinogenic, or antimutagenic and anti-inflammatory effects (Rice-Evans et al., 1997; Huang et al., 2010). Phenolic compounds are present in all plant part such as bark, stalks, leaves, fruits, roots, flowers, pods, seeds, stems, latex, hull etc. (Kalpna et al., 2011). Structurally they contain one or more aromatic rings along with one or more hydroxyl groups in their basic structure (Balasundram et al., 2006). These can be classified into flavonoids (flavonols, flavanones, flavones, flavanonols, isoflavones, flavanols, and anthocyanidins), tannins, stilbenes, phenolic acids and lignans (Robbins, 2003). For extraction of phenolic compound many novel extraction techniques have been developed including ultrasound-assisted extraction, supercritical fluid extraction, microwave-assisted extraction, and accelerated solvent extraction (Wang and Weller, 2006).

Cereal barn is rich in ferulic, vanillic, p-coumaric, caffeic, and chlorogenic acids (Adom et al., 2005). Wang et al. (Wang et al., 2008b) optimized the ultrasound-assisted extraction of phenolic compounds from wheat bran. In order to obtain the optimal conditions for phenol extraction, ultrasound-assisted extraction parameters such as the solvent, extraction temperature and extraction time were optimized using response surface methodology (RSM), by employing central composite rotatable design (CCRD). Ethanol concentration, 64%, extraction temperature, 60 °C, and extraction time, 25 min were the optimal conditions obtained by RSM for the extraction of phenolic compounds under ultra-sonication. Rind, peel, and seeds of fruits and vegetables are the rich source of phenolic compounds (Friedman, 1997). Potato peel is a major source of phenolic which is 50% of the total bioactives (Friedman, 1997). Luthria (Luthria, 2012) evaluated the influence of different extraction parameters to optimize phenolic acids extraction by using a pressurized liquid extractor. Solvent composition, extraction time, particle size, flush volume, temperature, pressure, and solid-to-solvent ratio are the studied parameters. Samarin et al. (Samarin et al., 2012) studied the extracts of five different solvents (water, ethanol, hexane, methanol, and acetone) and two solvent extraction methods (Solvent and ultrasound-assisted) for extraction of phenolic compounds. Choi et al. (2016) analyzed the bioactives in whole potatoes, peels, and pulps of "Superior" variety of Korean potato. Phenolic extraction from sweet potato peels was analyzed by modeling and optimization by response surface modeling and artificial neural network. Thus, this model studied the effect of solvent to solid ratio, time and temperature on the extraction of phenolic compounds (Anastácio et al., 2016).

Wastes from citrus juice industries include peels and seeds. These wastes are known to have high natural antioxidant properties. Citrus seeds possessed greater antioxidant activity than peels. Citrus peel contains high amounts of flavanone glycosides (hesperidin, neohesperidin, narirutin, naringin), lower amounts of polymethoxylated flavones (sinensetin, tangeretin, nobiletin), and traces of flavonols, glycosylated flavones, and hydrocinnamic acid (Kawaii et al., 1999; M'hiri et al., 2014). A phenolic compound extracted from peel by many methods such as conventional solvent extraction, supercritical CO_2 extraction ($SC-CO_2$), subcritical water extraction (SWE), pressurized fluid extraction (PFE), ultrasound-assisted extraction (UAE), microwave-assisted extraction (MAE) and combined approach (M'hiri et al., 2014). Phenolic composition of some citrus peels and seeds was described by Bocco et al. (Bocco et al., 1998). They studied the antioxidant activity of several citrus peels and seed extracts obtained by methanol extraction for free phenolic compounds and by alkaline hydrolysis for bound phenolic compounds in a model system based on accelerated citronellal oxidation. They reported methanol extracts are richer in flavones and glycosylated flavanones, whereas hydrolyzed extracts contained mainly phenolic acids. A review on extraction methods of citrus peel phenolic compounds by M'hiri et al. (M'hiri et al., 2014) concluded that conventional solvent extraction method can give reasonable yield, but it causes degradation of the thermolabile compounds of the extract. Other methods such as high-hydrostatic-pressure extraction, high temperatures (50–200 °C), and high pressures (10–15 MPa) although time savor but led to degradation of the phenolic compounds and represented especially by flavonoids, (characterized by their antioxidant activities). This review compiles the different extraction methods for phenolic compounds from citrus peel and shows that conventional solvent extraction method can give reasonable yield, but it causes degradation of the thermolabile compounds of the extract. They suggested ultrasound-assisted extraction for thermolabile components and microwave-assisted extraction to separate both polar and nonpolar phenolic compounds. It is a less matrix-dependent method and it needs less selective conditions by using a wide spectrum of organic solvents. This review shows that there is a lot to explore in terms of extraction process parameters and technologies, but few articles deal with a multifactor optimization extraction method of citrus peel phenols content. This review suggested a need for multifactor optimization of an extraction method for better extraction efficiency and preservation activities of phenolic compounds.

Tavares et al. (Tavares et al., 2015) analyzed the aqueous waste obtained after hydrodistillation of lemongrass (CcHD) in *Cymbopogon citratus* (Cc) (Lemongrass) oil industry. The CcHD is rich in phenolic acids and flavonoids, namely luteolin and apigenin derivatives. Banana peel contained large amounts of dopamine and L-dopa, catecholamines with a significant antioxidant activity (González-Montelongo et al., 2010). Evaluation of phenolic compound in peel and pulp or flesh of pomegranate and peaches reveled the higher phenolic compounds in peel than the pulp or flesh. (Li et al., 2006; Chang et al., 2000).

The grape seed from wine industry showed the highest amount of phenolic compounds such as gallic acid, catechin and epicatechin, and a wide variety of procyanidins whereas, skins revealed the highest levels of anthocyanins and p-coumaric acid hexoside (Peixoto et al., 2018; Maier et al., 2009). Ultrasound-assisted extraction technique was used by Ghafoor et al. (Ghafoor et al., 2011) for extraction of anthocyanins and phenolic compounds from grape peel Another extraction method i.e., microwave-assisted extraction of polyphenolic antioxidants from grape seeds was also studied and response surface methodology (RSM) was used to evaluate the effect of microwave power, solvent concentration, extraction time, and their interactions by krishnaswamy (Krishnaswamy et al., 2013).

Mango kernel is rich in gallic acid, ellagic acid, gallates, gallotannins, condensed tannins (Kumar et al., 2017). Mango peel contains anthocyanins, quercetin-glycosides, kaempferol-glycoside, xanthone-glycosides, cyanidin 3-O-galactoside anthocyanidin hexoside, γ-tocopherol, quercertin, mangiferin pentodise syringic, ellagic, gallic acid, condensed tannins etc. Mangiferin present in leaves of *M. indica* has been extracted by microwave assisted extraction using water as a solvent by Kulkarni and Rathore (Kulkarni and Rathod, 2016). Apple peel and pomace are the rich in the many phenolic compounds these include catechins, procyanidins, phloridzin, phloretin glycosides, caffeic acid, and chlorogenic acid; the peel possesses all of these compounds and has additional quercetin glycosides and cyanidin glycosides. The air-dried and freeze-dried apple peels had the highest total phenolic, flavonoid, and anthocyanin contents these includes contains neochlorogenic acid, 3-p-coumaroylquinic acid, chlorogenic acid, quercetin glucoside, and rutinoside, kaempferol-rutinoside, isorhamnetin-rutinoside, quercetin, kaempferol, isorhamnetin, anthocyanin (Kołodziejczyk et al., 2013).

3. Utilization of bioactive compound from food wastes

A bioactive compound extracted from the food wastes had a lot of potential as a nutraceutical, functional foods, and food additives (Baiano, 2014; Gil-ch et al., 2013). Bioactives present in the food wastes are known to have antioxidant and radical scavenging properties. These properties are responsible for delaying and inhibiting the oxidation of DNA, proteins, and lipids, therefore, lowered the risk of the development of diseases like cancer, Alzheimer, cataracts, and Parkinson etc. Some of the examples of the utilization of bioactives are as follows:

Oilcake a waste from the oil industry is highly popular feed utilized in dairy and poultry industries. The olive by-product, so-called "pâté," a natural source of bioactives characterized by the presence of hydroxytyrosol, β-hydroxyverbascoside, oleoside derivative, luteolin etc., is a potential ingredient for nutraceuticals preparations or feed industry (Sanchez et al., 2017). Protein isolates and protein concentrates from various oilcake meal of major and minor oilseeds such as soybean, peanut, rapeseed, sunflower, almond, groundnut, and walnut utilized as a food supplement. Protein degrading enzymes from the food wastes could also be utilized in meat or brewing industries.

Pectin is another bioactive compound, extracted from the fruit pomace and now being used as gelling agents in jams, fillings, sweets, etc. Besides this, the fruit pomace is also used to extract many food additives such as dietary fibers, lactic acid, pigments, vinegar, natural sweeteners and cellulose (Nawirska and Kwa'sniewska, 2018). Mangiferin is a bioactive compound obtained from the peel and leaves of mango is now been used to treat the cancer cell. A single administration of mangiferin or in combination other anticancerous drugs have shown the potential benefits in brain, lung, cervix, breast and prostate cancers, and leukemia besides its antioxidant and anti-inflammatory properties (Núñez Selles et al., 2016). Lignan concentrates obtained from the flaxseed oil cake is also rich in anti-cancer, antioxidant, antibacterial, antiviral, and anti-inflammatory properties (Zanwar et al., 2011). This lignans rich flaxseed meal is used in the preparation of bread, muffins and other bakery products. Ferulic, vanillic, and syringic acids and other phenolic compounds available in cereal bran provide resistance against free radical damage, cancer and cardiovascular diseases (Kim et al., 2006). γ-Oryzano obtained from rice bran is commercially available and is used as a cardioprotective and reduce the menopausal symptoms. β-glucans extracted from cereals especially barley flour used the production of pasta, noodles, breakfast cereals, and dairy products. Addition of flavonoids and saponins from black bean seed coat to whole wheat bread formulation was resulted in retention of added flavonoids and saponins as well as anthocyanins in bread after baking with many health benefits of prepared bread (Chávez-santoscoy et al., 2016).

As compared to the number of food wastes produced and its potential in food, health, and pharmaceutical industries it seems that food wastes are not fully utilized, due to non-availability of cost-effective efficient methods with less harmful waste production. The production cost of these compounds is much higher than the output.

Subsequently, there is a need to utilize novel extraction innovations fit for lessening dissolvable utilization, consequently equipped for expanding the general eco-supportability of the sustenance life cycle and diminish the creation cost. Therefore, there is a need to use novel extraction technologies with less solvent consumption and capable of increasing the overall eco-sustainability with low the production cost.

4. Conclusion and future aspects

Survival of any species on earth depends on the continuous supply of its food. To achieve sustainability in food production and consumption, utilization of food waste is one of the most important steps. Foods processing industrial unit that achieve the goal of getting these components from food waste always have better returns on their investments. It is the need of the era to have the most effective research environment for utilization of food waste using the principles of sustainability. The recent discovery of synthetic enzymes could be a starter point of this kind of research, besides biomimeticing the saprophytes waste utilization machinery. It should also be kept in mind that while utilizing waste, there should be minimum impact on the environment. In the end, we can say that the goal of zero or near zero food waste can be achievable by carefully designing of food processing units which can utilized food wastes directly into the useful commercial compounds at a lower cost.

Acknowledgments

Authors are thankful to the Department of Biochemistry, University of Missouri (MO) U.S.A. for its resources.

Disclosure of potential conflicts of interest

Authors declare no potential conflicts of interest.

References

Aachary, A.A., Thiyam, U., 2012. A pursuit of the functional nutritional and bioactive properties of canola proteins and peptides. Crit. Rev. Food Sci. Nutr. 52 (11), 965–979.

Adom, K.K., Sorrells, M.E., Liu, R.H., 2005. Phytochemicals and antioxidant activity of milled fractions of different wheat varieties. J. Agric. Food Chem. 53 (6), 2297–2306.

Ajila, C.M., Prasada Rao, U.J.S., 2013. Mango peel dietary fibre: composition and associated bound phenolics. J. Funct. Foods. 5 (1), 444–450.

Ajila, C.M., Bhat, S.G., Prasada Rao, U.J.S., 2007. Valuable components of raw and ripe peels from two Indian mango varieties. Food Chem. 102 (4), 1006–1011.

Alashi, A.M., Blanchard, C.L., Mailer, R.J., Agboola, S.O., Mawson, A.J., et al., 2014. Blood pressure lowering effects of Australian canola protein hydrolysates in spontaneously hypertensive rats. Food Res. Int. 55, 281–287.

Alimon, A.R., 2005. The nutritive value of palm kernel cake for animal feed. Palm Oil Dev. 40 (40), 12–14.
Anastácio, A., Silva, R., Carvalho, I.S., 2016. Phenolics extraction from sweet potato peels: modelling and optimization by response surface modelling and artificial neural network. J. Food Sci. Technol. 53 (12), 4117–4125.
Anderson, J.W., Randles, K.M., Kendall, C.W.C., Jenkins, D.J.A., 2004. Carbohydrate and fiber recommendations for individuals with diabetes: a quantitative assessment and meta-analysis of the evidence. J. Am. Coll. Nutr. 23 (1), 5–17.
Arefieva, O.D., Zemnukhova, L.A., Kovshun, A.A., Kovekhova, A.V., 2017. Processing methods of alkaline hydrolysate from rice husk. Rice Sci. 24 (4), 235–240.
Arifin, B., Bono, A., Farm, Y.Y., Ling, A.L.L., Fui, S.Y., 2009. Protein extraction from palm kernel meal. J. Appl. Sci. 9 (17), 2996–3004.
Ayala-Zavala, J.F., Rosas-Domínguez, C., Vega-Vega, V., González-Aguilar, G.A., 2010. Antioxidant enrichment and antimicrobial protection of fresh-cut fruits using their own byproducts: looking for integral exploitation. J. Food Sci. 75 (8), R175–R181.
Aydemir, L.Y., Gökbulut, A.A., Baran, Y., Yemenicioğlu, A., 2014. Bioactive, functional and edible film-forming properties of isolated hazelnut (*Corylus avellana* L.) meal proteins. Food Hydrocoll. 36, 130–142.
Baiano, A., 2014. Recovery of biomolecules from food wastes - a review. Molecules 19 (9), 14821–14842.
Balasundram, N., Sundram, K., Samman, S., 2006. Phenolic compounds in plants and agri-industrial by-products: antioxidant activity, occurrence, and potential uses. Food Chem. 99 (1), 191–203.
Balogun, O.O., 1982. Extraction of nitrogen from palm kernel meal and evaluation of digestibility of protein isolate from the meal by the in vitro method. J. Agric. Food Chem. 30 (6), 1163–1166.
Bau, H.M., Mohtadi-Nia, D.J., Mejean, L., Debry, G., 1983. Preparation of colorless sunflower protein products: effect of processing on physicochemical and nutritional properties. J. Am. Oil Chem. Soc. 60 (6), 1141–1148.
Benítez, V., Mollá, E., Martín-Cabrejas, M.A., Aguilera, Y., López-Andréu, F.J., et al., 2013. The impact of pasteurisation and sterilisation on bioactive compounds of onion by-products. Food Bioprocess Technol. 6 (8), 1979–1989.
Berkenkamp, J., Nennich, T., 2015. Beyond Beauty: The Opportunities and Challenges of Cosmetically Imperfect Produce. Report 1: Survey Results from Minnesota Produce Growers.
Beuchat, L.R., Cherry, J.P., Quinn, M.R., 1975. Physicochemical properties of peanut flour as affected by proteolysis. J. Agric. Food Chem. 23 (4), 616–620.
Bocco, A., Cuvelier, M.-E., Richard, H., Berset, C., 1998. Antioxidant activity and phenolic composition of citrus peel and seed extracts. J. Agric. Food Chem. 46 (6), 2123–2129.
Bos, C., Airinei, G., Mariotti, F., Benamouzig, R., Berot, S., et al., 2007. The poor digestibility of rapeseed protein is balanced by its very high metabolic utilization in humans. J. Nutr. 137 (3), 594–600.
Brienzo, M., Carvalho, A.F., Figueiredo, F.C.A., Oliva Neto, P d, 2016. Sugarcane bagasse hemicellulose properties, extraction technologies and xylooligosaccharides production. In: Riley, G.L. (Ed.), Food Waste. Nova Science Publishers, Inc, pp. 156–188.
Brown, L., Rosner, B., Willett, W.W., Sacks, F.M., 1999. Cholesterol-lowering effects of dietary fiber: a meta-analysis. Am. J. Clin. Nutr. 69 (1), 30–42.

Byg, I., Diaz, J., Øgendal, L.H., Harholt, J., Jørgensen, B., et al., 2012. Large-scale extraction of rhamnogalacturonan I from industrial potato waste. Food Chem. 131 (4), 1207–1216.

Campbell, L., Rempel, C., Wanasundara, J., 2016. Canola/rapeseed protein: future opportunities and directions—workshop proceedings of IRC 2015. Plants 5 (2), 17.

Celus, I., Brijs, K., Delcour, J.A., 2006. The effects of malting and mashing on barley protein extractability. J. Cereal Sci. 44 (2), 203–211.

Celus, I., Brijs, K., Delcour, J.A., 2007. Enzymatic hydrolysis of Brewers' spent grain proteins and technofunctional properties of the resulting hydrolysates. J. Agric. Food Chem. 55 (21), 8703–8710.

Chan, W.-M., Ma, C.-Y., 1999. Acid modification of proteins from soymilk residue (okara). Food Res. Int. 32 (2), 119–127.

Chang, S., Tan, C., Frankel, E.N., Barrett, D.M., 2000. Low-density lipoprotein antioxidant activity of phenolic compounds and polyphenol oxidase activity in selected clingstone peach cultivars. J. Agric. Food Chem. 48 (2), 147–151.

Chantaro, P., Devahastin, S., Chiewchan, N., 2008. Production of antioxidant high dietary fiber powder from carrot peels. LWT - Food Sci. Technol. 41 (10), 1987–1994.

Chau, C.-F., Hauun, L., 2003. Comparison of the chemical composition and physicochemical properties of different fibers prepared from the peel of Citrus sinensis L. Cv. Liucheng., 51, 2615–2618.

Chávez-santoscoy, R.A., Lazo-vélez, M.A., Serna-sáldivar, S.O., 2016. Delivery of flavonoids and saponins from black bean (Phaseolus vulgaris) seed coats incorporated into whole wheat bread. Int. J. Mol. Sci. 17 (222).

Chee, K.L., Ling, H.K., Ayob, M.K., 2012. Optimization of trypsin-assisted extraction, physico-chemical characterization, nutritional qualities and functionalities of palm kernel cake protein. LWT - Food Sci. Technol. 46 (2), 419–427.

Chinapongtitiwat, V., Jongaroontaprangsee, S., Chiewchan, N., Devahastin, S., 2013. Important flavonoids and limonin in selected Thai citrus residues. J. Funct. Foods. 5 (3), 1151–1158.

Choi, S.-H., Kozukue, N., Kim, H.-J., Friedman, M., 2016. Analysis of protein amino acids, non-protein amino acids and metabolites, dietary protein, glucose, fructose, sucrose, phenolic, and flavonoid content and antioxidative properties of potato tubers, peels, and cortexes (pulps). J. Food Compos. Anal 50, 77–87.

Constenla, D., Ponce, A.G., Lozano, J.E., 2002. Effect of pomace drying on apple pectin. LWT - Food Sci. Technol. 35 (3), 216–221.

Del Valle, M., Cámara, M., Torija, M.E., 2006. Chemical characterization of tomato pomace. J. Sci. Food Agric. 86 (8), 1232–1236.

Deng, Q., Penner, M.H., Zhao, Y., 2011. Chemical composition of dietary fiber and polyphenols of five different varieties of wine grape pomace skins. Food Res. Int. 44 (9), 2712–2720.

Dhingra, S., Kapoor, A.C., 1985. Nutritive value of mango seed kernel. J. Sci. Food Agric. 36 (8), 752–756.

Dhingra, D., Michael, M., Rajput, H., Patil, R.T., 2012. Dietary fibre in foods: a review. J. Food Sci. Technol. 49 (3), 255–266.

Paola Di, D., Annarita, P., Valentina, T., Barbara, N., 2015. Polysaccharides: applications in biology and biotechnology/polysaccharides from bioagro-waste new biomolecules-life. Polysaccharides, 1–2241 pp.

Djilas, S., Čanadanović-Brunet, J., Ćetković, G., 2009. By-products of fruits processing as a source of phytochemicals. Chem. Ind. Chem. Eng. Q. 15 (4), 191–202.

Dumay, J., Morançais, M., 2016. Proteins and Pigments. Elsevier Inc, 275–318 pp.
Elegbede, J.A., Achoba, I.I., Richard, H., 1995. Nutrient composition of mango (Mangnifera indica) seed kernel from Nigeria. J. Food Biochem. 19 (5), 391–398.
Ernster, J.H., 1986. Process for Extracting Proteins from Brewers Grain.
Ervin, V., Alli, I., Smith, J.P., Li, Z., 1989. Extraction and precipitation of proteins from brewer's spent grain. Can. Inst. Food Sci. Technol. J. 22 (3), 216–221.
Ezieshi, E., Olomu, J., 2007. Nutritional evaluation of palm kernel meal types: 1. Proximate composition and metabolizable energy values. Afr. J. Biotechnol. 6 (21), 2484–2486.
FAO, 2011. Rotein Quality Evaluation in Human Nutrition, 1–79 pp.
Fărcaş, A.C., Socaci, S.A., Mudura, E., Dulf, F.V., Vodnar, D.C., et al., 2017. Exploitation of brewing industry wastes to produce functional ingredients. In: Rijeka, M.K. (Ed.), Brewing Technology. IntechOpen.
Femenia, A., Lefebvre, A.C., Thebaudin, J.Y., Robertson, J.A., Bourgeois, C.M., 1997. Physical and sensory properties of model foods supplemented with cauliflower fiber. J. Food Sci. 62 (4), 635–639.
Ferrentino, G., Asaduzzaman, M., Scampicchio, M.M., 2018. Current technologies and new insights for the recovery of high valuable compounds from fruits by-products. Crit. Rev. Food Sci. Nutr. 58 (3), 386–404.
Friedman, M., 1997. Chemistry, Biochemistry, and dietary role of potato polyphenols. A review. J. Agric. Food Chem. 45 (5), 1523–1540.
Galanakis, C.M., 2012. Recovery of high added-value components from food wastes: conventional, emerging technologies and commercialized applications. Trends Food Sci. Technol. 26 (2), 68–87.
Garna, H., Mabon, N., Robert, C., Cornet, C., Nott, K., et al., 2007. Effect of extraction conditions on the yield and purity of apple pomace pectin precipitated but not washed by alcohol. J. Food Sci. 72 (1), C001–C009.
Ghafoor, K., Hui, T., Choi, Y.H., 2011. Optimization of ultrasonic-assisted extraction of total anthocyanins from grape peel using response surface methodology. J. Food Biochem. 35 (3), 735–746.
Gil-ch, G.J., Villa, A., Ayala-zavala, J.F., Heredia, J.B., Sepulveda, D., et al., 2013. Technologies for extraction and production of bioactive compounds to be used as nutraceuticals and food Ingredients : an overview, 12 (Harvey 2008), 5–23.
Golob, P., Farrell, G., Orchard, J.E., 2008. Crop Post-Harvest: Science and Technology, Principles and Practice. Wiley, 568 pp.
González-Centeno, M.R., Rosselló, C., Simal, S., Garau, M.C., López, F., Femenia, A., 2010. Physico-chemical properties of cell wall materials obtained from ten grape varieties and their byproducts: grape pomaces and stems. LWT - Food Sci. Technol. 43 (10), 1580–1586.
González-Montelongo, R., Gloria Lobo, M., González, M., 2010. Antioxidant activity in banana peel extracts: testing extraction conditions and related bioactive compounds. Food Chem. 119 (3), 1030–1039.
Gonzalez-Perez, S., Vereijken, J.M., van Koningsveld, G.A., Gruppen, H., Voragen, A.G.J., 2005. Physicochemical properties of 2S albumins and the corresponding protein isolate from sunflower (Helianthus annuus). J. Food Sci. 70 (1), 98–103.
Gorinstein, S., Martín-Belloso, O., Park, Y.S., Haruenkit, R., Lojek, A., et al., 2001. Comparison of some biochemical characteristics of different citrus fruits. Food Chem. 74 (3), 309–315.

Gumul, D., Ziobro, R., Noga, M., Sabat, R., 2011. Characterisation of five potato cultivars according to their nutritional and pro-health components. Acta Sci. Pol. Technol. Aliment. 10 (1), 77–81.

Gunders, D., 2012. NRDC Wasted Food Report.

Gustavsson, J., Cederberg, C., Sonesson, U., van Otterdijk, R., Meybeck, A., 2011. Global Food Losses and Food Waste Global Food Losses and Food Waste, 37 pp.

Hadnadjev, M., Dapcevic-Hadnadjev, T., Pojic, M., Saric, B., Misan, A., et al., 2017. Progress in vegetable proteins isolation techniques: a review. Food Feed Res. 44 (1), 11–21.

Haghdan, S., Renneckar, S., Smith, G.D., 2015. Sources of lignin. Lignin Polym. Compos. 1–11.

Happi Emaga, T., Robert, C., Ronkart, S.N., Wathelet, B., Paquot, M., 2008. Dietary fibre components and pectin chemical features of peels during ripening in banana and plantain varieties. Bioresour. Technol. 99 (10), 4346–4354.

Hipsley, E.H., 1953. Dietary "fibre" and pregnancy toxaemia. Br. Med. J. 2 (4833), 420–422.

Hoffman, J.R., Falvo, M.J., 2004. Protein - which is best? J. Sport. Sci. Med. 3 (3), 118–130.

Huang, W.-Y., Cai, Y.-Z., Zhang, Y., 2010. Natural phenolic compounds from medicinal herbs and dietary plants: potential use for cancer prevention. Nutr. Canc. 62 (1), 1–20.

Hwang, J.K., Kim, C.J., Kim, C.T., 1998. Extrusion of apple pomace facilitates pectin extraction. J. Food Sci. 63 (5), 841–844.

Ikram, S., Huang, L.Y., Zhang, H., Wang, J., Yin, M., 2017. Composition and nutrient value proposition of brewers spent grain. J. Food Sci. 82 (10), 2232–2242.

Imai, S., 2015. Soybean and processed soy foods ingredients, and their role in cardiometabolic risk prevention. NÃO; É BLOQUEADO. Recent Pat. Food. Nutr. Agric. 7 (2), 75–82.

Iriondo-DeHond, M., Miguel, E., Del Castillo, M.D., 2018. Food byproducts as sustainable ingredients for innovative and healthy dairy foods. Nutrients 10 (10).

Jaime, L., Mollá, E., Fernández, A., Martín-Cabrejas, M.A., López-Andréu, F.J., Esteban, R.M., 2002. Structural carbohydrate differences and potential source of dietary fiber of onion (Allium cepa L.) tissues. J. Agric. Food Chem. 50 (1), 122–128.

Joshi, V.K., Kumar, A., Kumar, V., 2012. Antimicrobial, antioxidant and phyto-chemicals from fruit and vegetable wastes: a review. Int. J. Food Ferment. Technol. 2 (2), 123–136.

Kalpna, R., Mital, K., Sumitra, C., 2011. Vegetable and fruit peels as a novel source of antioxidants. J. Med. Plants Res. 5 (1), 63–71.

Kammerer, D., Claus, A., Schieber, A., Carle, R., 2005. C : food chemistry and toxicology a novel process for the recovery of polyphenols from grape (Vitis vinifera L.) pomace. Food Chem. Toxicol. 70 (2), 0–8.

Kawaii, S., Tomono, Y., Katase, E., Ogawa, K., Yano, M., 1999. Effect of citrus flavonoids on HL-60 cell differentiation. Anticancer Res. 19 (2A), 1261–1269.

Kay, R.M., 1982. Dietary fiber. J. Lipid Res. 23, 221–242.

Keenan, J.M., Pins, J.J., Frazel, C., Moran, A., Turnquist, L., 2002. Oat ingestion reduces systolic and diastolic blood pressure in patients with mild or borderline hypertension: a pilot trial. J. Fam. Pract. 51 (4), 369.

Kennedy, J., 2009. Bananas and people in the homeland of genus Musa: not just pretty fruit. Ethnobot. Res. Appl. 7, 179–198.

Kibler, K.M., Reinhart, D., Hawkins, C., Motlagh, A.M., Wright, J., 2018. Food waste and the food-energy-water nexus: a review of food waste management alternatives. Waste Manag. 74, 52–62.

Kim, K., Tsao, R., Yang, R., Cui, S.W., 2006. Food Chemistry Phenolic acid profiles and antioxidant activities of wheat bran extracts and the effect of hydrolysis conditions, 95, 466–473.

Klockeman, D.M., Toledo, R., Sims, K.A., 1997. Isolation and characterization of defatted canola meal protein. J. Agric. Food Chem. 45 (10), 3867–3870.

Kołodziejczyk, K., Sójka, M., Abadias, M., Vi, I., Guyot, S., Baron, A., 2013. Polyphenol composition, antioxidant capacity , and antimicrobial activity of the extracts obtained from industrial sour cherry pomace. Ind. Crops Prod. 51, 279–288.

Kosmala, M., Milala, J., Kołodziejczyk, K., Markowski, J., Mieszczakowska, M., et al., 2009. Characterization of cell wall polysaccharides of cherry (Prunus cerasus var. Schattenmorelle) fruit and pomace. Plant Foods Hum. Nutr. 64 (4), 279–285.

Koubala, B.B., Mbome, L.I., Kansci, G., Tchouanguep Mbiapo, F., Crepeau, M.J., et al., 2008. Physicochemical properties of pectins from ambarella peels (Spondias cytherea) obtained using different extraction conditions. Food Chem. 106 (3), 1202–1207.

Kratchanova, M., Pavlova, E., Panchev, I., 2004. The effect of microwave heating of fresh orange peels on the fruit tissue and quality of extracted pectin. Carbohydr. Polym. 56 (2), 181–185.

Krishna Kumar, N.S., Yea, M.K., Cheryan, M., 2004. Ultrafiltration of soy protein concentrate: performance and modelling of spiral and tubular polymeric modules. J. Membr. Sci. 244 (1–2), 235–242.

Krishnaswamy, K., Orsat, V., Gariépy, Y., Thangavel, K., 2013. Optimization of microwave-assisted extraction of phenolic antioxidants from grape seeds (Vitis vinifera), 6, 441–455.

Kulkarni, V.M., Rathod, V.K., 2014. Extraction of mangiferin from Mangifera indica leaves using three phase partitioning coupled with ultrasound. Ind. Crops Prod. 52, 292–297.

Kulkarni, V., Rathod, V., 2016. Green process for extraction of mangiferin from *Mangifera indica* leaves. J. Biol. Act. Prod. Nat. 6 (5–6), 406–411.

Kumar, K., Yadav, A.N., Kumar, V., Vyas, P., Dhaliwal, H.S., 2017. Food waste: a potential bioresource for extraction of nutraceuticals and bioactive compounds. Bioresour. Bioprocess. 4 (1), 18.

Lairon, D., Arnault, N., Bertrais, S., Planells, R., Clero, E., et al., 2005. Dietary fiber intake and risk factors for cardiovascular disease in French adults. Am. J. Clin. Nutr. 82 (6), 1185–1194.

Lapornik, B., Prošek, M., Golc Wondra, A., 2005. Comparison of extracts prepared from plant by-products using different solvents and extraction time. J. Food Eng. 71 (2), 214–222.

Lawther, J.M., Sun, R., Banks, W.B., 1996. Effects of extraction conditions and alkali type on yield and composition of wheat straw hemicellulose. J. Appl. Polym. Sci. 60 (11), 1827–1837.

Lee, Y., Choi, H., Eun, J.-B., 2015. Protein Extraction and Purification of Soybean Flakes and Meals Using a Lime Treatment Followed by Ultrafiltration, vol. 5, pp. 7–15.

Lemes, A.C., Sala, L., Ores, J. da C., Braga, A.R.C., Egea, M.B., Fernandes, K.F., 2016. A review of the latest advances in encrypted bioactive peptides from protein-rich waste. Int. J. Mol. Sci. 17 (6).

Li, Y., Guo, C., Yang, J., Wei, J., Xu, J., Cheng, S., 2006. Evaluation of antioxidant properties of pomegranate peel extract in comparison with pomegranate pulp extract. Food Chem. 96 (2), 254–260.

Li, S., Zhu, D., Li, K., Yang, Y., Lei, Z., Zhang, Z., 2013. Soybean curd residue: composition, utilization, and related limiting factors. ISRN Ind. Eng. 2013, 1–8.

Li, X., He, X., Lv, Y., He, Q., 2014. Extraction and functional properties of water-soluble dietary fiber from apple pomace. J. Food Process. Eng. 37 (3), 293–298.

Liu, S., Stampfer, M.J., Hu, F.B., Giovannucci, E., Rimm, E., et al., 1999. Whole-grain consumption and risk of coronary heart disease: results from the Nurses' Health Study. Am. J. Clin. Nutr. 70 (3), 412–419.

Liu, Y., Shi, J., Langrish, T.A.G., 2006. Water-based extraction of pectin from flavedo and albedo of orange peels. Chem. Eng. J. 120 (3), 203–209.

Llobera, A., Cañellas, J., 2007. Dietary fibre content and antioxidant activity of Manto Negro red grape (Vitis vinifera): pomace and stem. Food Chem. 101 (2), 659–666.

Llobera, A., Cañellas, J., 2008. Antioxidant activity and dietary fibre of Prensal Blanc white grape (Vitis vinifera) by-products. Int. J. Food Sci. Technol. 43 (11), 1953–1959.

Luthria, D.L., 2012. Optimization of extraction of phenolic acids from a vegetable waste product using a pressurized liquid extractor. J. Funct. Foods. 4 (4), 842–850.

Lynch, K.M., Steffen, E.J., Arendt, E.K., 2016. Brewers' spent grain: a review with an emphasis on food and health. J. Inst. Brew. 122 (4), 553–568.

M'hiri, N., Ioannou, I., Ghoul, M., Boudhrioua, N.M., 2014. Extraction methods of citrus peel phenolic compounds. Food Rev. Int. 30 (4), 265–290.

Ma, C.-Y., Liu, W.-S., Kwok, K.C., Kwok, F., 1996. Isolation and characterization of proteins from soymilk residue (okara). Food Res. Int. 29 (8), 799–805.

Ma, H., Wang, Q., Qian, D., Gong, L., Zhang, W., 2009. The utilization of acid-tolerant bacteria on ethanol production from kitchen garbage. Renew. Energy. 34 (6), 1466–1470.

Maier, T., Schieber, A., Kammerer, D.R., Carle, R., 2009. Residues of grape (Vitis vinifera L.) seed oil production as a valuable source of phenolic antioxidants. Food Chem. 112 (3), 551–559.

Mandalari, G., Bennett, R.N., Bisignano, G., Saija, A., Giacomo, D., CR, B., Faulds, C.B., Waldron Keith, W.|, 2006. Characterization of flavonoids and pectins from bergamot (Citrus bergamia Risso) peel, a major byproduct of essential oil extraction. J. Agric. Food Chem. 54 (1), 197–203.

Meneses, N.G.T., Martins, S., Teixeira, J.A., Mussatto, S.I., 2013. Influence of extraction solvents on the recovery of antioxidant phenolic compounds from brewer's spent grains. Separ. Purif. Technol. 108, 152–158.

Michel, D., Bachelier, B., Drean, J.-Y., Harzallah, O., 2013. Preparation of cellulosic fibers from sugarcane for textile use. Conf. Pap. Eng. 2013, 1–6.

Min, B., Lim, J., Ko, S., Lee, K.-G., Lee, S.H., Lee, S., 2011. Environmentally friendly preparation of pectins from agricultural byproducts and their structural/rheological characterization. Bioresour. Technol. 102 (4), 3855–3860.

Minjares-Fuentes, R., Femenia, A., Garau, M.C., Meza-Velazquez, J.A., Simal, S., Rossello, C., 2014. Ultrasound-assisted extraction of pectins from grape pomace using citric acid: a response surface methodology approach. Carbohydr. Polym. 106, 179–189.

Mirabella, N., Castellani, V., Sala, S., 2014. Current options for the valorization of food manufacturing waste: a review. J. Clean. Prod. 65, 28–41.

Mohapatra, D., Mishra, S., Sutar, N., 2010. Banana and its by-product Utilisation: an overview. J. Sci. Ind. 69 (May), 323–329.

Montgomery, K.S., 2003. Soy protein. Education 12 (3), 42–45.

Montonen, J., Knekt, P., Jarvinen, R., Aromaa, A., Reunanen, A., 2003. Whole-grain and fiber intake and the incidence of type 2 diabetes. Am. J. Clin. Nutr. 77 (3), 622–629.

Murray, E.D., 1998. Oil Seed Protein Extraction. United States Patent (19),5,844,086.

Mussatto, S.I., Roberto, I.C., 2006. Chemical characterization and liberation of pentose sugars from brewer's spent grain. J. Chem. Technol. Biotechnol. 81 (3), 268–274.

Mussatto, S.I., 2013. Brewer's spent grain: a valuable feedstock for industrial applications. J. Sci. Food Agric. 94 (7), 1264–1275.

Nawirska, A., Kwa'sniewska, M., 2018. Food Chemistry Dietary fibre fractions from fruit and vegetable processing waste. Food Chem. 91 (2), 221–225.

Ng, K.L., Mohd Khan, A., 2012. Enzymatic preparation of palm kernel expeller protein hydrolysate (PKEPH). Int. Food Res. J. 19 (2), 721–725.

Ng, K.L., Ayob, M.K., Said, M., Osman, M.A., Ismail, A., 2013. Optimization of enzymatic hydrolysis of palm kernel cake protein (PKCP) for producing hydrolysates with antiradical capacity. Ind. Crops Prod. 43 (1), 725–731.

Núñez Selles, A.J., Daglia, M., Rastrelli, L., 2016. The potential role of mangiferin in cancer treatment through its immunomodulatory, anti-angiogenic, apoptopic, and gene regulatory effects. BioFactors 42 (5), 475–491.

Nzikou, J.M., K, A., M, L., Loum, B., Pambou, N.P.G., 2010. Extraction and characteristics of seed kernel oil from mango (Mangifera indica). Res. J. Environ. Earth Sci. 2 (1), 31–35.

Oecd/Fao, 2015. OECD-FAO Agricultural Outlook 2015–2024, 136 pp.

Ong, H.R., Khan, R., Prasad, D.M.R., Yousuf, A., Chowdhury, M.N.K., 2018. Palm kernel meal as a melamine urea formaldehyde adhesive filler for plywood applications. Int. J. Adhes. Adhesive. 85, 8–14.

Onwudike, O.C., 1986. Palm kernel meal as a feed for poultry. 1. Composition of palm kernel meal and availability of its amino acids to chicks. Anim. Feed Sci. Technol. 16 (3), 179–186.

Oosterveld, A., Beldman, G., Schols, H.A., Voragen, A.G.J., 1996. Arabinose and ferulic acid rich pectic polysaccharides extracted from sugar beet pulp. Carbohydr. Res. 288, 143–153.

Oosterveld, A., Beldman, G., Schols, H.A., Voragen, A.G.J., 2000. Characterization of arabinose and ferulic acid rich pectic polysaccharides and hemicelluloses from sugar beet pulp. Carbohydr. Res. 328 (2), 185–197.

Östergren, K., Gustavsson, J., Bos-Brouwers, H., Timmermans, T., Hansen, J., et al., 2014. Fusions Definitional Framework for Food Waste. Full report. 134 pp.

Padam, B.S., Tin, H.S., Chye, F.Y., Abdullah, M.I., 2014. Banana by-products: an underutilized renewable food biomass with great potential. J. Food Sci. Technol. 51 (12), 3527–3545.

Pagán, J., Ibarz, A., Llorca, M., Pagán, A., Barbosa-Cánovas, G.V., 2001. Extraction and characterization of pectin from stored peach pomace. Food Res. Int. 34 (7), 605–612.

Panchami, P.S., Gunasekaran, S., 2017. Extraction and characterization of pectin from fruit waste. Int. J. Curr. Microbiol. Appl. Sci. 6 (8), 943–948.

Panchev, I., Kirchev, N., Kratchanov, C., 1988. Improving pectin technology . 11 . Extraction using ultrasonic treatment. Int. J. Food Sci. Technol. 337–341.

Panouillé, M., Ralet, M.-C., Bonnin, E., Thibault, J.-F., 2007. Recovery and reuse of trimmings and pulps from fruit and vegetable processing. Food Process. 417–447. Handb. Waste Manag. Co-Product Recover.

Papargyropoulou, E., Lozano, R., Steinberger, K., Wright, N., Ujang, Z.B., 2014. The food waste hierarchy as a framework for the management of food surplus and food waste. J. Clean. Prod. 76, 106–115.

Parfitt, J., Barthel, M., MacNaughton, S., 2010. Food waste within food supply chains: quantification and potential for change to 2050. Philos. Trans. R. Soc. B Biol. Sci. 365 (1554), 3065–3081.

Peixoto, C.M., Dias, M.I., Alves, M.J., Calhelha, R.C., Barros, L., et al., 2018. Grape pomace as a source of phenolic compounds and diverse bioactive properties. Food Chem. 253, 132–138.

Pérez, S.G., Vereijken, J.M., 2007. Sunflower proteins: overview of their physicochemical, structural and functional properties. J. Sci. Food Agric. 87, 2173–2191.

Petruzziello, L., Iacopini, F., Bulajic, M., Shah, S., Costamagna, G., 2006. Review article: uncomplicated diverticular disease of the colon. Aliment. Pharmacol. Ther. 23 (10), 1379–1391.

Pinheiro, E.R., Silva, I.M.D.A., Gonzaga, L.V., Amante, E.R., Teófilo, R.F., et al., 2008. Optimization of extraction of high-ester pectin from passion fruit peel (Passiflora edulis flavicarpa) with citric acid by using response surface methodology. Bioresour. Technol. 99 (13), 5561–5566.

Praznik, W., Loeppert, R., Viernstein, H., Haslberger, A.G., Unger, F.M., 2015. Dietary fiber and prebiotics. In: Ramawat, K.G., Mérillon, J.-M. (Eds.), Polysaccharides: Bioactivity and Biotechnology. Springer International Publishing, Cham, pp. 891–925.

Ptichkina, N.M., Markina, O.A., Rumyantseva, G.N., 2008. Pectin extraction from pumpkin with the aid of microbial enzymes. Food Hydrocolloids 22 (1), 192–195.

Quist, E.E., Phillips, R.D., Saalia, F.K., 2009. The effect of enzyme systems and processing on the hydrolysis of peanut (*Arachis hypogaea* L.) protein. LWT - Food Sci. Technol. 42 (10), 1717–1721.

Rackis, J.J., Honig, D.H., Sessa, D.J., Cavins, J.F., 1971. Soybean whey proteins-recovery and amino acid analysis. J. Food Sci. 36 (1), 10–13.

Rashid, S., Rakha, A., Anjum, F.M., Ahmed, W., Sohail, M., 2015. Effects of extrusion cooking on the dietary fibre content and Water Solubility Index of wheat bran extrudates. Int. J. Food Sci. Technol. 50 (7), 1533–1537.

ReFED Collaborative, 2016. A Roadmap to Reduce U.S. Food Waste.

Rice-Evans, C., Miller, N., Paganga, G., 1997. Antioxidant properties of phenolic compounds. Trends Plant Sci. 2.

Rizzo, G., Baroni, L., 2018. Soy, soy foods and their role in vegetarian diets. Nutrients 10 (1).

Robbins, R.J., 2003. Phenolic acids in Foods: an overview of analytical methodology. J. Agric. Food Chem. 51 (10), 2866–2887.

Robertson, J.A., 1972. Sunflowers: America's neglected crop. J. Am. Oil Chem. Soc. 49 (4), 239–244.

Rodríguez, R., Jiménez, A., Fernández-Bolaños, J., Guillén, R., Heredia, A., 2006. Dietary fibre from vegetable products as source of functional ingredients. Trends Food Sci. Technol. 17 (1), 3–15.

Rudra, S.G., Jakhar, N., Kaur, C., Nishad, J., 2015. Food industry waste: Mine of nutraceuticals. Int. J. Sci. Environ. Technol. 4 (1), 205–229.

Russo, M., Bonaccorsi, I., Inferrera, V., Dugo, P., Mondello, L., 2015. Underestimated sources of flavonoids, limonoids and dietary fiber: availability in orange's by-products. J. Funct. Foods. 12, 150–157.

Sagar, N.A., Pareek, S., Sharma, S., Yahia, E.M., Lobo, M.G., 2018. Fruit and vegetable waste: bioactive compounds, their extraction, and possible utilization. Compr. Rev. Food Sci. Food Saf. 17 (3), 512–531.

Salihu, A., Muntari, B., 2011. Brewer's spent grain: a review of its potentials and applications. Afr. J. Biotechnol. 10 (3), 324–331.

Samarin, A.M., Poorazarang, H., Hematyar, N., Elhamirad, A., 2012. Phenolics in potato peels: extraction and utilization as natural antioxidants. World Appl. Sci. J. 18 (2), 191–195.

Sanchez, J., Bendini, A., Lecce, G.D., Valli, E., 2017. Macro and micro functional components of a spreadable olive by-product (pâté) generated by new concept of two-phase decanter. Eur. J. Lipid Sci. Technol. 119, 1–9, 1600096.

Sangnark, A., Noomhorm, A., 2003. Effect of particle sizes on functional properties of dietary fibre prepared from sugarcane bagasse. Food Chem. 80 (2), 221–229.

Saulnier, L., Vigouroux, J., Thibault, J.F., 1995. Isolation and partial characterization of feruloylated oligosaccharides from maize bran. Carbohydr. Res. 272 (2), 241–253.

Schwencke, K.V., Noordman, T.R., Mulder, H., Bruijn, P.J.M., Snip, O.C., et al., 2005. Method of Isolating a Protein Concentrate and a Fibre Concentrate from Fermentation Residue. WO 2005/029974 A1.

Schwencke, V., 2006. Sustainable, cost-effective, and feasible solutions for the treatment of brewers' spent grains. Tech. Q. - Master Brew. Assoc. Am. 43 (3), 199–202.

Sharoba, A.M., Farrag, M.A., El-Salam, A., 2013. Utilization of some fruits and vegetables waste as a source of dietary fiber and its effect on the cake making and its quality attributes. J. Agroaliment. Process. Technol. 19 (4), 429–444.

Smith, A.K., Nash, A.M., Eldridge, A.C., Wolf, W.J., 1962. Soybean by-products, recovery of soybean whey protein with edible gums and detergents. J. Agric. Food Chem. 10 (4), 302–304.

Socaci, S.A., Farcas, A.C., Vodnar, D.C., Tofana, M., 2017. Food wastes as valuable sources of bioactive Molecules. Superfood Funct. Food - Dev. Superfoods Their Roles Med.

Steffen, L.M., Jacobs, D.R.J., Stevens, J., Shahar, E., Carithers, T., Folsom, A.R., 2003. Associations of whole-grain, refined-grain, and fruit and vegetable consumption with risks of all-cause mortality and incident coronary artery disease and ischemic stroke: the Atherosclerosis Risk in Communities (ARIC) Study. Am. J. Clin. Nutr. 78 (3), 383–390.

Sun, R.C., Tomkinson, J., 2002. Characterization of hemicelluloses obtained by classical and ultrasonically assisted extractions from wheat straw. Carbohydr. Polym. 50 (3), 263–271.

Sun, R., Tomkinson, J., Wang, S., Zhu, W., 2000. Characterization of lignins from wheat straw by alkaline peroxide treatment. Polym. Degrad. Stabil. 67 (1), 101–109.

Tavares, F., Costa, G., Francisco, V., Liberal, J., Figueirinha, A., et al., 2015. Cymbopogon citratus industrial waste as a potential source of bioactive compounds. J. Sci. Food Agric. 95 (13), 2652–2659.

Tejada-Ortigoza, V., Garcia-Amezquita, L.E., Serna-Saldívar, S.O., Welti-Chanes, J., 2016. Advances in the functional characterization and extraction processes of dietary fiber. Food Eng. Rev. 8 (3), 251–271.

Tejada-Ortigoza, V., García-Amezquita, L.E., Serna-Saldívar, S.O., Welti-Chanes, J., 2017. The dietary fiber profile of fruit peels and functionality modifications induced by high hydrostatic pressure treatments. Food Sci. Technol. Int. 23 (5), 396–402.

Tomás-Barberan, F.A., Ferreres, F., Gil, M.I., 2000. Antioxidant phenolic metabolites from fruit and vegetables and changes during postharvest storage and processing. Stud. Nat. Prod. Chem. 23, 739–795.

Treimo, J., Aspmo, S.I., Eijsink, V.G.H., Horn, S.J., 2008. Enzymatic solubilization of proteins in brewer's spent grain. J. Agric. Food Chem. 56 (13), 5359–5365.

Tzeng, Y.-M., Diosady, L.L., Rubin, L.J., 1990. Production of canola protein materials by alkaline extraction, precipitation, and membrane processing. J. Food Sci. 55 (4), 1147–1151.

Venkat, K., 2011. The climate change and economic impacts of food waste in the United States. Int. J. Food Syst. Dyn. 2 (4), 431–446.

Vergara-Valencia, N., Granados-Pérez, E., Agama-Acevedo, E., Tovar, J., Ruales, J., Bello-Pérez, L.A., 2007. Fibre concentrate from mango fruit: characterization, associated antioxidant capacity and application as a bakery product ingredient. LWT - Food Sci. Technol. 40 (4), 722–729.

Vishwanathan, K.H., Singh, V., Subramanian, R., 2011. Influence of particle size on protein extractability from soybean and okara. J. Food Eng. 102 (3), 240–246.

Wahlström, R., Rommi, K., Willberg-Keyriläinen, P., Ercili-Cura, D., Holopainen-Mantila, U., et al., 2017. High yield protein extraction from brewer's spent grain with novel carboxylate salt - urea aqueous deep eutectic solvents. ChemistrySelect 2 (29), 9355–9363.

Wanasundara, J.P.D., 2011. Proteins of brassicaceae oilseeds and their potential as a plant protein source. Crit. Rev. Food Sci. Nutr. 51 (7), 635–677.

Wang, H.L., Cavins, J.F., 1989. Yield and amino acid composition of fractions obtained during tofu production. Cereal Chem. 66 (5), 359–361.

Wang, L., Weller, C.L., 2006. Recent advances in extraction of nutraceuticals from plants. Trends Food Sci. Technol. 17 (6), 300–312.

Wang, J., Sun, B., Cao, Y., Tian, Y., Li, X., 2008a. Optimisation of ultrasound-assisted extraction of phenolic compounds from wheat bran. Food Chem. 106 (2), 804–810.

Wang, Y.C., Chuang, Y.C., Hsu, H.W., 2008b. The flavonoid, carotenoid and pectin content in peels of citrus cultivated in Taiwan. Food Chem. 106 (1), 277–284.

Wang, L., Liu, H.M., Xie, A.J., Zhu, C.Y., Qin, G.Y., 2018. Dietary fiber extraction from defatted corn hull by hot-compressed water. Pol. J. Food Nutr. Sci. 68 (2), 133–140.

Whelton, S.P., Hyre, A.D., Pedersen, B., Yi, Y., Whelton, P.K., He, J., 2005. Effect of dietary fiber intake on blood pressure: a meta-analysis of randomized, controlled clinical trials. J. Hypertens. 23 (3), 475–481.

Wu, H., Wang, Q., Ma, T., Ren, J., 2009. Comparative studies on the functional properties of various protein concentrate preparations of peanut protein. Food Res. Int. 42 (3), 343–348.

Xing, Q., de Wit, M., Kyriakopoulou, K., Boom, R.M., Schutyser, M.A.I., 2018. Protein enrichment of defatted soybean flour by fine milling and electrostatic separation. Innov. Food Sci. Emerg. Technol. 50, 42–49.

Xiros, C., Christakopoulos, P., 2012. Biotechnological potential of brewers spent grain and its recent applications. Waste Biomass Valoriz. 3 (2), 213–232.

Yahia, E.M., Ornelas-Paz, J., Emanuelli, T., Jacob-Lopes, E., Zepka, L.Q., Cervantes-Paz, B., 2017. Chemistry, stability, and biological actions of carotenoids. In: Fruit and Vegetable Phytochemicals. Wiley-Blackwell, pp. 285–346.

Yan, H., Kerr, W.L., 2012. Total phenolics content, anthocyanins, and dietary fiber content of apple pomace powders produced by vacuum-belt drying. J. Sci. Food Agric. 93 (6), 1499–1504.

Yang, Y.Y., Ma, S., Wang, X.X., Zheng, X.L., 2017. Modification and application of dietary fiber in foods. J. Chem. 2017.

Yapo, B.M., 2009. Lemon juice improves the extractability and quality characteristics of pectin from yellow passion fruit by-product as compared with commercial citric acid extractant. Bioresour. Technol. 100 (12), 3147–3151.

Yılmaz, F.M., Görgüç, A., Karaaslan, M., Vardin, H., Ersus Bilek, S., et al., 2018. Sour cherry by-products: compositions, functional properties and recovery potentials – a review. Crit. Rev. Food Sci. Nutr. 0 (0), 1–15.

Yoon, K.Y., Cha, M., Shin, S.R., Kim, K.S., 2005. Enzymatic production of a soluble-fibre hydrolyzate from carrot pomace and its sugar composition. Food Chem. 92 (1), 151–157.

Yu, J., Ahmedna, M., Goktepe, I., 2007. Peanut protein concentrate: production and functional properties as affected by processing. Food Chem. 103 (1), 121–129.

Zanwar, A.A., Hegde, M.V., Bodhankar, S.L., 2011. Cardioprotective activity of flax lignan concentrate extracted from seeds of Linum usitatissimum in isoprenalin induced myocardial necrosis in rats. Interdiscip. Toxicol. 4 (2), 90–97.

Zee, J.A., Boudreau, A., Bourgeois, M., Breton, R., 1988. Chemical composition and nutritional quality of faba bean (Vicia faba L. Minor) based tofu. J. Food Sci. 53 (6), 1772–1774.

Zhang, L., Zhu, M., Shi, T., Guo, C., Huang, Y., et al., 2017. Recovery of dietary fiber and polyphenol from grape juice pomace and evaluation of their functional properties and polyphenol compositions. Food Funct. 8 (1), 341–351.

Zhao, X., Chen, J., Du, F., 2012. Potential use of peanut by-products in food processing: a review. J. Food Sci. Technol. 49 (5), 521–529.

Zou, T.B., Xia, E.Q., He, T.P., Huang, M.Y., Jia, Q., Li, H.W., 2014. Ultrasound-assisted extraction of mangiferin from mango (Mangifera indica L.) leaves using response surface methodology. Molecules 19 (2), 1411–1421.

Zykwinska, A., Boiffard, M.-H., Kontkanen, H., Buchert, J., Thibault, J.-F., Bonnin, E., 2008. Extraction of green labeled pectins and pectic oligosaccharides from plant byproducts. J. Agric. Food Chem. 56 (19), 8926–8935.

CHAPTER

Bioflavonoids: synthesis, functions and biotechnological applications

3

Madan L. Verma[a,b], Sneh Sharma[a], Raj Saini[c], Varsha Rani[d], Rekha Kushwaha[e]

Department of Biotechnology, Dr Y. S. Parmar University of Horticulture and Forestry, Neri Campus, Himachal Pradesh, India[a]; Centre for Chemistry and Biotechnology, Deakin University, Geelong Campus, VIC, Australia[b]; Department of Basic Sciences, Dr Y. S. Parmar University of Horticulture and Forestry, Himachal Pradesh, India[c]; Department of Biotechnology, Shoolini University Solan, Himachal Pradesh, India[d]; Department of Biochemistry, University of Missouri, Columbia, MO, USA[e]

Chapter outline

Biotechnological Production of Bioactive Compounds. https://doi.org/10.1016/B978-0-444-64323-0.00003-5

1. Introduction

Flavonoids are a group of bioactive compounds which are extensively found in foodstuffs of plant origin (Alvesa et al., 2018; Corell et al., 2018; Perez-Vizcaino and Fraga, 2018; Shang et al., 2018; Thakur and Verma, 2017; Puri and Verma, 2013; Puri et al., 2011). Plants have been used throughout the history, for the production of secondary metabolites having health promoting effects (Thakur et al., 2017). Regular consumption of flavonoids reduces the risk of a number of chronic diseases, including cancer, cardiovascular disease, diabetes, arthrosclerosis, neurodegenerative disorders, anti-ageing, anti-inflammatory, antiallergic, antiviral, and free radical scavenging. Among dietary sources of flavonoids, there are fruits, vegetables, nuts, seeds and spices. Consumption of these substances with diet appears to be safe and seems that a diet rich in flavonoids is beneficial and its promotion is thus justifiable (Vu et al., 2018; Wang et al., 2018; Kozłowska and Wegierek, 2014)

Flavonoids are plant pigments that are synthesized from phenylalanine and generally display marvelous colors to the flowering parts of plants. Flavonoids comprise a large group of poly phenolic compounds, characterized by a benzo-*y*-pyrone structure, which is ubiquitous in vegetables and fruits (Thakur et al., 2017). Besides their relevance in plants, flavonoids are very important for human health because of their high pharmacological activities. Flavonoids are thought to have health-promoting properties *in vivo* as well as *in vitro* systems, due to their high antioxidant capacity. The functionality in human health is supported by the ability of the flavonoids to induce human protective enzyme systems, and by the protective effects against cardiovascular diseases, cancers, and other age-related diseases (Yang et al., 2018; Sarian et al., 2017; Salvamani et al., 2014). Quercetin is one of the most abundant dietary flavonoids which can be obtained from apples, onions, broccoli and red wine. It is the most studied flavonoid and has been reported to exhibit wide range of biological activities including anticancer activity and direct scavenging of small free radicals such as HOO^-, NO^- and O_2^- which are produced by cell metabolic processes as well as endogenous and exogenous impacts (Vu et al., 2018; Sarian et al., 2017).

This chapter discusses chemistry, classification and sources of flavonoids prevalent in the food and biotechnology industries. The present chapter discusses the various methods of flavonoid extraction, chemical and physical methods of extraction, that includes microwave-assisted extraction method, ultrasound-assisted extraction method, accelerated solvent extraction method, and supercritical fluid extraction. Health benefits of flavonoids finds plethora of application due to unique properties of the bioactive compounds such as free radical scavenging, anti-tumour, anti-atherosclerotic, and anti-diabetic.

2. Chemistry and sources of flavonoids

Flavonoids constitute a large class of aromatic polyphenolic secondary metabolites widely existing in plants. More than 9000 flavonoids have been reported in the literature and are present in different types and parts of plants such as vegetables, fruits, grains, legumes, beans, herbs, roots, leaves, seeds etc (Wang et al., 2018). Flavonoids are also abundantly found in foods and beverages of plant origin, such as fruits, vegetables, tea, cocoa and wine; hence they are termed as dietary flavonoids (Panche et al., 2016; Williamson, 2017). Flavonoids were chemically characterized in various plants and synthesized in laboratory in the early 20th century. Early research was mainly carried on the anthocyanins and their role as pigments.

It was during late 1930s that Albert Szent-Györgyi focused on the health benefits of flavonoids and advanced their potential health-promoting activity in his Nobel lecture in 1937 (Perez-Vizcaino and Fraga, 2018).He and his coworkers pointed out that scurvy resulted from a combined deficiency of vitamin C and flavonoids, and coined the term 'Permeabilitäts-Vitamin'or 'vitamin P' for a mixture of citrus flavonoids; the term "vitamin P″ was discontinued in the 1950s. In the 1990s flavonoids were mainly considered as the active components of medicinal plants, while from 2000 onwards, they switched to be mainly regarded as bioactive food ingredients (Perez-Vizcaino and Fraga, 2018).

2.1 Chemistry of flavonoids

The core structure of flavonoids has a three-ring diphenylpropane (C6 –C3 –C6) unit, a fifteen-carbonskeleton (Fig. 3.1). The flavonoid contains two benzene rings (A ring and B ring) which are connected by a C3 moiety (Yang et al., 2015). The C3 moiety forms an aliphatic chain or a six-membered heterocyclic ring (ring C) attached to ring A.

The C ring may be connected to the second aromatic ring (ring B, benzenoid substituent) at the C-2, C-3 or C-4 positions forming flavone, iso-flavone or neoflavone respectively. In chalcone and aurone, six-membered heterocyclic pyran ring (ring C) is replaced by an acyclic moiety or a five membered heterocyclic furan ring respectively. Six-membered ring condensed with the benzene ring is either a 4-Pyrone (or γ-pyrone) as in flavones, flavonols and isoflavones or its dihydro-derivatives such as flavanones and flavanols.

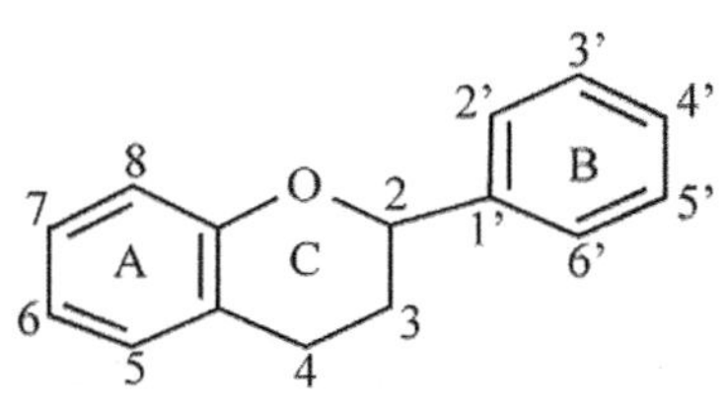

FIG. 3.1 Core structure of flavonoids.

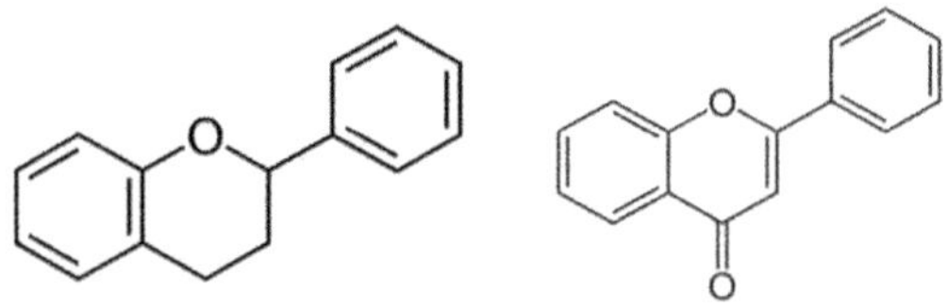

Structure of flavan backbone Structure of the flavone backbone

FIG. 3.2 Structure of flavan and flavone backbone.

Flavonoids of natural origin are also referred as bioflavonoids that encompass both flavonoids and flavanoids. The skeletal structures of flavonoids such as flavones and flavonols, iso-flavones and neo-flavones are based upon flavone backbone whereas flavanoids including flavanones, flavanols, and flavanonols have flavan backbone (Fig. 3.2). Flavonoids differ from flavanoids mainly by a C2–C3 double bond.

This basic structure allows a large number of substituition patterns leading to several subclasses of flavonoids such as anthocyanidins, flavanols, flavanones, flavones, flavonols, isoflavones and chalcones (Sebastian et al., 2017; Hossain et al., 2016) as in Table 3.1. Flavonoids are generally hydroxylated phenolic substances, often hydroxylated in positions 3, 5, 6, 7, 3′, 4′, and 5’, and therefore referred as plant polyphenols. They are present as flavonoid aglycones and with common substitution patterns such as methylation, glycosylation and prenylation which afford flavonoid complexity and diversity in structure (Yang et al., 2018; Hossain et al., 2016). These compounds are biosynthesized in plants through a combination of the shikimate-derived phenylpropanoid (– > C6–C3) pathway and the acetate/mevalonatepolyketide (– > C6) pathway (Rudrapal and Chetia, 2017).

A number of factors may lead to the quantitative variation of a specific flavonoid in a plant such as attack of pests, climatic stress, ultraviolet radiation, and others sources like cultivar, growing location, agricultural practices, harvesting and storage conditions, and processing and preparation methods, as well as analytical variability; among these, both location and cultivar account for 25% to 33% of the variability (Haytowitz et al., 2013).

2.2 Classification and sources of flavonoids

Based upon their chemical structure, flavonoids may be categorized into subclasses flavones, flavanols, flavanones, flavonols, isoflavones, anthocyaninsand chalcones (Table 3.1) which are found in a wide variety of plants. The flavonoids present in selected plant species such as medicinal plants, herbs, spices, fruits, berries, vegetables and legumes has been given in Table 3.2.

2.2.1 Flavonols

Flavonols were the most common flavonoids in many plants, such as fruits and vegetables and the main representatives are quercetin and kaempferol. They are flavonoids with a ketone group and also include fisetin, myricetin and isorhamnetin.

Table 3.1 Classification of flavonoids in plants with selected examples.

Flavonoids classes	Flavonoids subclasses	References
Flavones	Apigenin, Luteolin, Chrysin, Orientin Acacetin, Nevadensin, Scutellarein Reflevone, Calycopterin Cirsimaritin, Isokaempferide Penduletin, Afzelin, Baicalein, Baicalin, Sinensetin, Nobiletin Tangeretin, Diosmin, Saponaretin	Ming et al. (2018); Gutiérrez-Grijalva et al. (2018a, 2018b); Alves et al. (2018); Wang et al. (2018); Zefang et al. (2016); Pan et al. (2018); Mok et al. (2013); Diniz et al. (2015); Gutiérrez-Oladimeji et al. (2018); Fattahi et al. (2013)
Flavonols	Rutin, Quercetin, Kaempferol, Myricetin, Fisetin, Isorhamnetin Quercitrin, Myricitrin, Mearnsetin Astragalin, Epimedin A, Icariin Baohuoside I, Sophorin, Astilbin Engeletin, Laricitrin	Russo et al. (2014); Corell et al. (2018); Zhang et al. (2018); Oladimeji et al. (2018); Wang et al. (2017a,b); Phan et al. (2013); Xi et al. (2014); Georgiev et al. (2014)
Isoflavones	Daidzein, Glycitein, Genistein, Biochanin A, Formononetin, Ononin, Barbigerone	Lazo-Vélez et al. (2018); Guzmán-Ortiz et al. (2017); Guardado-Félix et al. (2017); Wang et al. (2014)
Flavanones	Naringin, Naringenin, Hesperetin, Eriodictyol, Butin, Poncirin, Sakuranin, Hesperidin, Neohesperidin, Eriocitrin, Narirutin, Didymin	Zhang et al. (2014a),b; Choi et al. (2016); Zhu et al. (2013); Zefang et al. (2016); Xi et al. (2014); Velisek (2013)
Flavanols	Monomers, Epigallocatechin-gallate, Epicatechin, Catechin, Gallocathechins, Oligomers, Proanthocyanidins, Procyanidins	Vu et al. (2018); Wojdyło et al. (2016); Taheri et al. (2013)
Anthocyanins	Cyanidin, Malvidin, Capensinidin Rosinidin, Hirsutidin, Delphinidin Pelargonidin, Peonidin, Petunidin	George et al. (2017) Georgiev et al. (2014)
Chalcones	Echinatin, Licochalcone A, Liquiritigenin, Phloretin, Phlorizin	Yang et al. (2018); Han et al. (2017)

They are present in abundance in a wide range of fruits like cherries, berries, *Ziziphus jujube* Mill, grapes, apples, mango, banana (Gündüz, 2016; Wojdyło et al., 2016; Georgiev et al., 2014; Hollands et al., 2017; Pierson et al., 2014; Pan et al., 2018; Passo Tsamo et al., 2015), vegetables like broccoli, tomatoes, onions (Kwak et al., 2017; Albishi et al., 2013; Silva-Beltrán et al., 2015; Neugart et al., 2018) medicinal plants, herbs and spices (Phan et al., 2013; Ghasemzadeh et al., 2016; Mok et al., 2013). Quercetin is the most common flavonol in berries. Flavonols are very diverse with methylation, hydroxylation and the different glycosylation patterns, forming the most common and largest subgroup of flavonoids.

Table 3.2 Flavonoids in selected plant species.

Description	Subclass/Flavonoid	References
Medicinal plants, spices and herbs		
Basil (*Ocimum basilicum*)	Flavanones: Hesperetin; Flavones: Apigenin, luteolin, Nevadensin (trimethoxyflavone); Flavonols: Isorhamnetin, Kaempferol, Quercetin, rutin; Flavan-3-ols: Catechins	Ghasemzadeh et al. (2016); Alhusainy et al. (2014)
Oregano *Hedeoma patens, Lippia graveolens, Lippia palmeri*	Flavonols: Quercetin-hexoside; Flavones: Scutellarein-hexoside, methylscutellarein; luteolin-glucoside-glucuronide, pentahydroxy dihydrochalcone derivative, luteolin trimethyl ether, apigenin	Gutiérrez-Grijalva et al. (2017), 2018a, 2018b
Azadirachta indica	Flavonols: Quercetin-3-O-β-D-glucoside	Tatke et al. (2014)
Scutia buxifolia	Flavonols: Quercetin, quercitrin, isoquercitrin, rutin	Boligon et al., 2012
Podophyllum hexandrum	Flavonols: Quercetin, kaempferol, rutin	Srivastava et al., 2014
Withania somnifera	Flavonols: Quercetin	Shah et al., 2015
Hippophae rhamnoides	Flavonols: Kaempferol, isorhamnetin, quercetin	Gou et al. (2017); Olas (2016) Ursache et al. (2017)
Dioclea virgata	Methoxyflavanone: 7-hydroxy-6-methoxyflavanone, 5,7-dihydroxy-6-methoxyflavanone, 3,7-dihydroxy-6-methoxyflavanone, 7-hydroxy-6-methoxyflavone Methoxyisoflavone: biochanin A; Flavone: luteolin	Alvesa et al., 2018
Dioclea reflexa	Flavone: Reflevone; Flavonol: Mearnsetin, Dioclins A, Dioclins B	Oladimeji et al., 2015, 2018
Australian flame tree *Brachychiton acerifolius*	Flavones: apigenin-7-O-α-rhamnosyl (1 → 2)-β-D-glucuronide, apigenin-7-O-β-D-glucuronide, apigenin-7-O-β-D glucoside, luteolin-7-O-β-D-glucuronide	Zeid et al. (2017)
Moringa oleifera	Flavone glycosides: Quercetin, kaempferol, isoquercetin, astragalin	Wang et al. (2017a); Vongsak (2014); Alhakmani (2013)
Dracocephalum kotschyi Boiss	Methoxylated Flavones: Apigenin, Cirsimaritin, Isokaempferid, Penduletin, Calycopterin, Xanthomicrol	Fattahi et al. (2013)

Table 3.2 Flavonoids in selected plant species. *Continued*

Description	Subclass/Flavonoid	References
Epimedium brevicornum	prenylated and glycosylated flavonols: epimedin A, epimedin B, epimedin C, icariin, baohuoside I	Phan et al. (2013)
Rhododendron mucronulatum for. *albiflorum*	glycosylated flavonols: Afzelin, quercitrin and myricitrin; aglycone flavonols: kaempferol, quercetin and myricetin	Mok et al. (2013)
Euphorbia neriifolia	2-(3,4-dihydroxy-5-methoxy-phenyl)-3,5-dihydroxy-6,7-dimethoxychromen-4-one	Sharma and Janmeda (2017)
Licorice (*Glycyrrhiza sp.*)	Chalcones: licochalcone A (LCA), LCB, LCC, LCD, LCE, Liquiritigenin, Isoliquintigenin and echinatin; Isoflavonoids: Isoangustone A; Isoflavans: Licoricidin, Glabridin, Licorisoflavan A, Dehydroglyasperin C, Dehydroglyasperin D	Fu et al. (2013) Yang et al. (2015); Yang et al. (2018)
Scutellaria baicalensis	Flavones- Baicalein, Baicalin	Moghaddam et al. (2014); Ming et al. (2018); Cheng et al. (2018)
Vegetables and legumes		
Onion (*Allium cepa* L.) Pearl, yellow, white, and red types	Flavonols: Isorhamnetin, Quercetin, kaempferol, myricetin, quercetin glycosides (quercetin 7,4′-diglucoside, quercetin 3-glucoside, quercetin 4′-glucoside), Quercetin aglycone- quercetin-3, 4′—O-diglucoside; Anthocyanins: Cyanidin glucosides {cyanidin-3-(6″-malonyl)-glucoside} and acylated glucosides of cyanidins	Albishi et al. (2013); Kwak et al. (2017); Manohar et al. (2017); Corell et al. (2018)
Tomatoes (*Solanum lycopersicum*)	Flavanone: chalconaringenin Flavonols: kaempferol 3-rutinoside, and rutin, dihydrochalcone phloretin 3′, 5′-di-C-beta-glucopyranoside, quercetin 3-O-(2″—O— beta-apiofuranosyl-6″—O-alpha-rhamnopyranosyl-beta-glucopyranoside	Silva-Beltrán et al. (2015)
Brassica spp.	Acylated and non-acylated quercetin and kaempferol glycosides: monoacylated	Neugart et al., 2018; Mageney et al., 2017

Continued

Table 3.2 Flavonoids in selected plant species. *Continued*

Description	Subclass/Flavonoid	References
	kaempferol-3-*O*-sinapoyl-sophoroside-7-*O*-glycoside, monoacylated kaempferol-3-*O*-feruloyl-sophoroside-7-*O*-glycoside, non-acylated kaempferol-3-*O*-sophoroside-7-*O*-glycoside, monoacylated kaempferol-3-*O*-caffeoyl-sophoroside-7-*O*-glycoside, kaempferol-3-*O*-hydroxyferuloyl-sophoroside-7-*O*-glycoside anthocyanin: (cyanidin-3-(sinapoyl)-(sinapoyl)diglycoside-5-glycoside, cyanidin-3-(sinapoyl)-(feruloyl)diglycoside-5-diglycoside, cyanidin-3-(sinapoyl)(feruloyl)diglycoside-5-glycoside)	
Chick pea (*Cicer arietinum*)	Isoflavones: biochanin A, biochanin B (formononetin), genistein, Ononin, Pseudobaptigenin, glycitein glucoside, isoformononetin glycoside, malonylated formononetin glycoside, 5-hydroxypseudobaptigenin, malonylated pseudobaptigenin glycoside, acetylated 5-hydroxypseudobaptigenin glycoside, biochanin A glycoside, malonylated biochanin A glycoside Isoflavanone-tetrahydroxyhomoisoflavanone	Megías et al. (2016); Gao et al. (2015) ; Guardado-Félix et al. (2017)
Soybean (*Glycine max*)	Isoflavones- genistein, daidzein, glycitein, malonylated genistein and malonylated daidzein glycosides	Lazo-Vélez et al. (2018) ; Guzmán-Ortiz et al. (2017)
Fruits and Berries		
Apples (*Malus domestica*)	Dihydrochalcones- Phloretin, phlorizin Flavonol-Kaempferol; Flavanol: monomers (−)-epicatechin, (+)-catechin and procyanidin dimers	Barreca et al. (2014) ; Han et al. (2017) ; Peng et al. (2016); Hollands et al. (2017)
Mandarin, or tangerine orange (*Citrus reticulata* Blanco)	Flavanone: Hesperidin; Flavonol: Rutin, Quercetin, Kaempferol, Myricetin, Fisetin, Isorhamnetin, Quercitrin, Myricitrin	Zhang et al. (2018) ; Zhang et al. (2014a,b)

Table 3.2 Flavonoids in selected plant species. *Continued*

Description	Subclass/Flavonoid	References
Citrus reticulata 'Chachi'	Flavanone: hesperidin; 6,7,8,3′,4′-pentamethoxyflavanone; 5,6,7,3′,4′-pentamethoxyflavanone; Polymethoxyflavones: sinensetin, nobiletin, tangeretin, 5-hydroxy-6,7,8,4′-tetramethoxyflavone; 5-hydroxy-6,7,8,3′,4′-pentamethoxyflavone; 3,5,6,7,8,3′,4′-heptamethoxyflavone; 5,6,7,8,3′,4′-hexamethoxyflavone; 5,6,7,8,4′-pentamethoxyflavone; 4,5,7,8-tetramethoxyflavone; 5-O-desmethyl nobiletinin	Duan et al. (2017); Fu et al. (2017, 2018)
kinnow (*Citrus reticulata* L.)	Flavanone: Hesperidin, Naringenin; Flavan-3-ols: Catechins, Epicatechins; Flavonols: Quercetin, Kaempferol	Safdar et al. (2017)
Pummelo (*Citrus grandis* L. Osbeck)	Flavanone- Eriocitrin, Hesperidin, Eriodictyol, Naringenin, Hesperitin, *Naringin,* neohesperidin, narirutin, *Didymin;* Flavone- Diosmin, Sinensetin, Nobiletin, Tangeretin; Flavonol-Rutin, sophorin	Zefang et al. (2016) ; Xi et al. (2014)
Grapefruit (*Citrus paradisi*)	Flavanone- Naringin, Neohesperidin, Narirutin, Eriocitrin, Hesperidin, Didymin, Naringenin Flavone- Hesperetin, Nobiletin, Tangeretin	Xi et al. (2014)
Lemon (*C. limon*)	Flavone: Apigenin 6,8-di-*C*-glucoside, Diosmetin 6,8-di-*C*-glucoside, Apigenin7-(malonylpiosyl)-glucoside; Flavanone: Eriocitrin, Hesperidin, Neoesperidin, Narirutin; Flavonol: Rutin	Russo et al. (2014)
Lime (*C. aurantifolia*)	Flavanone: Naringenin/4′,5, 7-Trihydroxyflavanone	
Oranges (*Citrus sinensis*)	Polymethoxylated flavones: nobiletin, tangeritin; C– or O-glycosylated flavones; Flavanone glycoside- narirutin, hesperidin; O-glycosylated flavanones; flavonols	Chen et al. (2017)

Continued

Table 3.2 Flavonoids in selected plant species. *Continued*

Description	Subclass/Flavonoid	References
Grapes (*Vitis vinifera*)	Anthocyanins: 3-*O*-monoglucosides or 3,5-*O*-diglucosides of malvidin, cyanidin, peonidin, delphinidin, pelargonidin and petunidin, as well as their acetyl-, p-coumaroyl- and/or caffeoyl-esters; Flavonols: 3-*O*-glycosides of quercetin, kaempferol, myricetin, laricitrin, isorhamnetin, syringetin; the acetylated and *p*-coumaroylated derivatives 3-*O*-glucosides of isorhamnetin, laricitrin and syringetin; Flavanols: catechin, epicatechin, epicatechin-3-*O*-gallate, 4′–*O*-*β*-glucosyl-(+)-catechin, 7-*O*-*β*-glucosyl-(+)-catechin; Flavanone: Naringenin; Dihydroflavonols: astilbin and engeletin; Proanthocyanidins	Georgiev et al., 2014; Zerbib et al., 2018; Favre et al., 2018
Banana (*Musa paradisiaca*)	Flavan-3-ols: catechin, gallocatechin, and epicatechin; Anthocyanidin: delphinidin, cyanidin, leucocyanidin, Flavonols: quercetin, myricetin, kaempferol Quercetin-deoxyhexose-hexoside, Quercetin -7-rutinoside, Quercetin -3-rutinoside, Quercetin-3/7-rutinoside-3/7-rhamnoside, Kaempferol-deoxyhexose-hexoside, Kaempferol-3-rutinoside, Kaempferol-7-rutinoside, Kaempferol-3/7-rutinoside-3/7-rhamnoside, Isorhamnetin-3- rutinoside, Isorhamnetin-3-rutinoside, Myricetin-deoxyhexose-hexoside, Myricetin-3-rutinoside, Laricitrin-3-rutinoside, Syringetin-3-rutinoside, Rutin	Vu et al. (2018); Singh et al. (2016); Rebello et al. (2014); Passo Tsamo et al. (2015)
Mango (*Mangifera indica*)	Flavonols: Quercetin, quercetin-3-O-β-D-xylopyranoside, quercetin-3-O-β-D-glucoside, quercetin-3-O-α-L-rhamnoside, quercetin-3-O-β-D-arabinoside, quercetin-3-O-β-D-galactoside, quercetin-4′–O-β-Dglucoside,	Pan et al. (2018); Pierson et al. (2014); Hoang et al. (2015)

Table 3.2 Flavonoids in selected plant species. *Continued*

Description	Subclass/Flavonoid	References
	taxifolin/Dihydroquercetin, kaempferol; Flavone: isovitexin/ *saponaretin;* isoswertisin; vitexin; amentoflavone, luteolin-7-O-β-D-glucoside; 3′,5′-dimethoxy-4,5,7-trihydroxyflavone; Flavanols: epicatechin-3-O-gallate	
Papaya *(Carica papaya)*	Flavonols: quercetin, kaempferol, rutin, quercetin 3-(2G-rhamnosylrutinoside), kaempferol 3-(2G-rhamnosylrutinoside), quercetin 3- rutinoside, myricetin 3-rhamnoside, kaempferol 3-rutinoside, kaempferol β-D *Mangifera indica)*-glucopyranoside, Flavones: Luteolin β -D-glucopyranoside	Nugroho et al. (2017); Nguyen et al. (2016);
Melon *(Cucumis melo L.)*	Flavones: Luteolin-7-glycoside, Apigenin-7-glycoside, Luteolin, Flavone, Amentoflavone Flavanone glycosides: Naringenin	Mallek-Ayadi et al. (2017, 2018)
Chokeberry (*Aronia melanocarpa*, *A. prunifolia*, *A. arbutifolia and A. mitschurinii*) black, purple, red Aronia and 'Viking'	Anthocyanin: cyanidin-3-galactoside Proanthocyanidins: catechin Flavonols: quercetin glycosides	Taheri et al. (2013)
Strawberry (*Fragaria ananassa*)	Anthocyanins: Cyanidin-3-glucoside, Cyanidin-3-rutinoside, Cyanidin-3-malonylglucoside, Cyanidin-3-malonylglucosyl-5-glucoside, Pelargonidin-3-galactoside, Pelargonidin-3-glucoside, Pelargonidin-3-rutinoside, Pelargonidin-3-arabinoside, Pelargonidin-3, 5-diglucoside, Pelargonidin-3-malylglucoside, Pelargonidin-3-malonylglucoside, Pelargonidin-3-acetylglucoside, Pelargonidin-disacharide (hexose þ pentose) acylated with acetic acid, 5-Pyranopelargonidin-3-glucoside Flavonols: Quercetin-3-glucuronide, Quercetin-3-malonyglucoside, Quercetin-rutinoside, Quercetin-glucoside, Quercetin-glucuronide, Kaempferol-3-glucoside,	Mandave et al. (2014); Gündüz (2016); Chaves et al., 2017; Kim et al. (2015)

Continued

Table 3.2 Flavonoids in selected plant species. *Continued*

Description	Subclass/Flavonoid	References
	Kaempferol-3-malonylglucoside, Kaempferol-coumaroyl-glucoside, Kaempferol-glucunoride, Rutin, Quercetin; Flavanols: Proanthocyanidin, Catechin	
Ziziphus jujube Mill.	Flavan-3-ols: (+)-catechin, Procyanidin B2, (-)-epicatechin, Polymeric proanthocyanidins Flavonols: Quercetin-3-O-rutinoside-7-O-hexoside, Quercetin-3-O-rutinoside-7-O-pentoside, Quercetin-3-O-robinobioside, Kaempferol-3-O-hexose-O-deoxy-hexose-O-pentoside, Quercetin-3-O-rutinoside, Quercetin-3-O-galactoside (or glucoside), Kaempferol-3-O-robinobioside, Quercetin-3-O-robinobioside, Kaempferol-3-O-rutinoside, Quercetin-3-O-rhamnoside, Quercetin-3-O-arabino-rhamnoside, Quercetin-3-O-xyloso-rhamnoside, Quercetin Flavanones: Eriodictyol derivative Dihydrochalcones: Phloretin-3′,5′-di-glucoside	Wojdyło et al. (2016); Bai et al. (2016)

2.2.2 Flavones

Flavones are a class of flavonoids based on backbone of 2-phenylchromen-4-one. The major flavones are apigenin, luteolin, baicalein, chrysin and their derivatives. The polymethoxylatedflavones tangeretin, nobiletin and sinensetin are present in oranges, grapefruit, pummel, *Citrus reticulata* 'Chachi' and othercitrus fruits (Chen et al., 2017; Xi et al., 2014; Zefang et al., 2016; Fu et al., 2017, 2018). Flavones are widely present in leaves, flowers and fruits as glucosides. Flavones from plants are typically conjugated as 7-O-glycosides, for example, in Australian flame tree as apigenin-7-O-α-rhamnosyl (1 → 2)-β-D-glucuronide, apigenin-7-O-β-D-glucuronide, apigenin-7-O-β-D glucoside, luteolin-7-O-β-D-glucuronide (Zeid et al., 2017) and apigenin7-(malonylpiosyl)-glucoside in *Citrus limon* (Russo et al., 2014).

They may also have acetyl or malonyl moieties as in parsley flavones apigenin 7-O-malonyl apiosylglucoside (malonylapiin) and apigenin 7- O-apiosylglucoside (apiin) (Hostetler et al., 2012; Ancuceanu et al., 2018) and lemon flavones apigenin 6,8-di-C-glucoside, diosmetin 6,8-di-C-glucoside (Russo et al., 2014) where flavone

C-glycosides are most commonly detected as 6-C and 8-C-glucosides. Flavones baicalein and baicalin have been reported in *Scutellaria baicalensis* (Cheng et al., 2018).

2.2.3 Flavanones

It is also called as dihydroflavones, hesperetin, naringenin, eriodictyol, isosakuranetin. Their respective glycosides are the major flavanones in plant species These are present mainly in all citrus fruits such as oranges, lemons, grapefruit, grapes, tomatoes and medicinal herbs belonging to the family *Rutaceae*, *Rosaceae*, and *Leguminosae* (Ninomiya and Koketsu, 2013; Zefang et al., 2016; Russo et al., 2014). Esperetin and naringenin are mainly found in the tomatoes and citrus fruit (Patel et al., 2018). Hesperetin and its derivatives are characteristic flavanones of sweet orange, tangelo, lemon and lime, while naringin is present in the grapefruits and oranges whereas neohesperidin in the sour oranges (Khan et al., 2014). These compounds are responsible for the bitter taste of the juice and peel of citrus fruits (Rajan et al., 2018).

2.2.4 Isoflavones

Isoflavones are predominantly produced by the members of the *Fabaceae* (i.e., *Leguminosae*, or bean) family and found in soyabeans and other leguminous plants. genistin, diadzin, and glycetin are glycoside conjugates of Isoflavones and their aglycone forms are genistein, diadzein, and glycetein (He and Chen, 2013). While isoflavones occur in many types of legumes, soybean contains the highest concentration of isoflavones. Genistein and daidzein are the principal soy isoflavones (Lazo-Vélez et al., 2018; Guzmán-Ortiz et al., 2017). Other isoflavones include biochanin A and biochanin B, formononetin found in chick pea, pea nut etc (Guardado-Félix et al., 2017; Bishial et al., 2015).

Isoflavones are biosynthesized via a branch of the general phenylpropanoid pathway, which begins from the amino acid phenylalanine naringenin, an intermediate of the pathway and is sequentially converted into the isoflavone genistein whereas naringenin chalcone, another intermediate is converted to the isoflavone daidzein by enzymatic action (He and Chen, 2013).

2.2.5 Flavanols

Flavanols are another very important group of flavonoids which occur as simple monomers of (+)- catechin or (-)- epicatechin, as well as in hydroxylated forms (gallocatechins)and their gallic acid esters. The most important among flavanols are catechin, epicatechin, epigallocatechin, epicatechin gallate and epigallocatechin gallate (EGCG). They are present in apples, grapes, berries and tea (Hollands et al., 2017; Zerbib et al., 2018; Gramza-Michałowska et al., 2018; Persson, 2013).When using the term 'flavanol' most scientists probably intend to refer to flavan-3-ols, and are not concerned with flavan-4-ols, compounds with low abundance in the average diet; and when specifying 'flavan-3-ols', they likely do not mean to include afzelechins, which are also low-level contributors, if any, to the average flavanol-3-ol intake (Ottaviani et al., 2018).

Proanthocyanidins are flavanol oligomers. Many are oligomers of catechin and epicatechin and their gallic acid esters. Major dietaryproanthocyanidins include procyanidins which are catechin oligomers (Verstraeten et al., 2013). Procyanidins including prodelphinidins (gallocatechin oligomers) and propelargonidins (afzelechin oligomers) can be found in many plants, most notably apples, cinnamon, aronia fruit, cocoa beans, grape seed, grape skin (Hollands et al., 2017; Teixeira et al., 2016; Taheri et al., 2013; Georgiev et al., 2014; Gültekin-Özgüven et al., 2016).

2.2.6 Anthocyanins

Anthocyanins are the water-soluble colored pigments widely distributed in terrestrial plants. They are responsible for the red, blue and purple coloration of many flowers, leaves, vegetables and fruits (Burton-Freeman et al., 2016).The color and stability of these pigments are influenced by pH, light, temperature, and structure. In acidic condition, anthocyanins appear as red but turn blue when the pH increases (Santos-Buelga and González-Paramás, 2018). Berries, currants, grapes, and some tropical fruits, red to purplish blue-colored leafy vegetables, onions, flowers, grains, roots, and tubers have a high level of anthocyanins (Shi et al., 2017; Kammerer, 2016; Corell et al., 2018; Neugart et al., 2018). Cyanidin-3-glucoside is the major anthocyanin found in most of the plants(Khoo et al., 2017).

Most anthocyanins consist of an aglycone (or anthocyanidins) derived from the 2-phenylbenzopyrylium (flavylium) skeleton, having a positive charge at the oxygen atom of the C-ring of basic flavonoid structure, diversely hydroxylated or methoxylated, which is linked to one or various sugar moieties that can be further acylated by aromatic or aliphatic organic acids (Santos-Buelga and González-Paramás, 2018). These naturally occurring anthocyanins are derived from their sugar-free counterparts which include aurantinidin, capensinidin, cyanidin, delphinidin, europinidin, hirsutidin, malvidin, pelargonidin, peonidin, petunidin, pulchellidin and rosinidin (Kristo et al., 2016). These compounds are common plant pigments that are mainly found in berries (Shi et al., 2017). In plants, the biosynthesis of anthocyanins involves the general precursor phenylalanine or tyrosine derived from the shikimate pathway which is converted to 4-coumaroyl-CoA. One molecule of 4-coumaroyl-CoA is condensed with three molecules of malonyl-CoA to form one molecule of naringenin chalcone, which is subsequently converted to naringenin. Naringenin, the major intermediate compound, undergoes various hydroxylations to form diverse anthocyanidins. Further glycosylation with sugar unit such as glucose, galactose, xylose and other modifications generate anthocyanins. (Zha and Koffas, 2017).

2.2.7 Chalcones

Chalcones are 1,3-diphenyl-2-propene-1-one in which two aromatic rings are linked by a three carbon α, β-unsaturated carbonyl system. This subclass of flavonoids is characterized by the absence of 'ring C' in the basic flavonoid skeleton structure, so referred as open-chain flavonoids. Chalcones are synthesized by Claisen–Schmidt condensation, which involves cross aldol condensation of appropriate aldehydes and ketones by base catalyzed or acid catalyzed reactions followed by dehydration (Ahmad et al., 2016). Some examples of chalcones include phloretin, phloridzin/phlorizin, arbutin, chalconaringenin, echinatin, licochalcone A, liquiritigenin occur in significant amounts in tomatoes, pears, apples, strawberries, licorice etc (Yang et al., 2018; Han et al., 2017; Lee et al., 2016; Wang et al., 2015). These are abundant in edible plants and are considered to be the precursors of flavonoids and isoflavonoids (Fu et al., 2013; Karimi-Sales, 2018).

3. Methods of flavonoid extraction

Many plants are currently in considerable significance due to their special characteristics as a large source of plant derived bioactive that may lead to the development of novel drugs. Through various technologies plants can be explored as reservoirs of bioactive molecules (Benchaachoua et al., 2018).

This has led to increased need for pre-extraction and the extraction procedures, which is an important step in the processing of the bioactive constituents from plant materials (Azwanida, 2015). Several advances have been made in the processing of the bioactive constituents from plant materials over traditional methods such as maceration and Soxhlet extraction. A variety of methods such as solvent extraction, alkaline extraction, electron beam-based extraction, enzyme assisted extraction, accelerated solvent extraction (ASE), microwave-assisted (MAE), ultrasound-assisted extraction (UAE) and supercritical fluid extraction (SFE) have been reported for the extraction of flavonoids to increase yield at lower cost (Table 3.3).

3.1 Chemical methods of extraction

Many of the pharmaceutical industries use selective solvents for the extraction of medicinally active portions of plant through standard procedures. The purpose of all extraction is to separate the soluble plant metabolites. Extraction of *Psidium guajava* L. leaves using ethanolic and hydroalcohol extracts (4:1 v/v) resulted in highest extraction yield with maximum presence of phytoconstituents (Arya et al., 2012).

Maceration process has been adopted and widely used in medicinal plants. The powdered material was extracted by adding boiling water and macerated for 2 h (Handa et al., 2008). Effect of different solvents using maceration at 1:10 w/v sample to solvent ratio for 1 h showed 70% acetone as efficient solvent for *Portucala oleracea* based on total phenolics, and 70% methanol as efficient solvent for flavonoids in *Cosmos caudatus* (Sulaiman et al., 2011). In the case of *Moringa oliefera* maceration with 70% ethanol powdered dried samples at 1:40 w/v exhibited highest phenolics and flavonoids content compared to Soxhlet extraction and percolation using similar solvent. This process has been found more applicable and convenient for small and medium enterprises compared to other methods (Dhanani et al., 2017). Leaves of *Ficus carica* were dried and extracted with 20 mL solvent for 3 days at room temperature 25°C. The higher flavonoids contents were obtained using 70% ethanol (Trifunschi and Ardelean, 2013).

Table 3.3 List of the flavonoid extraction methods.

Type of extraction method	Source of flavonoids	Name of extraction method	Applications	References
Chemical Method	*Silybum marianum* leaves	Maceration and decoction methods.	Extraction of phenolic compounds	Benchaachoua et al. (2018)
	Matricaria chamomilla inflorescence	Maceration method	Extraction of total flavonoid contents	Pereira et al. (2018)
	Euphorbia neriifolia leaves	Soxhlet extraction method	Extraction of flavonoids	Sharma and Janmeda (2017)
	Eryngium creticum L. leaves	Maceration and Soxhlet methods	Extraction of phenolic compounds and flavonoids	Hijazi et al.(2015)
Physical Method	*Cyclocarya paliurus* leaves	Microwave or ultrasound-assisted extraction Method	Extraction of flavonoids	Shang et al. (2018)
	Jussara (*Euterpe edulis M.*) and blueberry (*Vaccinium myrtillus*) fruits	Ultrasound-assisted extraction process	Extraction of phenolic compounds	De Rocha et al. (2018)
	Leaves of *Raphanus sativus* L.	Ultrasonic assisted extraction method	Quercetin extraction	Sharifi et al. (2017)
	Fresh peeled south *Algerian* red potatoes cultivars (*Solanum tuberosum* L)	Soxhlet extraction and ultrasonic-assisted extraction combines with maceration	Extraction of phenolic compounds and total flavonoids contents	Lanez and Ben haoua (2017)
	Passiflora species	Accelerated solvent extraction method	Extraction of five flavonoids	Gomes et al. (2017)
	Spina gleditsiae	Supercritical fluid extraction method	Flavonoid extraction	Liu and Gao (2018)
	Exotic *Gordonia axillaris* fruit	Microwave-assisted extraction method	Extraction of phenolic compounds	Li et al. (2017)

Dried peel powder of *Citrus decumana* was extracted using chloroform, ethyl acetate and methanol, respectively at room temperature over a period of 24 h. The extracts were filtered and concentrated under vacuum and stored for further studies (Sood et al., 2009). Shade dried leaves of *Euphorbia neriifolia* were powdered and Soxhlet extracted by 70% ethanol. After drying in hot air oven at 40–45°C these were extracted with benzene, chloroform, ethanol and ethyl acetate and finally macerated with distilled water. The extracts were further analyzed by TLC to determine the presence of high concentrations of flavonoid contents (Sharma and Janmeda, 2017).

The extraction efficiency was estimated by quantifying the total polyphenol contents using extract of *Silybum marianum* leaves by maceration and decoction method. The higher percentage yield was recorded 24.68% with highest polyphenol contents in the aqueous methanol i.e. 18.75 mg gallic acid equivalent/g, whereas highest flavonoids were obtained using ethanol as solvent by decoction method (Benchaachoua et al., 2018).

The total flavonoid contents and antioxidant activity was estimated maximum in 1 h extraction at 900 rpm, solvent ratio of 36.8% and 74.7% ethanol solution at temperature of 69 °C using dynamic maceration of *Matricaria chamomilla* inflorescence (Pereira et al., 2018).

3.2 Physical methods of extraction

3.2.1 Microwave-assisted extraction method

Microwave-assisted extraction (MAE) has received a great attention due to reduced extraction time, solvent volume and high extraction rate at low cost as compared to conventional methods like maceration and Soxhlet extraction (Spigno and Faveri, 2009) Using MAE method improved recoveries of analytes and reproducibility were observed but this method is limited to small molecule phenolic compounds like in case of citrus mandarin peels (Hayat et al., 2009). Extraction of tannins and anthocyanins may not be possible through MAE method because they were potentially subjected to thermal degradation (Kaufmann and Christen, 2002).

Using MAE method, high flavonoid contents were observed using 70% ethanol in *F. carica* (Trifunschi and Ardelean, 2013). Microwave assisted extraction for four pentacyclic triterpenes was evaluatedin *Centella asiatica*. MAE was capable of increasing the yield twice as compared to heat reflux method with absolute ethanol as solvent at75°C and irradiation power of 600 W for four irradiation cycles (Puttarak and Panichayupakaranant, 2013). An increase in yield was observed in *Dioscorea hispida* using MAE with 85% ethanol as solvent and irradiation power at 100W for 20 min. It was also recorded that optimum value of each parameters like extraction time and irradiation power is a critical factor for increase in yield (Kumoro and Hartati, 2015). Efficient extraction of phenolic compounds from blueberry leaves was reported using microwave in comparison with the extraction performed for 24 h at room temperature (Routray and Orsat, 2014).

A novel and ecofriendly extraction method based on microwave or ultrasound-assisted extraction with deep eutectic solvent extraction was investigated for the extraction of flavonoids from *Cyclocarya paliurus* leaves and the antioxidant activities of theses flavonoids were evaluated at DES water content 30% and extraction time 30 min, at 60°C (Shang et al., 2018).

3.2.2 Ultrasound-assisted extraction method

Ultrasound-assisted extraction (UAE) is a process that uses high intensity, high frequency sound waves ranging from 20 kHz to 2000 kHz and solvents to extract compounds from various plant resources (Handa et at., 2008; Chemat et al., 2008). The benefits of this method have proved significant in many vegetable crops due to reduction in extraction time and solvent consumption. The mechanic effect of cavitation disrupts the cell walls, which allow the extraction of compounds from the plant cell matrices (Trusheva et al., 2007).

UAE was found to be most suitable method for extraction of thermolabile compounds such as tannins and anthocyanins in flower of *Mirabilis jalab* L (Ebrahim et al., 2014). Extraction yield of *Withania somnifera* root using water, ethanol and water-ethanol showed maximum yield i.e. 11.85% at different extraction periods ranging from 5 to 20 min (Dhanani et al., 2017). Maximum phenol contents were obtained in the extract prepared with ethanol i.e. 35.93 mg/g followed by water-ethanol 21.15 mg/g and water 17.63 mg/g.

The UAE of phenolic antioxidants from orange peels with ethanol-water as solvent indicates that sonication power is the most important factor followed by temperature and pressure for better recoveries of cell contents. It is further suggested that this method of extraction can be used an environment friendly technique for the preparation of extracts rich in natural antioxidants by replacing synthetic antioxidants (Puri et al., 2012).

Ultrasound-assisted extraction of phenolic compounds from *Cratoxylum formosum* ssp. Formosum leaves using central composite design was evaluated maximum 50.33% at 65°C and extraction time of 15 min with ethanol as solvent by providing shorter extraction time and reduced energy consumption. These conditions resulted in higher efficiency of extraction of phenol compounds as compared to the conventional methods. This compound can br further used as valuable antioxidant source (Yingngam et al., 2014).

The ultrasound-assisted extraction process for extraction of phenolic compounds from jussara (*Euterpe edulis M.*) and blueberry (*Vaccinium myrtillus*) fruits was optimized using central composite design. Comparative analysis of extraction efficiency for the ultrasound-assisted extraction and the conventional extraction performed with ethanol and water. It was observed that there were no significant differences between the analyzed parameters for both extraction methods using the same solvent, for both fruits, except for the total anthocyanin content of the jussara extract obtained with ethanol solution.

This study suggests that efficient extraction of phenolic compounds can be obtained from jussara and blueberry fruits rapidly using ultrasound-assisted extraction rather than the conventional extraction (De Rocha et al., 2018).

3.2.3 Accelerated solvent extraction method

Accelerated solvent extraction method (ASE) is an efficient extraction method as compared to maceration and Soxhlet extraction method used for the extraction of plant materials. This technology is able to control temperature and pressure for each sample and requires less time for extraction like other solvent techniques.

Cyclohexane-acetone ratio on bixin extraction yield from *Bixa orellana* was recovered at the ratio of 6:4 with 5 min heating at 50 °C (Rahmalia et al., 2015). Highest yield of flavonoids 94% from *Rheum palmatum* were recorded using methanol as solvent by ASE (Tan et al., 2014). The optimum extraction efficiency of isoflavonoids from aerial parts of five *Trifolium* L. species was recorded using methanol-water (75:25, v/v) as an extraction solvent in a 125 °C oven. ASE offered significant benefits at a lower cost of reagents as well as a relatively high precision and accuracy compared with conventional solvent extraction and UAE (Blicharski and Oniszczuk, 2017).

Accelerated solvent extraction method was exploited for quantification of five flavonoids in *Passiflora* species and the optimal extraction was achieved at an extraction temperature of 80 °C, 64% (w/w) ethanol and five number of extraction cycles (Gomes et al., 2017).

3.2.4 Supercritical fluid extraction

Supercritical (Sc) fluid extraction is an excellent method for extraction of bioactive compounds from plant cells. It shares the physical properties of both gas and liquid that makes them suitable for the extraction.SC-solvents strength can be easily manipulated by changing the temperature, pressure. High pressure flavonoids extraction was performed using dried peel and edible fruit powder from *Citrus sulcata* and mixed with 40% ethanol for 30 min extraction time (Wang et al., 2011). Optimum yield of wedelolactone from *Wedelia calendulacea* was achieved at 25 °C, 25 MPa with 10% modifier concentration at 90 min extraction time. This method is very effective due to low cost and low toxicity but the major drawback is the high cost of the equipment (Patil et al., 2013).

There is a great emphasis on the recovery of high value-added products by using sustainable methods of extraction and one of the reasonable methods is to apply SFE for the separation of components with high feed to solvent ratio. This method was mainly focused on extractions of nutraceuticals and other phytochemicals from plant materials and the most common solvent used as a supercritical fluid is carbon dioxide (Oman et al., 2013).

The influence of supercritical fluid extraction from berry seeds on chemical composition and recovery of antioxidant compounds was achieved highest at temperature 80 °C (Gustinelli et al., 2018). Supercritical fluid extraction method was used for flavonoid extraction from *Spina gleditsiae* and the optimum yield i.e. 112% was obtained at extraction temperature of 45 °C, extraction pressure of 40 Mpa and extraction time of 1 h (Liu and Gao, 2018).

4. Health benefits of flavonoids

Flavonoids have many health properties due to the unique properties of the bioactive compounds. Various applications of flavonoids are discussed in this section as follows (Duenas et al., 2011).

4.1 Free radical scavenging

The best described property of almost every group of flavonoids is their capacity to act as antioxidants. The flavones and catechins seem to be the most powerful flavonoids for protecting our body against reactive oxygen species. Reactive Oxygen Species is a phrase used to describe number of reactive molecules and free radicals derived from molecular oxygen. These molecules, produced as by products during the mitochondrial electron transport of aerobic respiration or by oxidoreductase enzymes and metal catalyzed oxidation. Reactive Oxygen Species have the potential to cause a number of deleterious events. Atomic oxygen has two unpaired electrons in separate orbits in its outer electron shell (Amic et al., 2017). This electron structure makes oxygen susceptible to radical formation such as, superoxide; hydrogen peroxide; hydroxyl radical; hydroxyl ion and nitric oxide (Fig. 3.3).

Free radicals and reactive oxygen species (produced during normal oxygen metabolism or are induced by exogenous damage) are damaging our body cells and tissues continuously. Mechanisms by which free radicals interfere with cellular functions are not fully understood, but one of the most important events seems to be lipid peroxidation, which results in cellular membrane damage. This cellular damage causes a shift in the net charge of the cell, changes osmotic pressure, leads to swelling and eventually cell death. Living organisms can protect themselves from the reactive oxygen species by enzymatic as well as nonenzymatic mechanisms. The antioxidant-defense mechanisms of the body can be achieved enzymatically via, superoxide dismutase, catalase, and glutatione peroxidise as well as vianonenzymatic counterparts such as glutathione, ascorbic acid, and α-tocopherol. Flavonoids can prevent injury caused by free radicals in various ways. One way is

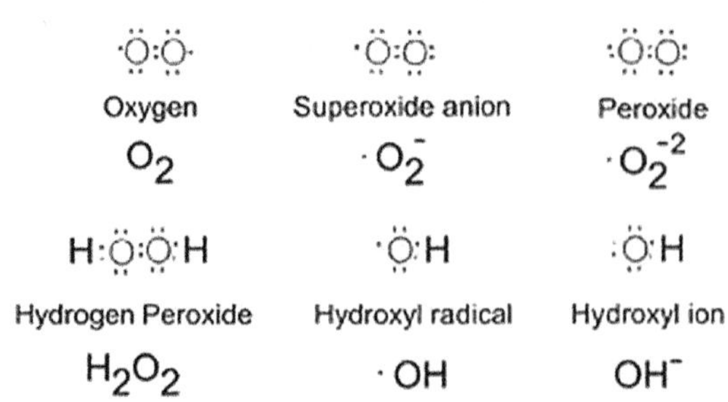

FIG. 3.3

Electron structures of common reactive oxygen species. The red • designates an unpaired electron.

the direct scavenging of free radicals. Flavonoids stabilize the reactive oxygen species by reacting with the reactive compound of the radical. Because of the high reactivity of the hydroxyl group of flavonoids, radicals are made inactive, as shown by the following equation (Nijveldtet al., 2001):

$$\text{Flavonoid(OH)} + \text{R}\cdot > \text{flavonoid(O}\cdot\text{)} + \text{RH(1)}$$

R• – free radical and O• – oxygen free radical.

Selected flavonoids can directly scavenge superoxides, whereas other flavonoids can scavenge the highly reactive oxygenderived radical called peroxynitrite. Flavonoids like epicatechin and rutin are powerful radical scavengers. The scavenging ability of rutin may be due to its inhibitory activity on the enzyme xanthine oxidase. By scavenging radicals, flavonoids can inhibit LDL oxidation *in vitro*. This action protects the LDL particles. Thus, flavonoids may have preventive action against atherosclerosis (Nijveldtet al., 2001). Free radical scavenging activity is reported in *Citrus paradisi* (commonly called as Grape fruit) as well as in its bioactive component naringin (Roghini et al., 2018). Different extracts of *Schima wallichii* is reported to scavenge different free radicals efficiently because of the presence of flavonoids and polyphenols which may be helpful in the treatment of free radical-induced diseases (Lalhminghlui and Jagetia, 2018).

4.2 Anti tumor effects

Flavonoids, the biologically active substances, found in foodstuffs may affect stages of carcinogenesis which are initiation, promotion and progression. Among the various mechanisms of flavonoid action, initiation and promotion stages include: inactivation of the carcinogen, inhibition of cell proliferation, enhancement of DNA repair processes and reduction of oxidative stress. In the progression phase, flavonoids may induce apoptosis, inhibit angiogenesis, exhibit antioxidant activity, and also cytotoxication against cancer cells. Prevention of metabolic activation of procarcinogensis related to flavonoid interaction with phase I enzymes that are responsible for metabolism of various endogenous and exogenous substrates. This results from inhibition of the cytochrome P450 enzymes, such as CYP1A1 and CYP1A2. Flavonoids thus protect against cellular damage arising from the activation of carcinogenic factors. Another mechanism of their action is relatedto reinforcement of mutagen detoxification through induction of the phase II enzymes, such as glutathione S-transferase (GST) and UDP-glucuronyl transferase, which detoxify and eliminate carcinogens from the body (Nijveldtet al., 2001).

Anticancer effects of flavonoids can also be explained by the cell cycle inhibition. Two classes of regulatory molecules are responsible for cell cycle progression: cyclins and cyclin-dependent kinases (CDKs), which are activated under the influence of mitogenic signals within the cell. Uncontrolled activation of CDKs plays a key role in the pathogenesis of cancer. Excessive CDKs activity through gene mutation leads to various types of cancer. Flavonoids can inhibit or modulate CDKs, such as: genistein, quercetin, daidzein, luteolin, kaempferol, apigenin, and epigallocatechin.

Certain flavonoids are reported to inhibit both cancer initiation and progression. Tangeretin (a flavonoid) administrated during tumor initiation reduced the number of precancerous lesions (Kozłowska and Wegierek et al., 2014).

Catechins contained in tea may protect against development of cancer. Flavonoids, luteolin and apigenin are also reported to provide effective protection against cancer development. These flavonoids seem to inhibit CDKs. The hydroxylation pattern of the B ring of the flavones and flavonols, such as luteolin and quercetin, seems to critically influence tumor cell activities, especially the inhibition of protein kinase activity and antiproliferation. It is reported by Ref. Li Chen et al. (2018) that extract of persimmon leaves inhibited liver tumor growth in mice with a tumor inhibition rate of up to 49.35%.

4.3 Anti arthrosclerotic effects

Atherosclerosis is the process of hardening and narrowing of the arteries and is generally associated with cardiovascular diseases such as strokes, heart attacks, and peripheral vascular diseases. Statins are widely used for the clinical treatment for atherosclerosis due to its excellent efficacy in reducing the low-density lipoprotein level. Since the usage of the synthetic drug, statins, leads to various side effects, the plants flavonoids with anti-artherosclerotic activity gained much attention and were proven to reduce the risk of atherosclerosis *in vitro* and *in vivo*. The flavonoids compounds also exhibit lipid lowering effects and anti-inflammatory and antiatherogenic properties. Flavonoids such as quercetin, kaempferol, myricetin, rutin, naringenin, catechin, fisetin, and gossypetin are reported to show antiarthrosclerotic effects. Table 3.4 is showing the summery of reported plants for various flavonoids with their uses (Salvamani et al., 2014).

4.4 Anti diabetic effects

Diabetes mellitus is the most prevalent metabolic syndrome world-wide with an incidence varying between 1% and 8%. Diabetes mellitus arises when insufficient insulin is produced, or when the available insulin does not function properly. Diabetes is characterized by hyperglycaemia which results various short-term metabolic changes in lipid and protein metabolism as well as long-term irreversible vascular changes. Long-term manifestation of diabetes results, in the development of some complications, broadly classified as micro vascular or macro vascular disease. Micro vascular complications include nerve damage, renal disease and vision disorders while macro vascular complications include heart disease, stroke and peripheral vascular disease, which can lead to ulcers, gangrene and amputation (Thakur et al., 2017; Thakur and Verma, 2017; Puri and Verma, 2013). Ethnobotanicals have been used from ancient times for the treatment of blood sugar abnormalities. In India, indigenous remedies have been used in the treatment of diabetes since the time of Charaka and Sushruta. Plants are the great source of bioactive compounds and many of the currently available drugs have been derived directly or indirectly from them (Grover et al., 2001).

Table 3.4 Summary of reported plants for various flavonoids.

Flavonoid	Bioavailability	Plant	References
Quercetin	Anti-inflammatory Antihypertensive Vasodilator effects Antiobesity Antihypercholesterolemic and antiatherosclerotic	*Morus alba* L, *Camellia chinensis, Allium fistulosum, Calamus scipionum*, *Moringa oleifera*, *Centella asiatica*, *Hypericum hircinum* and *Hypericum perforatum*	Sultana and Anwar (2008); Enkhmaa et al. (2005); Hertog et al. (2003); Miean and Mohamed (2001); Siddhuraju and Becker (2003); Bajpai et al. (2005); Chimenti et al. (2006).
Kaempferol	Enhances endothelium vasorelaxation Protective effects against endothelial damage Reduce oxidative stress Antiatherosclerotic Antihyperlipidemic	*Moringa oleifera*, *Centellaasiatica, Ginkgo biloba, Equisetum spp., Tilia spp., Sophora japonica, propolis, Euonymusalatus* and *Kaempferiagalangal*	Sultana and Anwar (2008); Siddhuraju and Becker (2003); Bajpai et al. (2005); Calderon-Montano et al. (2011); Fang et al. (2008); Sulaiman (2008).
Myricetin	Antiplatelet Cytoprotective effects Antihypertensive Antiobesity and antihyperlipidemic Antiartherosclerotic	*Calamus scipronum*, *Moringa oleifera, Aloe vera*, *Ampelopsis cantoniensis, Myrica cerifera* L, *Chrysobalanusicaco* L	Sultana and Anwar (2008); Miean and Mohamed (2001); Ha et al. (2007); Paul et al. (1974); Barbosa et al. (2006).
Rutin	Anti-inflammatory Improves capillary fragility and antihypertensive Suppresses oxidative stress and antiobesity Antiartherosclerotic Antihypercholesterolemic	*Dimorphandramollis, Floshippocastani*, *Rutagraveolens, Rhuscotinus* and *Phyllanthus amarus*	Kanashiro et al. (2009); Buszewski et al. (1993); Afshar and Delazar (2009); Shukla et al. (2009)
Naringenin	Antihypercholesterolemic Antiatherogenic and antiobesity Anti-inflammatory	*Typha angustata*, *Solanum lycopersicum*and *citrus fruits, Mentha aquatica*L, *Citrus aurantium* and *Acacia podalyriifolia*	Lee et al. (2012); Krause and Galensa (1992); Kawaii et al. (1999); Liu et al. (2008); Andrade et al. (2010)

Continued

Table 3.4 Summary of reported plants for various flavonoids.*Continued*

Flavonoid	Bioavailability	Plant	References
Catechin	Antiplatelet and anti-inflammatory Cardioprotective effects Antiatherosclerotic Antihypercholesterolemic Antihypertensive	*Camellia sinensis, Betula pubescens and Betula pendula*, *Cocos nucifera, Argania spinosa* and *Cassia fistula*	Stangl et al. (2007); Khan and Mukhtar (2007); Widlansky et al. (2007); Ossipovet al., 1996; Kirszberg et al. (2003); Charrouf and Guillaume (2007); Daisy et al. (2010)
Fisetin	Antioxidative Anti-inflammatory Antiproliferative Antiobesity Antiatherosclerotic	*Butea frondosa, Gleditsia triacanthos,* and *Quebracho Colorado, Curcuma longa, Rhusverniciflua*, *Acacia greggii*and *Acacia berlandieri*	Gabor and Eperjessy (1966); Lako et al. (2007); Lee et al. (2009); Forbes and Clement (2010)
Gossypetin	Suppresses LDL oxidation Reduces oxidative stress Antihyperlipidemic Antiatherosclerotic	*Hibiscus* spp., *Hibiscus sabdariffa, Hibiscus vitifolius, Hibiscus esculentus, Empetrum nigrum,* and*Acacia constricta, Hibiscus rosa-sinensis, Chiranthodendronpentadactylon, Fremontiacalifornica, Thespesia populnea,* and *Fagoniacretica.*	Lin et al. (2011); Chen et al. (2013); Harborne (1969); Bendz (2013)

Bio-flavonoids are a group of phenolic secondary plant metabolites which are widespread in nature. Bio-flavonoids are well-known for their multi-directional biological activities including anti-diabetic efficacy. Intraperitoneal administration of prunin (naringenin 7-*O*-β-D-glucoside) is reported to produce a significant hypoglycaemic effect in diabetic rats. Anti-hyperglycemic effects have also been demonstrated for various flavonoids including chrysin and its derivatives, silymarin, isoquercetrin and rutin. Flavonoids like, Apigenin (5,7,4′-trihydroxyflavone), luteolin, hesperidin and naringin are also found to lower the blood glucose level. It is suggested by Ref. Hui Xu et al. (2018) that higher intakes of flavonoids and subclasses of flavonoids like anthocyanidins, flavan-3-ols, flavonols, and isoflavones are associated with the lower risk of Type 2 diabetes mellitus.

5. Conclusion and future directions

Flavonoids are an important class of natural products; particularly belonging to a class of plant secondary metabolites with polyphenolic structure and are widely found in fruits, vegetables and certain beverages. Their biochemical and antioxidant effects are associated with various diseases such as cancer, alzheimer's disease, and atherosclerosis. Flavonoids are associated with a broad spectrum of health-promoting effects and are used in nutraceutical, pharmaceutical, medicinal and cosmetic applications. These applications of flavonoids in various industries is due to their antioxidative, anti-inflammatory, anti-mutagenic and anti-carcinogenic properties coupled with their capacity to modulate key cellular enzyme functions.

Flavonoids are the attractive target of world-wide research. Many industries have been engaged in the discovery and commercialization of bioflavonoids. Currently, research is primarily focused on the development of novel methods of flavonoids extraction. However, flavonoids bioactive research needs to be more focus on the *in vitro*, and *in silico* studies. More molecular docking studies need to be explored on the novel flavonoids bioactives that will help to understand the mechanism of bioactives molecules on the receptor molecules at the disease sites. Innovative bioprospecting of flavonoids molecules from the nature is the still need to explored for the selection of hypeproducer source of these bioactive compounds.

References

Afshar, D.J., Delazar, A., 1994. Rutin from *Rutagraveolens* L., *DARU*. J. Pharm. Sci. 4, 1–12.

Ahmad, M.R., Sastry, V.G., Bano, N., Anwar, S., 2016. Synthesis of novel chalcone derivatives by conventional and microwave irradiation methods and their pharmacological activities. Arab. J. Chem. 9, S931–S935.

Albishi, T., John, J.A., Al-Khalifa, A.S., Shahidi, F., 2013. Antioxidative phenolic constituents of skins of onion varieties and their activities. J. Funct. Foods 5, 1191–1203.

Alhakmani, F., Kumar, S., Khan, S.A., 2013. Estimation of total phenolic content, in-vitro antioxidant and anti-inflammatory activity of flowers of *Moringa oleifera*. Asian Pac. J. Trop. Biomed. 3, 623–627.
Alhusainy, W., Williams, G.M., Jeffrey, A.M., Latropoulos, M.J., Taylor, S., Adams, T.B., Rietjens, I.M., 2014. The natural basil flavonoid nevadensin protects against a methyleugenol-induced marker of hepatocarcinogenicity in male F344 rat. Food Chem. Toxicol. 74, 28–34.
Alvesa, C.Q., David, J.M., David, J.P., Kijjoa, A., 2018. Flavonoids and other compounds from *Dioclea virgata* (rich.) Amsh. Biochem. Syst. Ecol. 78, 43–45.
Amic, A., Lucic, B., Stepanic, V., Markovic, Z., Markovic, S., Markovic, J.M.D., Amic, D., 2017. Free radical scavenging activity of quercetin catecholic colonic metabolites: thermodynamics of 2 $H^+/2e^-$ processes. Food Chem. 218, 144–151.
Ancuceanu, R., Anghel, A.I., Hovaneţ, M.V., Dinu, M., Olaru, O.T., Dune, A., Ciolea, M., Stoicescu, C.S., Popescu, C., 2018. Variation of iron contents, polyphenols and flavonoids in *Petroselinum crispum* (Mill.) Fuss (*Apiaceae*). Farmacia 66 (2), 275–281.
Andrade, C.A., Carvalho, J.L.D.S., Cunico, M.M., Lordello, A.L.L., Higaskino, C.E.K., Almeida, S.C.C., Dias, J.F.G., Kerber, V.A., Miguel, M.D., Miguel, O.G., 2010. Antioxidant and antibacterial activity of extracts, fractions and isolated substances from the flowers of Acacia podalyriifolia A. Cunn. ex G. Don. Braz. J. Pharmaceut. Sci. 46, 715–721.
Arya, V., Thakur, N.M., Kashyap, C., 2012. Preliminary phytochemical analysis of the extracts of *Psidium* leaves. J. Pharmacogn. Phytochem. 1, 1–5.
Azwanida, N.N., 2015. A review on the extraction methods use in medicinal plants, principle, strength and limitation. Med. Aromatic Plants 4, 196.
Bai, L., Zhang, H., Liu, Q., Zhao, Y., Cui, X., Guo, S., Zhang, L., Ho, C.T., Bai, N., 2016. Chemical characterization of the main bioactive constituents from fruits of *Ziziphus jujube*. Food and Function 7, 2870–2877.
Bajpai, M., Pande, A., Tewari, S.K., Prakash, D., 2005. Phenoliccontents and antioxidant activity of some food and medicinalplants. Int. J. Food Sci. Nutr. 56, 287–291.
Barbosa, W.L.R., Peres, A., Gallori, S., Vincieri, F.F., 2006. Determination of myricetin derivatives in *Chrysobalanusicaco* L., (Chrysobalanaceae). Braz. J. Pharmacog. 16, 333–337.
Barreca, D., Bellocco, E., Laganà, G., Ginestra, G., Bisignano, C., 2014. Biochemical and antimicrobial activity of phloretin and its glycosilated derivatives present in apple and kumquat. Food Chem. 160, 292–297.
Bendz, G., 2013. Chemistry in botanical classification: medicine and natural sciences. Nobel Symposia Medicine and Natural Science. Academic Press, New York.
Benchaachoua, A., Bessam, H.M., Saidi, I., 2018. Effects of different extraction methods and solvents on the phenolic composition and antioxidant activity of Silybum Marianum leaves extracts. Int. J. Med. Sci. Clin. Invent. 5 (3), 3641–3647. https://doi.org/10.18535/ijmsci/v5i3.16.
Bishi, S.K., Lokesh, K., Mahatma, M.K., Khatediya, N., Chauhan, S.M., Misra, J.B., 2015. Quality traits of Indian peanut cultivars and their utility as nutritional and functional food. Food Chem. 167, 107–114.
Blicharski, T., Oniszczuk, A., 2017. Extraction methods for the isolation of isoflavonoids from plant material. Open Chem. 15, 34–45.
Boligon, A.A., Sagrillo, M.R., Machado, L.F., de Souza Filho, O., Machado, M.M., da Cruz, I.B., Athayde, M.L., 2012. Protective effects of extracts and flavonoids isolated from Scutia buxifolia Reissek against chromosome damage in human lymphocytes exposed to hydrogen peroxide. Molecules 17 (5), 5757–5769.

Burton-Freeman, B., Sandhu, A., Edirisinghe, I., 2016. Anthocyanins. In: Nutraceuticals-Efficacy, Safety and Toxicity, pp. 489–500.

Buszewski, B., Kawka, S., Suprynowicz, Z., Wolski, T., 1993. Simultaneous isolation of rutin and esculin from plant material and drugs using solid-phase extraction. J. Pharm. Biomed. Anal. 11, 211–215.

Calderon-Montano, J.M., Burgos-Moron, E., Lopez-Lazaro, M., 2011. A review on the dietary flavonoid kaempferol. Mini Rev. Med. Chem. 11, 298–344.

Charrouf, Z., Guillaume, D., 2007. Phenols and polyphenols from *Argania spinosa*. Am. J. Food Technol. 2, 679–683.

Chaves, V.C., Calvete, E., Reginatto, F., 2017. Quality properties and antioxidant activity of seven strawberry (Fragaria x ananassa duch) cultivars. Sci Hort 225, 293–298. https://doi.org/10.1016/j.scienta.2017.07.013.

Chemat, F., Tomao, V., Virot, M., 2008. Ultrasound-assisted Extraction in Food Analysis, Handbook of Food Analysis Instruments by Semih Otles. CRC press, Boca Raton, Florida, USA, pp. 85–103.

Chen, J.H., Tsai, C.W., Wang, C.P., Lin, H.H., 2013. Anti-atherosclerotic potential of gossypetin via inhibiting LDL oxidation and foam cell formation. Toxicol. Appl. Pharmacol. 272, 313–324.

Chen, X.M., Tait, A.R., Kitts, D.D., 2017. Flavonoid composition of orange peel and its association with antioxidant and anti-inflammatory activities. Food Chem. 218, 15–21.

Chen, L., Wei, Y., Zhao, S., Zang, M., Yan, X., Gao, X., Li, J., Gao, Y., Xang, A., Gao, Y., 2018. Antitumor and immunomodulatory activities of total flavonoids extract from persimmon leaves in H_{22} livertumor-bearing mice. Sci. Rep. 8, 1–10.

Cheng, C.S., Chen, J., Tan, H.Y., Wang, N., Chen, Z., Feng, Y., 2018. *Scutellaria baicalensis* and cancer treatment: recent progress and perspectives in biomedical and clinical studies. Am. J. Chin. Med. 46, 25–54.

Chimenti, F., Cottiglia, F., Bonsignore, L., Casu, L., Casu, M., Floris, C., Secci, D., Bolasco, A., Chimenti, P., Granese, A., Befani, O., Turini, P., Alcaro, S., Ortuso, F., Trombetta, G., Loizzo, A., Guarino, I., 2006. Quercetin asthe active principle of *Hypericum hircinum* exerts a selective inhibitory activity against MAO-A: extraction, biological analysis and computational study. J. Nat. Prod. 69, 945–949.

Choi, H.S., Kim, M.K., Choi, Y.K., Shin, Y.C., Cho, S.G., Ko, S.G., 2016. *Rhus verniciflua* Stokes (RVS) and butein induce apoptosis of paclitaxel-resistant SKOV-3/PAX ovarian cancer cells through inhibition of AKT phosphorylation. BMC Complement. Altern. Med. 16, 122.

Corell, L., Armenta, S., Esteve-Turrillas, F.A., la Guardia, M., 2018. Flavonoid determination in onion, chili and leek by hard cap espresso extraction and liquid chromatography with diode array detection. Microchem. J. 140, 74–79.

Daisy, P., Balasubramanian, K., Rajalakshmi, M., Eliza, J., Selvaraj, J., 2010. Insulin mimetic impact of catechin isolated from *Cassia fistula* on the glucose oxidation and molecular mechanisms of glucose uptake on Streptozotocin induced diabetic Wistar rats. Phytomedicine 17, 28–36.

De Rocha, J.C.G., Procopio, F.R., Mendonca, A.C., Vieira, L.M., Perrone, I.T., de Barros, F.A.R., Stringheta, P.C., 2018. Optimization of ultrasound-assisted extraction of phenolic compounds from jussara (*Euterpe edulis M.*) and blueberry (*Vaccinium myrtillus*) fruits. Food Sci. Technol, Campinas 38, 45–53.

Dhanani, T., Shah, S., Gajbhiye, N.A., Kumar, S., 2017. Effect of extraction methods on yield, phytochemical constituents and antioxidant activity of *Withania somnifera*. Arab. J. Chem. 10, S1193–S1199.

Diniz, T.C., Silva, J.C., de Lima-Saraiva, S.R., de Almeida Ribeiro, F.P.R., Pacheco, A.G., de Freitas, R.M., Quintans-Júnior, L.J., de Souza Siqueira Quintans, J., Mendes, R.L., da Silva Almeida, J.R., 2015. The role of flavonoids on oxidative stress in epilepsy. Oxid. Med. Cell. Long. 171756.

Duan, L., Dou, L.-L., Yu, K.-Y., Guo, L., Bai-Zhong, C., Li, P., Liu, E.-H., 2017. Polymethoxyflavones in peel of *Citrus reticulata* 'Chachi' and their biological activities. Food Chem. 234, 254–261.

Duenas, M., Laos, F.S., Manzano, S.G., Paramas, A.M.G., Buelga, C.S., 2011. Antioxidant properties of major metabolites of quercetin. Eur. Food Res. Technol. 232, 103–111.

Ebrahim, N., Kershi, M., Butnariub, M., 2014. Anti oxidant activity and anthocyanin content in flower of *mirabilis jalab* l collected from Yemen. World Appl. Sci. J. 29, 247–251.

Enkhmaa, B., Shiwaku, K., Katsube, T., Kitajima, K., Anuurad, E., Yamasaki, M., Yamane, Y., 2005. Mulberry (*Morusalba* L.) leaves and their major flavonol quercetin 3-(6-malonylglucoside) attenuate atherosclerotic lesion developmentin LDL receptor-deficient mice. J. Nutr. 135, 729–734.

Fang, X.K., Gao, J., Zhu, D.N., 2008. Kaempferol and quercetinisolated from *Euonymus alatus* improve glucose uptake of 3T3-L1 cells without adipogenesis activity. Life Sci. 82, 615–622.

Fattahi, M., Nazeri, V., Torras-Claveria, L., Sefidkon, F., Cusido, R.M., Zamani, Z., Palazon, J., 2013. A new biotechnological source of rosmarinic acid and surface flavonoids: hairy root cultures of *Dracocephalum kotschyi* Boiss. Ind. Crops Prod. 50, 256–263.

Favre, G., González-Neves, G., Piccardo, D., Gómez-Alonso, S., Pérez-Navarro, J., Hermosín-Gutiérrez, I., 2018. New acylated flavonols identified in *Vitis vinifera* grapes and wines. Food Res. Int. 112, 98–107.

Forbes, T.D.A., Clement, B.A., 2010. Chemistry of *Acacia's* from South Texas. Texas A and M Agricultural Research and Extension Center, pp. 4–14.

Fu, Y., Chen, J., Li, Y.J., Zheng, Y.F., Li, P., 2013. Antioxidant and anti-inflammatory activities of six flavonoids separated from licorice. Food Chem. 141, 1063–1071.

Fu, M.Q., Xu, Y.J., Chen, Y.L., Wu, J.J., Yu, Y.S., Zou, B., An, K., Xiao, G.S., 2017. Evaluation of bioactive flavonoids and antioxidant activity in pericarpium *Citri reticulatae* (*Citrus reticulata* 'Chachi') during storage. Food Chem. 230, 649–656.

Fu, M., An, K., Xu, Y., Chen, Y., Wu, J., Yu, Y., Zou, B., Xiao, G., Ti, H., 2018. Effects of different temperature and humidity on bioactive flavonoids andantioxidant activity in pericarpium *Citri Reticulata* (*Citrus reticulate* 'Chachi'). LWT – Food Sci. Technol. 93, 167–173.

Gabor, M., Eperjessy, E., 1966. Antibacterial effect of fisetin and fisetinidin. Nature 212, 1273.

Gao, Y., Yao, Y., Zhu, Y., Ren, G., 2015. Isoflavone content and composition in chickpea (*Cicer arietinum* L.) sprouts germinated under different conditions. J. Agric. Food Chem. 63, 2701–2707.

George, V.C., Dellaire, G., Rupasinghe, H.P.V., 2017. Plant flavonoids in cancer chemoprevention: role in genome stability. J. Nutr. Biochem. 45, 1–14.

Georgiev, V., Ananga, A., Tsolova, V., 2014. Recent advances and uses of grape flavonoids as nutraceuticals. Nutrients 6, 391–415.

Ghasemzadeh, A., Ashkani, S., Baghdadi, A., Pazoki, A., Jaafar, H.Z., Rahmat, A., 2016. Improvement in flavonoids and phenolic acids production and pharmaceutical quality of sweet basil (*Ocimum basilicum* L.) by ultraviolet-B irradiation. Molecules 21 pii: E1203.

Gomes, S.V.F., Portugal, L.A., dos Anjos, J.P., de Jesus, O.N., 2017. Accelerated solvent extraction of phenolic compounds exploiting a Box-Behnken design and quantification of five flavonoids by HPLC-DAD in *Passiflora* species. Microchem. J. 132, 28–35.

Gramza-Michałowska, A., Sidor, A., Kulczyński, B., 2017. Berries as a potential anti-influenza factor –A review. J. Funct. Foods 37, 116–137.

Grover, J.K., Vats, V., Rathi, S.S., Dewar, R.J., 2001. Traditional Indian anti-diabetic plants attenuate progression of renal damage in streptozotocin induced diabetic mice. J. Ethnopharmol. 76, 233–238.

Guardado-Félix, D., Serna-Saldivar, S.O., Cuevas-Rodríguez, E.O., Jacobo-Velázquez, D.A., Gutiérrez-Uribe, J.A., 2017. Effect of sodium selenite on isoflavonoid contents and antioxidant capacity of chickpea (*Cicer arietinum* L.) sprouts. Food Chem. 226, 69–74.

Gültekin-Özgüven, M., Berktaş, I., Özçelik, B., 2016. Change in stability of procyanidins, antioxidant capacity and i*n-vitro* bioaccessibility during processing of cocoa powder from cocoa beans. LWT-Food Sci. Technol. 72, 559–565.

Gündüz, K., 2016. Strawberry: phytochemical composition of strawberry (*Fragaria* × *ananassa*). In: Nutritional Composition of Fruit Cultivars. https://doi.org/10.1016/B978-0-12-408117-8.00030-1.

Guo, R., Guo, X., Li, T., Fu, X., Liu, R.H., 2017. Comparative assessment of phytochemical profiles, antioxidant and antiproliferative activities of sea buckthorn (Hip*pophaë rhamnoides L.*) berries. Food Chem. 221, 997–1003.

Gustinelli, G., Eliasson, L., Svelander, C., Andlid, T., Lundin, L., Ahrne, L., Alminger, M., 2018. Supercritical fluid extraction of berry seeds: chemical composition and antioxidant activity. J. Food Qual. 2018, 6046074, p. 10. https://doi.org/10.1155/2018/6046074.

Gutiérrez-Grijalva, E.P., Angulo-Escalante, M.A., León-Félix, J., Heredia, J.B., 2017. Effect of in vitro digestion on the total antioxidant capacity and phenolic content of 3 species of Oregano (*Hedeoma patens, Lippia graveolens, Lippia palmeri*). J. Food Sci. 82, 2832–2839.

Gutiérrez-Grijalva, E.P., Picos-Salas, M.A., Leyva-López, N., Criollo-Mendoza, M.S., Vazquez-Olivo, G., Heredia, J.B., 2018a. Flavonoids and phenolic acids from Oregano: occurrence, biological activity and health benefits. Plants 7, 2.

Gutiérrez-Grijalva, E.P., Antunes-Ricardo, M., Acosta-Estrada, B.A., Gutiérrez-Uribe, J.A., Heredia, J.B., 2018b. Cellular antioxidant activity and in vitro inhibition of α-glucosidase, α-amylase and pancreatic lipase of oregano polyphenols under simulated gastrointestinal digestion. Food Res. Int. https://doi.org/10.1016/j.foodres.2018.08.096. Article (in press).

Guzmán-Ortiz, F.A., San Martín-Martínez, E., Valverde, M.E., Rodríguez-Aza, Y., Berríos, J.J., Mora-Escobedo, R., 2017. Profile analysis and correlation across phenolic compounds, isoflavones and antioxidant capacity during germination of soybeans (Glycine max L.). CyTA – J. Food 15, 516–524.

Ha, D.A., Thuang, P.T., Thuan, N.D., 2007. Protective action of *Ampelopsis cantoniensis* and its major constituent myricetin against LDL oxidation. J. Chem. 45, 768–771.

Han, L., Fang, C., Zhu, R., Peng, Q., Li, D., Wang, M., 2017. Inhibitory effect of phloretin on α-glucosidase: kinetics, interaction, mechanism and molecular docking. Int. J. Biol. Macromol. 95, 520–527.

Handa, S.S., Khanuja, S.P.S., Longo, G., Rakesh, D.D., 2008. Extraction Technologies for Medicinal and Aromatic Plants. no. 66, first ed. United Nations Industrial Development Organization and the International Centre for Science and High Technology, Italy.

Harborne, J.B., 1969. Gossypetin and herbacetin as taxonomic markersin higher plants. Phytochemistry 8, 177–183.
Hayat, K., Sarfraz, H., Shabbar, A., Umar, F., Baomiao, D., Shuqin, X., Chengsheng, J., Xiaoming, Z., Wenshui, X., 2009. Optimized microwave-assisted extraction of phenolic acids from citrus Mandarin peels and evaluation of antioxidant activity *in vitro*. Separ. Purif. Technol. 70, 63–70.
Haytowitz, D.B.S., Bhagwat, S., Holden, J.M., 2013. 36th National nutrient databank conference-sources of variability in the flavonoid content of foods. Proc. Food Sci. 2, 46–51.
He, F.J., Chen, J.Q., 2013. Consumption of soybean, soy foods, soy isoflavones and breast cancer incidence: differences between Chinese women and women in western countries and possible mechanisms. Food Sci. Human Wellness 2, 146–161.
Hertog, M.G.L., Hollman, P.C.H., Van de Putte, B., 1993. Content of potentially anticareinogenic flavonoids of tea infusions, wines, and fruit juices. J. Agric. Food Chem. 41, 1242–1246.
Hijazi, A., Al Masri, D.S., Farhan, H., Nasser, M., Rammal, H., Annan, H., 2015. Effect of different ethanol concentrations, using different extraction techniques, on the antioxidant capacity of Lebanese *Eryngium creticum*. J. Pharm. Chem. Biol. Sci. 3, 262–271.
Hoang, V.L.T., Innes, D.J., Shaw, P.N., Monteith, G.R., Gidley, M.J., Dietzgen, R.G., 2015. Sequence diversity and differential expression of major phenylpropanoid-flavonoid biosynthetic genes among three mango varieties. BMC Genomics 16, 561.
Hollands, W.J., Voorspoels, S., Jacobs, G., Aaby, K., Kroon, P.A., 2017. Development, validation and evaluation of an analytical method for the determination of monomeric and oligomeric *procyanidins* in *apple* extracts. J. Chromatogr. A 1495, 46–56.
Hossain, M.K., Dayem, A.A., Han, J., Yin, Y., Kim, K., Saha, S.K., Yang, G.M., Choi, H.Y., Cho, S.G., 2016. Molecular mechanisms of the anti-obesity and anti-diabetic properties of flavonoids. Int. J. Mol. Sci. 17, 569.
Hostetler, G.L., Riedl, K.M., Schwartz, S.J., 2012. Endogenous enzymes, heat, and pH affect flavone profiles in parsley (*Petroselinum crispum* var. neapolitanum) and celery (*Apium graveolens*) during juice processing. J. Agric. Food Chem. 60, 202–208.
Kammerer, D.R., 2016. 3 anthocyanins. In: Handbook on Natural Pigments in Food and Beverages-Industrial Applications for Improving Food Color, pp. 61–80.
Kanashiro, A., Andrade, D.C., Kabeya, L.M., Turato, W.M., 2009. Modulatoryeffects of rutin on biochemical and hematological parameters in hypercholesterolemic Golden Syrian hamsters. An. Braz. Acad. Sci. 81, 67–71.
Karimi-Sales, E., Mohaddes, G., Alipour, M.R., 2018. Chalcones as putative hepatoprotective agents: preclinical evidence and molecular mechanisms. Pharmacol. Res. 129, 177–187.
Kaufmann, B., Christen, P., 2002. Recent extraction techniques for natural products: microwave-assisted extraction and pressurized solvent extraction. Phytochem. Anal. 13, 105–113.
Kawaii, S., Tomono, Y., Katase, E., Ogawa, K., Yano, M., 1999. Quantitation of flavonoid constituents in citrus fruits. J. Agric. Food Chem. 47, 3565–3571.
Khan, N., Mukhtar, H., 2007. Tea polyphenols for health promotion. Life Sci. 81, 519–533.
Khan, M.K., Zill-E-Huma, Dangles, O., 2014. A comprehensive review on flavanones, the major citrus polyphenols. J. Food Compos. Anal. 33, 85–104.
Khoo, H.E., Azlan, A., Tang, S.T., Lim, S.M., 2017. Anthocyanidins and anthocyanins: colored pigments as food, pharmaceutical ingredients, and the potential health benefits. Food Nutr. Res. 61, 1361779.

Kim, S.K., Kim, D.S., Kim, D.Y., Chun, C., 2015. Variation of bioactive compounds content of 14 oriental strawberry cultivars. Food Chem. 184, 196–202.

Kirszberg, C., Esquenazi, D., Alviano, C.S., Rumjanek, V.M., 2003. The effect of a catechin-rich extract of *Cocos nucifera* on lymphocytes proliferation. Phytother Res. 17, 1054–1058.

Kozlowska, A., Wegierek, D.S., 2014. Flavonoids-food sources and health benefits. Rocz. Panstw. Zakl. Hig. 65, 79–85.

Krause, M., Galensa, R., 1992. Determination of naringenin and naringenin-chalcone in tomato skins by reversed phase HPLC after solid phase extraction. ZeitschriftfürLebensmittel-Untersuchung and Forschung 194, 29–32.

Kristo, A.S., Klimis-Zacas, D., Sikalidis, A.K., 2016. Protective role of dietary berries in cancer. Antioxidants 5, 37.

Kumoro, C., Hartati, I., 2015. Microwave assisted extraction of dioscorin from gadung (*Dioscorea Hispida* Dennst) tuber flour. Proc. Chem. 14, 47–55.

Kwak, J.H., Seo, J.M., Kim, N.H., Arasu, M.V., Kim, S.M.K., Kim, S.J., 2017. Variation of quercetin glycoside derivatives in three onion (*Allium cepa* L.) varieties Saudi. J. Biol. Sci. 24, 1387–1391.

Lako, J., Trenerry, V.C., Wahlqvist, M., Wattanapenpaiboon, N., Sotheeswaran, S., Premier, R., 2007. Phytochemical flavonols, carotenoids and the antioxidant properties of awide selection of Fijian fruit, vegetables and other readily available foods. Food Chem. 101, 1727–1741.

Lalhminghlui, K., Jagetia, G.C., 2018. Evaluation of the free-radical scavenging and antioxidant activities of Chilauni, *Schima wallichii* Korth *in vitro*. Future Sci. 4.

Lanez, T., Ben Haoua, K., 2017. The effect of soxhlet and ultrasonic-assisted extraction on antioxidant components and antioxidant properties of selected south algerian red potatoes cultivars. Sci. Study Res. 18, 435–448.

Lazo-Vélez, M.A., Guardado-Félix, D., Avilés-González, J., Romo-López, I., Serna-Saldívar, S.O., 2018. Effect of germination with sodium selenite on the isoflavones and cellular antioxidant activity of soybean (*Glycine max*). LWT – Food Sci. Technol. 93, 64–70.

Lee, J.D., Huh, J.E., Jeon, G., Yang, H.R., Woo, H.S., Choi, D.Y., Park, D.S., 2009. Flavonol-rich RVHxRfrom *Rhusverniciflua*stokes and its major compound fisetininhibits inflammation-related cytokines and angiogenic factor in rheumatoid arthritic fibroblast-like synovial cells and *in vivo*models. Int. Immunopharmacol. 9, 268–276.

Lee, J.J., Yi, H., Kim, I.S., Kim, Y., Nhiem, N.X., Kim, Y.H., Myung, C.S., 2012. (2S)-Naringenin from *Typhaangustata* inhibits vascular smooth muscle cell proliferation via a G0/G1 arrest. J. Ethnopharmacol. 139, 873–878.

Lee, S.H., Cho, J.Y., Hang, Y.J., Da, E.J., Kim, D., Cho, S.Y., Kim, W.S., Moon, J.H., 2015. Comparison of bioactive compound contents and *in vitro* and *ex vivo* antioxidative activities between peel and flesh of pear (*Pyrus Pyrifolia* Nakai). Food Sci. Biotechnol. 24, 207–216.

Li, Y., Li, S., Lin, S.J., Zhang, J.J., Zhao, C.N., Li, H.B., 2017. Microwave-assisted extraction of natural antioxidants from the exotic *Gordonia axillaris* fruit: optimization and identification of phenolic compounds. Molecules 22, 1481.

Lin, H.H., Chen, J.H., Wang, C.J., 2011. Chemo preventive properties and molecular mechanisms of the bioactive compounds in *Hibiscus sabdariffa* linne. Curr. Med. Chem. 18, 1245–1254.

Liu, X., Gao, X., 2018. Optimization of supercritical CO_2 fluid extraction conditions of flavonoids from *Spina Gleditsiae*. Earth Environ. Sci. 186.

Liu, L., Shan, S., Zhang, K., Ning, Z.Q., Lu, X.P., Cheng, Y.Y., 2008. Naringenin and hesperetin, two flavonoids derived from *Citrusaurantium* up-regulate transcription of adiponectin. Phytother Res. 22, 1400–1403.

Mageney, V., Neugart, S., Albach, D.C., 2017. A guide to the variability of flavonoids in *Brassica oleracea*. Molecules 22 (2), 252.

Mallek-Ayadi, S., Bahloul, N., Kechaou, N., 2017. Characterization, phenolic compounds and functional properties of *Cucumis melo* L. peels. Food Chem. 221, 1691–1697.

Mallek-Ayadi, S., Bahloul, N., Kechaou, N., 2018. Chemical composition and bioactive compounds of *Cucumis melo* L. seeds: potential source for new trends of plant oils. Process Saf. Environ. Protect. 113, 68–77.

Mandave, P.C., Pawar, P.K., Ranjekar, P.K., Mantri, N., Kuvalekar, A.A., 2014. Comprehensive evaluation of in vitro antioxidant activity, total phenols and chemical profiles of two commercially important strawberry varieties. Sci. Hortic. 172, 124–134.

Manohar, C.M., Xue, J., Murayyan, A., Neethirajan, S., John, S., 2017. Antioxidant activity of polyphenols from Ontario grown onion varieties using pressurized low polarity water technology. J. Funct. Foods 31, 52–62.

Megías, C., Cortés-Giraldo, I., Alaiz, M., Vioque, J., Girón-Calle, J., 2016. Isoflavones in chickpea (*Cicer arietinum*) protein concentrates. J. Funct. Foods 21, 186–192.

Miean, K.H., Mohamed, S., 2001. Flavonoid (myricetin, quercetin, kaempferol, luteolin, and apigenin) content of edible tropical plants. J. Agric. Food Chem. 49, 3106–3112.

Ming, J., Zhuoneng, L., Guangxun, Z., 2018. Protective role of flavonoid baicalin from Scutellaria baicalensis in periodontal disease pathogenesis: a literature review. Complement. Therapy Med. 38, 11–18.

Moghaddam, E., Teoh, B.T., Sam, S.S., Lani, R., Hassandarvish, P., Chik, Z., Yueh, A., Abubakar, S., Zandi, K., 2014. Baicalin, a metabolite of baicalein with antiviral activity against dengue virus. Sci. Rep. 4, 5452.

Mok, S.Y., Lee, S., 2013. Identification of flavonoids and flavonoid rhamnosides from *Rhododendron mucronulatum* for. *albiflorum* and their inhibitory activities against aldose reductase. Food Chem. 136, 969–974.

Neugart, S., Baldermann, S., Hanschen, F.S., Klopsch, R., Wiesner-Reinhold, M., Schreiner, M., 2018. The intrinsic quality of brassicaceous vegetables: how secondary plant metabolites are affected by genetic, environmental, and agronomic factors. Sci. Hortic. 233, 460–478.

Nguyen, T.T., Parat, M.O., Hodson, M.P., Pan, J., Shaw, P.N., Hewavitharana, A.K., 2016. Chemical characterization and in vitro cytotoxicity on squamous cell carcinoma cells of *Carica papaya* leaf extracts. Toxins 8 pii: E7.

Nijveldt, R.J., Nood, E.V., Hoorn, D.E.C.V., Boelens, P.G., Nooren, K.V., Leeuwen, P.A.M.V., 2001. Flavonoids: a review of probable mechanisms of action and potential applications. Am. J. Clin. Nutr. 74, 418–425.

Ninomiya, M., Koketsu, M., 2013. Minor flavonoids (chalcones, flavanones, dihydrochalcones, and aurones). In: Ramawat, K.G., Mérillon, J.-M. (Eds.), Natural Products. Springer, New York, USA, pp. 1867–1900.

Nugroho, A., Heryani, H., Choi, J.S., Park, H.J., 2017. Identification and quantification of flavonoids in *Carica papaya* leaf and peroxynitritescavenging activity. Asian Pac. J. Trop. Biomed. 7, 208–213.

Oladimeji, A.O., Oladosu, I.A., Ali, M.S., Ahmed, Z., 2015. Flavonoids from the roots of dioclea reflexa (hook F.). Bull. Chem. Soc. Ethiop. 29, 441–448.

Oladimeji, A.O., Oladosu, I.A., Ali, M.S., Lateef, M., 2018. Dioclins A and B, new antioxidant flavonoids from *Dioclea reflexa*. Nat. Prod. Res. 32 (17).

Olas, B., 2016. Sea buckthorn as a source of important bioactive compounds in cardiovascular diseases. Food Chem. Toxicol. 97, 199–204.

Oman, M., Mojca Skerget, M., Knez, Z., 2013. Application of supercritical fluid extraction for the separation of nutraceuticals and other phytochemical from plant material. Maced. J. Chem. Chem. Eng. 32, 183–226.

Ossipov, V., Nurmi, K., Loponen, J., Haukioja, E., Pihlaja, K., 1996. High-performance liquid chromatographic separation and identification of phenolic compounds from leaves of *Betula pubescens* and *Betula pendula*. J. Chromatogr. A 721, 59–68.

Ottaviani, J.I., Heiss, C., Spencer, J.P.E., Kelm, M., Schroeter, H., 2018. Recommending flavanols and procyanidins for cardiovascular health: revisited. Mol. Aspect. Med. 61, 63–75.

Pan, J., Yi, X., Zhang, S., Cheng, J., Wang, Y., Liu, C., He, X., 2018. Bioactive phenolics from mango leaves (*Mangifera indica* L.). Ind. Crops Prod. 111, 400–406.

Panche, A.N., Diwan, A.D., Chadra, S.R., 2016. Flavonoids: an overview. J. Nutr. Sci. 5, e47.

Passo Tsamo, C.V., Herent, M.F., Tomekpe, K., Happi Emaga, T., Quetin-Leclercq, J., Rogez, H., Andre, C.M., 2015. Effect of boiling on phenolic profiles determined using HPLC/ESI-LTQ-Orbitrap-MS, physico-chemical parameters of six plantain banana cultivars (*Musa* sp). J. Food Compos. Anal. 44, 158–169.

Patel, K., Kumar, V., Rahman, M., Verma, A., Patel, D.K., 2018. New insights into the medicinal importance, physiological functions and bioanalytical aspects of an important bioactive compound of foods 'Hyperin': health benefits of the past, the present, the future. Beni-Suef Univ. J. Basic Appl. Sci. 7, 31–42.

Patil, Sachin, B.S., Wakte, P.S., Shinde, D.B., 2013. Optimization of supercritical fluid extraction and HPLC identification of wedelolactone from *Wedelia calendulacea* by orthogonal array design. J. Adv. Res. 5, 629–635.

Paul, B.D., Rao, G.S., Kapadia, G.J., 1974. Isolation of myricadiol, myricitrin, taraxerol, and taraxerone from *Myrica cerifera* L. root bark. J. Pharm. Sci. 63, 958–959.

Peng, X., Zhang, Liao, Y., 2016. Deming gong inhibitory kinetics and mechanism of kaempferol on α-glucosidase. Food Chem. 190, 207–215.

Pereira, S.V., Reis, R.A.S.P., Garbuio, D.C., Pedro de Freitas, L.A., 2018. Dynamic maceration of *Matricaria chamomilla* inflorescences: optimal conditions for flavonoids and antioxidant activity. Revista Brasileira de Farmacognosia 28, 111–117.

Perez-Vizcaino, F., Fraga, C.G., 2018. Research trends in flavonoids and health. Arch. Biochem. Biophys. 646, 107–112.

Persson, I.A.L., 2013. Tea flavanols: an overview. In: Tea in Health and Disease Prevention, pp. 73–78.

Phan, M.A.T., Wang, J., Tang, J., Yan Zhi Lee, Y.Z., Ng, K., 2013. Evaluation of α-glucosidase inhibition potential of some flavonoids from *Epimedium brevicornum*. LWT – Food Sci. Technol. 53, 492–498.

Pierson, J.T., Monteith, G.R., Roberts-Thomson, S.J., Dietzgen, R.G., Gidley, M.J., Shaw, P.N., 2014. Phytochemical extraction, characterisation and comparative distribution across four mango (*Mangifera indica* L.) fruit varieties. Food Chem. 149, 253–263. https://doi.org/10.1016/j.foodchem.2013.10.108.

Puri, M., Verma, M., 2013. Innovation in technology development with reference to enzymatic extraction of Flavonoids. In: Ghosh, D., Das, S., Bagchi, D., Smarta, R.B. (Eds.), Innovation in Healthy and Functional Foods. CRC Press, Boca Raton, Fl, pp. 331–342.

Puri, M., Verma, M.L., Mahale, K., 2011. Processing of citrus peel for the extraction of flavonoids for biotechnological applications. In Flavonoids: Dietary sources, Properties and Health Benefits. Nova Publishers, USA, pp. 443–459.

Puri, M., Verma, M.L., Mahale, K., 2012. Processing of citrus peel for the extraction of flavonoids for biotechnological applications. In: Yamane, K., Kato, Y. (Eds.), Handbook on Flavonoids: Dietary Sources, Properties, and Health Benefits. Nova Science Publishers, Hauppauge, New York, pp. 443–459.

Puttarak, P., Panichayupakaranant, P., 2013. A new method for preparing pentacyclic triterpene rich *Centella asiatica* extracts. Nat. Prod. Res. 27, 7.

Rahmalia, W., Fabre, J.F., Mouloungui, Z., 2015. Effects of cyclohexane/acetone ratio on bixin extraction yield by accelerated solvent extraction method. Proc. Chem. 14, 455–464.

Rajan, V.K., Muraleedharan, K., Hussan, K.P.S., 2018. Structural evaluation and toxicological study of a bitter masking bioactive flavanone, 'eriodictyol'. In: Polyphenols: Prevention and Treatment of Human Disease, second ed., pp. 45–60.

Rebello, L.P.G., Ramos, A.M., Pertuzatti, P.B., Barcia, M.T., Castillo-Muñoz, N., Hermosín-Gutiérrez, I., 2014. Flour of banana (Musa AAA) peel as a source of antioxidant phenolic compounds. Food Res. Int. 55, 397–403.

Roghini, R., Vijayalakshmi, K., 2018. Free radical scavenging activity of ethanolic extract of *Citrus paradise* and Naringin an *in vitro* study. Int. J. Pharmacog. Phytochem. Res. 10, 11–16.

Rudrapal, M., Cheti, D., 2017. Plant Flavonoids as Potential Source of Future Antimalarial leads. Sys. Rev. Pharm. 8 (1), 13–18.

Routray, W., Orsat, V., 2014. MAE of phenolic compounds from blueberry leaves and comparison with other extraction methods. Ind. Crops Prod. 58, 36–45.

Russo, M., Bonaccorsi, I., Torre, G., Sarò, M., Dugo, P., Mondello, L., 2014. Underestimated sources of flavonoids, limonoids and dietary fibre: availability in lemon's by-products. J. Funct. Foods 9, 18–26.

Safdar, M.N., Kausar, T., Jabbar, S., Mumtaz, A., Ahad, K., Saddozai, A.A., 2017. Extraction and quantification of polyphenols from kinnow (*Citrus reticulate* L.) peel using ultrasound and maceration techniques. J. Food Drug Anal. 25, 488–500.

Salvamani, S., Gunasekaran, B., Shaharuddin, N.A., Ahmed, S.A., Shukor, M.Y., 2014. Antiartherosclerotic effects of plant flavonoids. BioMed Res. Int. 1–11.

Santos-Buelga, C., González-Paramás, A.M., 2018. Anthocyanins. In: Reference Module in Food Science.

Sarian, M.N., Ahmed, Q.U., Soad, S.Z.M., Alahassan, A.M., Murugesu, S., Perumal, V., Mohamad, S.N.A.S., Khatib, A., Latip, J., 2017. Antioxidant and anti diabetic effects of flavonoids: a structure activity relationship-based study. BioMed Res. Int. 1–14.

Sebastian, R.S., Enns, C.W., Goldman, J.D., Steinfeldt, L.C., Martin, C.L., Clemens, J.C., Murayi, T., Moshfegh, A.J., 2017. New, publicly available flavonoid data products: valuable resources for emerging science. J. Food Compos. Anal. 64, 68–72.

Shah, N., Singh, R., Sarangi, U., Saxena, N., Chaudhary, A., Kaur, G., Kaul, S.C., Wadhwa, R., 2015. Combinations of Ashwagandha leaf extracts protect brain-derived cells against oxidative stress and induce differentiation. PLoS One 10 (3), e0120554. https://doi.org/10.1371/journal.pone.0120554. PubMed PMID: 25789768; PubMed Central PMCID: PMC4366112.

Shang, X., Tan, J.N., Du, Y., Liu, X., Zhang, Z., 2018. Environmentally-friendly extraction of flavonoids from *Cyclocarya paliurus* (Batal.) iljinskaja leaves with deep eutectic solvents and evaluation of their antioxidant activities. Molecules 23.

Sharifi, N., Mahernia, S., Amanlau, M., 2017. Comparision of different methods in quercetin extraction from leaves of *Raphanus sativa* L. Pharmaceut. Sci. 23, 59–65.

Sharma, V., Janmeda, P., 2017. Extraction, isolation and identification of flavonoids from *Euphobia nerrifolia* leaves. Arab. J. Chem. 10, 509–514.

Shi, M., Loftus, H., McAinch, A.J., Su, X.Q., 2017. Blueberry as a source of bioactive compounds for the treatment of obesity, type 2 diabetes and chronic inflammation. J. Funct. Foods 30, 16–29.

Shukla, P., Gopalkrishna, B., Shukla, P., 2012. Isolation of rutin from *Phyllanthus amarus*. Int. J. Pharm. Sci. Res. 3, 1198–1201.

Siddhuraju, P., Becker, K., 2003. Antioxidant properties of various solvent extracts of total phenolic constituents from three different agroclimatic origins of drumstick tree (*Moringa oleifera* Lam.) leaves. J. Agric. Food Chem. 51, 2144–2155.

Silva-Beltrán, N.P., Ruiz-Cruz, S., Cira-Chávez, L.A., Estrada-Alvarado, M.I., de Jesús Ornelas-Paz, J., López-Mata, M.A., Del-Toro-Sánchez, C.L., Ayala-Zavala, J.F., Márquez-Ríos, E., 2015. Total phenolic, flavonoid, tomatine, and tomatidine contents and antioxidant and antimicrobial activities of extracts of tomato plant. Int. J. Anal. Chem. 2015. Article ID 284071, 10 pages.

Singh, B., Singh, J.P., Kaur, A., Singh, N., 2016. Bioactive compounds in banana and their associated health benefits-A review. Food Chem. 206, 1–11.

Sood, S., Arora, B., Bansal, S., Muthuraman, A., Gill, N., Arora, R., Bali, M., Sharma, P., 2009. Antioxidant, anti-inflammatory and analgesic potential of the *Citrus decumana* L. peel extract. Injlammopharmacology 17, 267–274.

Spigno, G., Faveri, D.M.,D., 2009. Microwave-assisted extraction of tea phenols: a phenomenological study. J. Food Eng. 93, 210–217.

Srivastava, N.N., Shukla, S.K., Yashavarddhan, M.H., Devi, M., Tripathi, R.P., Gupta, M.L., 2014. Modification of radiation-induced DNA double strand break repair pathways by chemicals extracted from *Podophyllum hexandrum*: an in vitro study in human blood leukocytes. Environ. Mol. Mutagen. 55, 436–448.

Stangl, V., Dreger, H., Stangl, K., Lorenz, M., 2007. Molecular targets of tea polyphenols in the cardiovascular system. Cardiovasc. Res. 73, 348–358.

Sulaiman, M.R., Zakaria, Z.A., Daud, I.A., Ng, F.N., Ng, Y.C., Hidayat, M.T., 2008. Antinociceptive and anti-inflammatoryactivities of the aqueous extract of *Kaempferiagalanga* leaves in animal models. J. Nat. Med. 62, 221–227.

Sulaiman, S.F., Sajak, A.A.B., Ooi, K.L., Supriatno, Seow, E.M., 2011. Effect of solvents in extracting polyphenols and antioxidants of selected raw vegetables. J. Food Compos. Anal. 24, 506–515.

Sultana, B., Anwar, F., 2008. Flavonols (kaempeferol, quercetin, myricetin) contents of selected fruits, vegetables and medicinal plants. Food Chem. 108, 879–884.

Taheri, R., Connolly, B.A., Brand, M.H., Bolling, B.W., 2013. Underutilized chokeberry (*Aronia melanocarpa, Aronia arbutifolia, Aronia prunifolia*) accessions are rich sources of anthocyanins, flavonoids, hydroxycinnamic acids, and proanthocyanidins. J. Agric. Food Chem. 61, 8581–8588.

Tan, Ji, T., Jiang, G., Hu, F., 2014. Simultaneous identification and quantification of five flavonoids in the seeds of *Rheum palmatum* L. by using accelerated solvent extraction and HPLC-PDA-ESI/MS. Arab. J. Chem. 31.

Tatke, P., Desai, S., Gabhe, S.Y., 2014. Isolation of quercetin-3-O-β-D-Glucoside from *Azadirachta indica*. Am. J. Phytomed. Clin. Ther. 2, 870–876.

Teixeira, N., Mateus, N., de Freitas, V., 2016. Updating the research on prodelphinidins from dietary sources. Food Res. Int. 85, 170–181.

Thakur, R., Verma, M.L., 2017. Food Bioactives with Special Reference to Himalayan Tartary Buckwheat. LAMBERT Academic Publishing, Saarbrücken, Germany.

Thakur, R., Kumar, S., Awasthi, C.P., Verma, M.L., 2017. Biochemical evaluation of tartary buckwheat (*Fagopyrum tataricum* Gaertn.) genotypes of cold desert of Himachal Pradesh. Biosci. Biotechnol. Res. Asia 14, 1–16.

Trifunschi, S.I., Ardelean, D.G., 2013. Flavonoid extraction from *Ficus carica* leaves using different techniques and solvents. J. Nat. Sci. Matica Srpska Novi Sad 125, 79–84.

Trusheva, B., Trunkova, D., Bankova, V., 2007. Different extraction methods of biologically active components from propolis: a preliminary study. Chem. Cent. J. 13.

Ursache, F.M., Ghinea, I.O., Turturică, M., Aprodu, I., Râpeanu, G., Stănciuc, N., 2017. Phytochemicals content and antioxidant properties of sea buckthorn (*Hippophae rhamnoides L.*) as affected by heat treatment-quantitative spectroscopic and kinetic approaches. Food Chem. 233, 442–449.

Velisek, J., 2013. The Chemistry of Food. John Wiley & Sons, 1128 pages.

Verstraeten, S.V., Jaggers, G.K., Fraga, C.G., Oteiza, P.I., 2013. Procyanidins can interact with Caco-2 cell membrane lipid rafts: involvement of cholesterol. Biochim. Biophys. Acta Biomembr. 1828, 2646–2653.

Vongsak, B., Sithisarn, P., Gritsanapan, W., 2014. Simultaneous HPLC quantitative analysis of active compounds in leaves of *Moringa oleifera* Lam. J. Chromatogr. Sci. 52, 641–645.

Vu, H.T., Scarlett, C.J., Vuong, Q.V., 2018. Phenolic compounds within banana peel and their potential uses: a review. J. Funct. Foods 40, 238–248.

Wang, G., Wang, F., Cao, D., Liu, Y., Zhang, R., Ye, H., Li, X., He, L., Yang, Z., Peng, A., Wang, A.Y., Zhou, M.Y., Lin, W.C., 2011. Antioxidative and anti-inflammatory properties of *Citrus sulcata* extracts. Food Chem. 124, 958–963.

Wang, G., Wang, F., Cao, D., Liu, Y., Zhang, R., Ye, H., Li, X., He, L., Yang, Z., Ma, L., Peng, A., Xiang, M., Wei, Y., Chen, L., 2014. Synthesis, structure–activity relationships and biological evaluation of barbigerone analogues as anti-proliferative and anti-angiogenesis agents. Bioorg. Med. Chem. Lett. 24 (14), 3158–3163.

Wang, L., Yang, R., Yuan, B., Liu, Y., Liu, C., 2015. The antiviral and antimicrobial activities of licorice a widely-used Chinese herb. Acta Pharm. Sin. 5, 310–315.

Wang, H., Guo, X., Hu, X., Tong Li, T., Fu, X., Liu, R.H., 2017a. Comparison of phytochemical profiles, antioxidant and cellular antioxidant activities of different varieties of blueberry (Va*ccinium* spp.). Food Chem. 217, 773–778.

Wang, Y., Gao, Y., Ding, H., Liu, S., Han, X., Gui, J., Liu, D., 2017b. Subcritical ethanol extraction of flavonoids from *Moringa oleifera* leaf and evaluation of antioxidant activity. Food Chem. 218, 152–158.

Wang, T., Li, Q., Bi, K., 2018. Bioactive flavonoids in medicinal plants: structure, activity and biological fate. Asian J. Pharma. Sci. 13, 12–23.

Widlansky, M.E., Hamburg, N.M., Anter, E., Holbrook, M., Kahn, D.F., Elliott, J.G., Keaney, J.F., Vita, J.A., 2007. Acute EGCG supplementation reverses endothelial dysfunction in patients with coronary artery disease. J. Am. Coll. Nutr. 26, 95–102.

Williamson, G., 2017. The role of polyphenols in modern nutrition. Nutr. Bull. 42, 226–235.

Wojdyło, A., Carbonell-Barrachina, A.A., Pilar Legua, P., Hernández, F., 2016. Phenolic composition, ascorbic acid content, and antioxidant capacity of Spanish jujube (*Ziziphus jujube* Mill.) fruits. Food Chem. 201, 307–314.

Xi, W., Fang, B., Zhao, Q., Jiao, B., Zhou, Z., 2014. Flavonoid composition and antioxidant activities of Chinese local pummelo (*Citrus grandis* Osbeck.) varieties. Food Chem. 161, 230–238.

Xu, H., Luo, X., Huang, X., 2018. Qien Wan, Flavonoids intake and risk of type 2 diabetes mellitus. Medicine 97.

Yang, R., Yuan, B.C., Ma, Y.S., Zhou, S., Liu, Y., 2017. The anti-inflammatory activity of licorice, a widely used Chinese herb. Pharmaceut. Biol. 55, 5–18.

Yang, B., Liu, H., Yang, J., Gupta, V.K., Jiang, Y., 2018. New insights on bioactivities and biosynthesis of flavonoid glycosides. Trends Food Sci. Technol. 79, 116–124.

Yingngam, B., Monschein, M., Brantner, A., 2014. Ultrasound-assisted extraction of phenolic compounds from *Cratoxylum formosum* ssp. Formosum leaves using central composite design and evaluation of its protective ability against H_2O_2-induced cell death. Asian Pac. J. Trop. Med. 7, S497–S505.

Zefang, L., Zhao, Z., Hongmei, W., Zhiqin, Z., Jie, Y., 2016. Phenolic composition and antioxidant capacities of Chinese local pummelo cultivars' peel. Horticult. Plant J. 2, 133–140.

Zeid, A.B.A., Farag, M.A., Hamed, M.A.A., Kandil, Z.A.A., El-Akad, R.H., El-Rafie, H.M., 2017. Flavonoid chemical composition and antidiabetic potential of *Brachychiton acerifolius* leaves extract. Asian Pac. J. Trop. Biomed. 7, 389–396.

Zerbib, M., Mazauric, J.P., Meudec, E., Le Guernevé, C., Saucier, C., 2018. New *flavanol* O-glycosides in grape and wine. Food Chem. 266, 441–448.

Zha, J., Koffas, M.A.G., 2017. Production of anthocyanins in metabolically engineered microorganisms: current status and perspectives. Synth. Syst. Biotechnol. 2, 259–266.

Zhang, J., Wu, Y., Zhao, X., Luo, F., Li, X., Zhu, H., Sun, C., Chen, K., 2014a. Chemopreventive effect of flavonoids from Ougan (*Citrus reticulata* cv. *Suavissima*) fruit against cancer cell proliferation and migration. J. Funct. Foods 10, 511–519.

Zhang, Y.M., Sun, Y., Xi, W.P., Shen, Y., Qiao, L.P., Zhong, L.Z., Ye, X.Q., Zhou, Z.Q., 2014b. Phenolic compositions and antioxidant capacities of Chinese wild Mandarin (*Citrus reticulate* Blanco) fruits. Food Chem. 145, 674–680.

Zhang, H., Yang, Y., Zhou, Z., 2018. Phenolic and flavonoid contents of Mandarin (*Citrus reticulata* Blanco) fruit tissues and their antioxidant capacity as evaluated by DPPH and ABTS methods. J. Integrat. Agri. 17, 256–263.

Zhu, X., Luo, F., Zheng, Y., Zhang, J., Huang, J., Sun, C., Li, X., Chen, K., 2013. Characterization, purification of poncirin from edible *citrus* ougan (*Citrus reticulate* cv. *Suavissima*) and its growth inhibitory effect on human gastric cancer cells SGC-7901. Int. J. Mol. Sci. 14, 8684–8697.

CHAPTER

Bioactive peptides: synthesis, functions and biotechnological applications

4

Vivek Chauhan, Shamsher Singh Kanwar

Department of Biotechnology, Himachal Pradesh University, Summer Hill, Shimla, India

Chapter outline

Biotechnological Production of Bioactive Compounds. https://doi.org/10.1016/B978-0-444-64323-0.00004-7

1. Introduction

Proteins referred to as the 'Building blocks' of body are important constituents of food, which not only serve as nutritional sources but are also responsible for various sensory and physicochemical properties of foods (Shahidi and Zhong, 2008). Previous studies have shown that dietary proteins have a rich content of bioactive peptides (BAPs), which are responsible for many of the functional and physiological properties of these proteins (Gobbetti et al., 2007; Shahidi and Zhong, 2008). BAPs (Shahidi and Zhong, 2008; Walther and Sieber, 2011) are defined as characteristic protein fragments that have a positive impact on body functions and biological processes and may ultimately lead to health benefits (Kitts and Weiler., 2003). 'Biopep' is a database which contains more than 1500 BAPs till date (Singh et al., 2014). Synthesized in cells as inactive large bioactive peptides, these are later cleaved and modified to gain an active state (Sharma et al., 2011). Most of the modifications are done in three ways: (a) hydrolysis using digestive enzymes, (b) hydrolysis done by proteolytic microorganisms and (c) through the action of proteolytic enzymes obtained from microorganisms or plants. (Korhonen and Pihlanto, 2006).The activity of bioactive peptides is based on their inherent amino acid composition and protein sequence. These sometimes large in size are generally 3−20 amino acids long (Shahidi and Zhong, 2008; Moller et al., 2008), show the presence of hydrophobic amino acids in addition to proline, lysine or arginine groups and are also known to exhibit resistance to the actions of digestion peptidases (Kitts and Weiler, 2003). BAPs are highly specific in their functions and choice of target (Wang et al., 2018). All proteins have active components or parts of the protein chain that actually perform the function of that protein. Protein functions only when its active site makes contact with specific target (Nardo et al., 2018). Hence when the BAP portion is separated from the rest of its protein chain, the function of that protein is preserved in the BAP (Fields et al., 2009). BAPs perform diverse functions ranging from specific hormone-like tasks to functions relating to cellular processes.

Gastrointestinal digestion of food proteins and fermentation of food materials with lactic acid bacteria contribute to *in vivo* production of physiologically active peptides (Korhonen and Pihlanto, 2006). BAPs are functional food component involved in regulatory activities in human beings (Gobbetti et al., 2002). When taken orally, bioactive peptides are known to affect major body systems namely, the digestive, immune, cardiovascular and nervous systems depending on their amino acid sequence. Both *in vitro* and *in vivo* studies reveal that bioactive peptides contribute to a large spectrum of biological functions ranging from immunomodulatory, antimicrobial, opioid-like, antithrombotic, antihypertensive, mineral-binding, antioxidative, hypocholesterolemic (Coda et al., 2012) and antitumor activities (Rizzello et al., 2015). Bioactive peptides make a notably high constituent in health-promoting functional foods, pharmaceutical preparations and dietary supplements (Coda et al., 2012). This has attracted the interest of many scientists lately toward the potential of distinct dietary peptide sequences to promote human health by boosting natural immune protection or reducing the risk of chronic diseases.

Many peptides are known to reveal multifunctional properties (Meisel and FitzGerald, 2003).

Proteins from plant and animal origins are potential sources of a wide range of BAPs encrypted in their structure (Carrasco-Castilla et al., 2012; Bhat et al., 2015a). Bioactive peptides produced from various animal resources have been studied extensively. Among them prominent BAPs are those extracted from caseins, which have high functional potential. The scientists have considered plants as potent sources for production of economically important bioactive peptides. Distinctive efforts are made to identify vegetable sources, which can produce good quantity of bioactive peptide. This is due to higher acceptance of vegetables as a food source amongst the consumers. Considered as a new generation of biologically active regulators, BAPs have shown useful applications in important fields like food preservation and medical science (Lemes et al., 2016). In present chapter sources, synthesis and uses of bioactive peptides have been discussed.

2. Sources

BAPs are vastly encrypted inside bioactive proteins (Meisel and Bockelmann, 1999). Till now cheese (Pritchard et al., 2010), dairy products (Choi et al., 2012) and bovine milk (Korhonen, 2009; Léonil, 2014; Mohanty et al., 2015, 2016) have been considered as the greatest sources of bioactive proteins and peptides derived from foods. BAPs can also be obtained from other animal sources such as bovine blood (Przybylski et al., 2016), gelatin (Lassoued et al., 2015), meat, eggs, fish species such as tuna, sardine, herring and salmon (Kouhdasht and Nasab, 2018). Some vegetable sources of BAPs and proteins are maize, rice (Selamassakul et al., 2016), soy (Singh et al., 2014), mushrooms, pumpkin, sorghum (Moller et al., 2008) and amaranth (Silva-Sanchez et al., 2008).

3. BAPs from animal sources

Among animal sources BAPs obtained from animal proteins have shown different health benefits (Bhat et al., 2015b). Blood is known to be rich in proteins (circa 20%), which make it a promising source of BAPs. This was highlighted in a study where different concentrations of trypsin was used to hydrolyze serum albumin and the peptide sequences were obtained that showed activities like dipeptidyl peptidase IV inhibition (glucose regulation), angiotensin-converting enzyme inhibition (antihypertensive activity) and antioxidation (Arrutia et al., 2016). Blood from the slaughterhouse is one such unexploited source that has a high potential for production of BAPs in bulk. In a study, blood sample obtained from the slaughterhouse was subjected to *in vitro* gastrointestinal digestion. This study helped in identification of 75 unique peptides using low-resolution liquid chromatography analysis and more than 950 unique peptides were identified under high-resolution liquid chromatography (Caron et al., 2016). Hemoglobin may also serve as a rich source for BAPs. The α- and β-globin chains subjected to proteolysis coupled with excretion can produce peptides similar to one found in animal tissue extracts (Ivanov et al., 1997).

4. BAPs from vegetal sources

BAPs extracted from the soybean and the soy milk has been widely studied (Singh et al., 2014). Soybean and the soymilk obtained from gastrointestinal tract digestion were used to produce bioactive peptides, which showed antimicrobial properties. Endogenous proteases were also used to investigate the presence of peptides (Singh et al., 2014). Peptides found in soy milk samples could be formed during food processing (Capriotti et al., 2016). Soy hydrolysate and the soy-fermented foods such as natto and tempeh were digested with a variety of endoproteases such as pronase, trypsin, Glu C protease, plasma proteases and kidney membrane proteases to generate oligopeptides, most likely derived from glycinin, a soy protein. These digests showed angiotensin-converting enzyme inhibitory activity and surface active properties (Gibbs et al., 2004).

Cereal grains are other rich sources of BAPs, grains such as rice, wheat, barley, millet, rye, sorghum, oat and corn have shown substantial qualitative production of active peptides (Malaguti et al., 2014). Studies have proved huge health benefits of consuming whole grains for preventing diseases such as cancer, diabetes and cardiovascular diseases. Oat and wheat grains showed the presence of angiotensin-converting enzyme inhibitory peptides, as well as peptides with hypotensive, antioxidant, antithrombotic and opioid activities. BAPs obtained from rice have also shown anticancer activity. Grains of wheat and barley showed the greatest diversity and abundance of peptides with potential biological activity amongst the cereal proteins (Malaguti et al., 2014). Further research is required to establish the mechanism to release the active peptides sequences from cereal grains. The 2, 5-diketopiperazines (DKPs) also acknowledged as cyclic dipeptides are present in various foods, particularly in cocoa, roasted coffee, chicken essence, roasted malt and fermented foods such as distillation residue of awamori, beer and aged sake have received considerable attention as bioactive compounds. They are formed of N-terminal amino acid residues of a linear peptide or protein and have shown antioxidant activities (Kumar et al., 2012). DKPs are also found in whey protein hydrolysate and some beverages; therefore are considered as a functional component. It has been reported that some DKPs found in the distillation residue of awamori showed antioxidant activity. Another compound cyclo(-Phe-Phe) present in chicken essence has shown many health related benefits. It acts as a dual inhibitor of the serotonin transporter and acetylcholinesterase resulting into food intake inhibition and body weight reduction in rats, which definitely provides a keen possibility that it might even influence human biological regulation (Yamamoto et al., 2016).

Many waste products of different food industries are also seen as potential sources of BAPs. For instance, olive oil extraction produces two different kinds of waste materials: aqueous liquor and the solid waste, which is a combination of olive pulp and stone. These residues are particularly polluting products that are not easily biodegradable and are difficult to treat. New strategies have been developed for the recovery of such waste proteins from olive seed with the potential to produce antioxidant and antihypertensive peptides (Esteve et al., 2015). Fruit and vegetable processing also generates a lot of waste products. Stones of many fruits as plum (*Prunus domestica* L.) are rich in proteins and may serve as a cheap source of BAPs that could be used in the food and pharmaceutical industries. In a study, extraction of proteins from a residual material of plum was done using a high intensity focused ultrasound waves (González-García et al., 2014). Further, upon digestion of these proteins, 13 peptides were obtained, which showed diverse antioxidant and antihypertensive activities.

5. Peptides from food sources

Amongst food crops the greatest sources of bioactive proteins and peptides are milk, cheese and other dairy products (Korhonen, 2009; El-Salam and El-Shibiny, 2013; Lemes et al., 2016; Mohanty et al., 2016). Presumably, this can be one of the primary reasons for milk to be required beyond nutrition in the first months of life (Moller et al., 2008). Milk proteins have a vide stretch of biological activities ranging from immunoglobulins, which have immunoprotective effects to the lactoferrin, which displays antibacterial activity. Growth factors and hormones that are abundant in colostrum play very significant role in post-natal development (Park and Nam, 2015). The major role of milk proteins is to supply amino acids and nitrogen to the young mammals and constitute an important part of dietary proteins for the adult.

BAPs can be released from milk by gastrointestinal digestion or food processing (Fitzgerald and Meisel, 2003; Meisel and Fitzgerald, 2003). Many opioid peptides present in dairy products have pharmacological properties similar to those of morphine and thus appear to play active role in the functioning of central nervous system (Haque et al., 2008). Large number of such medium and low molecular weight BAPs (opioid and phosphopeptides) have been recognized in human milk using tandem mass spectrometry and high performance liquid chromatography-mass spectrometry (HPLC-MS) techniques. A complete chain of gastrointestinal reactions takes place resulting in casein digestion. This stepwise degradations process is very hard to create *in-vitro* condition, which depicts the importance of maternal milk and demonstrates the fact, that how difficult it is to reproduce it artificially. As a consequence of the dynamic nature of maternal milk, a succession of potential BAP is produced in the intestine, which is hard to reproduce in artificial products (Ferranti et al., 2004). Another BAP present in all mammal milks is lactoferrin, an iron-binding glycoprotein, which has shown antimicrobial properties, immunomodulating effects and is also an important growth factor. Milk derived BAPs have been reported to influence immune and inflammatory processes of the body by effecting cytokine production (Moller et al., 2008). The use of lactic acid bacteria for milk production is also widely used to generate functional foods enriched with BAPs given the low cost and positive nutritional image associated with fermented milk products (Hayes et al., 2007). In a previous study, peptides (Tyr-Pro and Lys-Val-Leu-Pro-Val-Pro-Gln) purified and characterized from fermented milk showed angiotensin-converting enzyme inhibitory activity in hypertensive rats (Jäkälä and Vapaatalo, 2010).

6. Milk

Milk of different animals such as goat, buffalo, sheep, camel, mare and yak has been reported to be a rich source of BAPs. Among them predominately, peptides isolated and characterized from cow milk proteins have shown mineral-binding, angiotensin-converting enzyme inhibitory, immunomodulatory, opioid, anticarcinogenic, cytotoxic, antibacterial and antithrombotic activities. Physiological properties and functional nature of BAPs vastly vary with the milk protein source (El-Salam and El-Shibiny, 2013). Daily use of fermented milk containing BAPs has a blood pressure lowering effect in hypertensive subjects. Whey protein concentrate produces β-lactoglobulin peptides predominated in the hydrolysate when treated with trypsin followed by separation using membrane ultrafiltration (Arrutia et al., 2016). Bovine colostrum is another rich source, which upon gastrointestinal digestion or dialysis produces BAPs possessing high health benefits (Jorgensen et al., 2010).

7. Eggs

Egg, a major source of valuable proteins in human nutrition has been identified as an important source of many important BAPs (Wu et al., 2010; Zambrowicz et al., 2011; *Bhat* et al., 2015a) with many applications in medicinal and food processing industries (Sun et al., 2016). The identification and characterization of BAPs from egg both *in vivo* and *in vitro* conditions have helped in changing the image of egg as a food. Now egg is seen as a new source containing biologically active ingredients for the development of functional foods, which have specific benefits for the prevention and treatment of different human diseases (Bhat et al., 2015a).

The studies have invariably proven that eggs contain numerous substances, which have therapeutic effects beyond supplying basic nutritional requirements (Zambrowicz et al., 2011). A novel egg-derived product is egg white protein powder (EWPP), which because of long shelf life is extensively used in food processing industries. Antioxidant BAPs have been derived from EWPP upon hydrolysis by different proteases followed by sequential fractionation by ultrafiltration (Lin et al., 2011). Health benefits of egg can be simply analyzed by the fact that boiled egg white hydrolysate showed a total of 63 peptides with novel antioxidant peptides (Remanan and Wu, 2014). Many BAPs obtained from eggs have potential applications for maintenance of dermal health and treatment of skin diseases (Sun et al., 2016). Today the BAPs are increasingly used in production of many different cosmetics aimed at treating dysfunctional and damaged human skin (Lima and Moraes (2018).

8. Meat

Though traditionally associated with increased risk of obesity, cancer and other diseases, meat and meat products are rich source of BAPs. Peptides derived from meat products are potentially able to get incorporated into functional foods and nutraceuticals with many benefits to humans health (Howell and Kasase, 2010; Stadnik and Keska, 2015). Peptides derived from meat and fish have shown antihypertensive effects *in vivo*, along with other bioactivities such as antioxidation, antiproliferative and antimicrobial (Lafarga and Hayes, 2014; Liu et al., 2016; Ryder et al., 2016). A major limitation to meat as a source of BAPs is that food products containing meat derived BAPs are commercially quite limited (Ryan et al., 2011).

9. Exogenous peptides

Highly dynamic marine environment ensures the huge variability in organisms those acts as reservoirs containing unlimited resource for production of compounds with potential use as bioactive products. BAPs obtained from many marine organisms (Ngo et al., 2012) have been employed for the treatment of various diseases (Kang et al., 2015; Manikkam et al., 2016). Marine organisms such as shellfish, crustaceans, mollusks, fish and marine processing waste are abundant sources of a myriad of structurally diverse bioactive organic compounds (Cheung et al., 2015; Jo et al., 2017; Kouhdasht and Nasab, 2018). Well-documented evidence of their potential for human health (Fan et al., 2014; Ngo et al.,2012) includes activities as antioxidant (Ngo and Kim, 2013), anticoagulant, antihypertensive (Kim et al.,2012a,b), immunostimulatory, antimicrobial (Kang et al., 2015; Falanga et al., 2016), calcium-binding, antidiabetic (Manikkam et al., 2016), anticancer, appetite suppression and hypocholesteremic have incentivized the interest of these compounds as functional food ingredients (Harnedy and Fitzgerald, 2012).

Proteins derived from these marine organisms are rich source of BAPs and thus they serve as raw material in many industries (Kim et al., 2012a,b; Ngo and Kim, 2013; Kang et al., 2015; Falanga et al., 2016). Many studies have been conducted to understand the biological activities and chemistry of BAPs obtained from different marine coelenterates, algae, tunicates, sponges, mollusks and ascidians (Cheung et al., 2015). Marine algae due to its high protein content (up to 47% of the dry weight) are favored by many scientists as a rich source of BAPs

(Fan et al., 2014). Marine algae have developed complex metabolic pathways for BAPs production. Some of these have been characterized with specific health benefits and disease-preventing properties like antihypertensive (Harnedy and Fitzgerald, 2012).

10. Production of BAPs

High demand of BAPs for application in physiological, chemical, physical, pharmacological, biochemical, and clinical studies has paved a way for their industrial production. But, lack of suitable technology and low yield is a big hindrance to their industrial production (Korhonen and Pihlanto, 2003, 2006; Pihlanto, 2006). Methods used for BAPs synthesis (Fig. 4.1) are further discussed. Chemical synthesis uses chemical reagents to mediate peptide bond formation (Andreu and Rivas, 2002), enzymatic synthesis causes peptide bond formation catalyzed by enzymes and the DNA recombinant technology synthesis is primarily based on the use of cloning and ribosomal techniques from biological systems for peptide formation (Sewald and Jakubke, 2002). However, most sources of natural peptides are poor in these compounds, thus preventing their isolation in sufficient quantities for research.

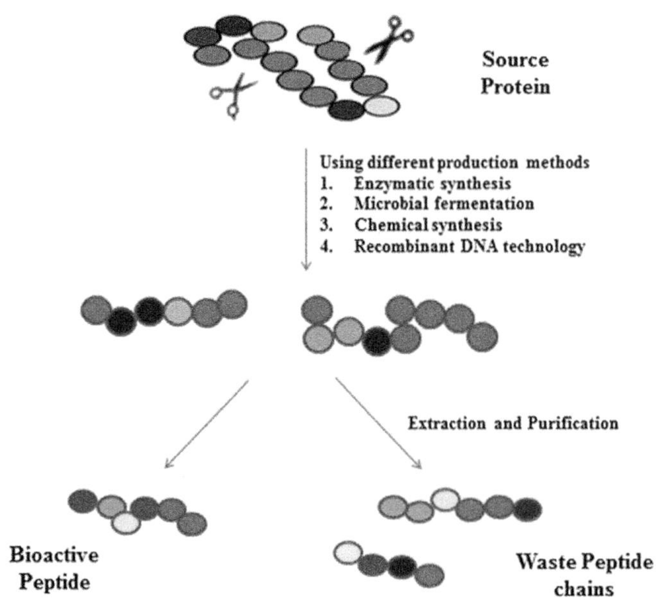

FIG. 4.1

An overview of production of bioactive peptides.

10.1 Enzymatic synthesis

This method utilizes enzymes (protease) in immobilized or free form for catalysis of peptide bond formation. It is a well-studied and most reliable method for production of BAPs (Bongers and Heimer, 1994; Boeriu et al., 2010). Very short peptides (2–5 oligomers) are produced using enzymatic synthesis and in the condensation of large peptide fragments. Most commonly used proteolytic enzymes for the degradation are chymotrypsin, pepsin, termolisin, papain, trypsin, subtilisin, among others (Ogino et al., 1999). Crude form of enzyme is preferred over the pure form as it reduces the overall cost of production of BAPs at commercial scale (Zarei et al., 2012).

This method has some noticeable advantages as regioselectivity and good stereoselectivity over other methods used for BAPs production. Though a very successful method it has certain shortcomings too. BAPs synthesis is thermodynamically unfavourable in water, which hinders the formation of large peptide bonds and thus serves as a disadvantage to enzymatic synthesis (So et al., 1998). Thus to use a protease for BAPs production, suitable conditions have to be identified, which allow peptide bond formation without mediating secondary hydrolysis (Bongers and Heimer, 1994). Transpeptidation and reverse hydrolysis reaction of amides are two of several methods that could be used for formation of peptide bonds by enzyme catalysis (Machado et al., 2004; Boeriu et al., 2010). Microscopic reversibility principle serves as the basis of the reverse hydrolysis reaction and it depicts that the hydrolysis reaction and peptide bond formation come from the same intermediate. Hence, the reaction conditions are manipulated to shift the equilibrium toward peptide bond formation, which results in BAPs production. On the other hand transpeptidation mechanism occurs when active acyl-enzyme intermediate is formed as a result of peptide bond break. This intermediate in the presence of a nucleophile (peptide or amino acid blocked in the α-carboxyl group) results in formation of a new peptide bond (Machado et al., 2004). Enzymatic synthesis has been successfully deployed for production of a broad range of antioxidant peptides and peptide mixtures (hydrolysates) from food sources like potato, soy, peanut, corn, egg, milk, meat and whey proteins. Also BAPs with antioxidant activity and angiotensin-converting enzyme inhibitory have been obtained from 'Thornback ray skin' gelatin upon hydrolysis with two different proteases (Lassoued et al., 2015).

10.2 Microbial fermentation

In this approach, microorganisms rather than purified enzymes are used in many countries as a traditional way for production of antioxidant peptides rich food products. Natto and tempeh are two best suited examples, which are produced by action of fungal proteases and are rich in antioxidant peptides (Wongputtisin et al., 2007). Nature of the used microorganism determines the type, amount and activity of the BAPs produced. In this method of synthesis, the source is worked upon by mixture of different enzymes in a series of reactions to produce BAPs. *Lactobacillus* sp. is among the most commonly used and commercially exploited microorganisms used for BAPs production *via* fermentation of milk (Hayes et al., 2007).

10.3 Chemical synthesis

This is the oldest method for production of BAPs with research first initiated about 30 years ago. However, the method has become more accessible over a period of time due to advancements in process efficiency, including the development and use of advanced coupling reagents, as well as the minimizing side reactions (Kishore, 2004; Udenigwe et al., 2013). Activation and protection are the main aspects of chemical synthesis with protection strategies mainly intended to provide chemical selectivity that is necessary for particular peptide sequence construction and activation. It refers to the chemical coupling, which is necessary for ensuring quantitative formation of each peptide bond in the sequence (Andreu and Rivas, 2002).

In this process, different chemical reagents are used for activating the carboxylic acid group (RCOOH) of the amino acid, which will in turn donate the acyl group (R—CO—) for peptide bond formation (Machado et al., 2004). Solid-phase synthesis and synthesis in solution (classical) are two types of chemical peptide synthesis (Saladino, 2012). In classical method, all reagents and reaction products are dissolved in a medium (Kent, 1988) whereas in solid-phase synthesis large quantities of peptides are produced on a solid support, which remain insoluble in the reaction medium (Shigeri et al., 2001). The solid support is mostly a polymeric resin that has a functional group on its surface (linker) that helps in stable bond formation. The solid phase peptide synthesis occurs *via* acylation of an amino acid to be linked to an insoluble support (resin) *via* a linker followed by removal of the protecting group from N-terminal, which allows the attachment of next amino acid to the complex 'peptide-linker-resin'. This cycle is repeated until the peptide with desired sequence is achieved. Finally, the separation of this complex 'peptide-linker-resin' is done by using a suitable cleavage reagent (Borgia and Fields, 2000).

Peptide chemical synthesis may be done by two different methods namely Boc (tert-butyloxycarbonyl) and Fmoc (9-fluorenylmethyloxycarbonyl), which are named based on type of protector present in the reactive group of the amino acids (N-terminal) involved in the synthesis. In first protocol, Boc is employed for *N*-amino protection (Borgia and Fields, 2000) while the second protocol uses Fmoc as the *N*-amino protecting group (Andreu and Rivas, 2002). This protocol provides a greater degree of chemoselectivity than the Boc protocol, since the Fmoc group is removed under basic conditions (piperidine in *N*, *N* methylpyrrolidone or dimethylformamide), without alteration of the acid-sensitive lateral chains. Many chemically synthesized BAPs have been deployed for biological studies (Smacchi and Gobbetti, 2000; Freidinger, 2003; Rajesh and Iqbal, 2006). In a study, the chemically synthesized BAPs were used to treat certain pathological conditions related to oxidation (Lalenti et al., 2001; Van Lancker et al., 2011). Another study showed use of a chemically synthesized peptide (Lys-Arg-Glu-Ser) upon oral administration, which helped in lowering LDL peroxidation and also reduced atherosclerosis in apoE-null mice (Navab et al., 2004). Peptides with active Pro-His-His fragment have been synthesized, which showed remarkable inhibition of lipid peroxidation.

10.4 Synthesis by recombinant DNA technology

Recombinant DNA technology makes use of modern methods of cloning and gene expression in microorganisms that result in production of a recombinant peptide or many different peptides simultaneously. Expression is done mostly in bacterial expression system with *Escherichia coli* being the most common host. Antimicrobial peptides produced *via* this process are destructive against the host and have a relative sensitivity to proteolytic degradation, due to this reason peptides are often expressed as fusion proteins, which neutralizes their innate toxic activity and also helps in increasing the expression levels (Wang et al., 2011). Over time, this method has gained a lot of popularity and offers most cost-effective alternative for large-scale peptide production (Li, 2011). The novel biotechnologies for the release of bioactive peptides from vegetable matrices, including microbial fermentation and the use of microbial enzymes have been thoroughly investigated.

11. Purification and recovery approaches

Purity is one of the major concerns which occur during BAPs synthesis. Peptides synthesized for structural and biological studies as well as for therapeutic and clinical research must have 95% or greater purity (Ridge and Hettiarachchi, 1998). However, some BAPs may have applications with purity level ranging between 70 and 95%. Purification of BAPs largely depends upon the separation technique deployed, which may be improved by studying the separation conditions on analytical scale prior to the execution of any preparative separation process (Sewald and Jakubke, 2002). After synthesis, BAPs are subjected to a separation procedure consisting of washing and centrifugation to remove residues of the reagents as well as the products of side reactions. After accomplishing the separation process, peptides are cleaved and subjected to filtration as well as lyophilization (Dagan et al., 2002). While reverse-phase high performance liquid chromatography (RP-HPLC) is the most commonly used methods for the purification of BAPs, other employed methods are capillary electrophoresis, size exclusion chromatography, ion exchange chromatography and affinity chromatography.

At industrial scale, many different technologies are employed for purification of BAPs. Of these membrane technology processes, have advantages as low energy expenditure, absence of phase, operation at low temperature and transition

(Conde et al., 2013). Membranes act as semipermeable barriers that split up two distinct phases and limit the movement of definite components in a selective way (Bazinet and Firdaous, 2013). Some elements are diffused through membrane while others are retained resulting in enrichment of permeate and the retentive in specific components (de Morais Coutinho et al., 2009). The diffusion of different compounds through the membrane can be assisted by applying a driving force (concentration gradient, pressure, temperature or electric potential). Membranes technologies include ultrafiltration, microfiltration, reverse osmosis and nanofiltration with ultrafiltration and nanofiltration being most widely used for industrial synthesis of bioactive peptides or proteins (Herrero et al., 2012; Szydłowska-Czerniak et al., 2011). Ultrafiltration is functional at pressure ranging from 2 to 30 bar (Nawaz et al., 2006) and it is helpful in retaining colloids and macromolecules with molar masses between 1 and 300 kDa, while allowing small molecules and water to permeate. On the other hand, NF is carried out at 10–40 bar with the separation capacity ranging between 0.35 and 1 kDa. The efficiency and performance of the separation are influenced by the properties of solution, charge of solution and ions and the type of membrane used (Langevin et al., 2012).

Recovery is an important stage in BAPs production. Most of the separation and purification processes for BAPs recovery used in laboratory scale procedures are too expensive and complex to be industrialized (Agyei et al., 2015). Some limitations during industrial scale production are making a reproducible product, lack of clinical trials to confirm bioactivity, efficacy and safety and food sensory changes such as bitterness and undesirable color (Carrasco-Castilla et al., 2012; Chaves-López et al., 2012). The peptide recovery has been applied to some protein hydrolysates, which show antioxidant activity against lipid and/or fatty acids peroxidation (Wang et al., 2007). Many different processes have been set up for proteins and peptides recovery from plants. Ultrafiltration is most widely used for recovery of BAPs from many different plant sources such as wheat gluten (Kong et al., 2008), rapeseed, soy flour and many more.

A method different from that used in production is then deployed to verify the purity of a peptide (Ridge and Hettiarachchi, 1998). Thus, the characterization is done by different methods of mass spectrometry. The ionization methods of Mass spectrometry that are used for BAPs characterization are matrix-assisted laser desorption/ionization mass spectrometry (MALDI-MS), fast atom bombardment mass spectrometry (FAB-MS) and electrospray ionization mass spectrometry (ESI-MS). Accurate chemical structure determination as well as the determination the molecular mass-charge ratio of the peptide with high resolution and sensitivity can be done using these mass spectrometry techniques (Sewald and Jakubke, 2002). Many new technologies are being worked upon for BAPs recovery from residual waste (Harnedy and Fitzgerald, 2012) and for their transformation into valuable products, as well as to promote large-scale recovery (Kitts and Weiler, 2003), which later can be used in pharmaceutical (Agyei and Danquah, 2011) and food industries (Korhonen and Pihlanto, 2007).

12. Broader functions and biotechnological applications of peptides

Peptides have a wide range of applications based on varying activities and properties they exhibit. The peptide activity depends upon its structure, *i.e.* the type of *N*– and C-terminal amino acid, the length of the peptide chain, charge character of the amino acids forming the peptide, the amino acid composition and the hydrophobic/hydrophilic characteristics of the amino acid chain (Li and Yu, 2015). A peptide is considered bioactive only when it has a potential positive effect on health of an individual where damaging effects such as mutagenicity, allergenicity and toxicity are not taken in consideration (Moller et al., 2008). Thus varying applications of BAPs may be depicted based upon the activities they exhibit (Fig. 4.2). Following is an attempt to demonstrate various functions of BAPs based on the activities they possess.

13. Antimicrobial BAPs

Most of the BAPs obtained from tissues and cell-types of a variety of animal, plant and invertebrate species possess antimicrobial activity by virtue of their ability to interact with cytoplasmic membrane of the microorganisms (Powers and Hancock, 2003). Many factors such as amino acid composition, cationic charge, size and amphipathicity allow BAPs to interact with and insert into membrane bilayers of microorganisms (Brogden, 2005). Studies have shown that hydrophobicity and flexibility of BAPs also serves as important factors in their interaction with many pathogenic microorganisms (Jenssen et al., 2006). The mentioned factors though vary from peptide to peptide are essential in determining antimicrobial nature of BAPs. Antimicrobial peptides are usually small ranging from 10 to 50 kDa, mostly

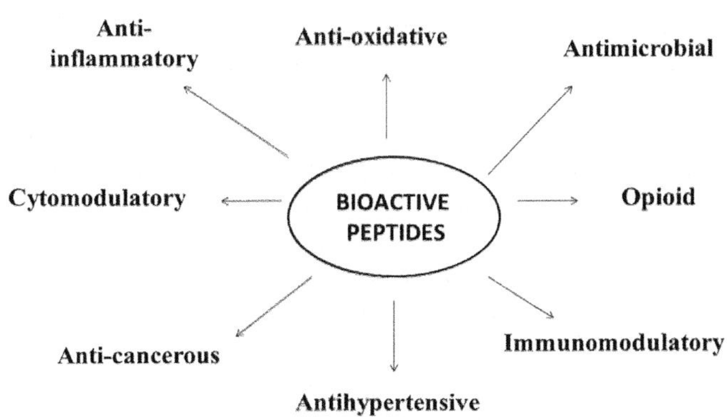

FIG. 4.2

An overview of important characteristics/functions of bioactive peptides (BAPs).

containing native protein precursors which can be generated *in vitro* by using enzymatic hydrolysis (Kim and Wijesekara, 2010).

Antimicrobial peptides have been used successfully as therapeutic agents (Cruz et al., 2014) against different pathogenic microbes. More than 150 peptides are being clinically tried for their therapeutic properties and about 60 peptide drugs have paved their way into market (Fosgerau and Hoffmann, 2015). Generally, antimicrobial peptides are classified in to three families based upon their structure. These are; i. Disulfide-bridged cyclic and open-ended cyclic peptides; ii. Peptides with high amount of glycine, proline or histidine in primary structures and iii. α-Helical linear peptides (Pimenta and De Lima, 2005). Antimicrobial peptides have been produced from many sources ranging from plants to animals. Many antimicrobial peptides have been produced from venoms of different organisms as primitive scorpions (Pimenta and De Lima, 2005), wasps and bees. Milk serves as a rich source for production of antimicrobial peptides. The lactoferrin, a milk whey protein was cleaved to produce a peptide, which served as antimicrobial agent against many naked and enveloped viruses such as adenovirus enterovirus and rotavirus (Ng et al., 2015). Casein-derived antibacterial peptides serve as production inhibitors to many pathogens such as *Streptococcus sanguis, Streptococcus mutans, Sthaphylococcus aureus, E. coli, Streptococcus sobrinus*, *Salmonella typhimurium* and *Porphyromonas gingivalis* (Mohanty et al., 2015). It seems, the antimicrobial peptides due to their effectiveness against wide microbial host range help in combating with briskly increasing incidences of multidrug-resistant infections as well as also serve as natural bio preservatives.

14. Antioxidative peptides

Presence of reactive oxygen species and free radicals can cause cell damage, consequently leading to problems like diabetes mellitus, cardiovascular, hypertension, neurodegenerative diseases, inflammatory diseases and cancer (Fan et al., 2012). Antioxidants are known to protect the body from these reactive moieties by scavenging reactive oxygen species and free radicals and by inhibiting lipid peroxidation reactions, which leads to oxidative damage. Antioxidant properties of BAPs have been established *via* many different studies conducted in past few years (Agyei and Danquah, 2011). BAPs (with the sequence identified as Glu-Ser-Thr-Val-Pro-Glu-Arg Thr-His-Pro- Ala-Cys-Pro-Asp-Phe-Asn) having antioxidant properties were identified from hoki frame protein hydrolysates using pepsin. Antioxidant peptides obtained from sweet potato were rich in both hydrophobic amino acids and antioxidant amino acids such as Tyr, Met, His, Phe and Cys and had protective effects against DNA damage *via* Fe^{2+}chelation and scavenging of hydroxyl radicals (Zhang et al., 2012).

Many milk protein-derived BAPs have shown antioxidant properties, which help in prevention of essential fatty acids peroxidation. Studies have revealed that casein produces phosphorylated peptides on digestion that exhibit both lipophilic and hydrophilic antioxidant activity due to metal ion sequestering and reactive oxygen species quenching (Clare and Swaisgood, 2000). BAPs obtained from protein hydrolysates of fish proteins were prepared using different enzymes (Elias et al., 2008). The antioxidant potential of obtained BAPs from fish protein hydrolysates depends on the manner in which the tertiary structure of parent protein disrupts and also on their amino acid composition.

Different food-derived BAPs like soy protein hydrolysate (Singh et al., 2014) and germinated black bean (Barrios et al., 2018) also showed antioxidant nature. BAPs derived after hydrolysis of soy protein showed antioxidant activities ranging from 28% to 65% (Peñta-Ramos and Xiong, 2002). The α- and β-lactoglobulin (Hernandez-Ledesma et al., 2008) and wheat germ protein (Zhu et al., 2006) enzymatic hydrolysate are known to possess free radical-scavenging and antioxidant activities. Egg (Sakanaka and Tachibana, 2006), gelatin (Je et al., 2005) and potato are the other food sources, which contain antioxidant proteins and peptides. Industrial waste like enzyme extract of sardinelle (*Sardina pilchardus*) was also used to produce antioxidation efficient peptides such as Leu-Ala-Arg-Leu, Leu-His-Tyr, Gly-Gly-Glu, Pro-His-Tyr-Leu and Gly-Ala-His. Gly-Ala-Trp-Ala peptide was obtained from industrial waste generated using crude enzyme extract from sardine (*S. pilchardus*) (Bougatef et al., 2010). Health promoting effects of dietary proteins in both animals and human beings may be due to presence of antioxidant peptides (Elias et al., 2008). *In vitro* studies have proved that peptide (2–4 amino acid residues) mixture obtained from specific food proteins using human digestive enzymes possess potent antioxidant activity (Zhu et al., 2008).

15. Cytomodulatory and immunomodulatory/ anti-inflammatory peptides

Many immunocompromised diseases like cancer and HIV have paved their way into the mankind and are on an increase with every passing day. Cytomodulatory BAPs are most suitable ones for the modulation of body immune cells. They are known to modulate viability (*e.g.* apoptosis, differentiation and proliferation) of different types of cells and in association with immunomodulatory peptides; they fight to control the development of tumor cells in body (El-Salam and El-Shibiny, 2013). Many BAPs from food sources are known to show cytomodulatory activities especially peptides obtained from waste whey of Mozzarella cheese showed antiproliferative effect when tested on human colorectal adenocarcinoma cell line (Simone et al., 2009). Another study showed cytomodulatory peptides to inhibit cancer cell growth besides stimulated activity of neonatal intestinal and immunocompetent cells (Meisel and FitzGerald, 2003).

On the other hand, immunomodulatory peptide, bolstered the functions of immune cell, helped to improve cytochrome regulations and antibody synthesis and also served to enhance the mucosal immunity in the gastrointestinal system (El-Salam and El-Shibiny, 2013). Milk serves as the most common and most efficient source for peptides with immunomodulatory properties, which are known to trigger either specific (antibody production, proliferation, cytokine expression and lymphocyte activation) and/or non-specific (functions of granulocytes, macrophages and natural killer cells) immune responses (Gauthier et al., 2006; Huang et al., 2010). BAPs derived from milk have been extensively studied for their immunodulatory properties. Hydrolysates and peptides derived from milk protein and caseins are known to enhance lymphocyte proliferation, immune cell functions, cytokine regulation and antibody synthesis (Gill et al., 2000). Different peptides released during milk fermentation using lactic acid bacteria are of special interest as their BAPs can modulate activities like proliferation of human lymphocytes, enhance phagocytic activities of macrophages and also help in down-regulation of certain cytokines production (Meisel and FitzGerald, 2003). A study showed a casein-derived immunopeptide to have a protective effect on resistance to microbial infection by *Klebsiella pneumoniae* (Migliore-Samour and Jollés, 1988). Immunomodulatory milk peptides are also known to alleviate allergic reactions in atopic humans and also enhance mucosal immunity of the gastrointestinal tract (Korhonen and Pihlanto, 2003). This could be the reason why immunomodulatory peptides might result in immune system development in new-born. Studies have shown that the immunomodulating peptides could be released from commercially available whey protein upon enzymatic digestion (Mercier et al., 2004).

Many non-specific immune defense systems are triggered by immunomodulatory peptides obtained from hydrolysates of soybean and rice proteins. These hydrolysates are known to stimulate reactive oxygen species. The anterior pituitary is known to contain a variety of BAPs as growth factors, posterior lobe peptides, brain-gut peptides, hypothalamic releasing factors, opioids, and various other peptides. Evidence have proved these peptides to be effective in enhancing many intracrine, autocrine and paracrine functions, regeneration and differentiation or controlling the release of growth hormones (Houben and Denef, 1994).

Body responds to nonlethal injury in form of inflammation, which is characterized by increased leakage of protein-rich exudates, infiltration of leukocytes into extravascular tissues and endothelial permeability. Though inflammation stands essential to resist microbial infections and wound healing, however the excessive and uncontrolled inflammatory changes might lead to chronic diseases such as cancer, asthma, obesity, cardiovascular disease, diabetes, osteoporosis, inflammatory bowel disease and neurological diseases like Parkinson's (Chakrabarti et al., 2014; Majumder et al., 2016). A recent study showed BAPs to improve diet-induced hepatic fat deposition and hepatocyte pro-inflammatory response when tested on a SAMP8 aging mice (Dumeusa et al., 2018). Many BAPs could modify intestinal barrier function (Martínez-Augustin et al., 2014) alterations, which were related to many non-inflammatory and inflammatory disorders. Corn, whey and soybean enzymatic hydrolysates have shown potent anti-inflammatory effects both in *in vitro* and *in vivo* studies (Shahi et al., 2012; Dia et al., 2014).

16. Antihypertensive peptides

Of all BAPs, antihypertensive peptides (also known as angiotensin-converting enzyme inhibitors) induced by food protein hydrolysates (Joffres et al., 2001) are most widely studied ones. According to worldwide statistics, there were 972 million adults suffering from hypertension in year 2000, and these cases have been predicted to increase up to 60% to a total of 1.56 billion by year 2025 (Chockalingam et al., 2006). Hypertension has thus become a high-priority global public health challenge that requires an urgent attention with prevention, detection, treatment and control. Angiotensin-converting enzyme, peptidyl di-peptide hydrolase (EC 3.4.15.1) an angiotensin I-converting enzyme plays a key role in stopping hypertension. It is a key enzyme of the renin-angiotensin system, which functions by regulating the arterial vasoconstriction and extracellular fluid volume either by inactivating the vasodilator bradykinin or by stimulating the conversion of angiotensin I to the vasoconstrictor angiotensin II. This Inhibition of angiotensin-converting enzyme results in a blood pressure decrease and thus, helps in controlling hypertension.

There are many potent synthetic peptides suggested to be used for the clinical treatment of heart failure and hypertension but most of them are costly and have numerous obnoxious side effects (Roy et al., 2010). Interestingly, the antihypertensive bioactive peptides emerge as a suitable alternative since they have none or very little side effects and have a relatively cheaper production cost (Agyei and Danquah, 2012). Antihypertensive peptides have a characteristic proline residue at the carboxyl terminal end and are usually shorter in size when compared to other BAPs. Presence of proline residues helps in prevention of enzymatic degradation of peptide in gastrointestinal carnal (Korhonen, 2009). The lactotri peptides valine-proline-proline (Val-Pro-Pro) and isoleucine proline-proline (Ile-Pro-Pro) obtained from sour milk are two well-known antihypertensive peptides (Korhonen and Pihlanto, 2006). Studies have shown that these two tripeptides have positive effect upon body by reducing the blood pressure in hypertensive patients (Marques et al. (2012).

Plant food proteins have served as a rich source for production of new and potent antihypertensive peptides (Kim et al., 2012a,b). Best suited examples could be tryptophan and valine rich peptides in which high level of antiangiotensin converting enzyme (anti-angiotensin-converting enzyme) activity was identified and for purification of these peptides alfalfa white protein concentrate hydrolysate was successfully used (Boudesocque et al., 2012). Upon subtilisin catalyzed digestion of rapeseed protein, four angiotensin-converting enzyme inhibitory peptides (Val-Trp, Val-Trp-Ile-Ser, Ile-Tyr and Arg-Ile-Tyr) were isolated. These peptides upon oral administration were found to lower down the blood pressure of hypertensive rats with maximum effect observed between 2 and 4 h (Marczak et al., 2003). These and other antihypertensive peptides may be thus useful for prevention and treatment of hypertension by being incorporated into pharmaceutical lead drugs and functional foods. Many other food sources as cereal grains (wheat, oat, barley) (Cavazos and

Gonzalez de Mejia, 2013), rice protein (Li et al., 2007) and milk (Tuomilehto et al., 2004; Moller et al., 2008) are extensively studied for isolation and purification of antihypertensive peptides.

Antihypertensive peptides also help in regulation of salt balance and fluids in mammals (El-Salam and El-Shibiny, 2013). Human trail of few BAPs have suggested that bioactive tripeptides with antihypertensive activities help in reducing the elevated levels of blood pressure in subjects with moderate hypertension (Cicero et al., 2013). Milk BAPs may be employed in prevention of metabolic syndrome risks and its complications by regulating blood pressure, control of food intake and the uptake of free radicals (Reddi et al., 2012).

17. Anticancerous peptides

Cancer has become one of the most dreaded, death causing diseases all around the world. Anticancer and antitumor drugs are amongst the highest developed drugs by pharmaceutical companies. Also the research in field of oncology is well advanced that over the years has enhanced our understanding of cancers. Besides all technological advances the therapeutic opportunities for cancer are still limited owing to the difficulty in selective targeting of cancer cells and leaving healthy ones. Survival and proliferation of cancerous cells may be modulated by the use of biomolecules, which target specific antigens associated with tumor expression present on cancer cells (Noguchi et al., 2012).

Food protein hydrolysate serves as a major source for production of anticancer peptides. Enzymatic hydrolysates of vegetable proteins as rice and soy have shown a high anticancer activity. Alcalase digest of rice bran proteins produces anticancer peptides (Kannan et al., 2010). An anticancer (Ala-Phe-Asn-Ile-His-Asn-Arg-Asn-Leu-Leu) peptide was isolated from protein of shellfish *Mytiluscoruscus*, which effectively induced cell death in breast, prostate and lung cancer cells but normal liver cells were spared (Kim et al., 2012a,b). *Styelaclava* proteins upon alcalase treatment resulted in production of anticancer peptides rich in arginine, tyrosine, phenylalanine, histidine and lysine. Upon purification of hydrolysate using gel-filtration, two peptide fractions with antioxidant and anticancer activities were obtained, which significantly inhibited the growth of colorectal adenocarcinoma cell line (DLD-1), HeLa cells and human stomach adenocarcinoma cell line (Jumeri and Kim, 2011). Lipopeptides have shown great results in controlling cancer (Meena et al., 2017). Bioactive peptides, due to their immense potential in killing cancerous cells or controlling their growth and proliferation possess significant health-promoting effects, which could possibly be harnessed into a pharmaceutical industry.

Hoskin and Ramamoorthy (2008) classified anticancerous peptides into 2 major groups: peptides with non-selective activity, which were active against carcinogenic cells, bacteria and healthy cells and the other were peptides with selective activity *i.e.* were only active against specific cancer causing antigen or protein.

18. Opioid peptides

Teschemacher et al. (1997) described opioid peptides as one that exists in dairy products and plays an active role in the functioning of nervous system. Opioid BAPs are pharmacological similarity to opium (morphine). These peptides have short sequences of amino acids, which mimic the effect of opiates in the brain. Mostly opioid peptides, originate from one of three precursor proteins (Hollt, 1983); proenkephalin (enkephalin), prodynorphin (dynorphins) and proopiomelanocortin (endorphins). Opioid peptides can be either produced by the body itself, as endorphins or absorbed from partially digested food (rubiscolins, casomorphins and exorphins).

BAPs with opioid activity (Yamamoto et al., 2003; Aldrich and McLaughlin, 2009) serve as the opioid receptor ligands with antagonistic or agonistic activities. Such peptides have common N-terminal sequence of Tyr-Gly-Gly-Phe-(Met or Leu) termed as *opioid motif* that is followed by different C-terminal extensions having peptides between 5 and 31 residues in length (Aldrich and McLaughlin, 2009).

Opioid BAPs react with opioid receptors present in the gastrointestinal tract, immune, endocrine, nervous system of mammals and influence the functioning of different organ systems. As in case of gastrointestinal track, orally administered opioid BAPs modulate absorption processes and thus influence the gastrointestinal functions either by affecting smooth muscles, which leads to reduction in transit time or by affecting the intestinal transport of electrolytes. Opioid peptide system of brain is known to play an important role in attachment behavior, motivation, the response to stress and pain, the control of food intake and emotions. Activity of opioid peptides is dependent upon their binding to specific receptors of the target cell while specific physiological effects is dependent upon the nature of each individual receptor as the s-receptor is for emotional behavior, m receptor is responsible for emotions and intestinal motility suppression and k-receptor is for food intake and sedation.

Casein has been identified as a rich source of opioid active BAPs. Upon hydrolysis by different digestive enzymes, various casein fragments result in production of peptides with opioid activity (Pihlanto-Leppä la et al., 1994). Interestingly, these opioid casein fragments were only found in plasma of infants. A commercially important, 1-casein-derived peptide commonly used in confectionery and soft drinks has been demonstrated to possess anxiolytic-like stress-relieving properties in animal model as well as human studies (Lefranc, 2001). Milk protein derived opioid BAPs may act as pheromones or exorphins (food hormones). Studies have also shown these BAPs to possess pharmacological properties similar to that of naloxone and morphine (Gobbetti et al., 2002). The α- and β-casomorphins act as opioid agonists, which could modulate social behavior, influence postprandial metabolism produce analgesia (Meisel and Schlimme, 1990). Wheat gluten upon digestion with pepsin serves as a good source of BAPs with opioid activity. These milk and wheat (exogenous sources) derived BAPs have shown to have similar structure to that of endogenous opioid peptides containing a tyrosine residue at the amino terminal or bioactive site (Kitts and Weiler, 2003; Yamamoto et al., 2003; Aldrich, 2009).

19. Biotechnology applications

Biotechnology is described as the field of biology where living systems, their derivatives and organisms interact with each other to produce or modify products. One major problem with BAPs is their unavailability. Due to a wide range of uses, demand of BAPs is growing with every passing day and BAPs exist only in a limited quantity within natural settings or nature. Hence to meet the demand at commercial scale, different biotechnological processes and methods are used to produce modified form of BAPs artificially (Danquah and Agyei, 2012). Synthetic peptide is the name given to such artificially produced BAPs, which have major advantage of the naturally occurring ones (Moser et al., 1985). These synthetic BAPs and proteins are well recognized today and have a big market share of billions of US dollars. Synthetic BAPs are known to positively influence metabolic functioning in human beings.

Protein engineering is the basic technique where using the process of recombinant DNA technology, the synthetic BAPs are produced at a large scale (Ryu et al., 2014). This involves insertion of chemically synthesised DNA into an organism to produce modified peptides. Here one or few amino acid sequence is/are altered, which result(s) in production of desired product. This leads in manufacturing of highly advanced bioactive peptides and enzymes at an industrial scale which was not possible traditionally using natural resources.

Though bacteria like *E. coli* is the best choice of host for DNA insertion, the yeast *Saccharomyces cerevisiae* has also been recognized as a host organism. This technology has paved a way to produce BAPs with multiple benefits as vaccines production for diseases like malarial, protozoa and hepatitis B virus.

Application of genetic engineering in protein industry has progressed so much that an entire new field has merged, called metabolic engineering. This field primarily deals with the modification of metabolic pathways *via* recruitment of peptides from different cells. It results in a change of both rate and pathway distribution. Metagenomics results in multiple expression systems in the host and thus the expression of a particular peptide or product is not limited upon a conserved metabolic pathway. This methodology results in an easy scale up of BAPs produce on an industrial level.

An innovation of biotechnology the peptide tag has seen its use in health industry (Einhauer and Jungbauer, 2001). Peptide tags can be manufactured for different type of proteins *i.e.* antigens, hormones *etc*. Tags can be deployed for different purposes. Most of the tags can be used to enable protein detection through ELISA, Western blot, ChIP, immunohistochemistry, immunocytochemistry and other such sensitive methods (Kimple et al., 2013). This property of peptide tags makes them the suitable candidate for detection of many diseases. These nano-sized tags not only help in rapid detection of a disease but also help in fast and efficient drug delivery (Lamla and Erdmann, 2014). Due to this property, the demand of peptide tags has increased drastically in health sector. Many peptide tags are currently utilized in the protein manufacture industry for protein purification. Peptide tags recognize the particular protein and then its extraction from the mixture can be done easily and efficiently.

20. Conclusion and future directions

The large amount(s) of BAPs are present entrapped within the complex chemical structure of proteins present in food matrices and some other natural resources. These are released during the breakdown of the food by chemical or enzymatic hydrolysis. The broad spectrum of BAP activities includes mineral binding, antimicrobial, antihypertensive, antithrombotic, antioxidative, immunomodulatory and opioid function, which makes BAPs as the most favored ingredient of pharmaceuticals and food products. BAPs are also used as potential drugs against many life style diseases, which include hypertension, diabetes type II and obesity.

Though BAPs are now recognized as beneficial products, their bulk availability is still a major issue. BAPs obtained from natural resources are not enough to thrive with the growing commercial demand. To tackle the problem of availability, BAPs are produced using the alternative methods and techniques of biotechnology. Using biotechnological tools and techniques the modified version of desired BAPs can be produced at a large scale, now known as synthetic BAPs. Immense benefits of BAPs have drawn scientific community toward them. Efficient and productive ways have been developed for identification, isolation and purification of BAPs, which will ultimately result in their extended benefits to the humanity.

Diverse functions of bioactive peptides have made them a hot topic of research. In near future bioactive peptides will be readily available as pharmaceutical drugs for many different diseases. Genetic modifications and advancement in biotechnology will result in production of more efficient BAPs which may be used extensively in protein targeting and signaling pathways. Pure food derived bioactive peptides will be available in the market to be sold as nutraceuticals. Further studies should be done on consumer acceptance as well as the safety considerations in order to make BAPs a commercial breakthrough and a common house hold name (Table 4.1).

Table 4.1 Major sources for production of BAPs.

Source(s)	Bioactivity	Reference(s)
Milk	Antimicrobial, antioxidative, immunomodulatory, opioid and anti-inflammatory	Meisel and Fitzgerald (2003)
Dairy products	Anti-inflammatory and antioxidative	Hayes et al. (2007)
Rice	Antihypertensive, anticancerous and opioid	Li et al. (2007)
Soy	Anti-inflammatory, antioxidative and anticancerous	Singh et al. (2014)
Potato	Antioxidative	Kudo et al. (2009)
Corn	Anti-inflammatory	Dia et al. (2014)
Whey	Anti-inflammatory	Shahi et al. (2012)
Wheat	Antihypertensive and opioid	Zhu et al. (2006)
Egg	Antioxidative	Bhat et al. (2015a)
Gelatine	Antioxidative	Kim SK et al. (2001).

Acknowledgments

This work has been funded by Council for Scientific and Industrial Research, New Delhi, under a CSIR-NET Junior Research Fellowship [File No.09/237(0161)/2017-EMR-1] awarded to one of the authors (VC). The authors are thankful to Department of Biotechnology, Ministry of Science and Technology, New Delhi for continuous financial support to the Bioinformatics Center, Himachal Pradesh University, Shimla (India).

Conflict of interest

Both the authors declare that they have no conflict of interest with the parent institute or among themselves.

References

Agyei, D., Danquah, M.K., 2011. Industrial-scale manufacturing of pharmaceutical-grade bioactive peptides. Biotechnol. Adv. 29, 272–277.

Agyei, D., Danquah, M.K., 2012. Rethinking food derived bioactive peptides for antimicrobial and immunomodulatory activities. Trends Food Sci. Technol. 23, 62–69.

Agyei, D., Potumarthi, R., Danquah, M.K., 2015. Food-derived multifunctional bioactive proteins and peptides: sources and production. Biotechnol. Bioact. Compds: Sources Appl. 483–505.

Aldrich, J., McLaughlin, J.P., 2009. Peptide kappa opioid receptor ligands: potential for drug development. AAPS J 11, 312.

Andreu, D., Rivas, L., 2002. Chemistry and applications of synthetic antimicrobial peptides. In: Peptide Antibiotics: Discovery, Modes of Action, and Applications. Marcel Dekker Inc, New York, pp. 15–45.

Arrutia, F., Puente, Á., Riera, F.A., Menéndez, C., González, U.A., 2016. Influence of heat pre-treatment on BSA tryptic hydrolysis and peptide release. Food Chem. 202, 40–48.

Barrios, L.L., Olea, E.H., Flores, D.G., Carrillo, E.P., Uribe, J.A.G., 2018. Bioactive peptides by *in vitro*Digestion of germinated bean digestion cotyledons extrudates. J. Food Res. 7, 1.

Bazinet, L., Firdaous, L., 2013. Separation of bioactive peptides by membrane processes: technologies and devices. Recent Pat. Biotechnol. 7, 9–27.

Bhat, Z.F., Kumar, S., Bhat, H.F., 2015a. Bioactive peptides from egg: a review. Nutr. Food Sci. 45, 190–212.

Bhat, Z.F., Kumar, S., Bhat, H.F., 2015b. Bioactive peptides of animal origin: a review. J. Food Sci. Technol. Mysore 52, 5377–5392.

Boeriu, C.G., Frissen, A.E., Boer, E., van Kekem, K., van Zoelen, D.-J., Eggen, I.F., 2010. Optimized enzymatic synthesis of C-terminal peptide amides using subtilisin A from *Bacillus licheniformis*. J. Mol. Catal. B Enzym. 66, 33–42.

Bongers, J., Heimer, E.P., 1994. Recent applications of enzymatic peptide synthesis. Peptides 15, 183–193.

Borgia, J.A., Fields, G.B., 2000. Chemical synthesis of proteins. Trends Biotechnol. 18, 243–251.

Boudesocque, L., Kapel, R., Paris, C., Dhulster, P., Marc, I., Renault, J.H., 2012. Concentration and selective fractionation of an antihypertensive peptide from an alfalfa white proteins hydrolysate by mixed ion-exchange centrifugal partition chromatography. J. Chromatogr. B Biomed. Sci. Appl. 905, 23–30.

Bougate, A., Arroume, N.N., Mannia, L., Ravallec, R., Barkiaa, A., Guillochon, D., Nasria, M., 2010. Purification and identification of novel antioxidant peptides from enzymatic hydrolysates of sardinelle (*Sardinella aurita*) byproducts proteins. Food Chem. 118, 559–565.

Brogden, K.A., 2005. Antimicrobial peptides: pore formers or metabolic inhibitors in bacteria? Nat. Rev. Microbiol. 3, 238–250.

Capriotti, A.L., Cavaliere, C., Piovesana, S., Samperi, R., Aldo Laganà, A., 2016. Recent trends in the analysis of bioactive peptides in milk and dairy products. Anal. Bioanal. Chem. 408, 2677–2685.

Caron, J., Chataigné, G., Gimeno, J.P., Duhal, N., Goossens, J.F., Dhulster, P., Cudennec, B., Ravallec, R., Flahaut, C., 2016. Food peptidomics of *in vitro*gastrointestinal digestions of partially purified bovine hemoglobin: low-resolution versus highresolution LC-MS/MS analyses. Electrophoresis 37, 1814–1822.

Carrasco-Castilla, J., Hernandez-Alvarez, A.J., Jimenez-Martınez, C., Gutierrez-Lopez, G.F., Davila-Ortiz, G., 2012. Use of proteomics and peptidomics methods in food bioactive peptide science and engineering. Food Eng. Rev. 4, 224–243.

Cavazos, A., Gonzalez de Mejia, E., 2013. Identification of bioactive peptides from cereal storage proteins and their potential role in prevention of chronic diseases. Compr. Rev. Food Sci. Food Saf. 12, 364–380.

Chakrabarti, S., Jahandideh, F., Wu, J., 2014. Food-derived bioactive peptides on inflammation and oxidative stress. BioMed Res. Int. 2014, 608979.

Chaves-López, C., Tofalo, R., Serio, A., Paparella, A., Suzzi, G., 2012. Yeasts from Colombian Kumis as source of peptides with angiotensin I converting enzyme (ACE) inhibitory activity in milk. Int. J. Food Microbiol. 59, 39–46.

Cheung, R.C.F., Ng, T.B., Wong, J.H., 2015. Marine peptides: bioactivities and applications. Mar. Drugs 13, 4006–4043.

Chockalingam, A., Campbell, N.R., Fodor, J.G., 2006. Worldwide epidemic of hypertension. Can. J. Cardiol. 22, 553–555.

Choi, J., Sabikhi, L., Hassan, A., Anand, S., 2012. Bioactive peptides in dairy products. Int. J. Dairy Technol. 65, 1–12.

Cicero, A., Aubin, F., Azais-Braesco, V., Borghi, C., 2013. Do the lactotripeptides isoleucine–proline–proline and valine–proline–proline reduce systolic blood pressure in European subjects? A meta-analysis of randomized controlled. Trials Am J Hypertens 26, 442–449.

Clare, D.A., Swaisgood, H.E., 2000. Bioactive milk peptides: a prospectus1. J. Dairy Sci. 83, 1187–1195.

Coda, R., Rizzello, C.G., Pinto, D., Gobbetti, M., 2012. Selected lactic acid bacteria synthesize antioxidant peptides during sourdough fermentation of cereal flours. Appl. Environ. Microbiol. 78, 1087–1096.

Conde, E., Reinoso, B.D., González-Muñoz, M.J., Moure, A., Domínguez, H., Parajó, J.C., 2013. Recovery and concentration of antioxidants from industrial effluents and from processing streams of underutilized vegetal biomass. Food Public Health 3, 69–91.

Cruz, J., Ortiz, C., Guzman, F., Fernandez-Lafuente, R., Torres, R., 2014. Antimicrobial peptides: promising compounds against pathogenic microorganisms. Curr. Med. Chem. 21, 2299–2321.

Dagan, A., Efron, L., Gaidukov, L., Mor, A., Ginsburg, H., 2002. *In vitro*antiplasmodium effects of dermaseptin S4 derivatives. Antimicrob. Agents Chemother. 46, 1059–1066.

Danquah, M.K., Agyei, D., 2012. Pharmaceutical applications of bioactive peptides. Biotechnology 29, 5.

de Morais Coutinho, C., Chiu, M.C., Basso, R.C., Ribeiro, A.P.B., Gonçalves, L.A.G., Viotto, L.A., 2009. State of art of the application of membrane technology to vegetable oils: a review. Food Res. Int. 42, 536–550.

De Simone, C., Picariello, G., Mamone, G., Stiuso, P., Dicitore, A., Vanacore, D., 2009. Characterisation and cytomodulatory properties of peptides from Mozzarella di Bufala Campana cheese whey. J. Pept. Sci. 15, 251–258.

Dia, V.P., Bringe, N.A., de Mejia, E.G., 2014. Peptides in pepsin–pancreatin hydrolysates from commercially available soy products that inhibit lipopolysaccharide-induced inflammation in macrophages. Food Chem. 152, 423–431.

Dumeusa, S., Shibua, M.A., Linc, W.T., Wang, M.F., Laie, C.H., Shenf, C.Y., Ling, Y.M., Viswanadhai, V.P., Kuoj, W.W., Huanga, C.Y., 2018. Bioactive peptide improves diet-induced hepatic fat deposition and hepatocyte proinflammatory response in SAMP8 ageing mice. Cell. Physiol. Biochem. 48, 1942–1952.

Einhauer, A., Jungbauer, A., 2001. The FLAG™ peptide, a versatile fusion tag for the purification of recombinant proteins. J. Biochem. Biophys. Methods 49, 455–465.

El-Salam, M.H.A., El-Shibiny, S., 2013. Bioactive peptides of buffalo, camel,goat, sheep, mare, and yak milks and milk products. Food Rev. Int. 29, 1–23.

Elias, R.J., Kellerby, S.S., Decker, E.A., 2008. Antioxidant activity of proteins and peptides. Crit. Rev. Food Sci. Nutr. 48, 430–441.

Esteve, C., Marina, M.L., García, M.C., 2015. Novel strategy for the revalorization of olive (Olea europaea) residues based on the extraction of bioactive peptides. Food Chem. 167, 272–280.

Falanga, A., Lombardi, L., Franci, G., Vitiello, M., Iovene, M.R., Morelli, G., 2016. Marine antimicrobial peptides: nature provides templates for the design of novel compounds against pathogenic bacteria. Int. J. Mol. Sci. 17, 785–803.

Fan, J., He, J., Zhuang, Y., Sun, L., 2012. Purification and identification of antioxidant peptides from enzymatic hydrolysates of tilapia (*Oreochromis niloticus*) frame protein. Molecules 17, 12836–12850.

Fan, X., Bai, L., Zhu, L., Yang, L., Zhang, X., 2014. Marine algae-derived bioactive peptides for human nutrition and health. J. Agric. Food Chem. 62, 9211–9222.

Ferranti, P., Traisci, M.V., Picariello, G., Nasi, A., Boschi, V., Siervo, M., 2004. Casein proteolysis in human milk: tracing the pattern of casein breakdown and the formation of potential bioactive peptides. J. Dairy Res. 71, 74–87.

Fields, K., Falla, T.J., Rodan, K., Bush, L., 2009. Bioactive peptides: signalling the future. J. Cosmet. Dermatol. 8, 8–13.

Fitzgerald, R.J., Meisel, H., 2003. Milk protein hydrolysates and bioactive peptides. In: Fox, P.F., McSweeney, P.L.H. (Eds.), Advanced Dairy Chemistry 1 Proteins. Springer US, Boston, pp. 675–698.

Fosgerau, K., Hoffmann, T., 2015. Peptide therapeutics: current status and future directions. Drug Discov. Today 20, 122–128.

Freidinger, R.M., 2003. Design and synthesis of novel bioactive peptides and peptidomimetics. J. Med. Chem. 46, 5563–5566.

Gauthier, S.F., Pouliot, Y., Saint-Sauveur, D., 2006. Immunomodulatory peptides obtained by the enzymatic hydrolysis of whey proteins. Int. Dairy J. 16, 1315–1323.

Gibbs, B.F., Zougman, A., Masse, R., Mulligan, C., 2004. Production and characterization of bioactive peptides from soy hydrolysate and soy-fermented food. Food Res. Int. 37, 123–131.

Gill, H.S., Doull, F., Rutherfurd, K.J., Cross, M.L., 2000. Immunoregulatory peptides in bovine milk. Br. J. Nutr. 84, 111–S117.

Gobbetti, M., Stepaniak, L., De Angelis, M., Corsetti, A., Di Cagno, R., 2002. Latent bioactive peptides in milk proteins: proteolytic activation and significance in dairy processing. Crit. Rev. Food Sci. Nutr. 42, 223–239.

Gobbetti, M., Minervini, F., Rizzello, C.G., 2007. Bioactivepeptides in dairy products. In: Handbook of Food Products Manufacturing, pp. 489–517.

González-García, E., Marina, M.L., García, M.C., 2014. Plum (Prunus Domestica L.) by-product as a new and cheap source of bioactive peptides: extraction method and peptides characterization. J. Funct. Foods 11, 428–437.

Haque, E., Chand, R., Kapila, S., 2008. Biofunctional properties of bioactive peptides of milk origin. Food Rev. Int. 25, 28–43.

Harnedy, P.A., FitzGerald, R.J., 2011. Bioactive proteins, peptides, and amino acids from macroalgae. J. Phycol. 47, 218–232.

Hayes, M., Ross, R.P., Fitzgerald, G.F., Stanton, C., 2007. Putting microbes to work: dairy fermentation, cell factories and bioactive peptides. Part I: overview. Biotechnol. J. 2, 426–434.

Hernandez-Ledesma, B., Recio, I., Amigo, L., 2008. Beta-lactoglobulin as source of bioactive peptides. Amino Acids 35, 257–265.

Herrero, M., Castro-Puyana, M., Rocamora-Reverte, L., Ferragut, J.A., Cifuentes, A., Ibáñez, E., 2012. Formation and relevance of 5-hydroxymethylfurfural in bioactive subcritical water extracts from olive leaves. Food Res. Int. 47, 31–37.

Hollt, V., 1983. Multiple endogenous opioid peptide. Trends Neurosci. 6, 24–26.

Hoskin, D.W., Ramamoorthy, A., 2008. Studies on anticancer activities of antimicrobial peptides. Biochim. Biophys. Acta Biomembr. 1778, 357–375.

Houben, H., Denef, C., 1994. Bioactive peptides in anterior pituitary cells. Peptides 15, 547–582.

Howell, N.K., Kasase, C., 2010. Bioactive peptides and proteins from fish muscle and collagen. In: Bioactive Proteins and Peptides as Functional Foods and Nutraceuticals, pp. 203–223.

Huang, S.M., Chen, K.N., Chen, Y.P., Hong, W.S., Chen, M.J., 2010. Immunomodulatory properties of the milk whey products obtained by enzymatic and microbial hydrolysis. Int. J. Food Sci. 45, 1061–1067.

Ivanov, V.T., Karelin, A.A., Philippova, M.M., Nazimov, I.V., Pletnev, V.Z., 1997. Hemoglobin as a source of endogenous bioactive peptides: the concept of tissue-specific peptide pool. Peptide Sci. 43, 171–188.

Jäkälä, P., Vapaatalo, H., 2010. Antihypertensive peptides from milk proteins. Pharmaceuticals 3, 251–272.

Je, J.Y., Park, P.J., Kim, S.K., 2005. Antioxidant activity of a peptide isolated from Alaska pollack (*Theragra chalcogramma*) frame protein hydrolysate. Food Res. Int. 38, 45–50.

Jenssen, H., Hamill, P., Hancock, R.E.W., 2006. Peptide antimicrobial agents. Clin. Microbiol. Rev. 19, 491–511.

Jo, C., Khan, F.F., Khan, M.I., Iqbal, J., 2017. Marine bioactive peptides: types, structures, and physiological functions. Food Rev. Int. 33, 44–61.

Joffres, M.R., Hamet, P., MacLean, D.R., L'italien, G.J., Fodor, G., 2001. Distribution of blood pressure and hypertension in Canada and the United States. Am. J. Hypertens. 14, 1099–1105.

Jorgensen, A.L.W., Juul-Madsen, H.R., Stagsted, J., 2010. Colostrum and bioactive, colostral peptides differentially modulate the innate immune response of intestinal epithelial cells. J. Pept. Sci. 16, 21–30.

Jumeri, Kim, S., 2011. Antioxidant and anticancer activities of enzymatic hydrolysates of solitary tunicate (*Styela clava*). Food Sci. Biotechnol. 20, 1075–1085.

Kang, H.K., Seo, C.H., Park, Y., 2015. Marine peptides and their anti-infective activities. Mar. Drugs 13, 618–654.

Kannan, A., Hettiarachchy, N.S., Lay, J.O., Liyanage, R., 2010. Human cancer cell proliferation inhibition by a pentapeptide isolated and characterized from rice bran. Peptides 31, 1629–1634.

Kent, S.B.H., 1988. Chemical synthesis of peptides and proteins. Ann. Rev. Biochem. 57, 957–989.

Kim, S.K., Wijesekara, I., 2010. Development and biological activities of marine-derived bioactive peptides: a review. J. Funct. Foods 2, 1–9.

Kim, S.K., Kim, Y.T., Byun, H.G., Nam, K.S., Joo, D.S., Shahidi, F., 2001. Isolation and characterization of antioxidative peptides from gelatin hydrolysate of Alaska pollack skin. J. Agric. Food Chem. 4, 1984–1989.

Kim, E.K., Joung, H.J., Kim, Y.S., Hwang, J.W., Ahn, C.B., Jeon, Y.J., 2012a. Purification of a novel anticancer peptide from enzymatic hydrolysate of *Mytilus coruscus*. J. Microbiol. Biotechnol. 22, 1381–1387.

Kim, S.K., Ngo, D.H., Vo, T.S., 2012b. Marine fish-derived bioactive peptides as potential antihypertensive agents. Adv. Food Nutr. Res. 65, 249–260.

Kimple, M.E., Brill, A.L., Pasker, R.L., 2013. Overview of affinity tags for protein purification. Curr. Protoc. Protein Sci. 73. Unit 9.9.

Kishore, R., 2004. β-Ala containing peptides: potentials in design and construction of bioactive peptides and protein secondary structure mimics. Curr. Protein Pept. Sci. 5, 435–455.

Kitts, D.D., Weiler, K., 2003. Bioactive proteins and peptides from food sources: applications of bioprocesses used in isolation and recovery. Curr. Pharmaceut. Des. 9, 1309–1323.

Kong, X., Zhou, H., Hua, Y., 2008. Preparation and antioxidant activity of wheat gluten hydrolysates (WGHs) using ultrafiltration membranes. J. Sci. Food Agric. 88, 920–926.

Korhonen, H., 2009. Milk-derived bioactive peptides: from science to applications. J. Func. Foods 1, 177–187.

Korhonen, H., Pihlanto, A., 2003. Food-derived bioactive peptides—opportunities for designing future foods. Curr. Pharmaceut. Des. 9, 1297–1308.

Korhonen, H., Pihlanto, A., 2006. Bioactive peptides: production and functionality. Int. Dairy J. 16, 945–960.

Korhonen, H., Pihlanto, A., 2007. Technological options for the production of health-promoting proteins and peptides derived from milk and colostrum. Curr. Pharmaceut. Des. 13, 829–843.

Kouhdasht, M.A., Nasab, M.M., 2018. Bioactive peptides derived from fish by-product collagen. Int. J. Environ. Sci. Nat. Res. 13 https://doi.org/10.19080/IJESNR.2018.13.555859.

Kudo, K., Onodera, S., Takeda, Y., Benkeblia, N., Shiomi, N., 2009. Antioxidative activities of some peptides isolated from hydrolyzed potato protein extract. J. Func. Foods 1, 170–176.

Kumar, S.N., Siji, J.V., Nambisan, B., Mohandas, C., 2012. Activity and synergistic antimicrobial activity between diketopiperazines against bacteria in vitro. Appl. Biochem. Biotechnol. 168, 2285–2296.

Lafarga, T., Hayes, M., 2014. Bioactive peptides from meat muscle and byproducts: generation, functionality and application as functional ingredients. Meat Sci. 98, 227–239.

Lalenti, A., Santagada, V., Caliendo, G., Severino, B., Fiorino, F., Maffia, P., Ianaro, A., Morelli, F., Di Micco, B., Cartenì, M., Stiuso, P., Metafora, V., Metafora, S., 2001. Synthesis of novel anti-inflammatory peptides derived from the amino-acid sequence of the bioactive protein SV-IV. Eur. J. Biochem. 268, 3399–3406.

Lamla, T., Erdmann, V.A., 2014. The Nano-tag, a streptavidin-binding peptide for the purification and detection of recombinant proteins. Protein Expr. Purif. 33, 39–47.

Langevin, M.E., Roblet, C., Moresoli, C., Ramassamy, C., Bazinet, L., 2012. Comparative application of pressure- and electrically-driven membrane processes for isolation of bioactive peptides from soy protein hydrolysate. J. Membr. Sci. 403, 15–24.

Lassoued, I., Mora, L., Barkia, A., Aristoy, M.C., Nasri, M., Toldra, F., 2015. Bioactive peptides identified in thornback ray skin's gelatin hydrolysates by proteases from Bacillus subtilis and Bacillus amyloliquefaciens. J. Proteomics 128, 8–17.

Lefranc, C., 2001. Cool, calm and collected. Dairy Ind. Int. 66, 36–37.

Lemes, A.C., Sala, L., Ores, J.D.C., Braga, A.R.C., Egea, M.B., Fernandes, K.F., 2016. A review of the latest advances in encrypted bioactive peptides from protein-rich waste. Int. J. Mol. Sci. 17, 950.

Léonil, J., 2014. Milk bioactive peptides: their interest for cardiovascular diseases and metabolic syndrome prevention. Medecine Des Maladies Metaboliques 8, 495–499.

Li, Y., 2011. Recombinant production of antimicrobial peptides in *Escherichia coli*: a review. Protein Expr. Purif. 80, 260–267.

Li, Y., Yu, J., 2015. Research progress in structure-activity relationship of bioactive peptides. J. Med. Food 18, 147–156.

Li, G.H., Qu, M.R., Wan, J.Z., You, J.M., 2007. Antihypertensive effect of rice protein hydrolysate with *in vitro*angiotensin I-converting enzyme inhibitory activity in spontaneously hypertensive rats. Asia Pac. J. Clin. Nutr. 16, 275–280.

Lima, T.N., Moraes, C.A.P., 2018. Bioactive peptides: applications and relevance for cosmeceuticals. Cosmetics 5, 21.

Lin, S., Guo, Y., You, Q., Yin, Y., Liu, J., 2011. Preparation of antioxidant peptide from egg white protein and improvement of its activities assisted by high-intensity pulsed electric field. J. Sci. Food Agric. 92, 1554–1561.

Liu, R., Xing, L., Fu, Q., Zhou, G.H., Zhang, W.G., 2016. A review of antioxidant peptides derived from meat muscle and by-products. Antioxidants 5, 32.

Machado, A., Liria, C.W., Proti, P.B., Remuzgo, C., Miranda, M.T.M., 2004. Sınteses qu'ımica e enzim'atica de pept'ıdeos: princ'ıpios básicos e aplicaç ̃oes. Quim. Nova 27, 781–789.

Majumder, K., Mine, Y., Wu, J., 2016. The potential of foodprotein derived anti-inflammatory peptides against various chronic inflammatory diseases. J. Sci. Food Agric. 96, 2303–2311.

Malaguti, M., Dinelli, G., Leoncini, E., Bregola, V., Bosi, S., Cicero, A., Hrelia, S., 2014. Bioactive peptides in cereals and legumes: agronomical, biochemical and clinical aspects. Int. J. Mol. Sci. 15, 21120–21135.

Manikkam, V., Vasiljevic, T., Donkor, O.N., Mathai, M.L., 2016. A review of potential marine-derived hypotensive and anti-obesity peptides. Crit. Rev. Food Sci. Nutr. 56, 92–112.

Marczak, E.D., Usui, H., Fujita, H., Yang, Y., Yokoo, M., Lipkowski, A.W., 2003. New antihypertensive peptides isolated from rapeseed. Peptides 24, 791–798.

Marques, C., Amorim, M., Pereira, J.O., Pintado, M.E., Moura, D., Calhau, C., 2012. Bioactive peptides – are there more antihypertensive mechanisms beyond ACE inhibition? Curr. Pharmaceut. Des. 18, 4706–4713.

Martínez-Augustin, O., Rivero-Gutiérrez, B., Mascaraque, C., Sánchez de Medina, F., 2014. Food derived bioactive peptides and intestinal barrier function. Int. J. Mol. Sci. 15, 22857–22873.

Meena, K.R., Sharma, A., Kanwar, S.S., 2017. Microbial lipopeptides and their medical Applications. Ann. Pharmacol. Pharma. 2, 1–5.

Meisel, H., Bockelmann, W., 1999. Bioactive peptides encrypted in milk proteins: proteolytic activation and thropho-functional properties. Antonie Van Leeuwenhoek Int. J. Gen. Mol. Microbiol. 76, 207–215.

Meisel, H., FitzGerald, R.J., 2003. Biofunctional peptides from milk proteins: mineral binding and cytomodulatory effects. Curr. Pharmaceut. Des. 9, 1289–1295.

Meisel, H., Schlimme, E., 1990. Milk proteins: precursors of bioactive peptides. Trends Food Sci. Technol. 1, 41–43.

Mercier, A., Gauthier, S.F., Fliss, I., 2004. Immunomodulating effects of whey proteins and their enzymatic digests. Int. Dairy J. 14, 175–183.

Migliore-Samour, D., Jollés, P., 1988. Casein prohormone with an immunomodulating role for the newborn. Experientia 44, 188–193.

Mohanty, D.P., Mohapatra, S., Misra, S., Sahu, P.S., 2015. Milk derived bioactive peptides and their impact on human health: a review. Saudi J. Biol. Sci. 23, 577–583.

Mohanty, D., Jena, R., Choudhury, P.K., Pattnaik, R., Mohapatra, S., Saini, M.R., 2016. Milk derived antimicrobial bioactive peptides: a review. Int. J. Food Prop. 19, 837–846.

Moller, N.P., Scholz-Ahrens, K.E., Roos, N., Schrezenmeir, J., 2008. Bioactive peptides and proteins from foods: indication for health effects. Eur. J. Nutr. 47, 171–182.

Moser, R., Klauser, S., Leist, T., Langen, H., Epprecht, T., Gutte, B., 1985. Applications of synthetic peptides. Angew Chem. Int. Ed. Engl. 24, 719–727.

Nardo, A.E., AñoÂn, M.C., Parisi, G., 2018. Large-scale mapping of bioactive peptides in structural and sequence space. PLoS One 13 (1), e0191063.

Navab, M., Anantharamaiah, G.M., Reddy, S.T., Van Lenten, B.J., Datta, G., Garber, D., Fogelman, A.M., 2004. Human apolipoprotein A-I and A-I mimetic peptides: potential for atherosclerosis reversal. Curr. Opin. Lipidol. 15, 645–649.

Nawaz, H., Shi, J., Mittal, G.S., Kakuda, Y., 2006. Extraction of polyphenols from grape seeds and concentration by ultrafiltration. Separ. Purif. Technol. 48, 176–181.

Ng, T.B., Cheung, R.C., Wong, J.H., Wang, Y., Ip, D.T., Wan, D.C., Xia, J., 2015. Antiviral activities of whey proteins. Appl. Microbiol. Biotechnol. 99, 6997–7008.

Ngo, D.H., Kim, S.K., 2013. Marine bioactive peptides as potential antioxidants. Curr. Protein Pept. Sci. 14, 189–198.

Ngo, D.H., Vo, T.S., Ngo, D.N., Wijesekara, I., Kim, S.K., 2012. Biological activities and potential health benefits of bioactive peptides derived from marine organisms. Int. J. Biol. Macromol. 51, 378–383.

Noguchi, T., Kato, T., Wang, L., Maeda, Y., Ikeda, H., Sato, E., 2012. Intracellular tumor-associated antigens represent effective targets for passive immunotherapy. Cancer Res. 72, 1672–1682.

Ogino, H., Yamada, M., Watanabe, F., Ichinose, H., Yasuda, M., Ishikawa, H., 1999. Peptide synthesis catalyzed by organic solvent-stable protease from *Pseudomonas aeruginosa* PST-01 in monophasic aqueous-organic solvent systems. J. Biosci. Bioeng. 88, 513–518.

Park, Y.W., Nam, M.S., 2015. Bioactive peptides in milk and dairy products: a review. Kor. J. Food Sci. Animal Res. 35, 831–840.

Peñta-Ramos, E.A., Xiong, Y.L., 2002. Antioxidant activity of soy protein hydrolysates in a liposomal system. J. Food Sci. 67, 2952–2956.

Pihlanto, A., 2006. Bioactive peptides: functionality and production. Agro Food Ind. Hi-Tech 17, 24–26.

Pihlanto-Leppä lä, A., Antila, P., Mä ntsä lä, P., Hellman, J., 1994. Opioid peptides produced by *in vitro*proteolysis of bovine caseins. Int. Dairy J. 4, 291–301.

Pimenta, A.M.C., De Lima, M.E., 2005. Small peptides, big world: biotechnological potential in neglected bioactive peptides from arthropod venoms. J. Pept. Sci. 11, 670–676.

Powers, J.-P.S., Hancock, R.E.W., 2003. The relationship between peptide structure and antibacterial activity. Peptides 24, 1681–1691.

Pritchard, S.R., Phillips, M., Kailasapathy, K., 2010. Identification of bioactive peptides in commercial Cheddar cheese. Food Res. Int. 43, 1545–1548.

Przybylski, R., Firdaous, L., Châtaigné, G., Dhulster, P., Nedjar, N., 2016. Production of an antimicrobial peptide derived from slaughterhouse by-product and its potential application on meat as preservative. Food Chem. 211, 306–313.

Rajesh, B.M., Iqbal, J., 2006. Recent advances in the synthesis of some bioactive peptides and peptidomimetics. Curr. Pharmaceut. Biotechnol. 7, 247–259.

Reddi, S., Kapila, R., Dang, A.K., Kapila, S., 2012. Evaluation of allergenic response of milk bioactive peptides using mouse mast cell. Milchwissenschaft Milk Sci. Int. 67, 189–191.

Remanan, M.K., Wu, J., 2014. Antioxidant activity in cooked and simulated digested eggs. Food Funct. 5, 1464–1474.

Ridge, S., Hettiarachchi, K., 1998. Peptide purity and counter ion determination of bradykinin by high-performance liquid chromatography and capillary electrophoresis. J. Chromatogr. 817, 215–222.

Rizzello, C.G., Lavecchia, A., Gramaglia, V., Gobbetti, M., 2015. Long-term fungal inhibitory activity of *Pisum sativum* flour hydrolysate during storage of wheat flour bread. Appl. Environ. Microbiol. 81, 4195–4206.

Roy, F., BoyeJ, I., Simpson, B.K., 2010. Bioactive proteins and peptides in pulse crops: pea, chickpea and lentil. Food Res. Int. 43, 432–442.

Ryan, J.T., Ross, R.P., Bolton, D., Fitzgerald, G.F., Stanton, C., 2011. Bioactive peptides from muscle sources: meat and fish. Nutrients 3, 765–791.

Ryder, K., Bekhit, A.E.D., McConnell, M., Carne, A., 2016. Towards generation of bioactive peptides from meat industry waste proteins: generation of peptides using commercial microbial proteases. Food Chem. 208, 42–50.

Ryu, J.S., Cho, A.Y., Seo, S.W., Min, H., 2014. Engineering bioactive peptide-based therapeutic molecules. Methods Mol. Biol. 1088, 35–50.

Sakanaka, S., Tachibana, Y., 2006. Active oxygen scavenging activity of egg-yolk protein hydrolysates and their effects on lipid oxidation in beef and tuna homogenates. Food Chem. 95, 243–249.

Saladino, R., 2012. Advances in the synthesis of bioactive unnatural amino acids and peptides. Mini Rev. Med. Chem. 12, 277–300.

Selamassakul, O., Laohakunjit, N., Kerdchoechuen, O., Ratanakhanokchai, K., 2016. A novel multi-biofunctional protein from brown rice hydrolysed by endo/endo-exoproteases. Food and Funct. 7, 2635–2644.

Sewald, N., Jakubke, H., 2002. Peptides: Chemistry and Biology. Wiley-VCH Verlag GmbH and Co. KGaA, Weinheim, p. 543.

Shahi, M.M., Rashidi, M.R., Mahboob, S., Haidari, F., Rashidi, B., Hanaee, J., 2012. Protective effect of soy protein on collageninduced arthritis in rat. Rheumatol. Int. 32, 2407–2414.

Shahidi, F., Zhong, Y., 2008. Bioactive peptides. J. AOAC Int. 91, 914–931.

Sharma, S., Singh, R., Rana, S., 2011. Bioactive peptides: a review. In. J. Bioautomotion 15, 223–250.

Shigeri, Y., Tatsu, Y., Yumoto, N., 2001. Synthesis and application of caged peptides and proteins. Pharmacol. Ther. 91, 85–92.

Silva-Sanchez, C., de la Rosa, A.P.B., Leon-Galvan, M.F., de Lumen, B.O., de Leon-Rodriguez, A., de Mejia, E.G., 2008. Bioactive peptides in amaranth (*Amaranthus hypochondriacus*) seed. J. Agric. Food Chem. 56, 1233–1240.

Singh, B.P., Vij, S., Hati, S., 2014. Functional significance of bioactive peptides derived from soybean. Peptides 54, 171–179.

Smacchi, E., Gobbetti, M., 2000. Bioactive peptides in dairy products: synthesis and interaction with proteolytic enzymes. Food Microbiol. 17, 129–141.

So, J.E., Kang, S.H., Kim, B.G., 1998. Lipase-catalyzed synthesis of peptides containing D-amino acid: facts and artefacts. Enzym. Microb. Technol. 23, 211–215.

Stadnik, J., Keska, P., 2015. Meat and fermented meat products as a source of bioactive peptides. Acta Sci. Pol. Technol. Aliment. 14, 181–190.

Sun, X., Chakrabarti, S., Fang, J., Yin, Y., Wu, J., 2016. Low-molecular weight fractions of Alcalase hydrolyzed egg ovomucin extract exert anti-inflammatory activity in human dermal fibroblasts through the inhibition of tumor necrosis factor-mediated nuclear factor κB pathway. Nutr. Res. 36, 648–657.

Szydłowska-Czerniak, A., Trokowski, K., Szłyk, E., 2011. Optimization of extraction conditions of antioxidants from sunflower shells (*Helianthus annuus* L.) before and after enzymatic treatment. Ind. Crops Prod. 33, 123–131.

Teschemacher, H., Koch, G., Brantl, V., 1997. Milk protein-derived opioid receptor ligands. Peptide Sci. 43, 99–117.

Tuomilehto, J., Lindström, J., Hyyrynen, J., Korpela, R., Karhunen, M.L., Mikkola, L., Jauhiainen, T., Seppo, L., Nissinen, A., 2004. Effect of ingesting sour milk fermented using Lactobacillus helveticus bacteria producing tripeptides on blood pressure in subjects with mild hypertension. J. Hum. Hypertens. 18, 795–802.

Udenigwe, C.C., Gong, M., Wu, S., 2013. *In silico* analysis of the large and small subunits of cereal RuBisCO as precursors of cryptic bioactive peptides. Process Biochem. 48, 1794–1799.

Van Lancker, F., Adams, A., De Kimpe, N., 2011. Chemical modifications of peptides and their impact on food properties. Chem. Rev. 111, 7876–7903.

Walther, B., Sieber, R., 2011. Bioactive proteins and peptides in foods. Int. J. Vitam. Nutr. Res. 81, 181–191.

Wang, J.S., Zhao, M.M., Zhao, Q.Z., Bao, Y., Jiang, Y.M., 2007. Characterization of hydrolysates derived from enzymatic hydrolysis of wheat gluten. Food Chem. 101, 1658–1663.

Wang, Q., Zhu, F., Xin, Y., Liu, J., Luo, L., Yin, Z., 2011. Expression and purification of antimicrobial peptide buforin IIb in *Escherichia coli*. Biotechnol. Lett. 33, 1–6.
Wang, J., Yin, T., Xiao, X., He, D., Xue, Z., Jiang, X.1, Wang, Y., 2018. StraPep: a structure database of bioactive peptides. Database 2018. https://doi.org/10.1093/database/bay038.
Wongputtisin, P., Khanongnuch, C., Pongpiachan, P., Lumyong, S., 2007. Antioxidant activity improvement of soybean meal by microbial fermentation. Res. J. Microbiol. 2, 577–583.
Wu, J., Majumder, K., Gibbons, K., 2010. Bioactive proteins and peptides from egg proteins. In: Bioactive Proteins and Peptides as Functional Foods and Nutraceuticals, pp. 247–263.
Yamamoto, N., Ejiri, M., Mizuno, S., 2003. Biogenic peptides and their potential use. Curr. Pharmaceut. Des. 9, 1345–1355.
Yamamoto, K., Hayashi, M., Murakami, Y., Araki, Y., Otsuka, Y., Kashiwagi, T., Shimamura, T., Ukeda, H., 2016. Development of LC-MS/MS analysis of cyclic dipeptides and its application to tea extract. Biosci. Biotechnol. Biochem. 80, 172–177.
Zambrowicz, A., Zelazko, M., Trziszka, T., 2011. Hen egg proteins: precursors of bioactive peptides. Med. Weter. 67, 157–161.
Zarei, M., Ebrahimpour, A., Abdul Hamid, A., Anwar, F., Saari, N., 2012. Production of defatted palm kernel cake protein hydrolysate as a valuable source of natural antioxidants. Int. J. Mol. Sci. 13, 8097–8111.
Zhang, M., Mu, T.H., Sun, M.J., 2012. Sweet potato protein hydrolysates: antioxidant activity and protective effects on oxidative DNA damage. Int. J. Food Sci. Technol. 47, 2304–2310.
Zhu, K., Zhou, H., Qian, H., 2006. Antioxidant and free radical-scavenging activities of wheat germ protein hydrolysates (WGPH) prepared with alcalase. Process Biochem. 41, 1296–1302.
Zhu, Y.P., Fan, J.F., Cheng, Y.Q., Li, L.T., 2008. Improvement of the antioxidant activity of Chinese traditional fermented okara (Meitauza) using Bacillus subtilis B2. Food Control 19, 654–661.

CHAPTER

Biotechnological production of phytosteviosides and their potential applications

5

Meenu Thakur[a], Anuj K. Chandel[b], Santosh Kumar[c], Madan L. Verma[d,e]

Department of Biotechnology, Shoolini Institute of Life Sciences and Business Management, Solan, Himachal Pradesh, India[a]; Department of Biotechnology, Engineering School of Lorena (EEL), University of São Paulo, Estrada Municipal do Campinho, Lorena, São Paulo, Brazil[b]; Department of Biochemistry, University of Missouri, Columbia, MO, USA[c]; Department of Biotechnology, Dr Y. S. Parmar University of Horticulture and Forestry, Neri Campus, Himachal Pradesh, India[d]; Centre for Chemistry and Biotechnology, Deakin University, Geelong Campus, VIC, Australia[e]

Chapter outline

Biotechnological Production of Bioactive Compounds. https://doi.org/10.1016/B978-0-444-64323-0.00005-9

1. Introduction

Demand of alternative sweeteners and healthy lifestyle resulted into the increased market potential for sweeteners. These can be broadly classified into two types: first one is low calorie/dietic segment whereas other one is alternative to sugar and other nutritive sweeteners (Gibbs and Mulligan, 1996). There has been increasing demand for natural sweeteners with low calorific value which can act as substitute for saccharine and aspartame (Rao et al., 2012). Diabetes has affected 246 million people worldwide and India has the largest number of diabetic patients approximately 40.9 million patients (Agarwal et al., 2010). Synthetic sweeteners have noted and adverse effects on humans so stevioside is gaining significant importance as a natural sweetener in various parts of World (Abou-Arab et al., 2010; Brahmchari et al., 2011; Chhya et al., 2012). Steviosides and rebaudioside A are two main types of diterpenoid glucosides present in leaf extracts of the *Stevia rebaudiana* Bertoni (Rao et al., 2012).

Stevia rebaudiana Bertoni is a herbaceous perennial plant of the Asteraceae family native to Paraguay (South America) (Yadav et al., 2011). Steviosides have much higher sweetener property and exhibit 300 times more sweetness as compared to Sucrose. Moreover, these steviosides have enormous therapeutic potential and can be used for the treatment of patients suffering from Diabetes mellitus, obesity and hypertension etc. Steviol glycosides have also wide applications in food, beverage, pharmaceuticals, wine making, cosmetics and other industries (Gregerson et al., 2004; Chatsudhipong and Muanprasat, 2009; Wolver-Rieck et al., 2010). Use of pure steviol glycosides (>95%) is allowed for human consumption (4 mg/kg) in USA and Europe (Wolver-Rieck et al., 2010). According to market reports and global sales of stevia derived sweeteners are forecast to increase from 3500 tonnes (2013) to 10000 tonnes (2017) with a growth trend of 186% (Yucesan et al., 2016). Despite all potential applications its uses are limited due to taste bitterness, unknown glycosides and other alkaloid impurities (JECFA Joint FAO/WHO Expert Committee on Food Additives, 2007). Moreover, there are so many production constraints using conventional methods such as seed infertility and lack of space etc. Use of biotechnological strategies such as tissue culture techniques of in vitro propagation, molecular marker technology, recombinant strategies and metabolic engineering can improve the production of phytosteviosides. In this chapter, all of these biotechnological strategies and applications has been discussed.

2. Structure and biochemical properties of steviosides

Stevioside is a diterpenoid glycoside, comprising an aglycone (steviol) and three molecules of glucose. In addition to stevioside, several other sweet compounds such as steviobioside, rebaudioside A, B, C, D, E and ducoside A were isolated from *S. rebaudiana* Bertoni leaf. All of these isolated diterpenoid glycosides have

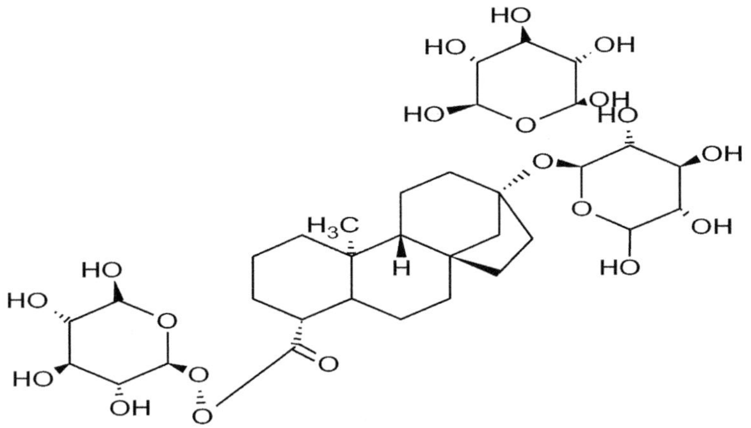

FIG. 5.1

Chemical structure of stevioside (Brandl et al., 1998).

the same chemical backbone structure (steviol) but differ in the residues of carbohydrate at positions C13 and C19 (Shibata et al., 1995). The major components of the leaf are stevioside (5%–10% of total dry weight), rebaudioside A (2%–4%), rebaudioside C (1%–2%) and dulcoside A (0.4%–0.7%) (Brandle et al., 1998). The chemical structure of stevioside is shown in Fig. 5.1.

There are eight diterpene glycosides present in leaf of stevia with gibberellic acid as precursor for biosynthesis. Stevioside is a glycoside which can be extracted from plants and gives more sweetness than sucrose. IUPAC name of glycoside is ent-13-hydroxykaur-16-en-1-oic acid. All the glycosides differ in their sugar moieties R1 chain and R2 chain of compound which result in difference of the fold of sweetness. Out of all these eight glycosides two are more significant one of them is stevioside and other is rebaudioside A (Yadav et al., 2011). Rebaudioside A has raised particular interest due to most desirable flavor profile (Dubois, 2000). Stevioside contributes to sweetness as it is 110–270 times sweeter than sugar. But stevioside (60% –70% of total glycoside content) has lingering effect of bitter after taste which is not acceptable by most of People. It has resulted into lesser acceptability. Whereas, Rebaudioside A (30%–40% of total glycosides) responsible for 180–400 times more sweetness than sugar without bitter aftertaste. So, the ratio of rebaudioside A to stevioside is the measure for the quality of sweetness (Yadav et al., 2011). The yield of sweetness and glycosides varies depending upon various factors such as method of propagation, daylength and other practices. Moreover, stevioside is stable at high temperature (100 °C) and over range of pH values. It is also non-calorific, non-fermentable and can be therefore used in the food and beverage industry (Brandle et al., 1998).

3. Estimation of steviosides

There are various techniques which has been used for estimation of steviosides ranging from simple such as thin layer chromatography (TLC) to more complex such as high-performance liquid chromatography (HPLC), reverse phase high performance liquid chromatography (RPHPLC), nuclear magnetic resonance (NMR) and HPTLC (Table 5.1). Silica gel 60 plate on metal was used as TLC plate for thin layer chromatography (Mali et al., 2015). The plates were developed using chloroform:methanol:water (6:2.5:1.5) as solvent system. These were air dried and sprayed with 50% sulfuric acid followed by saturated with iodine vapors which showed deep brown color on reacting with stevioside (Mali et al., 2015).

Karim et al. (2015) have used HPLC method for identification of stevioside with tissue culture derived *Stevia rebaudiana* leaves. Extraction of stevioside was

Table 5.1 Methods of extraction and estimation for stevioside.

Method of extraction and purification of steviosides	Method of estimation	References
Drying, defatting with petroleum ether, KOH added steam bath and extracted with chloroform and methanol (2:1)	High pressure liquid chromatography	Giridhar et al. (2010)
Pressurized hot water nano filters and ultrafilters	High pressure liquid chromatography with NH_2 column	Rao et al. (2012)
Ethanol/aqueous extraction with ethanol, soxhlet with 70% ethanol and column with color absorbing resin calcium hydroxide	Thin layer Chromatography	Mali et al. (2015)
Methanol with sonication	High pressure liquid chromatography	Karim et al. (2015)
Methanol with dried leaves powder using shaker for 48 h	High pressure liquid chromatography with C18 column	Sharma et al. (2015)
Microwave technology with 20% ethanol as solvent with 120 s of irradiation	High pressure liquid chromatography with C18 column	Javad et al. (2016)
Methanol extraction	Liquid Chromatography-Mass Spectrometry	Olsson et al. (2016)
Methanol extraction	High pressure liquid chromatography	Yucesan et al. (2016)
n-butyl alcohol with chloroform extraction	Thin layer Chromatography	Kumari et al. (2017)

performed with methanol along with sonication. HPLC was performed using crest-pak C18 column (Karim et al., 2015). Whereas in another study concentration of stevioside present in leaf extract were quantified by HPLC using waters analytical column. Solvent used was acetonitrile with 0.1 mM sodium phosphate buffer for elution of steviosides. The column temperature was maintained at 27–28 °C and UV detection was adjusted to 210 nm (Huang et al., 2010; Gardana et al., 2010). In another study analysis of stevioside has been performed using HPLC with acetonitrile: water (3:2) with a flow rate of 0.5 mL/min and spectrophotometric determination at 258 nm (Giridhar et al., 2010).

Liquid chromatography mass spectrometry (LC-MS) analysis was performed using C18 column packed with 1.7 μm particles, 130 Å pore size, temperature 35 °C coupled with quantum access spectrometer with electrospray ionization operation. Mobile phases used were water with 0.1% formic acid and acetonitrile for elution of the stevioside. In another similar study extracted stevioside was characterized through H NMR spectroscopy by comparison with a stevioside standard (Puri et al., 2012).

4. Extraction of steviosides

Traditional separation and purification techniques are available for effective extraction of stevioside (Table 5.1) from *Stevia rebaudiana* Bertoni (Rao et al., 2012). There are various reports on published and patented literature of previous work on stevioside extraction. Number of patents around 22 has been granted for different extraction and purification methods for steviosides.

These extraction methods range from conventional methods such as solvent extraction, column chromatography and ion-exchange methods whereas novel methods include enzymatic extraction, high speed counter current chromatography, microwave assisted extraction and ultra-nano membrane processes (Zhang et al., 2000; Pol et al., 2007; Teo et al., 2010; Puri et al., 2012). Out of all the methods, membrane-based methods such as microfiltration, ultrafiltration and nanofiltration techniques are preferred due to less yield losses of biological activity associated with other methods (Fuh and Chiang, 1990; Zhang et al., 2000; Li and Chase., 2010; Vanneste et al., 2011). But different extraction procedures have some of the constraints such a complexity and length of extraction procedures, expansive and costly processes, alkaloids as intermediates which can adversely affect human health. Moreover, most of the solvents used in the extraction procedures are toxic in nature. So, one of the major challenges in extraction of stevioside is to search for simple, inexpensive and environment friendly method (Liu et al., 2011).

Three different extraction procedures were used for stevioside extraction (Mali et al., 2015). In the first method, aqueous and ethanolic extraction were tried and extracts were dried (The ayurved pharmacopeia of India, 2008). In second method, soxhlet extraction was used with 70% ethanol as extraction solvent. Whereas in third method, column extraction was used by color removal resin. Filtrated extract 100 mL was passed through 50 mL of absorbent resin in a column (Rougfu et al., 2002).

Utilization of different membranes singly or in combination of other separation techniques can also be used for better extraction and retain biological activity (Rao et al., 2012). In this process has been developed on extraction of stevioside using pressurized hot water followed by purification using ultra and nanofilter membrane for filtration of high purity steviosides. This method has not only increased the purification yield but can be considered as green method due to less use of solvents (Rao et al., 2012). In one another study solvent based method using petroleum ether, chloroform and methanol (2:1) was utilized by Giridhar and co-workers (2010).

Enzymes can catalyze and lead to easy extraction of steviosides from tissue of *Stevia rebaudiana* leaves. Different concentrations of various enzymes such as cellulase, pectinase and hemicellulase were used for extraction of stevioside (Puri et al., 2012). Hemicellulase was observed to extract highest yield of stevioside (369.2 μg) in 1 h whereas second higher yield was obtained with cellulase (359 μg). It has resulted in 35 times more yield than control experiment. Interactive effects of various parameters on stevioside yield such as temperature, time and enzyme concentration has been studied using response surface methodology. This has resulted in 3 times more yield than previous experiment. One of the promising technologies for extraction of stevioside has been proposed (Puri et al., 2012).

High speed counter current chromatography was applied for isolation of stevioside, rebaudioside A and rebaudioside C from leaf extracts of *Stevia rebaudiana* Bertoni and different levels of purification was achieved (98.3%, 98.5% and 97.6%) respectively (Kumari et al., 2017). Extraction of steviol glycosides was performed using methanol (Rajab et al., 2009; Abou-arab et al., 2010). All the filtrates were combined and refluxed with another solvent chloroform with gentle heating and kept overnight in the refrigerator. The settled mass was filtered and washed with methanol. Most of the colors and less polar compounds were removed with chloroform treatment (Kumari et al., 2017). Microwave assisted extraction has been performed with ethanol as solvent during production of stevioside (Javad et al., 2016).

5. Conventional approaches for phytostevioside production

Natural methods for stevioside production are not sufficient for effective production of steviosides. Some of the major reasons are poor seed germination (Yadav et al., 2011), self-incompatibility leading to lack of fertilization (Midmore and Rank, 2002), small and infertile seeds (Kawatani et al., 1977), requirement of more time to establish seedlings (Yadav et al., 2011). Moreover, there is wide difference in glycoside content due to plants propagation using seeds. Other cultivation constraints include yield effected by rainfall during pollination, requirement of short day for cultivation. Seeds can be stored at 0 °C but under low temperature germination percentage declines by 50% over three years (Madan et al., 2010). Stevioside accumulation can be increased by delaying flowering period.

Due to above mentioned problems related with seeds and germination conventional plant breeding approaches were used for selection and crossbred among desirable genotypes for highly cross-pollinated *Stevia rebaudiana* (Yadav et al., 2011). Various plant types such as RSIT 94-1306, RSIT 94-75, RSIT 95-166 have been patented using breeding approach. Mutation breeding and polyploidy breeding has been checked and reviewed for production of *Stevia rebaudiana* (Yadav et al., 2011).

6. Biotechnological production of stevioside

Biotechnological approaches have tremendous potential to enhance the accumulation of steviosides. There are various strategies which can be explained as follows (Fig. 5.2).

7. Using tissue culture technology

Vegetative propagation of *Stevia rebaudiana* has resulted into poor level accumulation of stevioside (~4%) in individual plants after using selection program

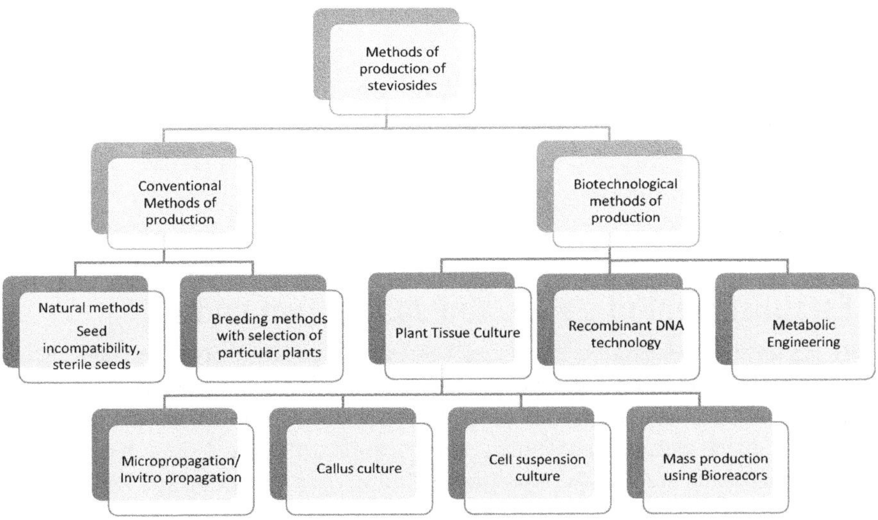

FIG. 5.2

Conventional and novel method of stevioside production.

(Giridhar et al., 2010). Tissue culture technology is the process which can be used for mass propagation. It ensures rapid and reliable multiplication for commercial production (Sharma et al., 2015). Micropropagation can be used for clonal production of elite plant varieties on large scale. It has benefit over other conventional approaches to achieve viable homogenous population containing similar content of steviosides. In vitro propagation of *Stevia rebaudiana* has been performed using various tissues as explants such as from nodes (Ahmed et al., 2007; Ibrahim et al., 2008), leaf (Sreedhar et al., 2008), shoot apex (Latha and Usha., 2003; Hossain et al., 2008) and directly from the stem cuttings etc.

Giridhar et al. (2010) have studied and developed rapid procedure for clonal propagation along with stevioside profiles. In their study they found Gamborg (B5) medium better than Murashige and Skoog (MS) medium for shoot induction from shoot tip and nodal explants of *Stevia rebaudiana.* The combination of different plant growth regulators such as benzoic acid BA (4.44 μM) and 1-Naphthaleneacetic acid (NAA, 0.80 μM) were produced more clones. The tissue culture propagation was followed by hardening of the plantlets. 70%−75% survival rate could be achieved which showed that this method can be used successfully for in vitro propagation of *Stevia rebaudiana* plant. Moreover, stevioside content was higher than plants grown with conventional methods.

In another study, a procedure was developed for in vitro propagation of Malaysian clone of *S. rebaudiana* with different plant growth regulators such as 6-benzylaminopurine and kinetin for formation of multiple shoots. Different combinations of Indole-3-Acetic acid (IAA), Indole 3-butyric acid (IBA) and 1-Naphthalene acetic acid (NAA) was used for root induction (Razak et al., 2014). All the studies and resulted were recorded and statistically significant combinations were used. For hardening polythene bags were used to increase the production of steviosides.

In similar study, plant regeneration protocol has been developed from leaf explants of *S. rebaudiana* (Abraham and Samrithi, 2016). Maximum percentage (82%) of induction was achieved with medium containing 2 mg/L kinetin and 0.5 mg/L IAA. Hardening studies showed 80% survival of the plantlets. This protocol can be used for commercial production of steviosides.

8. Establishment of callus and suspension cultures

Callus culture in *Stevia rebaudiana* has been achieved from organs (Singh et al., 2017). Callus is important for rapid mass multiplication to obtain cell suspension culture, preservation of cell line and secondary metabolite production. Callus culture has been produced through leaf, nodal and intermodal tissues (Uddin et al., 2006). Internodal segments are known to produce callus earlier than node and leaf. Callusing was enhanced by using combinations of NAA and 2,4-D (Gupta et al., 2010). The callus obtained from leaves are shiny and green and in case of callus obtained from other tissues is hard and brown (Singh et al., 2017).

Due to increase in population and limited natural resources there is raising demand for renewable products. As the agricultural land is also shrinking one of the alternatives can be suspension culture (Dicosmo and Misawa, 1995). Moreover, conditions for production of secondary metabolite can be controlled unlike soil and climate conditions. Cell suspension culture is effective method for enhancement of stevioside production on commercial level. Though some of the workers have reported suspension culture but this is one of the recent successful methods for production of secondary metabolites. Steviosides has been produced using cell suspension cultures from callus of *Stevia rebaudiana* (Javad et al., 2016). Twenty-two different combinations of plant growth regulators were used for initiating callus and calli obtained were compared for texture, color and proliferation rates. Callus cultures produced on BAP/NAA supplemented media did not proliferate on subculturing while callus with 2,4-D was very hard. Friable green calli were obtained using combination of all three regulators with maximum stevioside production (33.87 mg/g) of plant material.

Most of the plants produce secondary metabolites in very less quantities which can be enhanced further by application of various biotic and abiotic elicitors. These elicitors enhance the secondary metabolite production and thus increase the overall product yield. Effect of different elicitors such as PEG and proline on stevioside production using cell suspension culture of *Stevia rebaudiana* has been studied (Gupta et al., 2015). It was concluded that applying stress on suspension culture induced better concentration of rebaudioside A than stevioside (Gupta et al., 2015). Significant increase in production of steviol glycosides (3.7–4.7 times higher) was observed than control in callus and suspension cultures using 7.5 mM proline and 5% PEG. In one similar study an efficient method of callus and cell suspension culture has been developed (Sharma et al., 2015). Molecular characterization has been performed in two stevia variants from Russia and India which revealed genetic variations among both the variants.

9. Stevioside production using bioreactors

In view of growing world population, increasing anthropogenic activities and rapidly eroding natural ecosystems the natural habitats for a large number of plants so cell suspension cultures and bioreactors can be used for production of these bioactive compounds on large scale. These biotechnological techniques will not only help in easier extraction and purification of steviosides but can also provide basis of technology development for production of bioactive compounds using bioreactors on large scale. Bioreactors provide nutritional and close controlled environment for optimum growth of plant cells in which cells perform biochemical transformation to synthesize bioactive compounds. Bioreactors have several advantages such as mass cultivation of plant cells and better control for scale up of cell suspension cultures under defined parameters. Few workers have worked on the production and extraction of other bioactive compounds using bioreactor (Fulzele and Heble, 1994). Production of ajmalicine from *Catharanthus roseus* cells using 20 L airlift bioreactor was developed (Fulzele and Heble, 1994). Very few reports of previous work were available on growing the multiple shoot *Stevia rebaudiana* B in bioreactor using temporary immersion bioreactor (Sreedhar et al., 2008; Vives et al., 2017).

Temporary immersion bioreactor (BIT®) clearly increased total fresh weight expressed as shoots with higher size leaves and steviol glycosides content. These results might be due to the efficacy of BIT® that combine ventilation of the plant tissues with intermittent contact between the entire surface of the tissue and the liquid medium. These two characteristics are not usually combined in other liquid and semisolid culture procedures (Berthouly and Etienne, 2005). On the other hand, in BIT®, there is a direct contact of the culture medium with the leaves, leading to the possibility of nutrient uptake by them, unlike in semisolid medium (Escalona et al., 1999, 2003; Ziv, 2005).

10. Production of steviosides with recombinant DNA technology

Recombinant DNA technology is one of the important biotechnological technique which can be employed to achieve enhanced production of steviosides. A patent has been granted for developing a recombinant process for production of rebaudioside D which is one of the glycoside sweeter than stevioside using recombinant hostssuch as recombinant microorganisms, plants, or plant cells (Olsson et al., 2016). This study relates to tools and methods for producing terpenoids by modulating the biosynthesis of terpenoid precursors of the squalene pathway. Some other constituents can produce off flavors which can effect the taste and flavor of steviosides. Provided herein has been a recombinant host, such as a microorganism, plant, or plant cell, comprising one or more biosynthesis genes whose expression results in production of steviol glycosides such as rebaudioside A, rebaudioside C, rebaudioside D,rebaudioside E, rebaudioside F, or dulcoside A. In particular, EUGTl 1, a uridine 5′-diphospho (UDP) glycosyl transferase described herein, can be used alone or in combination with one or more other UDP glycosyl transferases such as UGT74G1, UGT76G1, UGT85C2, and UGT91D2e, to allow the production and accumulation of rebaudioside D in recombinant hosts or using in vitro systems. A method has been developed of producing a steviol glycoside. The method includes growing any of the hosts described herein in a culture medium, under conditions in which the genes are expressed; and recovering the steviol glycoside produced by the host (WO 2013/022989 A2).

11. Production of steviosides using metabolic engineering

Metabolic engineering is one of the most significant method for enhancement of stevioside production. Steviol glycoside biosynthesis pathway is multienzymatic pathway whose gene flux can be diverted to increase the desired product in lucrative way. Metabolic engineering has paved a way for new opportunities in agriculture, environmental application, production of chemicals and medicines. The plants altered in their profile of terpenoids and precursor pools can make important contribution to fundamental studies of terpenoids biosynthesis and its regulation (Aharoni et al., 2003). The success in metabolic engineering of diterpenoids will led to future studies on biological activities of transgenic plants engineered for terpenoid pathway. Diterpenoids like steviol glycosides present in leaves of *Stevia rebaudiana* are important due to their applications in industrial products such as commercial sweeteners flavoring agents, pharmaceuticals and antimicrobial agents. The biochemical pathway of steviol glycosides is closely related to the gibberellins with which they share part of their biosynthetic pathway (Fig. 5.3). In *Stevia*

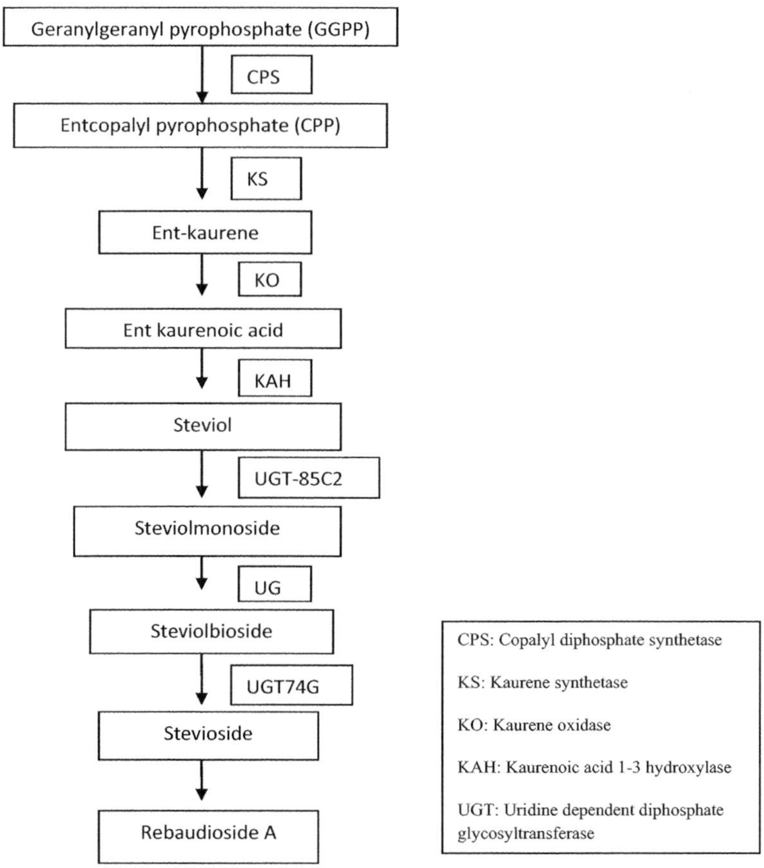

FIG. 5.3

Metabolic engineering steps for stevioside production.

Adapted from Olsson et al. (2016).

kaurenoic acid, an intermediate in gibberellic acid biosynthesis is converted into the tetracyclic diterpene steviol (Jain et al., 2014). Gibberelic acid and steviol are synthesized from the precursor geranyl geranyl diphosphate which is formed by the deoxyxululose -5-phosphate pathway. Two terpene cyclases Copalyl diphosphate Synthase (CPS) and Kaurene Synthase (KS) forms of Kaurene which is then oxidized by Kaurene Oxidase (KO) to form Kaurenic acid. Kaurene oxidase belongs to cytochrome P450 family which catalyses the three-step oxidation of Kaurene to form Kaurenoic acid. In *Stevia* leaves kaurenoic acid is then hydroxylated to form steviol (Kim et al., 1996). Steviol is than glycosylated by glycosyltransferase (UGTs). Three of which have been identified have been identified and characterized (UGT 85C2, UGT 74G1, UGT76G1) (Richman, 2005).

The glucosyltransferase UGT76G1 from *Stevia rebaudiana* is targeted enzyme for biosynthesis of the next-generation stevia sweeteners, rebaudioside D and rebaudioside M (Olsson et al., 2016). In the glycosylation of steviosides eight different reactions has been involved for production of rebaudioside D and rebaudioside M. Out of all the reactions four reactions result in formation of side products. Mutants have been identified by screening of mutant library and mutants with positive effect on accumulation of Reb D and Reb M were selected. The effect of the introduced mutations on other reactions was characterized. and variants were identified, such as UGT76G1Thr146Gly and UGT76G1His155Leu, which diminished accumulation of unwanted side-products and gave enhanced accumulation of the desired Reb D or Reb M sweeteners specifically (Olsson et al., 2016). This improvement in a key enzyme of the biosynthesis pathway represents an important step toward the commercial production of next-generation stevia sweeteners.

12. Applications of phytosteviosides

A number of studies have suggested that, beside sweetness, stevioside along with related compounds, which include rebaudioside A (second most abundant component of *S. rebaudiana* leaf), steviol and isosteviol (metabolic components of stevioside) may also offer therapeutic benefits, as they have anti-hyperglycemic, anti-hypertensive, anti-inflammatory, anti-tumor, anti-diarrheal, diuretic, and immunomodulatory actions (Fig. 5.4). Moreover, it is worth to note that stevioside exerts their effect only when level is abnormal with higher range. As steviol can interact with drug transporters, its role as a drug modulator is proposed.

13. Application of stevioside as substitute for table sugar

Stevioside is considered to be a sugar substitute and commercial sweetener, both in the form of stevioside and stevia extract (Kinghorn and Soejarto, 1985; Brandle and Rosa, 1998). They are used in variety of foods and products, such as pickled

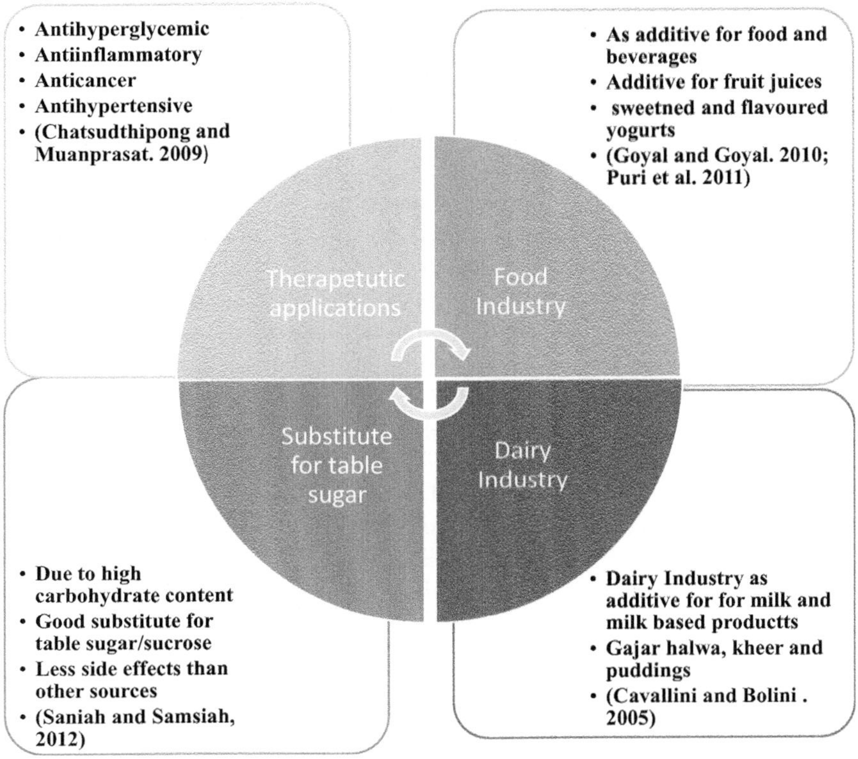

FIG. 5.4

Commercial applications of steviosides.

vegetables, dried seafood, soysauce, beverages, candies, chewing gum, yogurt and ice cream, as well as in toothpaste and mouth wash. Stevia extract and stevioside are officially approved as food additives in Brazil, Korea and Japan (Choi et al., 2002; Mizutani and Tanaka, 2002) and in the United States, they are permitted as a dietary supplement. They have not yet been approved by the European Commission due to safety concern. In 2006, the meeting of the Joint FAO/WHO Expert Committee on Food Additive (JECFA) to evaluate certain food additives and ingredients, flavoring agents, and natural constituent of food announced a temporary accepted daily intake (ADI) of stevioside of up to 5.0 mg/kg body weight (BW) (JECFA, 2016).

14. Application of steviosides in food industry and beverages

Sweeteners are the most important ingredients in the food industry. Stevioside is an intense sweetener, and the extract of *Stevia* finds extensive use in countries such as Japan, China, Russia, Korea, Paraguay, Argentina, Indonesia, Malaysia, Australia, New Zealand, South America, and others, to sweeten local teas, medicines, food, and beverages (Elkins, 1997). *Stevia* extract could be a preferred sweetener among other low-calorie sweeteners (saccharin, aspartame) as its source is natural, which might be appealing to health-conscious consumers. *Stevia* sweetener is heat-stable up to200 °C, acid-stable, and not fermentable (Kroyer, 2010), which make it suitable for use in different food products. Additionally, its organoleptic characteristics were found acceptable (Prakash et al., 2008), which suggest that it could be used as a substitute for sucrose. In another study, stevioside and sucrose were added into water and peach juice and compared in terms of sweetness, sweet, and bitter after tastes. Results revealed that 160 mg/L of stevioside successfully replaced 34 g/L of sucrose in juice without negatively affecting the sensory characteristics of the product (Parpinello et al., 2001). As such the functional food sector may be able to make use of stevioside in the preparation of dietetic and other low-calorie foods (Puri et al., 2012); however, more research is needed to determine the effects of long-term consumption of steviosideon the human health. There are many international food companies using stevia in their products among them Coca-Cola which uses *Stevia* in Japan for its Diet Coke and have filed patents applications in 2007 concerning extracting the tastiest parts of the *Stevia* plant. It is seeking exclusive rights to develop and market "rebina" for use in its drinks (Prakash et al., 2008; Prakash and Upreti, 2008). The stability of stevioside at elevated temperature during different processing and storage conditions has been evaluated. Stevioside at elevated temperature in tea and coffee beverages for 1 h showed good stability up to 120 °C. In aqueous solution, stevioside is remarkably stable in the pH range of 2—10. This revelation seems to be essential for its effective application in hot coffee and tea beverages (Kroyer, 2010). The bakery industry may also benefit from the use of *Stevia*. All cooked and baked food items such as puddings and cakes can be sweetened with only very small quantities of *Stevia* leaf powder. Stevioside is nonfermentable, and it does not undergo any kind of browning reaction while cooking. This further widens its area of application in baking, enhancing the quality and safety of the bakery industry may also benefit from the use of *Stevia*. All cooked and baked food items such as puddings and cakes can be sweetened with only very small quantities of *Stevia* leaf powder. A mere fragment of the leaf is enough to sweeten the mouth for an hour (De et al., 2013). The confectionery industry has yet to reap the benefit of Stevia, which can replace sugar as a sweetener. Stevia can be used in chocolates and candies, not only to meet market demand by diabetics but also to harvest the added advantage of this herb's actions against tooth decay (Lindsay, 2007). In addition, stevioside can also be used in chewing gum, mints, mouth refreshers, toothpaste, and some cosmetics. This low-calorie natural sweetener is used extensively in various food products such as biscuits, jams, chocolates, ice-creams, baked foods, soft drinks and fruit drinks (Goyal and Goyal, 2010; Jayaraman et al., 2008), sauces, sweet corn, delicacies, pickles (Koyama et al., 2003), yoghurt, soju, soysauce (Hossain et al., 2010), candies, sea foods (Goyal and Goyal, 2010; Koyama et al., 2008) and the common beverages like diptea, coffee, and herbal tea. Recently, several researchers have reported that purified stevioside from *S. rebaudiana* has potential to reduce the number of pathogenic bacteria in vitro (Puri and Sharma, 2011). Puri and Sharma (2011) documented *S. rebaudiana* solvent extracts (1000 mg/mL) displayed antibacterial activity using different solvents. Ghosh et al. (2008) reported (250 mg/mL) it is sufficient enough to inhibit the growth of test pathogenic bacteria completely in petriplates. These pathogenic bacteria such as *Bacillus cereus*, *Klebsiella pneumoniae*, and *Pseudomonas aeruginosa* are the root cause of many food-borne diseases such as enteric fever and diarrhea (Puri and Sharma, 2011). Further research is needed to assess the role of stevioside as an antimicrobial agent when used in different foods products (Gasmalla et al., 2014).

A similar investigation sensorially evaluated industrialized pineapple juice, sweetened with sucralose, aspartame, *Stevia* (pure commercial extract), cyclamate/saccharin and sucrose. The flavor profile obtained showed that juice made with stevioside had a sour and bitter aftertaste, demonstrating that stevioside may not be the appropriate sweetener for this particular juice formulation. In addition, residual sweetness, sweet flavor, spiciness and metallic taste were cited for these samples, but by very few tasters. These same sweeteners were also studied in guava juice (Fernandes et al., 2009; Koguishi et al., 2008). As regards sweetness, guava juice sweetened with sucrose or aspartame achieved similar acceptance ratings (81%), while samples sweetened with stevioside (extract obtained from leaves) and cyclamate/saccharine were given higher scores for negative attributes, positioning them in the rejection zone asper the hedonic scale used. It has been found that the matrix (guava nectar) did not affect the perception of sweetness of any of the sweeteners. Similar results were reported by Cavallini and Bolini (2005) who compared the temporal perception of sweetness, bitterness and flavor in reconstituted mango juice sweetened with sucrose, cyclamate:saccharin 2:1, aspartame, sucralose and stevioside. It has been concluded that, of all the sweeteners, aspartame showed a temporal profile closest to that of sucrose for the majority of the parameters tested (Gonzales et al., 2014). The juice made with stevioside, however, was the least similar to sucrose. From these studies, it can be seen that many products sweetened with *Stevia* extract tend to have a bitter residual taste. This may be attributed to the high proportions of stevioside glycoside contained in the mixture of leaf components obtained during the extraction process. This negative element is eliminated when plant varieties with higher concentrations of rebaudioside A are used, or simply by using steviol glycosides in pure form, due to their superior flavor profile (Dacome et al., 2005). A similar study, but this time applied to cake, assessed the rheological and microstructural properties and the final quality of cakes made by replacing sugar with stevioside (sweetener containing 80% stevioside) and liquid sorbitol (Manisha et al., 2012).

A central composite design and response surface methodology were used to establish the optimum level of *Stevia* sweetener as a replacement of sucrose in carbonated drinks. In this study, *Stevia*-sucrose combinations in the range of 0.2%–0.5% (*Stevia*) and 0%–54% (sucrose) were the independent variables and their effects on sensory acceptability and physico-chemical profile of product were investigated. Based on the response surface and superimposed plots, the desired sensory quality of orange flavored carbonated drink was obtained by incorporating 0.43% *Stevia* and 33.13% sucrose in syrup (Sainah and Samsiah, 2012).

15. Application of steviosides in dairy industry

There have been several investigations undertaken aimed at evaluating the potential of stevioside as a sweetener of specific products. For example, Lisak et al. (2012) prepared strawberry-flavored yogurt sweetened with either sucrose, stevioside or equal proportions of sucrose and stevioside, this last a pure extract with 90% stevioside content, at three different concentrations. According to the results given by the sensory panel judges, the authors determined that 6 g of stevioside is equivalent to 1000 g of sucrose. In taste tests, the yogurt sweetened with 4.5 g 100 g^{-1} of equal parts of sucrose and stevioside obtained the highest sensory acceptance scores. The apparent viscosity of the finished products was not affected by the addition of stevioside. Furthermore, after 7 days of cold storage, the degree of sweetness of all the different formulations tested remained the same as that of fresh samples (made the day before) of products prepared with sucrose. Another study evaluating stevioside-sweetened yogurt was carried out by (Guggisberg et al., 2011). Yogurt made with 8% sugar was replaced by stevioside, and combinations with other sweeteners. Neither stevosiden or the other two commercial sweeteners used had any negative effects on the yogurt making process or the pH, and did not significantly change the fermentation time or the generation of the casein network. However, the yogurt made with stevioside only had an unpleasant taste and thus could not be recommended (Guggisberg et al., 2011). Nevertheless, the yogurt made with a combination of Actilight (a commercial mixture of short-chain fructo-oligosaccharides) and stevioside showed a similar profile to the variant containing 8% sucrose. The authors concluded that low-calorie yogurts could be manufactured using commercial sweeteners including stevioside without modifying standard technological procedures. The addition of hydrocolloids, emulsifiers and debittered fenugreek seed powder was also considered. The results showed that the addition of stevioside did not change the amylographic viscosity of wheat flour batter during heating and cooling, unlike sucrose, which increases this property. Another study investigated the physical properties of cocoa powder drinks prepared with different fat contents and different sweeteners including *Stevia* extract (Belscak et al., 2010). The authors evaluated the bioactive content (content of polyphenols and antioxidant capacity) and the sensory properties of prepared cocoa drinks. The results showed that the type of sweetener used did not affect the polyphenolic constituents of the cocoa mixtures prepared. The results of the sensory evaluation revealed a preference for cocoa drinks made with the sweeteners (aspartame/acesulfame K and Stevia extract) than control. Finally, Mogran and Dashora (2009) has to optimize the concentration of stevioside to produce a degree of sweetness equivalent to sugar. Once the amounts required to achieve the desired sweetness had been determined, eleven products were elaborated: milk, coffee, tea, local Indian sweets *gajar halwa*, *milkshake*, *kheer*, curd, lemon water, custard, *halwa* and *lapsi*, replacing the sugar in the recipes with either Stevia extract or one of the other commercial sweeteners. The results revealed that 1.5 mL of Stevia extract in 100 mL of liquid was equivalent to 5 g of sugar. The recipes prepared using Stevia were more acceptable than the other sweeteners tested, as indicated by the fact that they were given the highest sensory acceptances cores by members of the panel (7.67–7.90), thus occupying first place ahead of the artificial sweeteners tested. They were also statistically comparable in taste to sugar (sugar scores: 7.47–8.47) ($p > 0.001\%$) in the case of coffee, halwa, milk drinks, *kheer* and *lapsi*, and were scored as tastier for the rest of the products evaluated. All of the above indicate the huge potential of *Stevia* as a realistic alternative to sugar in the products investigated. Not least because in addition to delivering similar physical and sensory properties, it also provides beneficial health effects for consumers (Gonzales et al., 2014).

16. Applications of steviosides as therapeutics

Steviosides have many therapeutic applications beyond acting as sweeteners because they have anti-hyperglycemic, anti-hypertensive, anti-inflammatory, antitumor, anti-diarrheal, diuretic, and immunomodulatory actions. It is of interest to note that their effects on plasma glucose level and blood pressure are only observed when these parameters are higher than normal. At present, there is a sharp increase in incidence of type 2 diabetes mellitus and obesity as a result of aging, dietary habits and decreasing physical activities. These metabolic syndromes have become major public health problems in industrialized and developing countries. Type 2 diabetes mellitus is a chronic metabolic disorder resulting from defects in both insulin secretion from β-cells of islets and insulin action (DeFronzo, 1988). Recently more interest has been raised for herbal or ayurvedic treatment for diabetes and extracts of *Stevia rebaudiana* has been used for this since long in South America (Kinghorn and Soejarto, 2002). In addition, stevioside, the major component of the extract, has a high sweetness with no calorie and only a small amount is needed for sweetening purposes. Therefore, it can be used as substitute of table sugar for diabetics. For anti hyperglycemic application their effect can be further segmented as effect on glucose absorption, effect on glucose synthesis and effect on insulin secretion and sensitivity.

The effects of stevioside and steviol on glucose absorption have been investigated using in vitro jejunal ring tissue and everted sac (Toskulkao et al., 1995) The effects of stevioside on glucose synthesis have been studied in two types of diabetic rats, type 1 (insulin dependent) and type 2. (insulin independent) (Chen et al., 2005). Studies in humans have also shown the effect of stevioside on cardiovascular system. Stevioside causes bradycardia and hypotension (Humboldt and Boech, 1977). Similarly, a slight hypotensive effect was observed in human subjects who received a tea prepared from *S. rebaudiana* (stevia extract) daily for 30 days. In these studies, stevioside was also suggested to have inotropic effect by shortening systole duration. This would reduce stroke Results from long-term clinical trials in humans with mild to moderate hypertension demonstrated that continued consumption of stevioside (750 mg/day) for one year reduces both systolic and diastolic blood pressure, whereas no significant side effect or alteration on lipid or fasting glucose was observed (Chan et al., 1998). Subsequent studies for up to 2 years with an increased dose of stevioside (1500 mg/day) showed that stevioside significantly decreases both systolic and diastolic blood pressure without any significant changes in body mass index, blood biochemistry values and left ventricular mass index (Hsieh et al., 2003). Boonkaewwan et al. (2006) measured the release by a human monocytic THP1 cell line of proinflammatory cytokines (TNF-α andIL-1β) and nitric oxide, all known to participate in the development of a number of inflammatory disorders, and found that stevioside (1 mM) moderately stimulates their release in unstimulated THP1 cells by interacting with toll-like receptor-4, a principal receptor for lipopolysaccharide (LPS) on gram-negative bacteria. At this level of monocyte stimulation, stevioside could be beneficial in healthy individuals as a result of its effect in enhancing innate immunity. Stevioside and steviol exert anti-inflammatory effects on colonic epithelial cells. Under physiological regulation, colonocytes not only function to form a barrier across which fluid and electrolyte are transported but also serve as an innate immune sensor of microbial pathogen and commensal organisms (Sartor, 2008).

17. Safety aspects and toxicity of steviosides

Proximate analysis of stevioside has shown that it contains carbohydrates and high protein. High ash content revealed presence of inorganic minerals. Potassium, calcium, magnesium, sodium and sulfur has been found in leaves in reasonable amount. Moreover, certain minerals such as zinc, calcium and potassium played important role in maintenance of blood glucose level. Tannins and other components have pronounced antioxidant activity which is responsible for antimicrobial activity of stevioside (Tadhani and Subhash, 2006).

As steviosides can be used in food industry the safety evaluation is also very important and revealed in terms of toxicity and teratogenicity by Chatsudthipong and Muanprasat (2009). In acute and chronic toxicity evaluations of stevioside ingestion investigated in mouse, rat and hamster, stevioside intake as high as15 g/

kg BW produces no acute toxicity (Akashi and Yokoyama, 1975; Mitsuhashi, 1976) Ingestion of stevioside (750 mg/day for 3 month) by healthy individuals or those with underlying diseases such as diabetic mellitus and hypertension produced no adverse effects or abnormalities in liver and renal function tests and serum electrolytes (Hsieh et al., 2003; Barriocanal et al., 2008). According to previous pharmacokinetic studies in humans, after oral administration of a single dose of 4.2 mg of stevioside/kg BW, mean maximal concentration of steviol glucuronide and free steviol in plasma is 1.89 μg/mL (3.7 μM) and 0.19 μg/mL (0.38 μM) respectively (Wheeler et al., 2008). The carcinogenic potential of stevioside is of particular concern and a number of investigations using different experimental models have been conducted to evaluate the mutagenic effects of stevioside and steviol. Bacterial genetic analysis revealed that stevioside is not mutagenic (Pezzuto et al., 1985; Klongpanichpak et al., 1997). However, from various mutagenic assays, the genotoxic potential of steviol remains inconclusive. In mutation assays using *Salmonella typhimurium* TM677, steviol showed genetic toxicity after metabolic activation by liver homogenate (Pezzuto et al., 1985; Terai et al., 2002). In vivo testing of stevioside carcinogenic potential has also been performed in both mouse and rat. Stevioside given orally does not increase the incidence of cancer in rat (Yamada et al., 1985; Toyoda et al., 1997; Sekihashi et al., 2002). No evidence of carcinogenicity of stevioside was obtained from in vivo studies in mice (Matsui et al., 1996; Yasukawa et al., 2002). A concern about antifertility and teratogenic effects of stevioside was raised after it was shown that stevia decoctions decrease live birth rate in rats (Planas and Kuc, 1968). After injection of stevioside or steviol into eggs, embryonic development undergoes normally without any decrease in embryonic mortality and body weight of hatchling, orany structural deformation. Taken together, most studies are in agreement demonstrating that oral stevioside, at an acceptable daily intake (5 mg/kg BW), is safe and not carcinogenic or teratogenic.

18. Current status and future prospects of steviosides

Steviosides hold an essential place under sweeteners segment for its low calorie and high protein content. The varied applications of stevioside as food ingredient include bakery, dairy, food products, beverages, packaged food products, dietary supplements, confectionary and others. Globally the sweetener market is estimated to be US$ 68.1 billion in 2014 and expected to reach US$ 95.9 billion by 2020 registering a compound annual growth rate of 5.7% during 2014–20. Whereas stevioside market is estimated to value at US$ 347.0 million in 2014 and expected to reach US$ 565.2 million by 2020 reflecting compound annual growth rate of 8.5% during the forecast period. In terms of volume consumption of stevioside it is expected to reach 8506.9 tonnes by the end of 2020 with annual growth of around 7%–8% during the forecast period (Hossain et al., 2017). Among all the segments packaged food products, beverages and table top sweeteners are collectively expected to account for around 72% of global stevia market (http://www.futuremarketinsight.com).

19. Future prospects

Stevioside is one of the most potential herbs with myriad of applications beyond acting as sweetener. There are so many therapeutic applications of this glycoside extracted from wonder herb *Stevia rebaudiana* Bertoni. As there is increasing demand of interest in ayurvedic and herbal medicine steviosides can tremendously help in treatment of patients of diabetes type II and blood pressure etc. Biotechnological production can not only enhance the stevioside content but can fulfill the requirement of renewable production due to shrinking natural resources. Bioreactors can provide controlled environment for production of steviosides and more studies can be conducted on this aspect in future. This can develop the bench level technology and later can be transferred to Industry for commercial production.

20. Conclusion

Steviosides is a potential constituent of herbal plant *Stevia rebaudiana* Bertoni having wide range of applications. In this chapter different structural biochemical property, extraction and estimation procedures followed by different researchers, conventional as well as biotechnological methods of stevioside production along with potential applications have been discussed. However various new biotechnological approaches can be employed to enhance the production of stevioside. Moreover, as steviosides have potential to be used in food industry in near future more work is needed to be done on its food application.

References

Abou-Arab, A.E., Abou-Arab, A.A., Abu-Salem, M.F., 2010. Physicochemical assessment to natural sweeteners steviosides produced from *Stevia rebaudiana* bertoni plant. African Food Sci. 4, 269–281.

Abraham, T., Samrithi, P.R., 2016. In vitro propagation of sweet herb, *Stevia rbaudiana* bertoni. Indian J. Plant Sci. 5 (4), 1–6.

Agarwal, V., Kochhar, A., Sachdeva, R., 2010. Sensory and nutritional evaluation of sweet milk products prepared using Stevia powder for diabetics. EthnoMed 4 (1), 9–13.

Aharoni, A., Giri, A.P., Deuerlein, S., Griepink, F., Kogel, W.J., Verstappen, F.W.A., 2003. Terpenoid metabolism in wild-type and transgenic Arabidopsis plants. Plant Cell. 15, 2866–2884.

Ahmed, M.B., Salahin, M., Karim, R., Razvy, M.A., Hannan, M.M., Sultana, R., Hossain, M., Islam, R., 2007. An efficient method for *in vitro* clonal propagation of a newly introduced sweetener plant (*Stevia rebaudiana* Bertoni) in Bangladesh. Am. Eura J. Sci. Res. 2, 121–125.

Akashi, H., Yokoyama, Y., 1975. Security of dried-leaf extracts of stevia: toxicological and its oxidative derivatives in *Salmonella typhimurium* TM677. Chem. Pharm. Bull. 50 (7), 1007–1010.

Barriocanal, L.A., Palacios, M., Benitez, G., Benitez, S., Jimenez, J.T., Jimenez, N., Rojas, V., 2008. Apparent lack of pharmacological effect of steviol glycosides used as sweeteners in humans. A pilot study of repeated exposures in some normotensive and hypotensive individuals and in Type 1 and Type 2 diabetics. Regul. Toxicol. Pharmacol. 51 (1), 37–41.

Belscak, A., Bonkovic, M., Komes, D., Bauman, I., Horzic, D., Dujmic, F., Matijasec, M., 2010. Physical properties and bioactive constituents of powdered mixtures and drinks prepared with cocoa and various sweeteners. J. Agric. Food Chem. 58, 7187–7195.

Berthouly, M., Etienne, H., 2005. Temporary immersion system: a new concept for use liquid medium in mass propagation. In: Liquid Culture Systems for In Vitro Plant Propagation. Springer, New York, pp. 165–195.

Boonkaewwan, C., Toskulkao, C., Vongsakul, M., 2006. Anti-inflammatory and immunomodulatory activities of stevioside and its metabolite steviol on THP-1 cells. J. Agric. Food Chem. 54 (3), 785–789.

Brahmachari, G., Mandal, L.C., Ray, R., Mondal, S., Brahmachari, A.K., 2011. Stevioside and related compounds—molecules of pharmaceutical promise: a critical overview. Chem. Life Sci. 1, 5–19.

Brandle, J.E., Rosa, N., 1998. Heritability for yield, leaf: stem ratio and stevioside content estimated from a landrace cultivar of *Stevia rebaudiana*. Retrieved from: http://res2.agr.ca/london/pmrc/faq/cjps-ste.html.

Brandle, J.E., Starratt, A.N., Gijzen, M., 1998. *Stevia rebaudiana*: its agricultural, biological and chemical properties. Can. J. Plant Sci. 527–536.

Cavallini, C.U., Bolini, H.M., 2005. Comparison of temporal perception of sweetness bitterness and fruitiness in mango juice sweetened with sucrose, cyclamate/saccharine 2:1, aspartame, sucralose and stevioside leaf extract. Bol. CEPPA. 23 (2), 361–382.

Chan, P., Xu, D.Y., Liu, J.C., Chen, Y.J., Tomlinson, B., Huang, W.P., 1998. The effect of stevioside on blood pressure and plasma catecholamines in spontaneously hypertensive rats. Life Sci. 63 (19), 1679–1684.

Chatsudhthipong, V., Muanprasat, C., 2009. Stevioside and related compounds:therapeutic benefits beyond sweeteners. Pharmacol. Ther. 121, 41–54.

Chen, T.H., Chen, S.C., Chan, P., Chu, Y.L., Yang, H.Y., Cheng, J.T., 2005. Mechanism of the hypoglycemic effect of stevioside, a glycoside of *Stevia rebaudiana*. Planta Med. 71 (2), 108–113.

Chhaya, Mondal, S., Majumdar, G.C., De, S., 2012. Clarification of stevia extract using cross flow ultrafiltration and concentration by nanofiltration. Separ. Purif. Technol. 89, 125–134.

Choi, Y.H., Kim, I., Yoon, K.D., Lee, S.J., Kim, C.Y., Yoo, K.P., Kim, J., 2002. Supercritical fluid extraction and liquid chromatographic-electrospray mass spectrometric analysis of stevioside from Stevia rebaudiana leaves. Chromatographia 55 (9–10), 617–620.

Dacome, A.S., Da Silva, C.C., Da Costa, C.E.M., Fontana, J.D., Adelmann, J., Da Costa, S.C., 2005. Sweet diterpenic glycosides balance of anew cultivar of *Stevia rebaudiana* Bertoni: isolation and quantitative distribution by chromatographic, spectroscopic and electrophoretic methods. Process Biochem. 44, 3587–3594.

De, S., Mondal, S., Banerjee, S., 2013. Stevioside: Technology, Applications and Health. Wiley, New York, pp. 185–196.

DeFronzo, R.A., 1988. The triumvirate: beta-cell, muscle, liver. A collusion responsible for NIDDM. Diabetes 37 (6), 667–687.

Dicosmo, F., Misawa, M., 1995. Plant cell and tissue culture: alternatives for metabolite production. Biotechnol. Adv. 13 (3), 425–453.

DuBois, G.E., 2000. Sweeteners: non-nutritive. In: Francis, F.J. (Ed.), Encyclopedia of Food Science and Technology, second ed., vol. 4. John Wiley and Sons, Inc., New York, NY, pp. 2245−2265.

Elkins, R., 1997. Stevia: Nature's Sweetener. Woodland Publishing, California, p. 64.

Escalona, M., Samson, G., Borroto, C., Desjardins, Y., 2003. Physiology of effects of temporary immersion bioreactors on micropropagated pineapple plantlets. In Vitro Cell. Dev. Biol. Plant 39, 651−656.

Escalona, M., Lorenzo, J.C., González, B., Daquinta, M., Borroto, C., González, J.L., Desjardines, Y., 1999. Pineapple micropropagation extracts. J. Food Hyg. Soc. Jpn. 26, 169−183.

Fernandes, A.G., Machado, P.H., Maia, G., Silva, D.S., Santos, S.M., 2009. Sensory evaluation of guava drinks sweetened with different stevioside sources. Food Sci. Technol. 36 (3), 413−420.

Fuh, W.S., Chiang, B.S., 1990. Purification of stevioside by membrane and ion exchange processes. J. Food Sci. 55, 1454−1457.

Fulzele, D., Heble, M.R., 1994. Large scale cultivation of *Catharanthus roseus* cells: production of ajmalicine in a 20L airlift bioreactor. Biotechnology 35 (1), 1−7.

Gardana, C.M., Scaglianti, M., Simonetti, P., 2010. Evaluation of steviol and its glycosides in *Stevia rebaudiana* leaves and commercial sweetener by ultra-high-performance liquid chromatography-mass spectrometry. J. Chromatogr. 1217, 1463−1470.

Gasmalla, M.A.A., Yang, R., Huz, X., 2014. *Stevia rebaudiana* bertoni: an alternative sugar replacer and its applications in food industry. Food Eng. Rev. 6 (4), 1−14.

Ghosh, S., Subudhi, E., Nayak, S., 2008. Antimicrobial assay of *Stevia rebaudiana* Bertoni leaf extracts against 10 pathogens. Int. J. Integr. Biol. 2 (1), 27−31.

Gibbs, B.F., Mulligan, C., 1996. Sweet and taste-modifying proteins: a review. Nutr. Res. 16, 1619−1630.

Giridhar, P., Sowmya, K.S., Ramakrishna, A., Ravishankar, G.A., 2010. Rapid clonal propagation and stevioside profiles of *Stevia rebaudiana* Bertoni. Int. J. Plant Dev. Biol. 4 (1), 47−52.

Gonzales, C., Tapia, M., Perez, E., Pallet, D., Dornier, M., 2014. Main properties of steviol glycosides and their potential in the food industry: a review. Fruits 69, 127−141.

Goyal, S., Goyal, R., 2010. Stevia (*Stevia rebaudiana*) a biosweetener: a review. Int. J. Food Sci. Nutr. 61, 1−10.

Gregersen, S., Jeppesen, P.B., Holst, J.J., Hermansen, K., 2004. Anti hyperglycemic effects of stevioside in type 2 subjects. Metabolism 53, 73−76.

Guggisberg, D., Piccinali, P., Schreier, K., 2011. Effects of sugar substitution with Stevia, Actilight and Stevia combinations or Palatinose on rheological and sensory characteristics of low fat and whole milk set yoghurt. Int. Dairy J. 21 (9), 636−644.

Gupta, P., Sharma, S., Saxena, S., 2010. Callusing in Stevia rebaudiana (natural sweetener) for steviol glycoside production. Int. J. Agric. Biol. Sci. 1 (1), 30−34.

Gupta, P., Sharma, S., Saxena, S., 2015. Biomass yield and steviol glycoside production in callus and suspension culture of *Stevia rebaudiana* treated with proline and polyethylene glycol. Appl. Biochem. Biotechnol. 176, 863−874.

Hossain, M.A., Kabir, A.H.M.S., Japhan, T.A., Hasan, M.N., 2008. Micropropagation of *stevia*. Int. J. Sustain. Crop. Prod. 3, 1−9.

Hossain, M.A., Siddique, A., Rhaman, M., Hossain, M.A., 2010. Chemical composition of the essential oils of *Stevia rebaudiana* bertoni leaves. Asian J. Trad. Med. 5 (2), 56−61.

Hossain, M.F., Islam, M.T., Islam, M.A., Akhtar, S., 2017. Cultivation and uses of stevia (*Stevia rebaudiana* bertoni): a review. Afr. J Food Agric. Nutr. Dev. 17 (4), 12745–12757.

Hsieh, M.H., Chan, P., Sue, Y.M., Liu, J.C., Liang, T.H., Huang, T.Y., Tomlinson, B., Chow, M.S., Kao, P.F., Chen, Y.J., 2003. Efficacy and tolerability of oral stevioside in patients with mild essential hypertension: a two-year, randomized, placebo-controlled study. Clin. Ther. 25 (11), 2797–2808.

Huang, X.Y., Fu, J.F., Di, D.L., 2010. Preparative isolation and purification of steviol glycosides from *Ste-via rebaudiana* bertoni using high-speed counter-current chromatography. Separ. Purif. Technol. 71, 220–224.

Humboldt, G., Boech, E.M., 1977. Efeito do edulcorante natural (stevioside) e sinte tico (sacarina) sobre o ritmo cardiaco em ratos. Arq. Bras. Cardiol. 30, 257–277.

Ibrahim, A.I., Nasr, M.I., Mohammded, B.R., El-Zefzafi, M.M., 2008. Plant growth regulators affecting *in vitro* cultivation of *Stevia rebaudiana*. Sugar Technol 10, 254–259.

Jain, P., Kacchwaha, S., Kothari, S.L., 2014. Biotechnology and metabolic engineering of *Stevia rebaudiana* Bertoni: perspectives and possibilities. Int. J. Life Sci. Biotechnol. Pharma Res. 3 (3), 1–23.

Javad, S., Naz, S., Ilyas, S., Aftab, A., 2016. Production of steviosides from callus and cell suspension cultures of *Stevia rebaudiana* Bertoni. JAPS 26 (5), 1374–1382.

Jayaraman, S., Manoharan, M.S., Illanchezian, S., 2008. In-vitro antimicrobial and antitumor activities of *Stevia rebaudiana*. Trop. J. Pharmaceut. Res. 7 (4), 1144–1151.

JECFA (Joint FAO/WHO Expert Committee on Food Additives), 2007. Evaluation of Certain Food Additives and Contominants. Sixty-Eight Report of the Joint FAO/WHO Expert Committee on Food Additives. World Health Organization (WHO), Geneva, pp. 50–54.

JECFA (2016) Compendium of food additive specifications. Joint FAO/WHO Expert Committee on Food Additives, 1–85. http://www.fao.org/3/a-a0675e.pdf.

Karim, M.Z., Uersugi, D., Nakayama, N., Hossain, M.M., Ishihara, K., Hamada, H., 2015. Identification of stevioside using tissue culture derived *Stevia* leaves. Biochem. Insights 8 (S2), 33–37.

Kawatani, T., Kaneki, Y., Tanabe, T., 1977. The cultivation of Kaa he-e (Stevia rebaudiana) II Seed germination with special reference to the optimum temperature and light sensitivity. Jpn. J. Trop. Agric. 20 (3), 137–142.

Kim, K.K., Yamashita, H., Sawa, Y., Shibata, H., 1996. A high activity of 3-hydroxy-3-methyl glutaryl coenzyme A in *Stevia rebaudiana* Bertoni. Biosci. Biotechnol. 60, 685–686.

Kinghorn (Ed.), 2002. Stevia, the Genus Stevia Medicinal and Aromatic Plants Industrial Profile, vol. 19. Taylor and Francis, London, pp. 178–195.

Kinghorn, A.D., Soejarto, D.D., 1985. Current status of stevioside as a sweetening agent for human use. In: Wagner, H., Hikino, H., Farnsworth, N.R. (Eds.), Economic and Medical Plant Research, vol. 1. Academic Press, London, pp. 1–52.

Kinghorn, A.D., Soerjarto, D.D., 2002. Discovery of terpenoid and phenolic sweeteners from plants. Pure Appl. Chem. 74 (7), 1169–1179.

Klongpanichpak, S., Temcharoen, P., Toskulkao, C., Apibal, S., Glinsukon, T., 1997. Lack of mutagenicity of stevioside and steviol in *Salmonella typhimurium* TA 98 and TA 100. J. Med. Assoc. Thail. 80 (1), 121–128.

Koguishi, C.A., Alves, V.H., Bolini, H.M., 2008. Equisweetness and sweetening power of guava nectars sweetened with different sweetening agents. Rev. Bras. Tecnol. Agroindustrial 2, 26–36.

Koyama, E., Kitazawa, K., Ohori, Y., Izawa, O., Kakegawa, K., Fujino, A., Ui, M., 2003. In vitro metabolism of the glycosidic sweeteners, stevia mixture and enzymatically modified stevia in human intestinal microflora. Food Chem. Toxicol. 41 (3), 359–374.

Kroyer, G., 2010. Stevioside and Stevia-sweetener in food: application, stability and interaction with food ingredients. J. Verbr. Lebensm 5, 225–229.

Kumari, N., Rana, R.C., Sharma, Y.P., Kumar, S., 2017. Extraction, purification and analysis of sweet compounds in *Stevia rebaudiana* Bertoni using chromatographic techniques. Indian J. Pharm. Sci. 79 (4), 617–624.

Latha, S., Usha, M., 2003. *In vitro* culture studies on *Stevia rebaudiana*. In Vitro Cell. Dev. Biol. Plant 39, 520–523.

Li, J., Chase, H.A., 2010. Applications of membrane techniques for purification of natural products. Biotechnol. Lett. 32, 601–608.

Lindsay, R., 2007. Coke Teams up with Cargill to Launch New Sweetener. Times.

Lisak, K., Jeličić, I., Tratnik, L., Božanić, R., 2012. Sensory evaluation of the strawberry flavored yoghurt with stevia and sucrose addition. Biotechnol. J. Nutr. 7, 39–43.

Liu, Y., Di, D., Bai, Q., Li, J., Chen, Z., Lou, S., Ye, H., 2011. Preparative separation and purification of rebaudi- oside A from steviol glycosides using mixed-mode macroporous adsorption resins. J. Agric. Food Chem. 59, 9629–9636.

Madan, S., Ahmand, S., Singh, G.N., Kohli, K., Kumar, Y., Singh, R., Garg, M., 2010. Stevia rebaudiana Bertoni: a review. Indian J. Nat. Prod. Res. 1 (3), 267–280.

Mali, A.B., Joshi, M., Kulkarni, V., 2015. Phytochemical screening and antimicrobial activity of *Stevia rebaudiana* leaves. Int. J. Curr. Microbiol. Appl. Sci. 4 (10), 678–685.

Manisha, G., Soumya, C., Indrani, D., 2012. Studies on interaction between stevioside, liquid sorbitol, hydrocolloids and emulsifiers for replacement of sugar in cakes. Food Hydrocoll. 29 (2), 363–373.

Matsui, M., Sofuni, T., Nohmi, T., 1996. Regionally-targeted mutagenesis by metabolically-activated steviol: DNA sequence analysis of steviol-induced mutants of guanine phosphoribosyltransferase (gpt) gene of *Salmonella typhimurium* TM677. Mutagenesis 11 (6), 565–572.

Midmore, D.J., Rank, A.H., 2002. A New Rural Industry- Stevia to Replace Imported Chemical Sweeteners. Project No: UCQ- 16A,1 1-3. RIRDC web publication.

Mitsuhashi, H., 1976. Report on Safety of Stevia. In: Safety of Stevioside. Tama Biochemical Co. Ltd., pp. 1–20

Mizutani, K., Tanaka, O., 2002. Use of *stevia rebaudiana* sweeteners in Japan. In: Kinghorn, A.D. (Ed.), Stevia, the Genus steviaMedicinal and Aromatic Plants Industrial Profile London, vol. 19. Taylor and Francis, pp. 178–195.

Mogran, R., Dashora, V., 2009. Exploring the use of *Stevia rebaudiana* as a sweetener in comparison with other sweeteners. J. Hum. Ecol. 25, 117–120.

Olsson, K., Carlsen, S., Semmler, A., Simon, E., Mikkelsen, M.D., Moller, B.L., 2016. Microbial production of next-generation stevia sweetness. Microb. Cell Factories 15 (207), 1–14.

Parpinello, G.P., Versari, A., Castellari, M., Galassi, S., 2001. Stevioside as a replacement of sucrose in peach juice: sensory evaluation. J. Sens. Stud. 16, 471–484.

Pezzuto, J.M., Compadre, C.M., Swanson, S.M., Nanayakkara, D., Kinghorn, A.D., 1985. Metabolically activated steviol, the aglycone of stevioside, is mutagenic. Proc. Natl. Acad. Sci. U. S. A. 82 (8), 2478–2482.

Planas, G.M., Kuc, J., 1968. Contraceptive properties of *Stevia rebaudiana*. Science 162–165.

Pól, P., Varadová Ostrá, E., Karásek, P., Roth, M., Bene- šová, K., Kotlaříková, P., Čáslavský, J., 2007. Comparison of two different solvents employed for pressurised fluid extraction of stevioside from Stevia rebaudiana: methanol versus water. Anal. Bioanal. Chem. 388, 1847–1857.

Prakash, I., Upreti, M., 2008. Stevioside Polymorphic and Amorphous Forms, Methods for Their Formulation, and Uses. WO Pattent 2,008,147,725.

Prakash, I., Upreti, M., Dubois, G., King, G., Klucik, J., San Miguel, R., 2008. Sweetness Enhancers, Sweetness Enhanced Sweetener Compositions, Methods for Their Formulations and Uses. WO Patent 2,008,147,727.

Puri, M., Sharma, D., 2011. Antibacterial activity of stevioside towards food-borne pathogenic bacteria. Eng. Life Sci. 11, 326–329.

Puri, M., Sharma, D., Barrow, C.J., Tiwary, A.K., 2012. Optimisation of novel method for the extraction of steviosides from *Stevia rebaudiana* leaves. Food Chem. 132, 1113–1120.

Rajab, R., Mohan Kumar, C., Murungan, K., Harish, M., Mohanan, P.U., 2009. Purification and toxicity studies of stevioside from *Stevia rebaudiana* Bertoni. Toxicol. Int. 16, 49–54.

Rao, A.B., Prasad, E., Roopa, G., Sridhar, S., Ravikumar, Y.V.L., 2012. Simple extraction and membrane purification process in isolation of steviosides with improved organoleptic activity. Adv. Biosci. Biotechnol. (3), 327–335.

Razak, U.N.N.A., Ong, C.B., Yu, T.S., Li, K.L., 2014. In vitro micropropagation of Stevia rebaudiana bertoni in Malaysia, 57 (1), 23–28.

Richman, A., 2005. Functional genomics uncovers three glucosyltransferases involved in the synthesis of the major sweet glucosides of *Stevia rebaudiana.* Plant J. 41, 56–67.

Rongfu, S., Mancai, X., Zuoqing, S., Yunge, F., Xianzhi, G., Yongning, L., Chunhong, W., Binglin, H., 2002. Synthesis of bifunctional polymeric adsorbent and its application in purification of stevia glycosides. React. Funct. Polym. 50, 107–116.

Saniah, K., Samsiah, S.M., 2012. The application of stevia as sugar substitute in carbonated drinks using response surface methodology. J. Trop. Agric. Food Sci. 40 (1), 23–24.

Sartor, R.B., 2008. Microbial influences in inflammatory bowel diseases. Gastroenterology 134 (2), 577–594.

Sekihashi, K., Saitoh, H., Sasaki, Y., 2002. Genotoxicity studies of stevia extract and steviol by the comet assay. J. Toxicol. Sci. 27 (1), 1–8.

Sharma, N., Dhurva, P., Gauchan, A.D., Luitel, A., Shakya, S., Shakya, R., 2015. Establishment of regenerative callus, cell suspension system and molecular characterization of *stevia rebaudiana* bertoni for the production of stevioside in *in vitro*. Int. J. Res. Appl. Sci. Eng. Technol. 3 (VIII), 133–144.

Shibata, H., Sawa, Y., Oka, T., Sonoke, S., Kim, K.K, Yoshioka, M., 1995. Steviol and steviol-glycoside. Glucosyltransferase activities in Stevia rebaudiana Bertoni. Purification and partial characterization. Arch. Biochem. Biophys 321, 390–396.

Singh, M., Saharan, V., Dayma, J., Rajpurohit, D., Sen, Y., Sharma, A., 2017. In vitro propagation of *Stevia rebaudiana* Bertoni: an overview. Int. J. Curr. Microbiol. App. Sci. 6 (7), 1010–1022.

Sreedhar, R.V., Venkatachalam, L., Thimmaraju, R., Bhagyalakshi, N., Narayan, M.S., Ravishankar, G.A., 2008. Direct organogenesis from leaf explants of *Stevia rebaudiana* and cultivation in bioreactor. Biol. Plant. 52, 355–360.

Tadhani, M.B., Subhash, R., 2006. Proximal composition and in vitro antimicrobial activity of *Stevia rebaudiana* Bertoni leaves. Trop. J. Pharmaceut. Res. 5, 557–560.

Tamura, Y., Nakamura, S., Fukui, H., Tabata, M., 1984. Comparison of Stevia plants grown from seeds, cuttings and stem tip cultures for growth and sweet diterpene glycosides. Plant Cell Rep. 3, 180–182.

Teo, C.C., Tan, S.N., Yong, J.W.H., Hew, C.S., Ong, E.S., 2010. Pressurized hot water extraction (PHWE). J. Chromatogr. A 1217, 2484–2494.

Terai, T., Ren, H., Mori, G., Yamaguchi, Y., Hayashi, T., 2002. Mutagenicity of steviol tests. Food Ind. 18, 34–43.

The Ayurvedic Pharmacopeia of India, 2008. Appendix 2. Part I, first ed., vol. I. Govt. Of India, Ministry of Health and Family Welfare, Dept. of Medicine and Homeopathy, New Delhi, pp. 142–156.

Toskulkao, C., Sutheerawattananon, M., Piyachaturawat, P., 1995. Inhibitory effect of towards food-borne pathogenic bacteria. Eng. Life Sci. 11, 326–329.

Toyoda, K., Matsui, H., Shoda, T., Uneyama, C., Takada, K., Takahashi, M., 1997. Assessment of the carcinogenicity of stevioside in F344 rats. Food Chem. Toxicol. 35 (6), 597–603.

Uddin, M.S., Chowdhury, M.S.H., Khan, M.M.M.H., Uddin, M.B., Ahmed, R., Baten, M.A., 2006. In vitro propagation of *Stevia rebaudiana* Bert in Bangladesh. Afr. J. Biotechnol. 5, 1238–1240.

Vanneste, J., Sotto, A., Courtin, C.M., Van Craeyveld, V., Bernaerts, K., Van Impe, J., Vandeur, J., Taes, S., Van der Bruggen, B., 2011. Application of tailormade membranes in a multi-stage process for the purification of sweeteners from *Stevia rebaudiana*. J. Food Eng. 103, 285–293.

Vives, K., Andujar, I., Lorenzo, J.C., Concepcion, O., Fernandez, M., Escalona, M., 2017. Comparison of different in vitro micropropagation methods of *Stevia rebaudiana* B. including temporary immersion bioreactor (BIT). Plant Cell Tissue Organ Cult. 131 (1), 195–199.

Wheeler, A., Boileau, A.C., Winkler, P.C., Compton, S.C., Prakash, J., Jiang, X., 2008. Pharmacokinetics of rebaudioside A and stevioside after single oral doses in healthy men. Food Chem. Toxicol. 46, S54–S60.

Wolwer-Rieck, U., Tomberg, W., Wawrzun, A., 2010. Investigations on the stability of stevioside and rebaudi-oside A in soft drinks. J. Agric. Food Chem. 58, 12216–12220.

Wood Jr., H.B., Allerton, R., Diehl, H.W., Fletcher Jr., H.G., 1955. Stevioside. I. The structure of glucose moieties. J. Org. Chem. 20, 875–883.

Yadav, A.K., Singh, S., Dhyani, D., Ahuja, P.S., 2011. A review on the improvement of stevia [Stevia rebaudiana (Bertoni)]. Can. J. Plant Sci. 91, 1–27.

Yamada, A., Ohgaki, S., Noda, T., Shimizu, M., 1985. Chronic toxicity of dietary stevia extracts. Shokun eiseigaku zaashi 26, 169–183.

Yasukawa, K., Kitanaka, S., Seo, S., 2002. Inhibitory effect of stevioside on tumor promotion by 12-O-tetradecanoylphorbol-13-acetate in two-stage carcinogenesis in mouse skin. Biol. Pharm. Bull. 25 (11), 1488–1490.

Yücesan, B., Büyükgöçmen, R., Mohammed, A., Sameeullah, M., Altuğ, C., Gürel, S., Gürel, E., 2016. An efficient regeneration system and steviol glycoside analysis of Stevia rebaudiana Bertoni, a source of natural high-intensity sweetener. Biol. Plant. 52, 330–337.

Zhang, S.O., Kumar, A., Kutowy, O., 2000. Membrane-based separation scheme for processing sweeteners from stevia leaves. Food Res. Int. 33, 617–620.

Ziv, M., 2005. Simple bioreactors for mass propagation of plants. In: Liquid Culture Systems for In Vitro Plant Propagation. Springer.

CHAPTER

Biotechnological application of health promising bioactive molecules

6

Pankaj Kumar[a,1], Ishani Shaunak[b,1], Madan L. Verma[a,c]

Department of Biotechnology, Dr Y. S. Parmar University of Horticulture and Forestry, Neri Campus, Himachal Pradesh, India[a]; Department of Biotechnology, Dr Y.S. Parmar University of Horticulture and Forestry, Solan, Himachal Pradesh, India[b]; Centre for Chemistry and Biotechnology, Deakin University, Geelong Campus, VIC, Australia[c]

Chapter outline

1. Introduction

Bioactive molecules are the phytochemicals found in fruits, vegetables, whole grains foods, medicinal plant/herbs that modulating the metabolic procedures and resulting in the advancement of better wellbeing (Marchetti et al., 2019; Chang et al., 2019). Different crop/plants species are the fundamental components of the biological system and the key part of metabolic pathways. Plant metabolism is considered as complex mechanism of physical and chemical events i.e. photosynthetic process, respiration, and organic compounds (biosynthesis and degradation) that takes place in the plant cell (Kumar et al., 2017; Mohanraj et al., 2018; Kumar and Shanker, 2018; Wieczorek and Jeleń, 2019). Primary metabolism contributes the essential components for plant life and synthesized primary metabolites which involved

[1]Equal contribution.

Biotechnological Production of Bioactive Compounds. https://doi.org/10.1016/B978-0-444-64323-0.00006-0

directly in the growth and metabolism (carbohydrates, lipids and proteins, purines and pyrimidines of nucleic acids and chlorophylls). Secondary metabolism is in connection to the primary metabolism by using building blocks and biosynthetic enzymes derived from primary metabolic processes and produce a huge number of specialized bioactive molecules (Steroids, alkaloids, terpenoids, flavonoids, sterols, lignins, tannins, curcumins, saponins, glucosides, phenolics and essential oils etc.) which aids different crop plant species such as horticultural/agricultural crops and medicinally important plants/herbs to live in its habitat and environment and also imparts health benefits (Fig. 6.1).

Crop plant-based foods are important source of nutrients like vitamins, minerals and also contained bioactive compounds which have been proven to be beneficial for

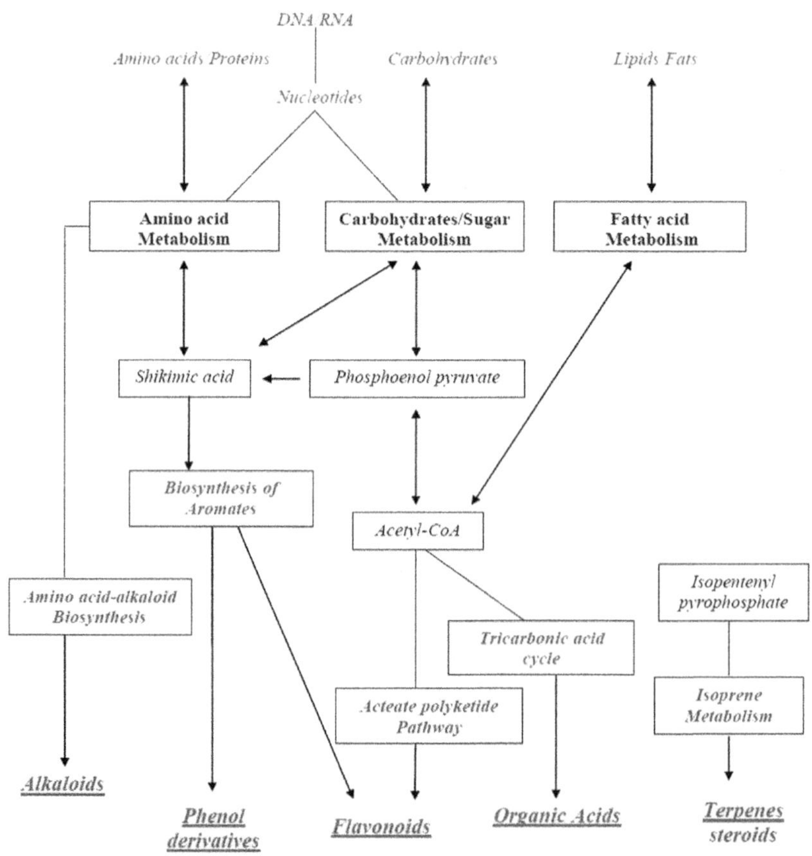

FIG. 6.1

Schematic representation of biosynthetic pathway and precursors for the major classes of bioactive compounds/secondary metabolites.

Adapted from Gutzeit and Ludwig-Muller (2014).

human health (Karalija et al., 2018; Jahurul et al., 2019). These bioactive molecules belong to the groups like phenolic compounds (flavonoids, phytoestrogens, and phenolic acids), phytosterols and phytostanols, carotenoid, organosulfur compounds i.e. allium compounds and glucosinolates. Phenolics are the chief category of phytochemicals and extensively distributed in the plant kingdom. Phenolic compounds play key role in plant defense response to various biotic and abiotic stresses by scavenging the free radicals produced (Lattanzio, 2013; Harsh et al., 2017). Flavonoids, phenolic acids, and polyphenols are the three central groups of dietary phenolics and have antioxidant, antibacterial and antiallergic applications etc. Whereas, glycosides and saponins are the bioactive molecules which reported to have antibacterial, antifungal properties, anti-inflammatory, antiviral plant defense applications (Maurya et al., 2008; Mohanraj et al., 2018). Alkaloids have analgesic, antispasmodic, antimalarial and diuretic activities. Terpenoids are the largest and most diverse class containing volatile molecules which gives plants and flowers their flavor and fragrance (Kumar and Shanker, 2018). Terpenoids as bioactive molecule remarkably contributed to various medicinal properties like anticarcinogenic, anti-viral, antibacterial, anti-malarial, antiulcer, hepaticidal, antimicrobial, diuretic activity and anti-inflammatory properties (Jan and Abbas, 2018). Due to the potential applications of bioactive medicinal molecules, recently large number of chemical compounds used in medicines is derived plant sources like atropine, aspirin, digoxin, ephedrine and colchicines (Moteriya et al., 2015).

Bioactive molecule extraction marks the initial step in compound analysis from the particular crops such as fruit crop, vegetable crops medicinal herbs/plants species (Gürbüz et al., 2018; Singh et al., 2019). For extract preparation from the desired crop/plant samples, suitable procedures should be undertaken such as prewashing, drying (freeze drying) and proper grinding etc. to maintain the bioactive constituents, which otherwise might be distorted or lost. Selection of proper solvent for extraction largely depends upon the nature and properties of bioactive compounds (nonpolar to polar; thermally labile). For extraction of hydrophilic compounds, polar solvents such as methanol, ethanol, or ethyl acetate are used, whereas for lipophilic molecule extraction dichloromethane or a mixture of dichloromethane/methanol (1:1) is used. Chlorophyll is also removed in some of the extraction procedures using hexane. Methods such as sonification, Soxhlet extraction, heating under reflux, microwave-assisted extraction, solid-phase extraction, Solid-phase microextraction, supercritical-fluid extraction, pressurized liquid extraction and surfactant-mediated techniques etc. are also utilized for extraction (Pharmacopeia, 2002; Cos et al., 2006).

Plants extract is mainly consisted of mixtures of bioactive molecules, so for identification and characterization of particular bioactive molecule, most critical step is bioactive compound separation. Large number separation techniques such as thin-layer chromatography (TLC), flash chromatography, column chromatography, high-performance liquid chromatography (HPLC) and high throughput compound separation techniques have been commercially exploited and the pure bioactive compounds are lined for structure and biological activity determination.

Phytochemicals screening assay, immunoassays (monoclonal antibodies based) and Fourier-transform infrared spectroscopy (FTIR), are some nonchromatographic techniques that has been also be utilized to acquire and initiate the bioactive molecule identification (Jan and Abbas, 2018).

The present chapter discusses the bioactive production in vegetable through biotechnological intervention. Biotechnological implication of medicinal plants research is critically discussed. Bioinformatics resources and "Omics" based Himalayan medicinal plants research along with scope of health promising bioactive compounds is particularly discussed.

2. Biotechnological intervention for improving bioactive molecules in vegetables

Vegetables are the important source of human's daily diet and provide nutrients, vitamins, antioxidants, phytosterols, and dietary fiber which impart health benefits. To strengthen the agricultural economy, vegetable farming is the most significant part in the developing nations (Srivastava et al., 2016). Nowadays, vegetable-based research is receiving attention worldwide due to bioactive medicinal molecules presents in vegetables such as polyphenols, phytoestrogens, carotenoids, anthocyanins, lycopene, resveratrol, tannins, indoles and glucosinolates which significantly contributed to health benefits i.e. prevention of cardiovascular disease, cancer and other diseases (Saha, 2016; Yalcin and Capar, 2017; Kumar et al., 2017). Health promising bioactive molecules present in vegetables have been discussed in detail in Table 6.1, which provide insight to researchers to work on it for its further genetic advancement having therapeutically potential. Recent biotechnological interventions such as marker assisted breeding using genetic map and quantitative trait loci analysis, genetic engineering approaches and advancements in plant genetics, plant tissue culture and plant molecular biology techniques in concurrence with traditional crop breeding programs have remarkable prospective to produce bioactive molecule enriched vegetable varieties (Kumar et al., 2017).

As per the recent research literature mining, it has been reported that synergistic interactions of bioactive molecules along with nutrients present in staple food crops, whole grains, fruits and vegetables, and plant-based foods resulted in health benefits (Liu, 2013). Therefore, end users should consume balanced diet enriched with vegetables, fruits, whole grains, and other plant foods for optimal nutrition, health, and well-being (Kumar et al., 2017). Tomato, Potato and brinjal are the foremost globally important vegetable and mostly studied as modal vegetable crop in vegetable breeding programs. Potatoes (low-fat foods) globally consumed vegetable having distinctive nutrients and bioactive molecules such as lutein, zeaxanthin, quercetin, kaempferol, chlorogenic acid and caffeic acid that imparts health benefits (Chakraborty et al., 2010). Brinjal is an economically important vegetable crop rich in medicinal molecule such as chlorogenic acid and nasunin which are having

Table 6.1 Health promising bioactive molecules in vegetables.

Vegetables	Bioactive molecule(s)	Health promising effects	References
Monocotyledonae			
Amaryllidaceae (Alliaceae)			
Onion (*Allium cepa*)	Allyl sulfides and Quercetin	Antioxidant, protection from cardiovascular disease and certain type of cancer	Beretta et al. (2017)
Garlic (*Allium sativum*)	Allicin, Alliin and Ajoene	Antibacterial and antioxidative etc.	
Leek (*Allium porrum*	Flavonols - Kaempferol or Quercetin derivatives	Protection from cardiovascular disease and cancer disease	
Liliaceae			
Asparagus (*Asparagus officinalis*)	Steroidal Saponins, Flavanoids, Glycosides	In ayurveda, effective in treating madhur rasam, madhur vipakam, seet-veeryam, som rogam, chronic fever and internal heat. Effective in problems related with female reproductive system.	Zhang et al. (2019)
Araceae			
Colocasia (*Colocasia esculenta*)	Alkaloids and Glycosides	Protection from cardiovascular disease etc.	Kumar and Sharma (2017)
Graminae			
Sweet corn (*Zea mays*)	Carotenoids, Phenolic compounds and Phytosterols	Preventing chronic diseases.	Sheng et al. (2018)
Dioscorecaeae			
Yam (*Dioscorea alata*)	Diosgenin	Used as a moderate laxative and vermifuge; treatment of fever, gonorrhea, leprosy, tumors and inflamed hemorrhoids.	Kanu et al. (2018)

Continued

Table 6.1 Health promising bioactive molecules in vegetables. *Continued*

Vegetables	Bioactive molecule(s)	Health promising effects	References
Dicotyledonae			
Cucurbitaceae			
Cucumber (*Cucumis sativus*)	Alkaloid, 9-beta-methyl-19-norlanosta-5- ene glycoside, Steroid, Saponin and Tannin	Antiulcer, diuretic and anthelmintic etc.	Rajasree et al. (2016); Saha (2016); Kumar et al. (2017)
Pumpkin (*Cucurbita moschata*)	Alkaloids, Flavonoids, and Palmitic, Oleic and Linoleic acids	Anti-diabetic, antioxidant, anti-carcinogenic, anti-inflammatory	
Water melon (*Citrullus lanatus*)	Lycopene, Cucurbitacin	Diuretic, cardiovascular activity, anti-inflammatory activity	
Summer squash (*Cucurbita pepo*)	Cucurbitacins, Avenasterol, Spinasterol, Triterpenoids, Sesquiterpenoids, Squalene, Tocopherols, Carotenoids	Antiandrogenic activity, immunological activity, antiviral activity, antifungal activity, cardiovascular activity, anti-inflammatory activity and hepatoprotective activity.	
Winter squash (*Cucurbita maxima*)	Flavonoids, Polyphenolics, Saponins, Cucurbitaxanthin and α-tocopherol	Antidiabetic, antitumor, antihypertensive, antiinflammatory and immunomodulatory etc.	
Bitter gourd (*Momordica charantia*)	Steroidal glycosides, Insulinomimeticlectins and Alkaloids	Antidiabetic, antitumor etc.	
Bottle gourd (*Lagenaria siceraria*)	β-carotene, Cucurbitacins, Saponins, Flavone-C-glycoside and Polyphenol	Antioxidant, antihepatotoxic, antitumor, immunoprotective and antipoliferative activity etc.	
Snake gourd (*Trichosanthes cucumerina*)	Triterpenoids, Saponins, Cucurbitacins	Anti-inflammatory activity, ant-diabetic activity etc.	
Sponge gourd (*Luffa cylindrica*)	Alkaloids, Saponins, Carotenoids, Terpenoids	Antioxidant, antimicrobial, anticancer, cardioprotective, gastroprotective, antidiabetic, hypolipidemic, hepatoprotective properties.	
Ridge gourd (*Luffa acutangula*)	Flavonoids, Saponins, Luffangulin, Sapogenin, Oleanolic acid and Cucurbitacin	Diuretic, antioxidant, expectorant, laxative, purgative, hypoglycemic agent etc.	

Cruciferae (Brassicacaeae)			
Broccoli (*Brassica oleraceae var. italica*)	Isothiocyanates, Glucosinolates, Sulforaphane, Lutein, Zeaxanthin, β-carotene, flavanoids	Antioxidant and anticancerous properties etc.	Srivastava et al. (2016); Kumar et al. (2017)
Cauliflower (*Brassica oleraceae var. botrytis*)	Indole-3-carbinol, Sulforaphane	Anti-inflammatory and anticancerous etc.	
Cabbage (*Brassica oleraceae var. capitata*)	Anthocyanins and Sulforaphane	Prevent cardiovascular disease and cancer etc	
Brussels sprout (*Brassica oleraceae var. gemmifera*)	Organosulfur compounds	Antioxidant properties	
Chinese cabbage (*Brassica Campestris var.pekinensis*)	Glucosinolates	Anti-tumor, immunomodulating, antioxidant etc.	
Turnip (*Brassica campestris var. rapa*)	Lutein and Glucosinolates	Anticancerous	
Mustard (*Brassica juncea*)	Allyl isothiocyanate	Cancer chemo-preventive etc.	
Raddish (*Raphanus sativus*)	Glucosinolate, Myrosinase, Isothiocyanate, Triterpenes, Alkaloids, Flavonoids, Tannins, Saponin and Coumarins	Anthelmintic, antifungal, antibacterial, antiscorbutic, diuretic, laxative, tonic, carminative, corrective, stomachic, cholagogue, lithotriptic, emmenagogue	
Compositae (Asteraceae)			
Lettuce (*Lactuca sativa*)	Lactucin, Lactucopicrin, Lactuerol	Antimalarial, analgesic and sedative properties	Sílvia et al. (2015)
Globe artichoke (*Cynara scolymus*)	Chlorogenic acid, Cynarin, Luteolin-7-rutinoside, and Cynaroside	Antioxidative, hepatoprotective, choleretic and anti-cholestatic effects etc	
Jerusalem artichoke (*Helianthus tuberosus*)	Sesquiterpene lactones, Phenolics	Antifungal, antioxidant, anticancer activities etc.	
Solanaceae			
Tomato (*Solanum lycopersicum*)	Lycopene, Carotenoids, Polyphenols, quercetin, naringenin, chalcone and chlorogenic acid	Antioxidant and anticancerous etc.	Saha (2016); Srivastava et al. (2016)

Continued

Table 6.1 Health promising bioactive molecules in vegetables. *Continued*

Vegetables	Bioactive molecule(s)	Health promising effects	References
Potato (*Solanum tuberosum*)	Carotene-ß, Cryto-xanthin-ß, Lutein-xanthin	Antiscorbutic, aperient, diuretic, galactagogue, stimulant, emollient, antidote, antispasmodic	
Brinjal (*Solanum melongena*)	Anthocyanins, Polyphenols, Alkaloids	Antioxidants, anticancerous, anti-inflammatory etc.	
Hot pepper (*Capsicum annuum var annuum*)	Carotenoids - capsanthin, capsorubin and cryptocapsin	Carminative, anti-inflammatory, antigenotoxic, anticarcinogenic, and antioxidant properties.	
Sweet pepper (*Capsicum annuum var grossum*)	Polyphenols, Carotenoids, Capsaicinoids and Ascorbic acid	Antioxidants, anticancerous, anti-inflammatory, antiallergic and antibacterial activities	
Leguminosae (Fabaceae)			
Cluster bean (*Cyamopsis tetragonoloba*)	Gallotannins, Gallic acid derivatives, Campesterol, Avenasterol, Stigmasterol	Acts as an appetizer, digestive aid and laxative	Wink (2013)
Indian bean/Hyacinth bean (*Dolichos lablab*)	Alkaloids, Tannins, Flavonoids, Saponins, Coumarins, Terpenoids, Glycosides, Anthnanoids	Antidiabetic, antiinflammatory, analgesic, antioxidant, cytotoxic, hypolipidemic, antimicrobial, insecticidal, hepatoprotective, antilithiatic, antispasmodic effects	
Lima bean (*Phaseolus lunatus*)	Lunatusin	Antibacterial, antifungal, antiproliferative etc.	
Kidney/snap/French bean (*Phaseolus vulgaris*)	Alkaloids, Anthocyanin, Catechin, Flavonoids, Glycosides, Phasine, Quercetin, Polyphenols, Saponins, Steroids, Tannins, and Terpenoids	Depurative, resolvent, carminative, diuretic, antidiarrheal, emollient, antioxidant, antidiabetic, osteoprotective properties etc.	
Cow pea (*Vigna sinensis*)	Folic Acid	Prevent neural tube defects in unborn babies.	
Winged bean/Goa bean (*Psophocarpus tetragonolobus*)	Sterols, folates, niacin	Antimicrobial, anti-inflammatory, anti-nociceptive, antioxidant properties etc.	
Sword bean (*Canavalia gladiata*)	Methyl gallate, Digalloyl hexoside and Digallic acid	Antioxidative, potential health benefits against oxidative stress-related chronic diseases.	
Methi/fenugreek (*Trigonella foenum-graecum*)	Galactomannan, 4-OH isoleucine, and Steroidal saponin	Treatments of diabetes, hypercholesterolemia, inflammation and cancers.	

Umbelliferae			
Carrot (*Daucus carota*)	Polyacetylenes (Falcarinol, Falcarindiol and Falcarindiol-3-acetate) and carotenoids (β-carotene and Lutein)	Treatment of leukemia etc	Altemimi et al. (2017)
Coriander (*Coriandrum sativum*)	Caffeic, Chlorogenic acid, Quercetin, Keampferol, Rhamnetin and Apigenin	Antibacterial, antifungal and antioxidant activities etc.	
Celery (*Apium graveolens*)	Limonene, Selinene, Frocoumarin glycosides, Flavonoids	Antioxidant, different healing effects etc.	
Chenopodiaceae			
Beet (*Beta vulgaris*)	Betaine and Betalain	Treat a wide variety of ailments, treatment for fevers and constipation etc.	Shivaranjani et al. (2014)
Spinach (*Spinacia oleracea*)	Flavonoid - 5,3‘,4‘-trihydroxy-3-methoxy-6:7-methylenedioxyflavone-4‘-glucuronide, Glycosides	Protection from eye disorders, oxidative stress, iron deficiency etc., reducing risks of specific diseases like diabetes, cancer and hepatotoxicity.	
Palak (*Beta vulgaris* var.bengalensis)	Betaine and betalain	Treat a wide variety of ailments, treatment for fevers and constipation etc	
Araliaceae			
Udo (*Aralia cordata*)	Saponins, Sesquiterpenes and Diterpene acids	Analgesic, antiinflammatory, carminative, diuretic, febrifuge, stomachic.	Saha (2016)
Aizoaceae			
New Zealand spinach (*Tetragonia expansa*)	Saponin, Vitamin C, E, K and Nitrates	Antiscorbutic, anticancerous etc.	Shivaranjani et al. (2014)
Convolvulaceae			
Sweet potato (*Ipomoea batatas*)	Anthocyanins, Phenolics, Coumarins, Phenolics, Caffeoylquinic acid derivatives etc	Anti-cancer, antidiabetic, and anti-inflammatory activities.	Saha (2016)
Euphorbiaceae			
Tapioca/cassava (*Manihot esculenta*)	Mandiocin - glucoside; Flavonoids, Saponins	Anti-inflammatory and antimicrobial activity etc	Altemimi et al. (2017)
Malvaceae			
Okra/Bhendi (*Abelmoschus esculentus*)	Phenolics, Quartering and Flavonol derivatives, Catechin oligomers and Hydroxycinnamic derivatives	Antioxidants activity, help to lower serum cholesterol, reducing the risk of heart diseases etc	Roy et al. (2014)

anti-carcinogenic, anti-diabetic and anti-obesity applications. Similarly, lycopene present in tomato, watermelon and carrot has been reported to fight against infection and anticancerous properties. Bioactive molecules present in bittergourd i.e. momordicin and charantin act as blood purifier, and having medical application in hypertension, dysentery and anathematic, and diabetes. Beta-carotene enriched vegetables such as carrot, sweet potato, pumpkin, cantaloupe and cauliflower are rich in antioxidants and betacyanins containing beet root are higher in antioxidants and shown free radical scavenging activities. Bioactive molecules i.e. anthocyanins pigments enriched vegetable crop such as cabbage, broccoli, brinjal, carrot, amaranth and bean provide protective effect on pancreatic cells and prevents from cardiovascular dysfunction. In vegetables like onion and garlic, the bioactive molecule allicin, allyl propyl disulfide, di-allyl disulfide significantly contributed toward protection against cancer and cardiovascular diseases and also acts as immunobooster. Quercetin is an important bioactive medicinal molecule in onion that impart health benefits by treating asthma, gout, diabetes, peptic ulcer, inflammation, viral infections and heart disease (Table 6.1).

Bioactive molecule i.e. glucosinolates that is attributed to anticancerous property, most scientifically studied bioactive molecule present in cruciferous vegetable such as cauliflower, and broccoli (Kumar and Srivastava, 2015, 2016). To enhance the bioavailability of glucosinolate, plant genetic engineering approaches has been successfully used to introgression chromosome segments from a wild ancestor, *Brassica villosa* in three high-glucoraphanin F1 broccoli hybrids that resulted in enhance glucosinolate levels (Traka et al., 2013). Lu et al. (2006) developed transgenic cauliflower enriched in β-carotene by transferring 'or' gene from wild cauliflower into cultivable varieties that converts the white color of curd tissue into distinct orange color, which shown increased level of β-carotene. Eady et al. (2009) also develop transgenic onion enriched with allicin using genetic engineering approaches. Using plant transgenic technologies, phytoene synthase (*CrtB*), phytoenedesaturase (*CrtI*) and lycopene beta-cyclase (*CrtY*) from *Erwinia* has been transferred in potato to produce higher β-carotene (Diretto et al. 2007). Maligeppagol et al. (2013) developed transgenic tomato enriched in anthocyanin pigment (70–100 fold) by transferring fruit specific expression *Delila* and *Rosea*1 (transcription factors) isolated from *Antirrhinum majus*.

Plant molecular breeding approaches have also been employed to produce bioactive compounds rich vegetable varieties/cultivars. Just et al. (2007) reported genetic mapping of carotenoid pathway (carotenoids; phytoene, α-carotene, β-carotene, zeta-carotene, and lycopene) structural genes and monogenic traits viz. *Rs, Mj-1, yel, cola, Y, Y2,* and *P1* in carrot. *Solanum lycopersicoides* aubergine (*Abg*) gene associated with carotenoids was introgressed using Marker assisted breeding program that resulted in strong and variegated pigmentation in tomatoes peel (Mes et al., 2008). QTLs for high β –carotene was reported in sweet potato by Cervantes-Flores et al. (2011) which possibly contributed in genetic advancement in root crops.

3. Biotechnological implication of medicinal plants research

Bioactive compounds take part in a vital role in high-value product development in the pharmaceutical industry. Medicinal plants/herbs used in traditional medicines are worldwide well known. India one of 17 mega biodiversity countries and contribute about seven percent of world biodiversity. India has 15 agro-climatic zones. Approximately, more than seven thousand plants species are estimated to have medicinal usage as per documented systems of medicine like Ayurveda, Unani, Siddha & Homoeopathy (AYUSH System of Medicine). In Ayurveda, siddha and unani systems of medicine have more than 90% formulations which are plant based. As per the recent update by National Medicinal Plant Board, domestic demand of medicinal plants has been estimated 195,000 MT for the year of 2014–15 and total consumption of herbal raw drug in the country for the year 2014–15 has been estimated 512,000 MT and the export demand of medicinal plants estimated about 134,500 MT with export value of 3211 crore during 2014–15.

Plants are highly specific to their natural growth habitat. Most of the plants grow in association with other plants and microbes that possibly resulted in increased biomass and improved growth characteristics. Due to continuous demand and over-exploitation of medicinal herb, the Himalayan plants/high altitude medicinal plants are facing threat of extinction, narrowing down their distribution range. In addition, various biotic and abiotic stress factors, loss in seed viability, poor seed germination and survival percentage also affect the overall growth and development of Himalayan plants. Biotechnological interventions such as plant tissue culture approaches, plant genetic engineering techniques, metabolic and pathway engineering etc. offer a sustainable way conservation and utilization at commercial scale (Fig. 6.2).

4. *In vitro* plant regeneration and micropropagation

For the cultivation of high-altitude medicinal plants, comprehensive understanding of the species biology is required. And there are very few published reports on the methods *in vitro* propagation of medicinal plants. For *in vitro* plant propagation, detailed standardization and optimization procedures (phytohormones, pH, temperature, aeration, agitation, light, etc.) is required for culture initiation, growth and multiplication and acclimatization (Briskin, 2000; Harsh et al., 2017). Once the reliable, genetically stable and highly efficient *in vitro* regeneration protocol has been optimized, then the most important step is acclimatization of the tissue culture raised plants. Successful acclimatization depends upon the optimization of humidity condition, light level, temperature and other environmental conditions as in the natural habitat/native environment of the cultured plant species. For the re-introduction of *in vitro* raised plants in their natural habitat is the most critical step in plant

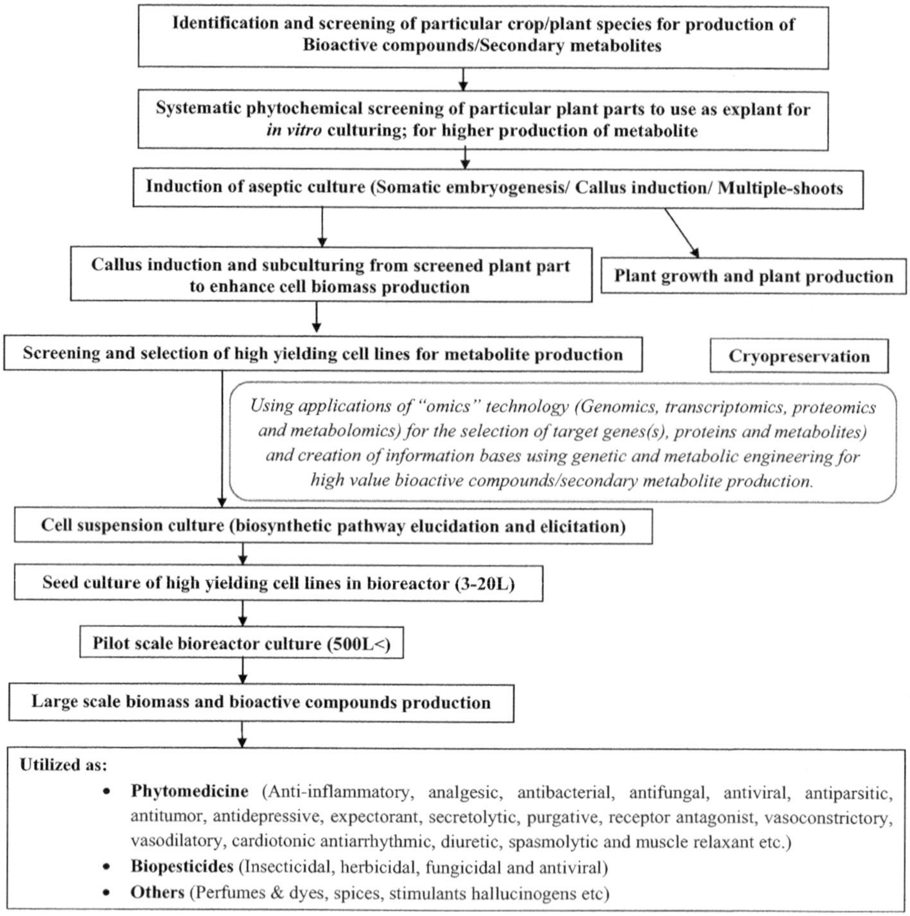

FIG. 6.2

Biotechnological approaches for the production of high value metabolites/bioactive compounds and its applications.

conservation strategy, which relies on successful acclimatization procedure (Chaturvedi et al., 2007). This process is time taking, tedious and huge efforts are required which might hinder the researcher's interest and that is the major obstacle in large scale quality plant production (Harsh et al., 2017).

Plant micropropagation is an important plant biotechnological process in which cells, tissues or organs are excised from the particular plant, then surface sterilized, and incubated in a growth-promoting aseptic environment to produce large number of clonally plantlets (Kumar and Srivastava, 2016). It is important tool to select, multiply, and screen the promising genotypes of Himalayan medicinal plants. *In vitro* culture is one of the key tools of plant biotechnology that exploits the totipotency nature of plant cells. As compared with conventional methods of vegetative propagation, this technique has many advantages such as the propagation of a large number of pathogen-free plants in a short period of time with high consistency. The *in vitro* plantlets derived from existing meristem yield plants that are genetically identical with the donor plants (Harsh et al., 2017; Jan and Abbas, 2018). *In vitro* propagation through meristem culture is the feasible means of virus elimination and produces a large number of plants in a very short time span and act as a robust tool for large-scale propagation of plants. Axillary or apical buds regeneration is an important technique of micropropagation in which nodal segments bearing these buds are involved in the recovery of multiple shoots without callus phase (Pati et al., 2006). This micropropagation technique plays a significant role in the regeneration of virus-free seeds from different plant species (Rout et al., 2006). By using this technique, plants can be grown and multiplied on a large scale without depending on their natural habitats, thus preserving the natural resources that are otherwise under the constant hazard of exploitation (Jan and Abbas, 2018). The important medicinal plants that have been raised via micropropagation from the shoot apical meristem have been reported in different studies like *C. roseus, C. ledgeriana, Digitalis* spp., *R. glutinosa, Rauvolfia serpentina, I. canariensis, Citrullus colocynthis, Zephyranthes bulbous, Plectranthus vettiveroids*, and *Glossocardia bosvallea* (Roy et al., 1994; Paek et al., 1995; Gayathri and Ramagopal, 2007; Meena and Patni, 2007). *Potentilla fulgens,* important Himalayan medicinal plant were also regenerated by the tissue culture from the axillary bud explants. Similarly, the mass multiplication of number of Himalayan threatened and endemic medicinal plants was also studied by this important technique of micropropagation, for example *Atropa acuminata, Bacopa monnieri, Coleus forskohlii, Decalepis hamiltonii, Eclipta alba, Gentiana kurroo, Ginkgo biloba, Janakia arayalpathra, Picrorhiza kurroa, Paederia foetida, Saussurea lappa, Salvadora persica, and Tylophora indica* (Mathur et al., 2003; Srivastava and Srivastava, 2004; Sood and Chauhan, 2009; Thangavel et al., 2014; Jan and Abbas, 2018).

Micropropagation of Himalayan medicinal plants by using the adventitious shoots and their conservation by establishing gene or tissue banks plays an important role for the rapid supply of good quality planting material. Thus, in order to retain the genetic makeup of high-yielding medicinal plants under the controlled environmental conditions, micropropagation plays a significant role for at least those plants where vegetative propagation by conventional methods is unsatisfactory (Chaturvedi et al., 2007). In fact, lots of protocol has been developed for *in vitro* propagation of Himalayan plants and the extraction of bioactive constituent from *in vitro* cultures but still lacking in the cost-benefit analysis and commercial exploitation (Moshelion and Altman, 2015). Till date, most of the medicinal plant based/pharmaceutical industries are still dependent upon the wild plants for their collection and extraction of bioactive compounds that resulted in narrowing down of biological diversity; and there is not any commercial technology for *in vitro* secondary metabolites/bioactive compounds extraction (Harsh et al., 2017).

5. Bioinformatics resources and "Omics" based Himalayan medicinal plants research

Bioinformatics resources and high throughput approaches such as science of omics (transcriptomics, proteomics, metabolomics, ionomics and phenomics) have revolutionized the biological research (Matsuura et al., 2018; Chakraborty, 2018; Wang et al., 2018; Saito, 2018). This give insight to reveal the bioactive compound chemistry and dynamics, complexity, nature, gene regulations and biosynthetic pathway elucidation for better understanding and to harness the potential in sustainable manner by saving time (Table 6.2). Identification, characterization and extraction of natural bioactive compounds from Himalayan medicinal plants have been commercially exploited for herbal medicine formulations and also act as significant resources for drug discovery (Kumar and Shankar, 2018). Due to utmost demand of plant materials supply, there is non-sustainable harvesting and overexploitation of Himalayan medicinal plants from their natural habitat which resulted in threatening and extinction of natural bio-resources (Kumar and Shankar, 2018). In order to fulfill the global demand supply, suitable conservation strategies, cultivation practices/ good agricultural practices and agro-technology turns out to be sustainable produce (Moshelion and Altman, 2015; Harsh et al., 2017; Tuyiringire et al., 2018). Controlled supply of plant based herbal medicines in sustainable manner depends upon the availability of a high-quality planting material sources. Various biotechnological interventions such as plant tissue culture, *in vitro* micropropagation techniques for the cultivation, rapid propagation, genetic engineering and metabolic engineering of Himalayan plants play significant role in constant supply of high quality material, large scale production of bioactive compounds and value addition to harness the true potential of Himalayan flora (Sarasan et al., 2006; Chaturvedi et al., 2007; Harsh et al., 2017; Jan and Abbas, 2018; Kumar and Shanker, 2018). Recently, genomic profiling, DNA fingerprinting and DNA barcoding has been used for identification and authentication of Himalayan medicinal herb/plant species based on species specific sequence variations in various nuclear DNA and Chloroplast DNA regions (Sucher et al., 2013) which otherwise traditionally used by practiced taxonomist with matching of microscopic and macroscopic type specimen descriptions. DNA sequence based "barcodes" generation for Himalayan medicinal plants considered to be determined significant attempt to define barcode for every species on earth which remarkably contribute to medicinal plant research community (Sucher et al., 2012, 2013).

For biological identification of species using short, standardized genomic regions as taxon barcodes (DNA barcoding) was first proposed by Hebert et al. (2003) and till now has been considered as indispensable tools to the plant taxonomist. Nuclear ribosomal marker ITS (internal transcribed spacer (ITS) or a subset of ITS i.e. (ITS2) and plastid markers/plastidial gene(s) i.e. *rbcL* (large subunit of ribulose 1,5 bisphosphate carboxylase/oxygenase), *matK* (maturase K encodes an intron maturase protein that splices introns) and the intergenic spacer trnH—psbA, has

Table 6.2 Bioinformatics resources i.e. biological databases and web resources for medicinal plant research.

Database(s)	Available resource link	Function	References
Indian Medicinal Plants Database	(http://www.medicinalplants.in/aboutfrlhtdb)	• Provides detailed information's on Indian medicinal plants i.e. traditional Indian medicine systems (Ayurveda, Siddha, Unani, Homeopathy, Sowa-Rigpa and Folk).	Kumar and Shanker (2018)
The Indian Medicinal Plants, Phytochemistry, and Therapeutics database (IMPPAT)	(https://www.imsc.res.in/~asamal/resources/imppat/home)	• Provides scientific information on phytochemicals constituents along with their therapeutic potential of commonly found medicinal plants in India.	Mohanraj et al. (2018)
The database of Indian Medicinal Plants for Diabetes (InDiaMed)	(http://www.indiamed.info)	• Provides information of Indian medicinal plants which are used for the treatment of diabetes. • Also comprises the polyherbal formulations list that can used for diabetes treatment in India.	Tota et al. (2013)
DiaMedBase	(http://www.progenebio.in/DMP/DMP.htm)	• Provides information about 389 medicinal plants that can be used in diabetes treatment.	Babu et al. (2006)
Medicinal Plants Database (MEDDB)	(http://www.ladydoakcollege.edu.in/meddb/home.html)	• Provide information on 110 different plants species, belonging to a total of 50 families, commonly used by Madurai, India region tribal community. • The disease search option displays Tamil vernacular name and the plants used for a particular disease.	Mary et al. (2012)
Natural Products CARE database (NPCARE)	(http://silver.sejong.ac.kr/npcare)	• Database of natural product and fractional extracts include information about 6578 natural compounds and 2566 fractional extracts with anticancer activities from range biological resources i.e. plants, bacteria, fungus, and sea creatures.	Choi et al. (2017)
Medicinal Materials DNA Barcode Database (MMDBD)	(http://www.cuhk.edu.hk/icm/mmdbd.htm)	• An integrated medicinal material DNA database provides information over	Lou et al. (2010)

Continued

Table 6.2 Bioinformatics resources i.e. biological databases and web resources for medicinal plant research. *Continued*

Database(s)	Available resource link	Function	References
		1000 species DNA sequences and key references of medicinal materials recorded in Chinese and American herbal pharmacopoeia with other related references.	
HerbMed	(http://www.herbmed.org)	• Provides detailed scientific information's related to the use of herbs for therapeutic purpose and health benefits.	Wootton (2002)
MPD3	(http://bioinform.info)	• Provides information about phytochemicals, activities, structural, and test target of medicinal plants at a single display place (genus, plant, activity, and phytochemical).	Mumtaz et al. (2017)
The Natural Medicines Comprehensive Database	(http://www.naturaldatabase.com)	• Scientific information on natural medicines. • Comprehensive and reliable natural evidence-based medicine sources, useful for healthcare professionals and patients	Hsu (2002)
Natural Products Alert Database (NAPRALERT)	(https://www.napralert.org/about)	• Provides scientific information on natural products, including ethnomedical, pharmacological and biochemical extracts of organisms. • Also contains information on secondary metabolites obtained from natural sources.	Loub et al. (1985)
Super Natural II Database	(http://bioinformatics.charite.de/supernatural)	• Have information's on natural compounds (about 326,000) with their corresponding two-dimensional structures and physicochemical properties. • Facilitates the pathways information related to synthesis and degradation of the natural products.	Banerjee et al. (2014)
Indonesian Medicinal Plants Database	(http://herbaldb.farmasi.ui.ac.id)	• Provides information of Indonesian medicinal plants with their three-dimensional (3D) structures of chemical compounds.	Yanuar et al. (2011)

been proved effective for the genomic authentication in a range of plant taxa and recently successfully employed in genomic profiling and species discrimination in large number of medicinal herbs/plant species (Sucher and Carles, 2008; Sucher et al., 2012, 2013; Kumar and Shankar, 2018). To study the potential of bioactive compounds, omics-based approach i.e. "metabolomics" significantly contributed in identification, screening and mapping of bioactive compounds/secondary metabolites through metabolite profiling process. Metabolomics investigations include metabolic profiling and metabolic fingerprinting (Aravindaram and Yang, 2013). In metabolic profiling, analysis of metabolites has been carried out to identify specific metabolite, specific metabolic pathway and classes of compound. Whereas metabolic fingerprinting techniques is not used to identify the metabolite, but to compare the metabolite pattern or "fingerprints" that change in response to environmental or genetic alteration, in contact to diseases/toxins etc. Metabolome analysis techniques include nuclear magnetic resonance and mass spectrometry coupled with gas or liquid chromatographic separation for analysis of a large number of metabolites/bioactive compounds in Himalayan plants (Shyur and Yang, 2008; Aravindaram and Yang, 2013). To utilize the potential of natural bioactive compounds and to sustained plant based medicinal research interdisciplinary approaches among bioinformatics, system and computational biology and chemometrics will be essential which ultimately benefits societal needs and public health (Kumar and Shanker, 2018).

6. Scope and limitation of health promising bioactive compounds

Recently, focus on commercialization of bioactive compounds has increased considerably all over the world (Puri et al., 2012; Puri and Verma, 2013; Thakur et al., 2017; Thakur and Verma, 2017). Bioactive compounds are extracted with well-established methodology from many sources including microorganisms, plants and animals (Banerjee et al., 2017). However, bioactive compounds from the medicinal plant provide unique platform for exploring the noble bioactive compounds of medicinal importance due to rich and varied biodiversity across the globe (Ares et al., 2018; Singh et al., 2017). In view of bioactive compounds, plants of medicine importance possess the chemical repertoire that capable to synthesize vast number of highly complex and unusual chemical compounds (Yuan and Grotewold, 2015). Asian continent, in particular to three countries India, china and Japan, is repertories for bioactive compound search. Till now, more than two hundred bioactive compounds are extracted from plants used in medicine throughout the world (Altemimi et al., 2017). The plant extracts are typically subjected to pharmacological and biochemical evaluation to determine the suitability of bioactive compounds. There has been a significantly increase in the production of Phyto active compounds-based medicines. The World Health Organization (WHO) estimates approximately

80% of traditional medicine that involves the use of plant extracts (Ekor, 2014). This means that approximately 4 billion people in the world rely on plants as sources of drugs. Rich biodiversity is yielding plant sources of various therapeutically valuable bioactive chemicals or their precursors, that are in great, demand in national as well as international drug and pharmaceutical industry.

There is a growing interest in correlating bioactive compounds of a plant with its pharmacological activity (Xi, 2017; Vardanega et al., 2014). Researchers have even started correlating the botanical properties of plants with their pharmacological activity (Wang et al., 2019). In future, more coordinated multidimensional research aimed at correlating botanical and phyto-chemical properties to specific pharmacological activities is expected. In terms of pharmacological activity, more attention has been paid to CNS-active, cyto-protective, immuno-modulators and chemotherapeutic plant products (Lele, 2010).

The prevalence of a variety of climatic conditions puts in a supreme position with respect to richness of medicinal flora. Bioactive prospecting should focus on process technology, standardization, quality control, research and development of purified and biochemically characterized safe bioactive molecules from plants (Gu et al., 2014). With the advances in cellular biology, a shift toward studying changes in cytosolic enzyme activities, DNA patterns and genetic control has been observed rather than concentrating merely on the gross effects induced by the plant drugs. In addition to the proper utilization of technological advances, a logical interpretation of the codified language of traditional medicine also becomes a necessity in order to further promote research in this field. The work done on rasayana group of plants is a good example of the above statement (Kuchewar et al., 2014). This understanding triggered the subsequent research work that was aimed at evaluating the immuno-stimulant potential of the rasayana group of plants and now, we have indigenous immuno-stimulants available, at affordable rates, in the global market.

However, there is a scary side too. There are very few studies involving evaluation of biochemically characterized immuno-modulating plant-based drugs/molecules. Any publication related to phyto-pharmacology should ideally provide data on the authentication and standardization of the plant products. Not all the research areas selected by the scientists were relevant to needs of current research. Infections like tuberculosis, malaria, diarrhea, AIDS and malnutrition are some of the major problems. However, these areas have not been extensively explored. Majority of the drugs are at the experimental stage and have to still undergo clinical trials (Katiyar et al., 2012). There is still a paucity of clinical studies, which are carried out in randomized, controlled, double blind manner. Today, concurrent consumption of drugs from different disciplines is a common finding. Very few studies, however, address the problem of drug interactions, if any. In view of this, the studies documenting interaction of plant drugs with modern medicine are important from the clinician's point of view (Thakur et al., 2017).

Similarly, the documented data on adverse drug reactions of plant drugs is also sparse (Hafner-Blumenstiel, 2011). A relevant point that arose while compiling this data and requires a special mention is the immense need for documentation of research activities related to medicinal plants and publications of results, whether positive or negative, in peer-reviewed journals. Although, work carried out on traditional medicine is presented in conferences and is available in the conference abstract books (data of which may or may not be complete) or published in local journals, these are not peer-reviewed and moreover this data is accessible to a select few (Thakur and Verma, 2017). This results in lacunae while judging the current status of research on traditional medicine and leads to a false impression that not enough research is being carried out in this field. Hence, networking of the available data sources generated by various research activities carried out by different scientists/research institutes has now become the need of the hour in order to provide information about and access to research work done in different laboratories of the country.

In order to meet requirement of the global plant-based drug market, herbal products should be standardized as per WHO guidelines. However, it is extremely important to establish and develop databases describing the medicinal plants, and linking such databases to national and international uplinks. Such an approach will definitely give bioactive compounds extracted from plants a global importance and shall further boost global economy.

7. Conclusion and future directions

Biotechnological interventions hold the significant promise in the production of novel bioactive compounds. Vegetable based bioactive compounds is considerably promote health benefits by prevention of various diseases. Bioprospecting of plant tissue culture technology is employed for conservation and propagation of valuable, rare and endangered plants in order to mass production of bioactive constituents via callus or suspension cultures. Improvement in bioactive yield production is feasible via genetic engineering and metabolic engineering for genetic makeup modification for higher biomass production. This health promising bioactive molecules present in vegetable crops and medicinal herbs/plants need further exploration to prior to commercialization.

Recently a couple of the research centers have developed appropriate technologies for the conservation, cultivation and applications of medically important plants. In order to push a significant player in the global plant-based drug market, herbal products need to be standardized as per WHO guidelines. Various academic and research institutes have also become primed with the need of production of safe plant/herbal drugs, conservation of important herbal/medicinal plants and have also adopted organized cultivation of medicinal plants. Establishment of a few herbal gardens besides arranging organized cultivation of medicinal plants indicates the concern of the research institutions. However, it is extremely important to establish and develop more databases describing the medicinal plants of Himalayan region, and linking such databases to national and international uplinks. Such an approach will definitely give medicinal plants of Himalayan a global importance and shall further boost bioeconomy.

Acknowledgments

The senior author (PK) thankfully acknowledges the award of National Post-Doctoral Fellowship, Science and Engineering Research Board, Department of Science & Technology, Government of India, New Delhi, India.

References

Aravindaram, K., Yang, N., 2013. Metabolomics in medicinal plant research. In: Chandra, et al. (Eds.), Biotechnology for Medicinal Plants. Springer-Verlag Berlin Heidelberg, pp. 275–294.

Ares, A.M., Valverde, S., Bernal, J.L., Nozal, M.J., Bernal, J., 2018. Extraction and determination of bioactive compounds from bee pollen. J. Pharm. Biomed. Anal. 147, 110–124.

Altemimi, A., Lakhssassi, N., Baharlouei, A., Watson, D.G., Lightfoot, D.A., 2017. Phytochemicals: extraction, isolation, and identification of bioactive compounds from plant extracts. Plants (Basel) 6 (4), 42.

Babu, P.A., Suneetha, G., Boddepalli, R., Lakshmi, V.V., Rani, T.S., Babu, Y.R., Srinivas, K., 2006. A database of 389 medicinal plants for diabetes. Bioinformation 1, 130–131.

Banerjee, P., Erehman, J., Gohlke, B.O., Wilhelm, T., Preissner, R., Dunkel, M., 2014. Super natural II-a database of natural products. Nucleic Acids Res. 43, D935–D939.

Banerjee, J., Singh, R., Vijayaraghavan, R., MacFarlane, D., Patti, A.F., Arora, A., 2017. Bioactives from fruit processing wastes: green approaches to valuable chemicals. Food Chem. 225, 10–22.

Beretta, H.V., Bannoud, F., Insani, M., Berli, F., Hirschegger, P., Galmarini, C.R., Federico, P., 2017. Relationships between bioactive compound content and the antiplatelet and antioxidant activities of six *Allium* vegetable species. Food Technol. Biotechnol. 55, 266–275.

Briskin, D.P., 2000. Medicinal plants and phytomedicines. Linking plant biochemistry and physiology to human health. Plant Physiol. 124, 507–514.

Cervantes-Flores, J.C., Sosinski, B., Pecota, K.V., Mwanga, R.O.M., Catignani, G.L., Truong, V.D., Watkins, R.H., Ulmer, M.R., Yencho, G.C., 2011. Identification of quantitative trait loci for dry-matter, starch, and β-carotene content in sweet potato. Mol. Breed. 28, 201–216.

Chakraborty, P., 2018. Genomics as tools for dissecting new metabolic pathways of unexplored medicinal plants and drug discovery. Biochim Open 6, 9–16.

Chakraborty, S., Chakraborty, N., Agrawal, A., Ghosh, S., Narula, K., Shekhar, S., Naik, P.S., Pande, P.C., Chakrborti, S., Datta, A., 2010. Next-generation protein-rich potato expressing the seed protein gene *AmA1* is a result of proteome rebalancing in transgenic tuber. PNAS USA 107, 17533–17538.

Chang, H.Y., Wu, J.R., Gao, W.Y., Lin, H.R., Chen, P.Y., Chen, C.I., Wu, M.J., Yen, J.H., 2019. The cholesterol-modulating effect of methanol extract of pigeon pea (*Cajanus cajan* (L.) Millsp.) leaves on regulating LDLR and PCSK9 expression in HepG2 cells. Molecules 24 (3) pii: E493.

Chaturvedi, H.C., Jain, M., Kidwai, N.R., 2007. Cloning of medicinal plants through tissue culture-a review. Indian J. Exp. Biol. 45, 937–948.

Choi, H., Cho, S.Y., Pak, H.J., Kim, Y., Choi, J.Y., Lee, Y.J., Gong, B.H., Kang, Y.S., Han, T., Choi, G., Cho, Y., 2017. NPCARE: database of natural products and fractional extracts for cancer regulation. J. Cheminf. 9, 2.

Cos, P., Vlietinck, A.J., Berghe, D.V., Maes, L., 2006. Anti-infective potential of natural products: how to develop a stronger in vitro 'proof-of-concept'. J. Ethnopharmacol. 106, 290–302.

Diretto, G., Al-Babili, S., Tavazza, R., Papacchioli, V., Beyer, P., Giuliano, G., 2007. Metabolic engineering of potato carotenoid content through tuber-specific overexpression of a bacterial mini-pathway. PLoS One 2, e350.

Eady, C., Hunger, A., Sarah, 2009. Alliums. https://doi.org/10.1002/9781405181099.k0608.

Ekor, M., 2014. The growing use of herbal medicines: issues relating to adverse reactions and challenges in monitoring safety. Front. Pharmacol. 4, 177.

Galanakis, C.M., 2017. Introduction. Nutraceut. Funct. Food Components 1–14. https://doi.org/10.1016/b978-0-12-805257-0.00001.

Gayathri, D., Ramagopal, G., 2007. *In vitro* propagation of *Amaryllis* and *Zephyranthes* bulbous species using household tissue culture medium. Plant Cell Biotechnol. Mol. Biol. 8, 131–136.

Gu, R., Wang, Y., Long, B., Kennelly, E., Wu, S., Liu, B., Li, P., Long, C., 2014. Prospecting for bioactive constituents from traditional medicinal plants through ethnobotanical approaches. Biol. Pharm. Bull. 37 (6), 903–915.

Gürbüz, N., Uluişik, S., Frary, A., Frary, A., Doğanlar, S., 2018. Health benefits and bioactive compounds of eggplant. Food Chem. 268, 602–610.

Gutzeit, H.O., Ludwig-Müller, J., 2014. Plant Natural Products: Synthesis, Biological Functions and Practical Applications. John Wiley & Sons, Weinheim.

Hafner-Blumenstiel, V., 2011. Herbal drug-drug interaction and adverse drug reactions. Ther. Umsch. 68 (1), 54–57.

Harsh, K.C., Anil, K.B., Indra, D.B., 2017. Impediments in the tissue culture of high-altitude medicinal plants. Adv. Biotechnol. Microbiol. 4, 555–631.

Hebert, P.D., Cywinska, A., Ball, S.L., deWaard, J.R., 2003. Biological identifications through DNA barcodes. Proc. Biol. Sci. 270, 313–321.

Hsu, P.P., 2002. Natural medicines comprehensive database. J. Med. Libr. Assoc. 90, 114.

Jahurul, M.H.A., Zaidul, I.S.M., Beh, L., Sharifudin, M.S., Siddiquee, S., Hasmadi, M., Sahena, F., Mansoor, A.H., Lee, J.S., Jinap, S., 2019. Valuable components of bambangan fruit (*Mangifera pajang*) and its co-products: a review. Food Res. Int. 115, 105–115.

Jan, S., Abbas, N., 2018. Himalayan Phytochemicals Sustainable Options for Sourcing and Developing Bioactive Compounds. Elsevier Ltds, Netherlands, ISBN 978-0-08-102227-6.

Just, B.J., Santos, C.A.F., Fonseca, M.E.N., Boiteux, L.S., Oloizia, B.B., et al., 2007. Carotenoid biosynthesis structural genes in carrot (*Daucus carota*): isolation, sequence-characterization, single nucleotide polymorphism (SNP) markers and genome mapping. Theor. Appl. Genet. 114, 693–704.

Kanu, A.N., Ezeocha, C.V., Ogunka, N.P., 2018. A review on bioactive compounds of Yam varieties for human disease management. Asian Food Sci. J. 1, 1–10.

Karalija, E., Parić, A., Dahija, S., Bešta-Gajević, R., Ćavar Zeljković, S., December 23, 2018. Phenolic compounds and bioactive properties of *Verbascum glabratum* subsp. bosnense (K. Malý) Murb., an endemic plant species. Nat. Prod. Res. 1–5.

Katiyar, C., Gupta, A., Kanjilal, S., Katiyar, S., 2012. Drug discovery from plant sources: an integrated approach. Ayu 33 (1), 10–19.

Kuchewar, V.V., Borkar, M.A., Nisargandha, M.A., 2014. Evaluation of antioxidant potential of Rasayana drugs in healthy human volunteers. Ayu 35 (1), 46–49.

Kumar, S., Shanker, A., 2018. Biological databases for medicinal plant research. In: Kumar, N. (Ed.), Biotechnological Approaches for Medicinal and Aromatic Plants. Springer Nature Singapore Pvt Ltd, pp. 655–665.

Kumar, V., Sharma, H.K., 2017. Process optimization for extraction of bioactive compounds from taro (*Colocasia esculenta*), using RSM and ANFIS modeling. J. Food Measure. Charac. 11, 704–718.

Kumar, P., Srivastava, D.K., 2015. High frequency organogenesis in hypocotyl, cotyledon, leaf and petiole explants of broccoli (*Brassica oleracea* L. var. *italica*), an important vegetable crop. Physiol. Mol. Biol. Plants 21, 279–285.

Kumar, P., Srivastava, D.K., 2016. Biotechnological advancement in genetic improvement of broccoli (*Brassica oleracea* L. var. *italica*), an important vegetable crop: a review. Biotechnol. Lett. 38, 1049–1063.

Kumar, P., Shaunak, I., Thakur, A.K., Srivastava, D.K., 2017. Health promising medicinal molecules in vegetable crops. J. Gene. Genom. 1, 1.

Lattanzio, V., 2013. Phenolic compounds: introduction. In: Natural Products. Springer, Berlin Heidelberg, pp. 1543–1580.

Lele, R.D., October 2010. Beyond reverse pharmacology: mechanism-based screening of Ayurvedic drugs. J. Ayurveda Integr. Med. 1 (4), 257–265.

Liu, R.H., 2013. Health-promoting components of fruits and vegetables in the Diet. Adv. Nutri. 4, 384–392.

Lou, S.K., Wong, K.L., Li, M., But, P.P., Tsui, S.K., Shaw, P.C., 2010. An integrated web medicinal materials DNA database: MMDBD (Medicinal Materials DNA Barcode Database). BMC Genomics 11, 402.

Loub, W.D., Farnsworth, N.R., Soejarto, D.D., Quinn, M.L., 1985. NAPRALERT: computer handling of natural product research data. J. Chem. Inf. Comput. Sci. 25, 99–103.

Lu, S., Eck, J.V., Zhou, X., Lopez, A.B., Halloran, M., Cosman, K.M., Conlin, B.J., Paolillo, D.J., Garvinm, D.F., Vrebalov, J., Kochian, L.V., Kupper, H., Earle, E.D., Cao, J., Li, L., 2006. The Cauliflower *or* gene encodes a *DnaJ* cysteine-rich domain-containing protein that mediates high levels of Beta-carotene accumulation. Plant Cell 18, 3594–3605.

Maligeppagol, M., Chandra, G.S., Navale, P.M., Deepa, H., Rajeev, P.R., Asokan, R., Babu, K.P., Barbu, C.S.B., Rao, V.K., Kumar, N.K.K., 2013. Anthocyanin enrichment of tomato (*Solanum lycopersicum* L.) fruit by metabolic engineering. Curr. Sci. 105, 72–80.

Marchetti, L., Pellati, F., Graziosi, R., Brighenti, V., Pinetti, D., Bertelli, D., 2019. Identification and determination of bioactive phenylpropanoid glycosides of *Aloysia polystachya* (Griseb. et Moldenke) by HPLC-MS. J. Pharm. Biomed. Anal. 166, 364–370.

Mary, J.A., Priyadharshini, K.C., Amal, G.P., Ramya, G., Nithya, R., Ambika, M.B., Shenbagarathai, R., 2012. MEDDB: a medicinal plant database developed with the information gathered from tribal people in and around Madurai, Tamil Nadu. Bioinformation 8, 391–393.

Mathur, S., Sharma, S., Gupta, M.M., Kumar, S., 2003. Evaluation of an Indian germplasm collection of the medicinal plant *Bacopa monnieri* (L.) Pennell by use of multivariate approaches. Euphytica 133, 255–265.

Matsuura, H.N., Malik, S., de Costa, F., Yousefzadi, M., Mirjalili, M.H., Arroo, R., Bhambra, A.S., Strnad, M., Bonfill, M., Fett-Neto, A.G., 2018. Specialized plant metabolism characteristics and impact on target molecule biotechnological production. Mol. Biotechnol. 60 (2), 169–183.

Maurya, R., Singh, G., Yadav, P.P., 2008. Antiosteoporotic agents from natural sources. Stud. Nat. Prod. Chem. 35, 517–548.

Meena, M.C., Patni, V., 2007. *In vitro* clonal propagation of *Citrullus colocynthis* (Linn.) Schrad: a threatened medicinal plant. Plant Cell Biotechnol. Mol. Biol. 8, 147–152.

Mes, P.J., Boches, P., Myers, J.R., Durst, R., 2008. Characterization of tomatoes expressing anthocyanin in the fruit. J. Am. Soc. Hort. Sci. 133, 262–269.

Mohanraj, K., Karthikeyan, B.S., Vivek-Ananth, R.P., Chand, R.B., Aparna, S.R., Mangalapandi, P., Samal, A., 2018. IMPPAT: a curated database of Indian medicinal plants, phytochemistry and therapeutics. Sci. Rep. 8, 1–17.

Moshelion, M., Altman, A., 2015. Current challenges and future perspectives of plant and agricultural biotechnology. Trends Biotechnol. 33, 337–342.

Moteriya, P., Satasiya, R., Chanda, S., 2015. Screening of phytochemical constituents in some ornamental flowers of Saurashtra region. J. Phcog. Phytochem. 3, 112–120.

Mumtaz, A., Ashfaq, U.A., Qamar, M.T., Anwar, F., Gulzar, F., Ali, M.A., Saari, N., Pervez, M.T., 2017. MPD3: a useful medicinal plants database for drug designing. Nat. Prod. Res. 31, 1228–1236.

Paek, K.Y., Yu, K.J., Park, S.I., Sung, N.S., Park, C.H., 1995. Micropropagation of *R. glutinosa* as medicinal plant by shoot tip and root segment culture. ISHS international symposium on medicinal and aromatic plants. Acta Hortic 390, 120–123.

Pati, P.K., Rath, S.P., Sharma, M., Sood, A., Ahuja, P.S., 2006. In vitro propagation of rose-a review. Biotechnol. Adv. 24, 94–114.

Pharmacopeia US, 2002. United States Pharmacopeia. The Standard. US Pharmacopeia 1995, USA.

Puri, M., Verma, M.L., 2013. Innovation in technology development with reference to enzymatic extraction of flavonoids. In: Ghosh, D., Das, S., Bagchi, D., Smarta, R.B. (Eds.), Innovation in Healthy and Functional Foods. CRC Press, Boca Raton, FL, pp. 331–342.

Puri, M., Verma, M.L., Mahale, K., 2012. Processing of citrus peel for the extraction of flavonoids for biotechnological applications. In: Yamane, K., Kato, Y. (Eds.), Handbook on Flavonoids: Dietary Sources, Properties, and Health Benefits. Nova Science Publishers, Hauppauge, New York, pp. 443–459.

Rajasree, R.S., Sibi, P., Francis, F., William, H., 2016. Phytochemicals of *Cucurbitaceae* family-a review. Int. J. Pharmacogn. Phytochem. Res. 8, 113–123.

Rout, G.R., Mohapatra, A., Jain, S.M., 2006. Tissue culture of ornamental pot plant: a critical review on present scenario and future prospects. Biotechnol. Adv. 24, 531–560.

Roy, S.K., Hossain, M.Z., Islam, M.S., 1994. Mass propagation of *Rauvolfia serpentina* by *in vitro* shoot tip culture. Plant Tissue Cult. 4, 69–75.

Roy, A., Shrivastava, S.L., Mandal, S.M., 2014. Functional properties of Okra *Abelmoschus esculentus* L. (Moench): traditional claims and scientific evidences. Plant Sci. Today 1, 121–130.

Saha, P., 2016. Biotechnology for improving bioactive compounds in vegetable crops. Biotech Res. Article. https://www.biotecharticles.com.

Saito, K., 2018. Development of plant metabolomics and medicinal plant genomics. Yakugaku Zasshi 138 (1), 1–18.

Sarasan, V., Cripps, R., Ramsay, M.M., Atherton, C., McMichen, M., et al., 2006. Conservation *In vitro* of threatened plants-progress in the past decade. In Vitro Cell. Dev. Plant 42, 206–214.

Sheng, S., Li, T., Liu, R.H., 2018. Corn phytochemicals and their health benefits. Food Sci. Human Wellness 7, 185–195.

Shivaranjani, V.L., Poornima, H., Umamaheswari, J., Devi, K.L., 2014. Preliminary phytochemical screening and quantification of bioactive compounds in the leaves of spinach (Spinaceae oleraceae L.). J. Pharm. Res. 8, 1113–1119.
Shyur, L.F., Yang, N.S., 2008. Metabolomics for phytomedicine research and drug development. Curr. Opin. Chem. Biol. 12, 66–71.
Bessada, S.M.F., João, C.M., Barreira, M., Beatriz, P.P., 2015. *Asteraceae* species with most prominent bioactivity and their potential applications: a review. Ind. Crops Prod. 76, 604–615.
Singh, M., Kumar, A., Singh, R., Pandey, K.D., 2017. Endophytic bacteria: a new source of bioactive compounds. 3 Biotech 7 (5), 315.
Singh, B.P., Yadav, D., Vij, S., 2019. Soybean bioactive molecules: current trend and future prospective. In: Mérillon, J.M., Ramawat, K. (Eds.), Bioactive Molecules in Food. Reference Series in Phytochemistry. Springer, Cham.
Sood, H., Chauhan, R.S., 2009. Development of a low cost micropropagation technology for an endangered medicinal herb (*Picrorhiza kurroa*) of North-Western Himalayas. J. Plant Sci. 4, 21–31.
Srivastava, S.K., Srivastava, N., 2004. *In vitro* multiplication of *Paedaria foetida* L.- a rare medicinal plant. J. Plant Biochem. Biotechnol. 13, 89–91.
Srivastava, D.K., Kumar, P., Sharma, S., Gaur, A., Gambhir, G., 2016. Genetic engineering for insect resistance in economically important vegetable crops. In: Ahmad, N., Anis, M. (Eds.), Plant Tissue Culture: Propagation, Conservation and Crop. Springer Publishing House, pp. 343–378.
Sucher, N.J., Carles, M.C., 2008. Genome-based approaches to the authentication of medicinal plants. Planta Med. 74, 603–623.
Sucher, N.J., Hennell, J.R., Carles, M.C., 2012. DNA fingerprinting, DNA barcoding, and next generation sequencing technology in plants. Methods Mol. Biol. 862, 13–22.
Sucher, N.J., Hennell, J.R., Carles, M.C., 2013. Genomic and transcriptomic profiling: tools for the quality production of plant-based medicines. In: Chandra, et al. (Eds.), Biotechnology for Medicinal Plants. Springer-Verlag Berlin Heidelberg, pp. 439–455.
Thakur, R., Verma, M.L., 2017. Food Bioactives with Special Reference to Himalayan Tartary Buckwheat. LAMBERT Academic Publishing, Saarbrücken, Germany.
Thakur, R., Kumar, S., Awasthi, C.P., Verma, M.L., 2017. Biochemical evaluation of tartary buckwheat (Fagopyrum tataricum Gaertn.) genotypes of cold desert of Himachal Pradesh. Biosci. Biotechnol. Res. Asia 14 (1), 1–16.
Thangavel, K., Ebbie, M.G., Ravichandran, P., 2014. Biotechnology and *in vitro* conservation of medicinal plants. Ann. Plant Sci. 3, 734–744.
Tota, K., Rayabarapu, N., Moosa, S., Talla, V., Bhyravbhatla, B., Rao, S., 2013. InDiaMed: a comprehensive database of Indian medicinal plants for diabetes. Bioinformation 9, 378.
Traka, M.H., Saha, S., Huseby, S., Mithen, R.F., 2013. Genetic regulation of glucoraphanin accumulation in Beneforte broccoli. New Phytol. 198, 1085–1095.
Tuyiringire, N., Tusubira, D., Munyampundu, J.P., Tolo, C.U., Muvunyi, C.M., Ogwang, P.E., 2018. Application of metabolomics to drug discovery and understanding the mechanisms of action of medicinal plants with anti-tuberculosis activity. Clin. Transl. Med. 7 (1), 29.
Vardanega, R., Santos, D.T., Meireles, M.A., 2014. Intensification of bioactive compounds extraction from medicinal plants using ultrasonic irradiation. Phcog. Rev. 8 (16), 88–95.
Wang, X., Zhang, J., He, S., Gao, Y., Ma, X., Gao, Y., Zhang, G., Kui, L., Wang, W., Wang, Y., Yang, S., Dong, Y., 2018. HMOD: an omics database for herbal medicine plants. Mol. Plant 11 (5), 757–759.

Wang, J., Xu, J., Gong, X., Yang, M., Zhang, C., Li, M., 2019. Biosynthesis, chemistry, and pharmacology of polyphenols from Chinese *Salvia* species: a review. Molecules 24 (1) pii: E155.

Wieczorek, M.N., Jeleń, H.H., 2019. Volatile compounds of selected raw and cooked *brassica* vegetables. Molecules 24 (3) pii: E391.

Wink, M., 2013. Evolution of secondary metabolites in legumes (*Fabaceae*). South Afr. J. Bot. 89, 164–175.

Wootton, J.C., 2002. Development of HerbMed®: an interactive, evidence-based herbal database. Adv. Phytomed. 1, 55–60.

Xi, J., 2017. Ultrahigh pressure extraction of bioactive compounds from plants-a review. Crit. Rev. Food Sci. Nutr. 57 (6), 1097–1106.

Yalcin, H., Capar, T.D., 2017. Bioactive compounds of fruits and vegetables. In: Yildiz, F., Wiley, R. (Eds.), Minimally Processed Refrigerated Fruits and Vegetables. Food Engineering Series. Springer, Boston, MA.

Yanuar, A., Munim, A., Lagho, A.B., Syahdi, R.R., Rahmat, M., Suhartanto, H., 2011. Medicinal plants database and three-dimensional structure of the chemical compounds from medicinal plants in Indonesia. Int. J. Comput. Sci. 8, 180–183.

Yuan, L., Grotewold, E., 2015. Metabolic engineering to enhance the value of plants as green factories. Metab. Eng. 27, 83–91.

Zhang, H., Birch, J., Pei, J., Ahmed, I.A.M., Yang, H., Dias, G., Abd El-Aty, A.M., Bekhit, A., 2019. Identification of six phytochemical compounds from *Asparagus officinalis* L. root cultivars from New Zealand and China Using UAE-SPE-UPLC-MS/MS: effects of extracts on H_2O_2-Induced oxidative stress. Nutrients 11, 107.

CHAPTER

7

Biotechnological production of high-valued algal astaxanthin and lutein under different growth conditions

Santosh Kumar[a,1], Balraj Singh Gill[b], Aruna Verma[c], Madan L. Verma[d,e], Rekha Kushwaha[a,1]

Department of Biochemistry, University of Missouri, Columbia, MO, USA[a]; Department of Higher Education, Shimla, Himachal Pradesh, India[b]; Department of Biosciences, Himachal Pradesh University Shimla, India[c]; Department of Biotechnology, Dr Y. S. Parmar University of Horticulture and Forestry, Neri Campus, Himachal Pradesh, India[d]; Centre for Chemistry and Biotechnology, Deakin University, Geelong Campus, VIC, Australia[e]

Chapter outline

[1] Contributed equally.

Biotechnological Production of Bioactive Compounds. https://doi.org/10.1016/B978-0-444-64323-0.00007-2

1. Introduction

Microalgae are the microscopic, unicellular autotrophs reside in fresh and marine water. Their ability to grow rapidly and adaptability under different and ever-changing environmental conditions explains their worldwide distribution (Katiyar et al., 2017; Coltelli et al., 2014). There are approximately 200,000–800,000 species of various phyla such as Cyanophyta (blue-green algae, cyanoprokaryotes, cyanobacteria), Chlorophyta (green algae), Rhodophyta (red algae), Cryptophyta, Haptophyta, Pyrrophyta, Streptophyta, Heterokontophyta (Gacheva and Gigova, 2014). As per size, they range from few μm to 100 μm. According to Chapman (Chapman, 2013), microalgae are the primary producers and are the main source of global oxygen (ca. 50% of all oxygen production).

Microalgae have an unusual proportion of carbon, hydrogen, nitrogen and phosphorous with the following composition: $CO_{0.48}H_{1.83}N_{0.11}P_{0.01}$ and are rich source of proteins (30%–40%), lipids (10%–20%) and carbohydrates (5%–15%) (Zhu, 2015). In addition, microalgae are equally rich in fibers, vitamins (A, C, B1, B2, B6, niacin), and minerals (iodine, potassium, iron, calcium, and magnesium). Therefore, these serve as a major food source in many Asian countries like China, Japan, and Korea (Bilal et al., 2017). In contrast to other energy resources, algae are known to have high nutritional index e.g. *Chlorella* has high protein content (58 gm/100 gm) compared with wheat (13 gm/100 gm), chicken (24 gm/100 gm) and fish (18–24 gm/100 gm).

15,000 novel compounds originated from algal biomass have been identified (Sun et al., 2016; Cardozo et al., 2007). Various bioactive compounds from these algae play the main role in the production of high valued product development in the biotechnology industry as well as an important source of the oxygen, food, fuel and many pharmaceutical compounds. Pharmaceutical compounds produced from these bioactive compound are used against drug-resistant bacterial strains, viruses (including Herpes Simplex and AIDS), cancers as well as the cosmetic industry (Chapman, 2013; Wells et al., 2017; Mimouni et al., 2012). Most of the microalgal species produce unique products like carotenoids, antioxidants, fatty acids, enzymes, polymers, peptides, toxins, and sterols. Xanthophylls, a major group of carotenoids, primarily include astaxanthin, β-cryptoxanthin, canthaxanthin lutein, and zeaxanthin (Kotake-Nara and Nagao, 2011; Zou et al., 2013; Fen et al., 2013). Astaxanthin (also called as Super Vitamin) and lutein are the most important as well as effective compounds against a wide range of ailments, such as cardiovascular problems, diabetes, chronic inflammatory diseases, different types of cancer, and some diseases of the immunological system as well as eyes disorders. This is due to their outstanding antioxidant activities.

Astaxanthin and lutein are available as nutritional supplement products in the nutraceutical market (Higuera-Ciapara et al., 2006; Zhao and Sweet, 2008). Despite high commercial as well as pharmaceutical values, these are available at very high prices in the market. This is due to their high cultivation cost as well deficiency of efficient tools for the successful extraction and purification processes (Ventura et al., 2017).

Therefore, biotechnologists are persistently working in devolvement of proficient conventions for their generation by utilizing different growth conditions. In the present chapter, an endeavor has been made to survey the recent literature for the biotechnological production of two noteworthy high valued bioactive compounds (astaxanthin and lutein) from their algal sources under different growth conditions. The present chapter is an attempt to summarize the sources, chemical structure therapeutic and pharmaceutical values and additionally the significance of this algal bioactive compound as food supplements.

2. Astaxanthin

2.1 Astaxanthin (sources and importance)

Astaxanthin (3, 3′-dihydroxy-β, β′-carotene-4, 4′-dione) a ketocarotenoid is a red fat-soluble pigment. This belongs to the xanthophyll family of carotenoids. Astaxanthin is present in many red-colored aquatic organisms. The natural sources of astaxanthin are algae, yeast, salmon, trout, krill, shrimp and crayfish (Ambati et al., 2014). Green algae *Haematococcus pluvialis*, *Chlorella zofingiensis*, and *Chlorococcum* sp are the main algal source. For the commercial purpose, it is naturally derived from microalgae *(Haematococcus)* and the yeast (*Xanthophyllomyces*) (Solymosi et al., 2015).

The United States Food and Drug Administration (USFDA) and European Commission have approved the use of astaxanthin as food and feed colorant (Ambati et al., 2014). USFDA also approved astaxanthin a food supplement in 1999 (Guerin et al., 2003).

Astaxanthin is known as "super vitamin E" due to its strong antioxidant activity (Miki, 1991; Fan and Chen, 2007; Murillo et al., 2016). It has free radical fighting capacity worth 500 times than that of vitamin E (Dufossé et al., 2005). According to the many reports, its antioxidant capacity is 65 times more powerful than vitamin C, 54 times stronger than β-carotene, and 10 times more potent than β-carotene, canthaxanthin, zeaxanthin, and lutein; and 100 times more effective than α-tocopherol (Miki, 1991; Borowitzka, 2013; Koller et al., 2014; Perez-lopez et al., 2016; Shah et al., 2016). According to Nishida et al. (2007), Astaxanthin is 8000 times stronger than vitamin C, 3000 times stronger than resveratrol, 800 times stronger than CoQ10, 560 times stronger than green tea catechins and 500 times stronger than vitamin E thus can be called as "King of antioxidants"(Yoshida et al., 2010). Astaxanthin is not converted to vitamin A in humans, which means excess intake will not cause hypervitaminosis A toxicity (Jyonouchi et al., 1995; Fassett and Coombes, 2011).

Astaxanthin reported having antioxidant (Naguib, 2000; Ranga Rao et al., 2010, 2013; Hussein et al., 2006; Liu and Osawa, 2007; Augusti et al., 2012; Kamath et al., 2008), anti-inflammatory properties (Liu and Lee, 2003; Bennedsen et al., 1999; Haines et al., 2011; Ohgami et al., 2003; Suzuki et al., 2006; Miyachi et al., 2015; Park et al., 2010; Coombes et al., 2016), antidiabetic activity (Uchiyama et al., 2002; Otton et al., 2010; Li et al., 2016; Nakano et al., 2008; Nishigaki et al., 2010; Hussein et al., 2007; Bhuvaneswari et al., 2010; Naito et al., 2004; Kim et al., 2009; Manabe et al., 2008; Zhu et al., 2018). These antioxidant and anti-inflammatory activities help astaxanthin to treat cancerous cells (Jyonouchi et al., 1995; Tanaka et al., 1994, 1995; Prabhu et al., 2009; Nakao et al., 2010; Maoka et al., 2012; Zhang and Wang, 2015; Davinelli et al., 2018) as well as work as an immuno-modulator and also prevent various cardiovascular diseases (Ambati et al., 2014). Because of its free radical, scavenging capacity astaxanthin has wide applications in the food, feed, cosmetic, aquaculture, nutraceutical, and pharmaceutical industries (Shah et al., 2016). The astaxanthin products are available in various forms in the market such as capsule, soft gel, tablet, powder, biomass, cream, energy drink, oil, and extract.

2.2 Structure

Astaxanthin is a metabolite of zeaxanthin and/or canthaxanthin. Astaxanthin contains both hydroxyl and ketone functional groups. The structure of astaxanthin was described in 1975 (Cooper et al., 1975). The molecular formula of astaxanthin is $C_{40}H_{52}O_4$ and molar mass is 596.84 g/mol. This is a carotenone that consists of beta, beta-carotene-4, 4′-dione bearing two hydroxy substituents at positions 3 and 3' (the 3S, 3′S diastereomer. Its systemic IUPAC nomenclature is 6S)-6-Hydroxy-3-[(1E,3E,5E,7E,9E,11E,13E,15E,17E)-18-[(4S)-4-hydroxy-2,6,6-trimethyl-3-oxo1cyclohexenyl]3,7,12,16-tetramethyloctadeca-1,3,5,7,9,11,13,15,17-nonaenyl]-2,4,4-trimethyl-1-cyclohex-2-enone) (Fig.7.1).

3. Lutein

3.1 Lutein sources and importance

Lutein is a xanthophyll synthesized by plants and is associated with photosystem II (PSII) in green plants and algae (Jahns and Holzwarth, 2012; Lin et al., 2015). With one or two fatty acids bound to the two hydroxyl-group, lutein is present as

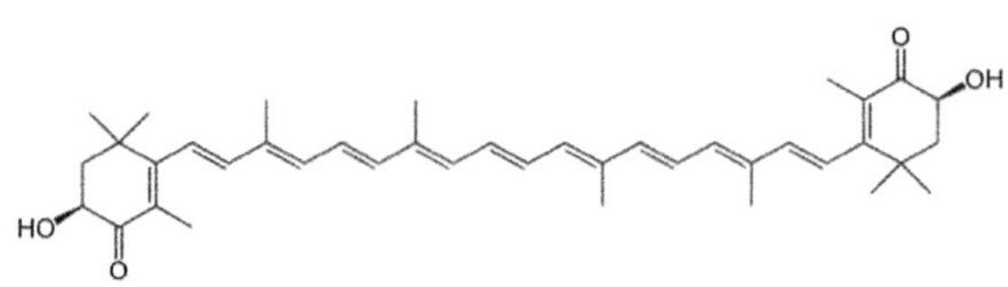

FIG. 7.1

Chemical structure of astaxanthin (ChemSpider ID4444636).

fatty-acid esters in the plants. Lutein protects plants from photo-oxidative damage due to excessive illumination (Frank and Cogdell, 1996). Most of the vegetables and fruits such as broccoli, spinach, kale, corn, orange pepper, kiwi fruit, grapes, orange juice, zucchini, and squash are rich in lutein (Perry et al., 2009). Marigold (*Tagetes erecta* L) flowers are the main source of commercial lutein production. These are grown for business purposes in Mexico, Peru, Ecuador, Spain, India or China (Šivel et al., 2014).

Microalgae have been popularly known as a potential source of naturally occurring lutein (Sun et al., 2016). Therefore, some of the microalgae particularly *Chlorella* has been utilized for lutein production (Shi et al., 1997). Lutein production from microalgae is cost-effective and does not require any conventional agricultural practice. Beside this growth rate of microalgae is higher and can be harvested throughout the year. Different microalgae species found to produce lutein with an intercellular lutein content that is between 6 and 15 times higher than marigold (Del Rio-Chanona et al., 2017). Several microalgae have been proposed as potentially adequate lutein sources, such as *Muriellopsis* sp. (Del Campo et al., 2001), *Chlorella zofingensis* (Del Campo et al., 2000), or *Chlorella protothecoides* (Shi et al., 1997, 2000, 2006; Shi, 2002), *Dunaliella* sp, (Kleinegris et al., 2011; García-González et al., 2005), *Scenedesmus almeriensis* (Sánchez et al., 2008a,b), *Chlorococcum citriforme* and *Neospongiococcus gelatinosum* (Del Campo et al., 2000).

Lutein some time also called "the eye vitamin' as it screen the retina from harmful blue light and protect it from photochemical light damage (Landrum et al., 1996). Beside this, lutein protects the eyes from an age-related disorder such as macular degeneration (AMD) and cataract by a significant decline in the plasma level factor D (Landrum et al., 1996; Kijlstra et al., 2012; Tian et al., 2015; Koushan et al., 2013). Lutein is very rich in antioxidant properties. Sindhu et al. (2010) and Suhler et al., (2008) described the protective role of these antioxidant properties of lutein to protect eyes against lipopolysaccharide (LPS) induced uveitis (common ophthalmic disorder which is responsible for approximately 10% of blindness in western countries (Chao et al., 2015). Kamoshita et al. (2016) reported the role of lutein in multiple antioxidant pathways in the photo-stressed retina. Anti-inflammatory and antitumor properties of lutein are responsible for treating different kinds of cancers (Astorg, 1997) such as prostrate (Rafi et al., 2015; Gunasekera et al., 2007; Chang et al., 2005), breast (Gong et al., 2018; Li et al., 2018; Mignone et al., 2009; Huang et al., 2007), colon (Reynoso-Camacho et al., 2011; Slattery et al., 2000; Männistö et al., 2007) etc. Selective role of lutein in inhibiting cancer cell growth and enhance the effect of chemotherapeutic agents through reactive oxygen species (ros)-mediated mechanisms explain its effectiveness against cancer (Gong et al., 2018).

Commercially lutein has been used for the pigmentation of animal tissues and products, as well as for the coloration of foods, drugs, and cosmetics (Thongkrajay et al., 2006).

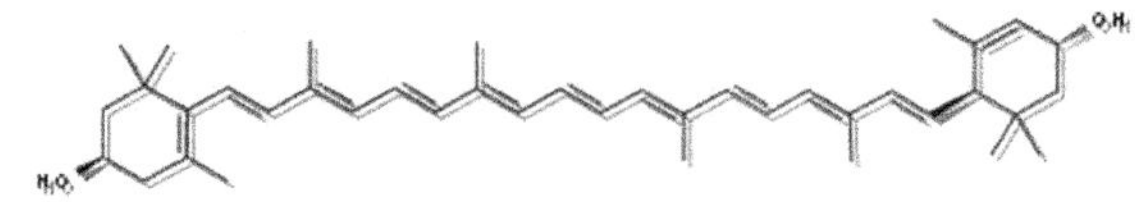

FIG. 7.2

Chemical structure of lutein (ChemSpider ID4444655).

3.2 Chemical structure

Lutein is isomeric with zeaxanthin. The only difference is in the placement of one double bond. Lutein and zeaxanthin can be interconverted in the body through an intermediate called meso-zeaxanthin. (3R,3′R,6′R)-beta, epsilon-carotene-3, 3′-diol is the natural stereoisomer of lutein. Being lipophilic molecule lutein is insoluble in water. Long chromophore of conjugated double bonds (polyene chain) is responsible for the light-absorbing properties of lutein. Heat and light make this polyene chain susceptible to oxidative degradation. This chain is chemically unstable in acids.

The molecular formula for lutein is $C_{40}H_{56}O_2$ and its molecular mass is 568.886 g/mol. The IUPAC name for lutein is (1R)-4--2-en-1-yl]-3,7,12,16-tetramethyloctadeca-1,3,5,7,9,11,13,15,17-nonaenyl]-3,5,5-trimethylcyclohex-3-en-1-ol (Fig.7.2).

4. Biotechnological production of algal high-valued compounds

Biotechnological production is the technological application that uses biological systems, living organisms, or derivatives thereof, to make or modify products or processes for the specific use (UN Convention on Biological Diversity, Art. 2). Although the term ‘Biotechnology’ is coined in 1919 by Hungarian engineer Ereky (1919), it’s not the initiation of biotechnology. Utilization of a biological system by the human being to make valuable products are as old as the history of beverages and bread. It was in 1970 since when the new biotechnology industry grew rapidly and this technique was extended to the manufacturing of many ingredients e.g. flavors and oligosaccharides. Since then, the industrial adoption of biotechnology has grown constantly. Production of bioactive compounds relies upon different growth conditions, including temperature, light, pH, availability, and source of nitrogen, salinity (or ionic strength) and presence of oxidizing substances (or redox potential) etc. These conditions can be obtained by the use of advanced technology with better and sophisticated instrumentation. Thus responsible for the great boost in biotechnological production of these bioactive compounds.

4.1 Astaxanthin

According to the ‘Astaxanthin - Global Market Outlook (2017–23)’ the Global astaxanthin market is expected to grow from $ 615.19 million in 2016 to reach $

1226.14 million by 2023 with a compound annual growth rate (CAGR) of 10.3%. This is due to rising demand for cosmetic products and customers varying preferences toward healthcare & nutraceutical products. Beside this, lack of raw material, inadequate advanced technologies are restricting the market. (https://www.researchandmarkets.com/reports/4426115/astaxanthin-global-market-outlook-2017-2023 # rela0-4605893). Nowadays, the market value of astaxanthin varies usually from \$2500–7000/kg. Most of the astaxanthin available in the market is synthetic astaxanthin due to its low production cost. Synthetic astaxanthin, synthesized from pasta-C_{15}-triarylphosphonium salt and the C_{10} -dialdehyde in a Wittig reaction (Al. & et al., 1973). The synthetic astaxanthin contributes 95% of total astaxanthin in the market. With the growing interest of society toward the natural astaxanthin, algal astaxanthin seems to be gaining potential in the market (Solymosi et al., 2015). Therefore, in this section, we are trying to include various study explaining the effect of various growth factors on astaxanthin production from *Haematococcus pluvialis*, *Chlorella zofingiens* (Table 7.1).

4.1.1 Haematococcus pluvialis

H. pluvialis is known to accumulate up to 3.8% of astaxanthin on the dry weight basis; this number is known to be the highest accumulation as compared to other algal sources. Thus, *H. pluvialis* serve as a main source of astaxanthin in astaxanthin production industries (Orosa et al., 2001). Astaxanthin is present in the resting cells, that are produced and accumulated rapidly when the environmental conditions are unfavorable to normal cell growth (Boussiba et al., 1999). Astaxanthin accumulation is known to be induced under unfavorable growth conditions such as nitrogen and phosphorus starvation (Boussiba et al., 1999; Boussiba and Vonshak, 1991) salt stress (Tjahjono et al., 1994; Sarada et al., 2002) and elevated temperature (Cordero et al., 1996) etc. Olaizola (2000b) presented *H. pluvialis* as the main candidate for the production of astaxanthin for commercial purpose by presenting a nine-month record of production data using the Aquasearch Growth Module (AGM), an enclosed 25,000 L computer-controlled outdoor photobioreactor. This study explained the significance of modification made in production parameters to the viability of commercial production of astaxanthin from high *H. pluvialis*.

Kobayashi et al., (1992) studied the astaxanthin formation in *Haematococcus pluvialis* on a Fe^{2+}-rich acetate medium under the influence of varying dark and light illumination cycles and showed that continuous illumination was most favorable for astaxanthin formation. Further, authors reported the enhanced astaxanthin accumulation in an algal cell by the addition of acetate plus Fe^{2+} than by the addition of acetate alone (Kobayashi et al., 1993). Effect of red cyst cell inoculation and iron (II) supplementation on autotrophic astaxanthin production under outdoor summer conditions studied by the Hong et al. (2016). Evaluation of the growth and astaxanthin production by *H. Pluvialis* in a 30(L)-airlift photobioreactor was performed by Harker et al. (1996). Researchers also evaluated the use of a conventional, reliable autotrophic system to minimize or eliminate the threat posed by potential predators and competitors. In order to promote astaxanthin production in the algae,

Table 7.1 Biotechnological production of astaxanthin from a various algal source.

Microalgae	Culture conditions				Reactor configuration	References
	Temp	pH	Light	Growth medium		
H. pluvialis	20	6.8	Flashing/ continuous red or blue light	Kobayashi's	–	Lababpour et al. (2005)
		7.0	–	–	Airlift column (30 L) indoor/batch	Harker et al. (1996)
	16–34	–	Day light cycle	Modified bold's	Outdoor tubular/ open pond (25,000 L)	Olaizola (2000a)
	25 °C	7.0	Continuous	BG11 & Nitrogen-deficient ($BG11_0$)	Indoor, tubular (50 L) semi continuous	Torzillo et al. (2003)
	25	7.0	Continuous	Basal inorganic	Indoor bubbling column (1.8 L), batch,	Del Río et al. (2005)
	23	7.5	Dark/Light 12:12 h	NIES-C	250-ML Erlenmeyer flask incubated in a photo incubator	Kang et al. (2005)
	20	8.0	–	Inorganic medium Free of Acetate	Airlift tubular (55 L)/ batch	López et al. (2006)
	25	6.4–7.0	Continuous illumination	BG-11	Indoor bubbling column (0.5 L)/ batch	Aflalo et al. (2007)
	24–26	6.4–7.0	Dark/Light 14:10h	BG-11	Outdoor tubular (200 L)/batch	
	23	7.5	Dark/Light 12:12 h	NIES-C	250-mililiters Erlenmeyer flask incubated in a photo Incubator	Kang et al. (2007)

	25	7.0–7.1	Continuous	Basal inorganic	Indoor bubbling column (1.8 L)/ batch	Del Río et al. (2008)
		6.8	Flashing	Kobayashi's basal	Indoor bubbling column (1 L)/batch	Ranjbar et al. (2008)
	20	8.0	Dark/Light 12:12 h	Standard culture	Continuous pilot Scale tubular photo -bioreactor operating under continuous conditions (50 L)	García-Malea et al. (2009)
	15–25	7–10	Light–Dark 14:10 h	BG-11	Open pond, indoor/ batch	Zhang et al. (2009)
	23	7.5	Dark/Light 12:12 h	NIES-C	CO_2-fed flat type photo bioreactor (1 L)/batch culture	Kang et al. (2010)
	22	–	Continuous	OHM	Airlift column indoor (0.5 L)/batch	Choi et al. (2011)
	23	7.5	Dark/Light 12:12 h	NIES-C	Airlift column indoor (6 L)/batch	Yoo et al. (2012)
	20	–	Continuous	Blue-Green Medium -11	Bubbling column (0.6 L) outdoor/ batch	Wang et al. (2013a)
	20	7–8	Continuous	Blue-Green Medium -11	Bubbling column (0.6 L) outdoor/ batch	Wang et al. (2013b)
	25	–	Continuous with varying Intensity	Blue-Green Medium -11	Attached cultivation system	Zhang et al. (2014)
	16–24	7.5	Dark/Light 12:12h	MCM	Photobioreactor (5 L)	Sun et al. (2015)

Continued

Table 7.1 Biotechnological production of astaxanthin from a various algal source.—*cont'd*

Microalgae	Culture conditions				Reactor configuration	References
	Temp	pH	Light	Growth medium		
	8–33	7.5–8.0	Dark/Light 12:12h	NIES-C	Indoor column bioreactor (1 L)	Wan et al. (2014a)
	2–18		Solar Cycle		Open pond	
	23	7.3–7.4	Light/Dark 14:10 h	Blue-Green Medium 11-H	Twin-layer photobioreactor	Kiperstok et al. (2017)
	25	–	12: 12 h	Bold's basal	Glass cylinder photobioreactors	Cheng et al. (2016)
	–	–	Continuous	MES-Volvox E	–	Christian et al. (2018)
	25	7.0	Light/Dark 16:8 h	NSIII	Glass photobioreactor column	Pan-utai et al. (2017)
	20	7.1	Continuous red LED	BG11	250-mililiters flask containing a 150 mililiters	Gao et al. (2017)
	25		Continuous Illumination	OHM	250-mL round-bottomed flask	Kim et al. (2018)
	28	–	Continuous	BAR	Batch (1 mililiters)	Barberà et al. (1993); Domínguez-Bocanegra et al. (2004)
	20	6.8	Effect of different LED with a different wavelength	Kobayashi's basal	200 mililiters Flask (50)/Batch	Katsuda et al. (2004)
	20	6.8	Intermittent flashing light	Kobayashi's basal	200 mililiters Flask (100 mililiters)/Batch	Katsuda et al. (2006)

C. zofingiensis	30	6.5	–	Bristol's	250-mL erlenmeyer flask (250 mililiters)/ batch	Ip et al. (2004)
	30	6.5	Total dark	Modified Bristol's	250-mL erlenmeyer flask (250 mililiters)/ batch	Ip & Chen (2005)
	20–28	–	–	Modified (Arnon et al., 1974)	Cylindrical flasks (1 L) (4.5/batch	Del Campo et al. (2004)
	25	6.5	Continuous	Kuhl's	250- mL erlenmeyer flask/Batch (50 mililiters)	Liu et al. (2013)
	25	6.5	Continuous	Kuhl's	250- mL erlenmeyer flask/Fed Batch (50 mililiters)	Liu et al. (2012)
	25	6.5	Total dark	Bristol's CZ-M1	200-mL erlenmeyer flask containing/ fed-batch (36 mililiters)	Sun et al. (2008)
	27	6.5	Total dark	Bristol's	Batch, fed-batch, and 2-Stage culture experiments/ erlenmeyer flask (50 mililiters)	Chen & Wang (2013)

Abbreviations: BAR- *Media by BAR Barberà et al. (1993);* BG-11 – *Blue-Green Medium; Nitrogen-deficient medium (BG11$_0$);* MCM-*marine culture medium;* MES - *2-(N-Morpholino)ethanesulfonic acid);* NIES *(National Institute for Environmental Studies)culture media;* OHM-*optimal* Haematococcus *medium* researchers suggested the addition of Sodium Chloride at the onset of the stationary phase of algal growth when levels of phosphate and especially nitrogen had been depleted.

In order to increase the cell concentration and the accumulation of astaxanthin Lababpour et al. (2005) used the fed-batch addition of 10-fold-concentrated medium to supply nutrients, as well as illumination with a blue light emitting diodes (LEDs). Significantly higher concentration of astaxanthin production by flashing light illumination than continuous light illumination was obtained by Katsuda et al. (2006). Authors also investigated the effects of the incident light intensity (2−12 μmol m^{-2} s^{-1}), duty cycle (17%−67%) and frequency (25−200 Hz) of flashing light on the cell growth and astaxanthin production. Katsuda et al. (2006) demonstrated flashing light from blue LEDs as a promising illumination method for indoor algal cultivation using photobioreactors. Katsuda et al. (2006) studied the effect of different wavelength light {(red ($\lambda_{max} = 625$ nm), green ($\lambda_{max} = 525$ nm), blue ($\lambda_{max} = 470$ nm), blue-purple ($\lambda_{max} = 410$ nm) and purple ($\lambda_{max} = 380$ nm)} on cell growth and astaxanthin accumulation from *H. pluvialis*.

Torzillo et al. (2003) studied the interplay between physiological and photochemical parameters changes and astaxanthin production in *Haematococcus pluvialis* cultures grown outdoors in photobioreactors under nitrogen deficiency. Authors also recorded the photochemical changes by photosynthetic oxygen evolution and chlorophyll fluorescence measurements.

Del Río et al. (2005) studied the accumulation of astaxanthin in *H. pluvialis* cells growing in a photochemostat under constant irradiance and dilution rate. These authors also studied the influence of nitrate availability on astaxanthin accumulation and productivity and identified the intensive variable controlling the behavior of the system as the specific nitrate input. Researchers also developed the mathematical model to links growth rate with both irradiance and specific nitrate input. This mathematical model was further validated by them (Del Río et al., 2008) by growing *H. pluvialis* cultures in photo-chemostats maintained in continuous growth regime under steady-state conditions, being operated at different forty combinations of irradiance, dilution rate and nitrate concentration in the feed medium and showed their models as helpful tools for efficient production of *Haematococcus* astaxanthin in the one-stop system.

Ranjbar et al. (2008) cultivated the H. *pluvialis* under photoautotrophic conditions in a bubble column with fed-batch addition of nutrients and showed that combination of fed-batch addition of nutrients and dilution of broth for nutrient deficiency enhanced the accumulation of astaxanthin After comparing the outdoor culture of *H. pluvialis* in a tubular photobioreactor and a bubble column under same conditions López et al. (2006) showed tubular reactors to be more suitable for the production of *H. pluvialis* biomass and/or astaxanthin, due to their higher light availability. Authors also showed the maximum astaxanthin production under conditions of nitrate saturation.

According to Aflalo et al. (2007) productivity of astaxanthin in the two-stage system is higher than the single step process, Thus, authors rejected the earlier reports in favor of a single-step process of astaxanthin production by *H. pluvialis*. Interestingly, García-Malea et al. (2009) after studying the feasibility of a one-step method for the continuous production of astaxanthin by the microalga *H. pluvialis* showed that astaxanthin can be efficiently produced outdoors in continuous mode with a suitable dosage of the specific nitrate as well as average irradiance inside the culture.

A new method named "two-stage growth one-step process" has been designed by Zhang et al. (2009). These authors used this simplified method to cultivate four strains of *Haematococcus* (*H. pluvialis* 26; *H. pluvialis* 30; *H. pluvialis* 34; *H. pluvialis* WZ). *pluvialis* for astaxanthin production by closely monitoring the cell density and medium pH, and also the astaxanthin content and yield of the four *strain*.

Kang et al. (2005) showed photoautotrophic induction is more effective than heterotrophic induction for astaxanthin synthesis in *H. pluvialis*. Another study on the effect of C/N ratio and light intensity on astaxanthin production by Kang et al. (2007) reported that the light intensity is more critical than the C/N ratio. These authors further proposed that the sequential photoautotrophic process (fed-batch culture process followed by subsequent photoautotrophic induction) has a high potential for the production of astaxanthin (Kang et al., 2010).

Astaxanthin productivities are correlated with growth phases, therefore, Choi et al. (2011) optimized the cellular phase for transferring into the second production stage and showed that senescent green phases could increase more than 10-fold productivity of Astaxanthin.

Yoo et al. (2012) optimized the slope of a V-shaped bottom of a photo-bioreactor, the volumetric flow rate (vvm) of air, the height/diameter (H/D) ratio, and the diameter of an air sparger for phototropic culturing of *H. pluvialis* and also evaluated the efficacy of the optimized factors for long-term culture test.

Wang et al. (2013a) studied the impact of initial biomass density on cell mortality, morphology, growth, pigment content, and productivity of biomass at the red stage growth in the outdoor glass columns. Authors further studied the effect of seasonal changes in light intensity and temperature on the relationship between initial biomass density and growth and astaxanthin production. In another study, these authors also explained the impact of initial biomass density and nitrogen concentration on growth and astaxanthin production of *H. pluvialis* and also optimized the initial biomass density under nitrogen limited (i.e., 4.4 mM nitrate) conditions which resulted in the highest astaxanthin productivity under outdoor glass column photobioreactor Wang et al., (2013b).

Zhang et al. (2014) cultivated *H. pluvialis* using an immobilized biofilm method, viz. "Attached Cultivation'. Authors also investigated the impact of inoculum density, light intensity, nitrogen quantity as well as the medium volume on growth and astaxanthin accumulation in immobilized *H. pluvialis.*

Wan et al. (2014b) studied the impact of temperature on cell growth and astaxanthin accumulation during a light–dark cyclic cultivation in *H. pluvialis* and suggested 23–28°C day temperature and below 28°C temperature for good production of astaxanthin.

Sun et al. (2015) investigated Fed-batch culture and the transformation conditions of *H. pluvialis* in a 5 L photobioreactor. Their application of mixed lights of blue and white at different ratios to determine the effects of lights on the process of astaxanthin-accumulation revealed that mixed lights were better than a single light. Kiperstok et al. (2017) used Biofilm cultivation of *H. pluvialis* that enables a highly productive one-phase process for astaxanthin production using high light intensities.

Recently, Christian et al. (2018) reported the enhanced astaxanthin accumulation in *H. pluvialis* by using high carbon dioxide concentration and light illumination without a medium replacement step.

Effect of nuclear irradiation and high concentration of carbon dioxide stress enhanced the growth rate and astaxanthin was studied by Cheng et al. (2016) and these authors reported 15% higher growth rate and astaxanthin production in mutant than that of the original strain.

Pan-utai et al. (2017) suggested the use of potassium acetate as an organic carbon source for use as an inducing agent in mixotrophic cultivations for astaxanthin accumulation by *H. pluviali*.

Gao et al. (2017) developed a Type II model of light distribution for green cells of the microalgae cultivated under red LED lights.

Interestingly, Kim et al. (2018) reported the direct impact of electrical treatment on the algal growth. By comparing the growth as different voltage these authors found 100 mA (voltage: 25 V) as an optimum current and periodic electrical treatment more effective than the single treatment.

Vo et al. (2016) explored the effects of 1-aminocyclopropane-1-carboxylic acid on growth stages of *H. pluvialis* and observed enhanced astaxanthin production whenever the cells were treated at the early red stage. This study also confirmed the combinatorial synergistic effect of ACC and light intensity.

4.1.2 *Chlorella zofingiens*

Although not much explored *C. zofingiens* is also a potential source of astaxanthin. *C. zofingiensis* is a fast growing alga and can be a culture in phototrophic heterotrophic as well as mixotrophic conditions besides this, it is easy to be cultured and scaled up both indoors and outdoors than *H. pluvialis* (Orosa et al., 2001; Liu et al., 2014).

Del Campo et al. (2004) represents *C. zofingiens* CCAP 211/14 strain a unique model system for advancing the knowledge of differential regulation during the synthesis of astaxanthin in response to a variety of nutritional as well as environmental factors. Authors also investigated the influence of temperature, light, nitrate concentration and salinity on the autotrophic growth and production of total and specific carotenoids including astaxanthins by *C. zofingiensis* CCAP 211/14, to evaluate its potential as a source of astaxanthin.

Ip et al. (2004) studied the effect of various concentrations of glucose and nitrate on the cultivation of *C. zofingiensis* in mixotrophic culture and showed the favorable effect of high glucose and low nitrate concentrations on the production of astaxanthin In another study, authors also showed higher production of astaxanthin by cultivating *C. zofingiens* in dark. Hence, researchers proposed this light-independent astaxanthin-producing ability of *C. zofingiensis* can be potentially employed for commercial production of astaxanthin on a large-scale (Ip and Chen, 2005).

Sun et al. (2008) studied the effect of various sugars such as glucose, mannose, fructose, sucrose, galactose, and lactose on cell growth as well as astaxanthin production in dark and reported glucose and mannose as the best carbon sources for the algal growth in the dark with high algal specific growth rates.

Liu et al. (2012, 2013) used the cane molasses in heterotrophic batch cultivation *Chlorella zofingiensis*, Thus proposed a cost-effective method for production of astaxanthin.

Chen and Wang (2013) identified and optimized the major factors affecting both cell growth and astaxanthin production of *Chlorella zofingiensis* under dark According to them glucose, $NaNO_3$, and $MgSO_4 \cdot 7H_2O$ were found to be the most important in affecting *C. zofingiensis*.

4.2 Lutein

Over the past few years, the lutein market has largely been driven by the growing demand for lutein for eye health supplements. This has been estimated to reach USD 263.8 million by 2017 and is projected to reach USD 357.7 million by 2022, growing at a compound annual growth rate of 6.3% from 2017(https://www.researchandmarkets.com/research/rhkmrk/global_lutein?w=5). As describe earlier (in section 2.5) there are many algal species used for the production of lutein but among these reported microalgae, only *Murielopsis sp.*, *S. almeriensis*, and *Chlorella* sp. have been tested in growth conditions that could be considered for mass production. Therefore, many researchers studied the effect of various factors such as temperature, pH, light, nutrients etc. on biomass growth and lutein yield in order to standardize the optimum conditions for algal growth and lutein yield (Table 7.2).

4.2.1 *Murielopsis sp*

Del Campo et al. (2000) compared the carotenoid profile (especially the lutein) in 15 strains of chlorophycean microalgae and reported highest lutein content (up to 35 mg L^{-1}culture), growth rate (up to 0.17–0.23 h^{-1}) and maximal standing cell density (up to 8×1010 cells 1–1culture) in the *Muriellopsis* sp. In this study, researchers also reported 20–40 mM sodium nitrate, 2–100 mM sodium chloride, 460 mmol photon $m^{-2}s^{-1}$, pH 6.5 and 28 °C are the best culture condition for maximum g level of lutein.

Del Campo et al. (2001) studied the effect of dilution rate, mixing and daily solar cycles on lutein and biomass productivity of the green unicellular alga *Muriellopsis* sp. throughout the year, in an outdoor tubular photobioreactor and reported the highest productivity in May and July. Authors also reported optimal dilution rate was found to be lower in May (0.06 h^{-1}) than in November (0.09 h^{-1}). No variations

Table 7.2 Biotechnological production of lutein from a various algal source.

Microalga source	Processing conditions				Reactor configuration	References
	Temp(°C)	pH	Light cycle	Culture medium		
Muriellopsis sp.	28	7	Continuous	Modified (Arnon et al., 1974)	Outdoor tubular photo bioreactor 55L/Batch culture	Del Campo et al. (2001)
		9	Continuous	Modified (Arnon et al., 1974)	batch culture 0.2 L	Del Campo et al. (2000)
	Not controlled	7.5–9.5	Not controlled	Modified media (Arnon et al., 1974) –	Outdoors, open tank, semi continuous culture, 100 L	Blanco et al. (2007)
Scenedesmus almeriensis	30–40	8	Day light	Modified (Mann and Myers, 1968)	Bubble column photo bioreactors (2.0 L) discontinuous and continuous mode	Sánchez et al. (2008a)
	35	8	Solar cycle	Modified (Mann and Myers, 1968)	Outdoors, tubular system, continuous culture, 4,000 L	Sánchez et al. (2008b)
Scenedesmus obliquus	20–30	6–8	Continuous illumination	Optimal *Haematococcus* medium (OHM)	Glass bioreactor (2 L) Batch culture	Guedes et al. (2011b)
Chlorella vulgaris UTEX 265	4–10	6.8	14–20 h light	Bold's basal (BBM)	Coiled tubular tree photo bioreactor	Gong & Bassi (2017)
Chlorella sorokiniana	28	–	Dark/Light 12:12h	Modified (Arnon et al., 1974)	Batch (1 L)	Cordero et al. (2011)
Chlorella protothecoides	28	6.5	Dark	Modified basal	Batch (16 L)	Wei et al. (2008)
	24, 28, 32	6.6	Dark	Modified basal	Batch (3.7- L and 30- L fermenters)	Shi (2002)
	26	5.0–8.0	Dark	Modified basal	16- L fermenters	Shi et al. (2006)
	24–40	6.6	Dark	Modified basal	3.7- L fermenters	

Chlorella zofingiensis ***Chlorococcum citriforme*** ***Neospongiococc us gelatinosum***	28	7	Continuous illumination	Modified (Arnon et al., 1974)	Batch (0.2 L)	Del Campo et al. (2000)
Coccomyxa onubensis	28	25	White fluorescent lamps	K9 (Silverman and Lundgren, 1959)	Batch (1 L)	Vaquero et al. (2014)
Dunaliella salina	25	8	Continuous illumination	Indoor- (Shaish et al., 1992) Outdoor-F2	Closed tubular photo bioreactor Semi-continous	García-González et al. (2005)
	–	6.5 and 7.5	–	Gg-8	Cylindrical bubble column 300 mililiter/Semi-continous	Fu et al. (2014)
Heterochlorella luteoviridis	22, 27 or 32	8.5	Dark/Light 12/12 h	f/2 Culture Media (Guillard, 1975)	Flat-plate airlift photo bioreactors (PBRs) with 2.4 L working volum	Menegol et al. (2017)

were found in photosynthetic efficiency (about 4%) values throughout the year. During the early hours of daylight authors reported a fast increase in lutein content with maximal lutein content (about 6 mg (g dry weight)$^{-1}$) being recorded at noon and decreasing slowly.

Another interesting research by Blanco et al. (2007) studied the effect of various nutritional and environmental factors on growth and accumulation of lutein in outdoor cultures of *Muriellopsis sp.* in open tanks agitated with a paddlewheel. These factors include the Effect of daily solar irradiance, seasonal solar irradiance, cell suspension depth (light path), pH value, sodium chloride etc.

4.2.2 Scenedesmus *sp.*

Sánchez et al. (2008b) investigated the optimum culture conditions for *S. almeriensis* in terms of temperature, pH, culture medium, external irradiance, and salinity. Authors reported 30–40 °C as optimal temperature and 8.0 as an optimal pH. According to them, low medium salinities (0–5g NaCl/L) are appropriate to promote growth rate. For better yield, authors suggested the culture medium proposed by Mann and Myers (1968). Authors did not see any photoinhibition even at high irradiances up to 1700 μE/m^{2}s by chlorophylls fluorescence. Thus, *S. almeriensis* is suitable to grow under outdoor conditions.

According to the Guedes et al. (2011a,b) , lutein production from microalgae *Scenedesmus obliquus* depends upon the various factors including temperature, light, pH, nitrogen source etc. According to this study, pH and temperature significantly affect biomass growth and antioxidant compound production. Biomass growth is favored by low pH and high temperature, whereas lutein production is enhanced by high pH (i.e., 30 °C and pH 8), i.e. 203.57 ± 1.41 mg mL^{-1} and 18.20 ± 0.33 mg mL^{-1}, respectively. Therefore, the novel strain studied holds a promising application as a functional ingredient for food or cosmetic/pharmaceutical purposes.

4.2.3 Chlorella *sp.*

Shi et al. (2000) investigated the influence of various nitrogen sources (nitrate, ammonium, and urea) on heterotrophic growth of *Chlorella protothecoides* and reported no significant inhibitory effect on the algal growth over a nitrogen concentration range of 0.85–1.7 g/l, but showed differences in specific growth rate and biomass production among various nitrogen compounds. Culture with urea as nitrogen source showed highest yields of both biomass and lutein.

Shi (2002) showed enhancement of lutein production as a result of a combination of nitrogen limitation and high-temperature stress in heterotrophically grown *Chlorella protothecoides. Chlorella protothecoides* was grown in batch mode in a 3.7-L fermenter containing 40 g/L glucose and 3.6 g/L urea.

Shi et al. (2006) investigated the influence of various temperature (24, 26, 28, 30, 35, and 40 °C) as well as pH (5–8) on the growth of *Chlorella protothecoides* and yield of lutein. Authors found larger cells at 35 or 40 °C; maximum biomass was reported at 24–30 °C *i.e.* 16.2–17.4 g/L dry cells. Authors further demonstrated 28 °C of temperature and 6.6 pH is good for better lutein yield and growth.

Wei et al. (2008) investigated the effects of two oxidant-forming reactive oxygen species (ROS) on the biomass concentration, and yield and content of lutein in batch culture of heterotrophic *Chlorella protothecoides*. According to, Wei et al. (2008), lower concentrations of ROS stimulate lutein production, whereas higher concentrations of ROS inhibit the lutein production.

Cordero et al. (2011) studied the effects of several nutritional and environmental factors on cell growth of *Chlorella sorokiniana* lutein accumulation and attained the maximal specific growth rate and lutein content at 690 μmol photons m^{-2} s^{-1}, 28 °C, 2 mM NaCl, 40 mM nitrate and under mixotrophic conditions. Authors also obtained high lutein yielding mutants by mutagen N-methyl-N′-nitro-nitrosoguanidine (MNNG). Selection of mutants were done based on their resistance to the carotenogenic pathway nicotine and norflurazon inhibitor Out of many selected mutants, mutant MR-16 has a 2.0-fold higher lutein content than that of the wild-type.

Recently, Gong and Bassi (2017) investigated the influence and interaction of light irradiance strength, lighting cycle, and temperature on cold tolerant *Chlorella vulgaris* UTEX 265 cultivation and lutein production efficiency at a low-temperature range in the Coiled Tree Photo-Bioreactor (CTPBR) and in flasks via response surface method (RSM). Authors proposed coiled tubular tree photobioreactor (CTPBR) configuration as an efficient photo-bioreactor for cultivating microalgae at low light at 22 C or at very low temperatures and at a light intensity of 470 μmol photons m^{-2} s^{-1}. Authors reported fast growth ($\mu = 1.50$ day^{-1}) and good lutein recovery (11.98 mg g^{-1} day^{-1}) in Coiled tree photo-bioreactor at the low irradiance stress condition and the low-temperature photo-inhibition conditions.

4.2.4 *Coccomyxa onubensis*

An interesting study by Vaquero et al. (2014) on *C. onubensis* an alga with acid-tolerant behavior showed the best growth and carbon uptake capacity at acidic pH.

4.2.5 *Dunaliella salina*

Fu et al. (2014) studied the impact of three abiotic stress variables (nitrate concentration, salinity, and light quality) on lutein production in batch cultures of *Dunaliella salina* by using response surface analysis. Authors further reported that *D. salina* was more tolerant to hyperosmotic stress than to hypo-osmotic stress. Besides this, authors showed the strong influence of interaction between nitrate concentration and salinity than between light quality and the other two on lutein production.

4.2.6 *Heterochlorella luteoviridis*

Menegol et al. (2017) evaluated the effect of different temperatures (22, 27 or 32 C) and sodium nitrate concentrations (12, 24, 36, 48 or 60 mg L-1 of N–NO3) in culture medium on *H. luteoviridis* biomass production and lutein contents. This study reported the high yield, high biomass concentration under autotrophic conditions.

5. Conclusion and future directions

In the present chapter, an attempt has been made to summarize the various studies performed to enhance the production of astaxanthin and lutein two high valued compounds from their different algal sources, under changing growth conditions such as temperature, light, nutrient supplements etc. Production of these compounds from their algal source has many advantages as these sources are renewable and production of valuable compounds are environmental friendly as well as it is easy to scale up the synthesis process by simply adjusting the growth conditions. However, the production cost of these compounds is still high and hence the market value.

In order to reduce the market price of these compounds, the production of biomass and the yield of these bioactive compounds must be improved. Therefore, scientists should focus their research on exploring new species with better yield and better adaptive strength in changing environment conditions. Beside this, algal species which are currently in use can be improved by genetic engineering for higher yield of bioactive compounds. To reduce losses during processing advances in processing techniques are necessary.

Acknowledgments

Dr. Balraj Singh Gill thanks department of higher education, Shimla, Himachal Pradesh, India for providing the necessary facilities to carry out the present work.

Disclosure of potential conflicts of interest

Authors declare no potential conflicts of interest.

References

Al, K., et al., 1973. United States Patent 19, pp. 4–8.

Aflalo, C., Meshulam, Y., Zarka, A., Boussiba, S., 2007. Determination of heat flux on dual bell nozzle by Monte Carlo method. Biotechnol. Bioeng. 98 (1), 300–305.

Ambati, R.R., Moi, P.S., Ravi, S., Aswathanarayana, R.G., 2014. Astaxanthin: sources, extraction, stability, biological activities and its commercial applications – a review. Mar. Drugs 12 (1), 128–152.

Arnon, D.I., McSwain, B.D., Tsujimoto, H.Y., Wada, K., 1974. Photochemical activity and components of membrane preparations from blue-green algae. I. Coexistence of two photosystems in relation to chlorophyll a and removal of phycocyanin. Biochim. Biophys. Acta Bioenerg. 357 (2), 231–245.

Astorg, P., 1997. Food carotenoids and cancer prevention: an overview of current research. Trends Food Sci. Technol. 8 (12), 406–413.

Augusti, P.R., Quatrin, A., Somacal, S., Conterato, G.M., Sobieski, R., et al., 2012. Astaxanthin prevents changes in the activities of thioredoxin reductase and paraoxonase in hypercholesterolemic rabbits. J. Clin. Biochem. Nutr. 51 (1), 42–49.

Barberà, E., Tomàs, X., Moya, M.-J., Ibàñez, A., Molins, M.-B., 1993. Significance tests in the study of the specific growth rate of Haematococcus lacustris: influence of carbon source and light intensity. J. Ferment. Bioeng. 76 (5), 403–405.

Bennedsen, M., Wang, X., Willen, R., Wadstrom, T., Andersen, L.P., 1999. Treatment of *H. pylori* infected mice with antioxidant astaxanthin reduces gastric inflammation, bacterial load and modulates cytokine release by splenocytes. Immunol. Lett. 70 (3), 185–189.

Bhuvaneswari, S., Arunkumar, E., Viswanathan, P., Anuradha, C.V., 2010. Astaxanthin restricts weight gain, promotes insulin sensitivity and curtails fatty liver disease in mice fed a obesity-promoting diet. Process Biochem. 45 (8), 1406–1414.

Bilal, M., Rasheed, T., Ahmed, I., Iqbal, H.M.N., 2017. High-value compounds from microalgae with industrial exploitability – a review. Front. Biosci. - Sch. 9 (3), 319–342.

Blanco, A.M., Moreno, J., Del Campo, J.A., Rivas, J., Guerrero, M.G., 2007. Outdoor cultivation of lutein-rich cells of Muriellopsis sp. in open ponds. Appl. Microbiol. Biotechnol. 73 (6), 1259–1266.

Borowitzka, M.A., 2013. High-value products from microalgae-their development and commercialisation. J. Appl. Phycol. 25 (3), 743–756.

Boussiba, S., Vonshak, A., 1991. Astaxanthin accumulation in the green alga haematococcus pluvialis1. Plant Cell Physiol. 32 (7), 1077–1082.

Boussiba, S., Bing, W., Yuan, J.P., Zarka, A., Chen, F., 1999. Changes in pigments profile in the green alga Haeamtococcus pluvialis exposed to environmental stresses. Biotechnol. Lett. 21 (7), 601–604.

Cardozo, K.H.M., Guaratini, T., Barros, M.P., Falcão, V.R., Tonon, A.P., et al., 2007. Metabolites from algae with economical impact. Comp. Biochem. Physiol. C Toxicol. Pharmacol.

Chang, S., Erdman, J.W.J., Clinton, S.K., Vadiveloo, M., Strom, S.S., et al., 2005. Relationship between plasma carotenoids and prostate cancer. Nutr. Canc. 53 (2), 127–134.

Chao, S.-C., Vagaggini, T., Nien, C.-W., Huang, S.-C., Lin, H.-Y., 2015. Effects of lutein and zeaxanthin on LPS-induced secretion of IL-8 by uveal melanocytes and relevant signal pathways. J. Ophthalmol. 2015, 152854.

Chapman, R.L., 2013. Algae: the world's most important "plants"—an introduction. Mitig. Adapt. Strategies Glob. Change 18 (1), 5–12.

Chen, T., Wang, Y., 2013. Optimized astaxanthin production in Chlorella zofingiensis under dark condition by response surface methodology. Food Sci. Biotechnol. 22 (5), 1–8.

Cheng, J., Zongbo Yang, K.L., Junhu, Z., Kefa Cen, S., 2016. Enhancing the growth rate and astaxanthin yield of Haematococcus pluvialis by nuclear irradiation and high concentration of carbon dioxide stress. Bioresour. Technol. 204, 49–54.

Choi, Y.E., Yun, Y.S., Park, J.M., Yang, J.W., 2011. Determination of the time transferring cells for astaxanthin production considering two-stage process of Haematococcus pluvialis cultivation. Bioresour. Technol. 102 (24), 11249–11253.

Christian, D., Zhang, J., Sawdon, A.J., Peng, C.A., 2018. Enhanced astaxanthin accumulation in Haematococcus pluvialis using high carbon dioxide concentration and light illumination. Bioresour. Technol. 256, 548–551.

Coltelli, P., Barsanti, L., Evangelista, V., Frassanito, A.M., Gualtieri, P., 2014. Water monitoring: automated and real time identification and classification of algae using digital microscopy. Environ. Sci. Process. Impacts. 16 (11), 2656–2665.

Coombes, J.S., Sharman, J.E., Fassett, R.G., 2016. Astaxanthin has no effect on arterial stiffness, oxidative stress, or inflammation in renal transplant recipients: a randomized controlled trial (the XANTHIN trial). Am. J. Clin. Nutr. 103 (1), 283–289.

Cooper, R.D., Davis, B., Leftwick, P., Price, C., Weedon, C., 1975. Carotenoids and related compounds. Part XXXII. synthesis of astaxanthin, phoenicoxanthin, hydroxyechinenone, and the corresponding diosphenols. J. Chem. Soc., Perkin Trans. 1, 2195–2204.

Cordero, B., Otero, A., Patiño, M., Arredondo, B.O., Fabregas, J., 1996. Astaxanthin production from the green alga Haematococcus pluvialis with different stress conditions. Biotechnol. Lett. 18 (2), 213–218.

Cordero, B.F., Obraztsova, I., Couso, I., Leon, R., Vargas, M.A., Rodriguez, H., 2011. Enhancement of lutein production in Chlorella sorokiniana (chorophyta) by improvement of culture conditions and random mutagenesis. Mar. Drugs 9 (9), 1607–1624.

Davinelli, S., Nielsen, M.E., Scapagnini, G., 2018. Astaxanthin in skin health, repair, and disease: a comprehensive review. Nutrients 10 (4), 1–12.

Del Campo, J.A., Moreno, J., Rodríguez, H., Angeles Vargas, M., Rivas, J., Guerrero, M.G., 2000. Carotenoid content of chlorophycean microalgae: factors determining lutein accumulation in Muriellopsis sp. (Chlorophyta). J. Biotechnol. 76 (1), 51–59.

Del Campo, J.A., Rodríguez, H., Moreno, J., Vargas, M.Á., Rivas, J., Guerrero, M.G., 2001. Lutein production by Muriellopsis sp. in an outdoor tubular photobioreactor. J. Biotechnol. 85 (3), 289–295.

Del Campo, J.A., Rodríguez, H., Moreno, J., Vargas, M.Á., Rivas, J., Guerrero, M.G., 2004. Accumulation of astaxanthin and lutein in Chlorella zofingiensis (Chlorophyta). Appl. Microbiol. Biotechnol. 64 (6), 848–854.

Del Río, E., Acién, F.G., García-Malea, M.C., Rivas, J., Molina-Grima, E., Guerrero, M.G., 2005. Efficient one-step production of astaxanthin by the microalga haematococcus pluvialis in continuous culture. Biotechnol. Bioeng. 91 (7), 808–815.

Del Río, E., Acién, F.G., García-Malea, M.C., Rivas, J., Molina-Grima, E., Guerrero, M.G., 2008. Efficiency assessment of the one-step production of astaxanthin by the microalga Haematococcus pluvialis. Biotechnol. Bioeng. 100 (2), 397–402.

Del Rio-Chanona, E.A., Fiorelli, F., Zhang, D., Ahmed, N.R., Jing, K., Shah, N., 2017. An efficient model construction strategy to simulate microalgal lutein photo-production dynamic process. Biotechnol. Bioeng. 114 (11), 2518–2527.

Domínguez-Bocanegra, A.R., Guerrero Legarreta, I., Martinez Jeronimo, F., Tomasini Campocosio, A., 2004. Influence of environmental and nutritional factors in the production of astaxanthin from Haematococcus pluvialis. Bioresour. Technol. 92 (2), 209–214.

Dufossé, L., Galaup, P., Yaron, A., Arad, S.M., Blanc, P., 2005. Microorganisms and microalgae as sources of pigments for food use: a scientific oddity or an industrial reality? Trends Food Sci. Technol 16 (9), 389–406.

Ereky, K., 1919. Biotechnologie der Fleisch-, Fett-, und Milcherzeugung im landwirtschaftlichen Grossbetriebe : für naturwissenschaftlich gebildete Landwirte verfasst Berlin (Catalog record from Hathi Library.

Fan, K.W., Chen, F., 2007. Production of high-value products by marine microalgae thraustochytrids. Bioprocess. Value-Added Prod. from Renew. Resour. 293–323.

Fassett, R.G., Coombes, J.S., 2011. Astaxanthin: a potential therapeutic agent in cardiovascular disease. Mar. Drugs 9 (3), 447–465.

Fen, Y., Hyeon, S., Jo, M., Bae, J., Duck, H., Gyoon, W., 2013. And zeaxanthin on cytochrome P450 enzyme activities. Food Chem. Toxicol. 59, 78–85.

Frank, H.A., Cogdell, R.J., 1996. Carotenoids in photosynthesis. Photochem. Photobiol. 63 (3), 257–264.

Fu, W., Paglia, G., Magnúsdóttir, M., Steinarsdóttir, E.A., Gudmundsson, S., Palsson, B., Andrésson, Ó.S., Brynjólfsson, S., 2014. Effects of abiotic stressors on lutein production in the green microalga Dunaliella salina. Microb. Cell Factories 13 (3), 1–9.

Gacheva, G.V., Gigova, L.G., 2014. Biological activity of microalgae can be enhanced by manipulating the cultivation temperature and irradiance. Cent. Eur. J. Biol. 9 (12), 1168–1181.

Gao, X., Wang, X., Li, H., Roje, S., Sablani, S.S., Chen, S., 2017. Parameterization of a light distribution model for green cell growth of microalgae: haematococcus pluvialis cultured under red LED lights. Algal Res 23, 20–27.

García-González, M., Moreno, J., Manzano, J.C., Florencio, F.J., Guerrero, M.G., 2005. Production of Dunaliella salina biomass rich in 9-cis-β-carotene and lutein in a closed tubular photobioreactor. J. Biotechnol. 115 (1), 81–90.

García-Malea, M.C., Gabriel Acién, F., Río, E Del, Fernández, J.M., Cerón, M.C., et al., 2009. Production of astaxanthin by haematococcus Pluvialis: taking the one-step system outdoors. Biotechnol. Bioeng. 102 (2), 651–657.

Gong, M., Bassi, A., 2017. Investigation of Chlorella vulgaris UTEX 265 cultivation under light and low temperature stressed conditions for lutein production in flasks and the coiled tree photo-bioreactor (CTPBR). Appl. Biochem. Biotechnol. 183 (2), 652–671.

Gong, X., Smith, J.R., Swanson, H.M., Rubin, L.P., 2018. Carotenoid lutein selectively inhibits breast cancer cell growth and potentiates the effect of chemotherapeutic agents through ROS-mediated mechanisms. Molecules 23 (4), 905.

Guedes, A.C., Amaro, H.M., Malcata, F.X., 2011a. Microalgae as sources of carotenoids. Mar. Drugs 9 (4), 625–644.

Guedes, A.C., Amaro, H.M., Pereira, R.D., Malcata, F.X., 2011b. Effects of temperature and pH on growth and antioxidant content of the microalga Scenedesmus obliquus. Biotechnol. Prog. 27 (5), 1218–1224.

Guerin, M., Huntley, M.E., Olaizola, M., 2003. Haematococcus astaxanthin: applications for human health and nutrition. Trends Biotechnol. 21 (5), 210–216.

Guillard, R.R.L., 1975. Culture of phytoplankton for feeding marine invertebrates. In: Smith, W.L., Chanley, M.H. (Eds.), Culture of Marine Invertebrate Animals: Proceedings — 1st Conference on Culture of Marine Invertebrate Animals Greenport. Springer US, Boston, MA, pp. 29–60.

Gunasekera, R.S., Sewgobind, K., Desai, S., Dunn, L., Black, H.S., et al., 2007. Lycopene and lutein inhibit proliferation in rat prostate carcinoma cells. Nutr. Canc. 58 (2), 171–177.

Haines, D.D., Varga, B., Bak, I., Juhasz, B., Mahmoud, F.F., et al., 2011. Summative interaction between astaxanthin, Ginkgo biloba extract (EGb761) and vitamin C in suppression of respiratory inflammation: a comparison with ibuprofen. Phytother Res. 25 (1), 128–136.

Harker, M., Tsavalos, A.J., Young, A.J., 1996. Autotrophic growth and carotenoid production of Haematococcus pluvialis in a 30 liter air-lift photobioreactor. J. Ferment. Bioeng. 82 (2), 113–118.

Higuera-Ciapara, I., Félix-Valenzuela, L., Goycoolea, F.M., 2006. Astaxanthin: a review of its chemistry and applications. Crit. Rev. Food Sci. Nutr. 46 (2), 185–196.

Hong, M.E., Choi, Y.Y., Sim, S.J., 2016. Effect of red cyst cell inoculation and iron(II) supplementation on autotrophic astaxanthin production by Haematococcus pluvialis under outdoor summer conditions. J. Biotechnol. 218, 25–33.

Huang, J., Zhang, M., Holman, C.D.J., Xie, X., 2007. Dietary carotenoids and risk of breast cancer in Chinese women. Asia Pac. J. Clin. Nutr. 16 (Suppl. 1), 437–442.

Hussein, G., Sankawa, U., Goto, H., Matsumoto, K., Watanabe, H., 2006. Astaxanthin, a carotenoid with potential in human health and nutrition. J. Nat. Prod. 69 (3), 443–449.

Hussein, G., Nakagawa, T., Goto, H., Shimada, Y., Matsumoto, K., et al., 2007. Astaxanthin ameliorates features of metabolic syndrome in SHR/NDmcr-cp. Life Sci. 80 (6), 522–529.

Ip, P.F., Chen, F., 2005. Production of astaxanthin by the green microalga Chlorella zofingiensis in the dark. Process Biochem. 40 (2), 733–738.

Ip, P.F., Wong, K.H., Chen, F., 2004. Enhanced production of astaxanthin by the green microalga *Chlorella zofingiensis* in mixotrophic culture. Process Biochem. 39 (11), 1761–1766.

Jahns, P., Holzwarth, A.R., 2012. The role of the xanthophyll cycle and of lutein in photoprotection of photosystem II. Biochim. Biophys. Acta Bioenerg. 1817 (1), 182–193.

Jyonouchi, H., Sun, S., Tomita, Y., Gross, M.D., 1995. Astaxanthin, a carotenoid without vitamin A activity, augments antibody responses in cultures including T-helper cell clones and suboptimal doses of antigen. J. Nutr. 125 (10), 2483–2492.

Kamath, B.S., Srikanta, B.M., Dharmesh, S.M., Sarada, R., Ravishankar, G.A., 2008. Ulcer preventive and antioxidative properties of astaxanthin from Haematococcus pluvialis. Eur. J. Pharmacol. 590 (1–3), 387–395.

Kamoshita, M., Toda, E., Osada, H., Narimatsu, T., Kobayashi, S., et al., 2016. Lutein acts via multiple antioxidant pathways in the photo-stressed retina. Sci. Rep. 6, 30226.

Kang, C.D., Lee, J.S., Park, T.H., Sim, S.J., 2005. Comparison of heterotrophic and photoautotrophic induction on astaxanthin production by Haematococcus pluvialis. Appl. Microbiol. Biotechnol. 68 (2), 237–241.

Kang, C.D., Lee, J.S., Park, T.H., Sim, S.J., 2007. Complementary limiting factors of astaxanthin synthesis during photoautotrophic induction of Haematococcus pluvialis: C/N ratio and light intensity. Appl. Microbiol. Biotechnol. 74 (5), 987–994.

Kang, C.D., Han, S.J., Choi, S.P., Sim, S.J., 2010. Fed-batch culture of astaxanthin-rich Haematococcus pluvialis by exponential nutrient feeding and stepwise light supplementation. Bioproc. Biosyst. Eng. 33 (1), 133–139.

Katiyar, R., Gurjar, B.R., Biswas, S., Pruthi, V., Kumar, N., Kumar, P., 2017. Microalgae: an emerging source of energy based bio-products and a solution for environmental issues. Renew. Sustain. Energy Rev. 72, 1083–1093.

Katsuda, T., Lababpour, A., Shimahara, K., Katoh, S., 2004. Astaxanthin production by Haematococcus pluvialis under illumination with LEDs. Enzym. Microb. Technol. 35 (1), 81–86.

Katsuda, T., Shimahara, K., Shiraishi, H., Yamagami, K., Ranjbar, R., Katoh, S., 2006. Effect of flashing light from blue light emitting diodes on cell growth and astaxanthin production of Haematococcus pluvialis. J. Biosci. Bioeng. 102 (5), 442–446.

Kijlstra, A., Tian, Y., Kelly, E.R., Berendschot, T.T.J.M., 2012. Lutein: more than just a filter for blue light. Prog. Retin. Eye Res. 31 (4), 303–315.

Kim, Y.J., Kim, Y.A., Yokozawa, T., 2009. Protection against oxidative stress, inflammation, and apoptosis of high-glucose-exposed proximal tubular epithelial cells by astaxanthin. J. Agric. Food Chem. 57 (19), 8793–8797.

Kim, J.Y., Lee, C., Jeon, M.S., Park, J., Choi, Y.E., 2018. Enhancement of microalga Haematococcus pluvialis growth and astaxanthin production by electrical treatment. Bioresour. Technol. 268, 815–819.

Kiperstok, A.C., Sebestyén, P., Podola, B., Melkonian, M., 2017. Biofilm cultivation of Haematococcus pluvialis enables a highly productive one-phase process for astaxanthin production using high light intensities. Algal Res 21, 213–222.

Kleinegris, D.M.M., Janssen, M., Brandenburg, W.A., Wijffels, R.H., 2011. Continuous production of carotenoids from Dunaliella salina. Enzym. Microb. Technol. 48 (3), 253–259.

Kobayashi, M., Kakizono, T., Nishio, N., Nagai, S., 1992. Effects of light intensity, light quality, and illumination cycle on astaxanthin formation in a green alga, Haematococcus pluvialis. J. Ferment. Bioeng. 74 (1), 61–63.

Kobayashi, M., Kakizono, T., Nagai, S., 1993. Enhanced carotenoid biosynthesis by oxidative stress in acetate-induced cyst cells of a green unicellular alga, Haematococcus pluvialis. Appl. Environ. Microbiol. 59 (3), 867–873.

Koller, M., Muhr, A., Braunegg, G., 2014. Microalgae as versatile cellular factories for valued products. Algal Res 6 (PA), 52–63.

Kotake-Nara, E., Nagao, A., 2011. Absorption and metabolism of xanthophylls. Mar. Drugs 9 (6), 1024–1037.

Koushan, K., Rusovici, R., Li, W., Ferguson, L.R., Chalam, K.V., 2013. The role of lutein in eye-related disease. Nutrients 5 (5), 1823–1839.

Lababpour, A., Shimahara, K., Hada, K., Kyoui, Y., Katsuda, T., Katoh, S., 2005. Fed-batch culture under illumination with blue light emitting diodes (LEDs) for astaxanthin production by Haematococcus pluvialis. J. Biosci. Bioeng. 100 (3), 339–342.

Landrum, J.T., Bonet, R.A., Kilburn, M.D., 1996. The macular pigment: a possible role in protection from age-related macular degeneration. Adv. Pharmacol. 38, 537–556.

Li, X., Qi, Z., Zhao, L., Yu, Z., 2016. Astaxanthin reduces type 2 diabetic-associated cognitive decline in rats via activation of PI3K/Akt and attenuation of oxidative stress. Mol. Med. Rep. 13 (1), 973–979.

Li, Y., Zhang, Y., Liu, X., Wang, M., Wang, P., et al., 2018. Lutein inhibits proliferation, invasion and migration of hypoxic breast cancer cells via downregulation of HES1. Int. J. Oncol. 52 (6), 2119–2129.

Lin, J.H., Lee, D.J., Chang, J.S., 2015. Lutein production from biomass: marigold flowers versus microalgae. Bioresour. Technol. 184, 421–428.

Liu, B.H., Lee, Y.K., 2003. Effect of total secondary carotenoids extracts from Chlorococcum sp on Helicobacter pylori-infected BALB/c mice. Int. Immunopharmacol. 3 (7), 979–986.

Liu, X., Osawa, T., 2007. Cis astaxanthin and especially 9-cis astaxanthin exhibits a higher antioxidant activity in vitro compared to the all-trans isomer. Biochem. Biophys. Res. Commun. 357 (1), 187–193.

Liu, J., Huang, J., Jiang, Y., Chen, F., 2012. Molasses-based growth and production of oil and astaxanthin by Chlorella zofingiensis. Bioresour. Technol. 107, 393–398.

Liu, J., Sun, Z., Zhong, Y., Gerken, H., Huang, J., Chen, F., 2013. Utilization of cane molasses towards cost-saving astaxanthin production by a Chlorella zofingiensis mutant. J. Appl. Phycol. 25 (5), 1447–1456.

Liu, J., Sun, Z., Gerken, H., Liu, Z., Jiang, Y., Chen, F., 2014. Chlorella zofingiensis as an alternative microalgal producer of astaxanthin: biology and industrial potential. Mar. Drugs 12 (6), 3487–3515.

López, M.C.G.-M., Sánchez, E.D.R., López, J.L.C., Fernández, F.G.A., Sevilla, J.M.F., et al., 2006. Comparative analysis of the outdoor culture of Haematococcus pluvialis in tubular and bubble column photobioreactors. J. Biotechnol. 123 (3), 329–342.

Manabe, E., Handa, O., Naito, Y., Mizushima, K., Akagiri, S., et al., 2008. Astaxanthin protects mesangial cells from hyperglycemia-induced oxidative signaling. J. Cell. Biochem. 103 (6), 1925–1937.

Mann, J.E., Myers, J., 1968. On pigments, growth, and photosynthesis of Phaeodactylum tricornutum. J. Phycol. 4 (4), 349–355.

Männistö, S., Yaun, S.-S., Hunter, D.J., Spiegelman, D., Adami, H.-O., et al., 2007. Dietary carotenoids and risk of colorectal cancer in a pooled analysis of 11 cohort studies. Am. J. Epidemiol. 165 (3), 246–255.

Maoka, T., Tokuda, H., Suzuki, N., Kato, H., Etoh, H., 2012. Anti-oxidative, anti-tumor-promoting, and anti-carcinogensis activities of nitroastaxanthin and nitrolutein, the reaction products of astaxanthin and lutein with peroxynitrite. Mar. Drugs 10 (6), 1391–1399.

Menegol, T., Diprat, A.,B., Rodrigues, E., Rosane, R., 2017. Effect of temperature and nitrogen concentration on biomass composition of Heterochlorella luteoviridis. Food Sci. Technol. 37, 28–37.

Mignone, L.I., Giovannucci, E., Newcomb, P.A., Titus-Ernstoff, L., Trentham-Dietz, A., et al., 2009. Dietary carotenoids and the risk of invasive breast cancer. Int. J. Cancer. 124 (12), 2929–2937.

Miki, W., 1991. Biological functions and activities of animal carotenoids. Pure Appl. Chem. 63 (1), 141–146.

Mimouni, V., Ulmann, L., Pasquet, V., Mathieu, M., Picot, L., et al., 2012. The potential of microalgae for the production of bioactive molecules of pharmaceutical interest. Curr. Pharmaceut. Biotechnol. 13 (15), 2733–2750.

Miyachi, M., Matsuno, T., Asano, K., Mataga, I., 2015. Anti-inflammatory effects of astaxanthin in the human gingival keratinocyte line NDUSD-1. J. Clin. Biochem. Nutr. 56 (3), 171–178.

Murillo, A.G., DiMarco, D.M., Fernandez, M.L., 2016. The potential of non-provitamin A carotenoids for the prevention and treatment of non-alcoholic fatty liver disease. Biology 5 (4).

Naguib, Y.M., 2000. Antioxidant activities of astaxanthin and related carotenoids. J. Agric. Food Chem. 48 (4), 1150–1154.

Naito, Y., Uchiyama, K., Aoi, W., Hasegawa, G., Nakamura, N., et al., 2004. Prevention of diabetic nephropathy by treatment with astaxanthin in diabetic db/db mice. Biofactors 20 (1), 49–59.

Nakano, M., Onodera, A., Saito, E., Tanabe, M., Yajima, K., et al., 2008. Effect of astaxanthin in combination with alpha-tocopherol or ascorbic acid against oxidative damage in diabetic ODS rats. J. Nutr. Sci. Vitaminol. 54 (4), 329–334.

Nakao, R., Nelson, O.L., Park, J.S., Mathison, B.D., Thompson, P.A., Chew, B.P., 2010. Effect of dietary astaxanthin at different stages of mammary tumor initiation in BALB/c mice. Anticancer Res. 30 (6), 2171–2175.

Nishida, Y., Yamashita, E., Miki, W., 2007. Quenching activities of common hydrophilic and lipophilic antioxidants against singlet oxygen using chemiluminescence detection System. Carotenoid Sci 11, 16–20.

Nishigaki, I., Rajendran, P., Venugopal, R., Ekambaram, G., Sakthisekaran, D., Nishigaki, Y., 2010. Cytoprotective role of astaxanthin against glycated protein/iron chelate-induced toxicity in human umbilical vein endothelial cells. Phytother Res. 24 (1), 54–59.

Ohgami, K., Shiratori, K., Kotake, S., Nishida, T., Mizuki, N., et al., 2003. Effects of astaxanthin on lipopolysaccharide-induced inflammation in vitro and in vivo. Investig. Ophthalmol. Vis. Sci. 44 (6), 2694–2701.

Olaizola, M., 2000a. Commercial production of astaxanthin from Haematococcus pluvialis using 25,000-liter outdoor photobioreactors. J. Appl. Phycol. 12 (3–5), 499–506.

Olaizola, M., 2000b. Commercial production of astaxanthin from. Growth 499–506.

Orosa, M., Valero, J.F., Herrero, C., Abalde, J., 2001. Comparison of the accumulation of astaxanthin in Haematococcus pluvialis and other green microalgae under N-starvation and high light conditions. Biotechnol. Lett. 23, 1079–1085, 23.

Otton, R., Marin, D.P., Bolin, A.P., Santos, R de CM dos, Polotow, T.G., et al., 2010. Astaxanthin ameliorates the redox imbalance in lymphocytes of experimental diabetic rats. Chem. Biol. Interact. 186 (3), 306–315.

Pan-utai, W., Parakulsuksatid, P., Phomkaivon, N., 2017. Effect of inducing agents on growth and astaxanthin production in Haematococcus pluvialis: organic and inorganic. Biocatal. Agric. Biotechnol. 12, 152–158.

Park, J.S., Chyun, J.H., Kim, Y.K., Line, L.L., Chew, B.P., 2010. Astaxanthin decreased oxidative stress and inflammation and enhanced immune response in humans. Nutr. Metab. 7, 18.

Perez-lopez, P., Gonzalez-garcia, S., Jeffryes, C., Spiros, N., Mchugh, E., et al., 2016. Life cycle assessment of the production of the red antioxidant carotenoid astaxanthin by microalgae : from lab to pilot scale to cite this version : HAL Id : hal-01284775, 64, 332–344.

Perry, A., Rasmussen, H., Johnson, E.J., 2009. Journal of Food Composition and Analysis Xanthophyll (lutein , zeaxanthin) content in fruits , vegetables and corn and egg. J. Food Compos. Anal. 22, 9–15.

Prabhu, P.N., Ashokkumar, P., Sudhandiran, G., 2009. Antioxidative and antiproliferative effects of astaxanthin during the initiation stages of 1,2-dimethyl hydrazine-induced experimental colon carcinogenesis. Fundam. Clin. Pharmacol. 23 (2), 225–234.

Rafi, M.M., Kanakasabai, S., Gokarn, S.V., Krueger, E.G., Bright, J.J., 2015. Dietary lutein modulates growth and survival genes in prostate cancer cells. J. Med. Food 18 (2), 173–181.

Ranga Rao, A., Raghunath Reddy, R.L., Baskaran, V., Sarada, R., Ravishankar, G.A., 2010. Characterization of microalgal carotenoids by mass spectrometry and their bioavailability and antioxidant properties elucidated in rat model. J. Agric. Food Chem. 58 (15), 8553–8559.

Ranga Rao, A., Baskaran, V., Sarada, R., Ravishankar, G.A., 2013. In vivo bioavailability and antioxidant activity of carotenoids from microalgal biomass — a repeated dose study. Food Res. Int. 54 (1), 711–717.

Ranjbar, R., Inoue, R., Shiraishi, H., Katsuda, T., Katoh, S., 2008. High efficiency production of astaxanthin by autotrophic cultivation of Haematococcus pluvialis in a bubble column photobioreactor. Biochem. Eng. J. 39 (3), 575–580.

Reynoso-Camacho, R., Gonzalez-Jasso, E., Ferriz-Martinez, R., Villalon-Corona, B., Loarca-Pina, G.F., et al., 2011. Dietary supplementation of lutein reduces colon carcinogenesis in DMH-treated rats by modulating K-ras, PKB, and beta-catenin proteins. Nutr. Canc. 63 (1), 39–45.

Sánchez, J.F., Fernández, J.M., Acién, F.G., Rueda, A., Pérez-Parra, J., Molina, E., 2008a. Influence of culture conditions on the productivity and lutein content of the new strain Scenedesmus almeriensis. Process Biochem. 43 (4), 398–405.

Sánchez, J.F., Fernández-Sevilla, J.M., Acién, F.G., Cerón, M.C., Pérez-Parra, J., Molina-Grima, E., 2008b. Biomass and lutein productivity of Scenedesmus almeriensis: influence of irradiance, dilution rate and temperature. Appl. Microbiol. Biotechnol. 79 (5), 719–729.

Sarada, R., Tripathi, U., Ravishankar, G.A., 2002. Influence of stress on astaxanthin production in Haematococcus pluvialis grown under different culture conditions. Process Biochem. 37 (6), 623–627.

Shah, M.M.R., Liang, Y., Cheng, J.J., Daroch, M., 2016. Astaxanthin-producing green microalga Haematococcus pluvialis: from single cell to high value commercial products. Front. Plant Sci. 7, 531.

Shaish, A., Ben-Amotz, A., Avron, M., 1992. Biosynthesis of β-carotene in Dunaliella. Methods Enzymol. 213 (C), 439–444.

Shi, X.M., 2002. High-yield production of lutein by the green microalga Chlorella protothecoides in heterotrophic fed-batch culture. Biotechnol. Prog. 18 (4), 723–727.

Shi, X.M., Chen, F., Yuan, J.P., Chen, H., 1997. Heterotrophic production of lutein by selected Chlorella strains. J. Appl. Phycol. 9, 445–450.

Shi, X.M., Zhang, X.W., Chen, F., 2000. Heterotrophic production of biomass and lutein by Chlorella protothecoides on various nitrogen sources. Enzym. Microb. Technol. 27 (3–5), 312–318.

Shi, X., Wu, Z., Chen, F., 2006. Kinetic modeling of lutein production by heterotrophic Chlorella at various pH and temperatures. Mol. Nutr. Food Res. 50 (8), 763–768.

Silverman, M.P., Lundgren, D.G., 1959. Studies on the chemoautotrophic iron bacterium Ferrobacillus ferrooxidans. I. An improved medium and a harvesting procedure for securing high cell yields. J. Bacteriol. 77 (5), 642–647.

Sindhu, E.R., Preethi, K.C., Kuttan, R., 2010. Antioxidant activity of carotenoid lutein in vitro and in vivo. Indian J. Exp. Biol. 48 (8), 843–848.

Šivel, M., Kráčmar, S., Fišera, M., Klejdus, B., Kubáň, V., 2014. Lutein content in marigold flower (Tagetes erecta L.) concentrates used for production of food supplements. Czech J. Food Sci. 32 (6), 521–525.

Slattery, M.L., Benson, J., Curtin, K., Ma, K.N., Schaeffer, D., Potter, J.D., 2000. Carotenoids and colon cancer. Am. J. Clin. Nutr. 71 (2), 575–582.

Solymosi, K., Latruffe, N., Morant-Manceau, A., Schoefs, B., 2015. Food colour additives of natural origin. Colour Addit. Foods Beverages 3–34.

Suhler, E.B., Lloyd, M.J., Choi, D., Rosenbaum, J.T., Austin, D.F., 2008. Incidence and prevalence of uveitis in veterans affairs medical centers of the Pacific northwest. Am. J. Ophthalmol. 146 (6), 890–896.

Sun, N., Wang, Y., Li, Y.T., Huang, J.C., Chen, F., 2008. Sugar-based growth, astaxanthin accumulation and carotenogenic transcription of heterotrophic Chlorella zofingiensis (Chlorophyta). Process Biochem. 43 (11), 1288–1292.

Sun, H., Kong, Q., Geng, Z., Duan, L., Yang, M., Guan, B., 2015. Enhancement of cell biomass and cell activity of astaxanthin-rich Haematococcus pluvialis. Bioresour. Technol. 186, 67–73.

Sun, Z., Li, T., Zhou, Z., Jiang, Y., 2016. Microalgae as a source of lutein: chemistry, biosynthesis, and carotenogenesis. In: Posten, C., Chen, S.F. (Eds.), Microalgae Biotechnology. Springer International Publishing, Cham, pp. 37–58.

Suzuki, Y., Ohgami, K., Shiratori, K., Jin, X.-H., Ilieva, I., et al., 2006. Suppressive effects of astaxanthin against rat endotoxin-induced uveitis by inhibiting the NF-kappaB signaling pathway. Exp. Eye Res. 82 (2), 275–281.

Tanaka, T., Morishita, Y., Suzui, M., Kojima, T., Okumura, A., Mori, H., 1994. Chemoprevention of mouse urinary bladder carcinogenesis by the naturally occurring carotenoid astaxanthin. Carcinogenesis 15 (1), 15–19.

Tanaka, T., Makita, H., Ohnishi, M., Mori, H., Satoh, K., Hara, A., 1995. Chemoprevention of rat oral carcinogenesis by naturally occurring xanthophylls, astaxanthin and canthaxanthin. Cancer Res. 55 (18), 4059–4064.

Thongkrajay, U., Poolsawat, N., Ratniyomchai, T., Kulworawanichpong, T., 2006. Unbalanced three-phase distribution power flow using alternative Newton-Raphson method. WSEAS Trans. Circuits Syst. 5 (3), 403–410.

Tian, Y., Kijlstra, A., Webers, C.A.B., Berendschot, T.T.J.M., 2015. Lutein and Factor D: two intriguing players in the field of age-related macular degeneration. Arch. Biochem. Biophys. 572, 49–53.

Tjahjono, A.E., Hayama, Y., Kakizono, T., Terada, Y., Nishio, N., Nagai, S., 1994. Hyper-accumulation of astaxanthin in a green algaHaematococcus pluvialis at elevated temperatures. Biotechnol. Lett. 16 (2), 133–138.

Torzillo, G., Goksan, T., Faraloni, C., Kopecky, J., Masojídek, J., 2003. Interplay between photochemical activities and pigment composition in an outdoor culture of Haematococcus pluvialis during the shift from the green to red stage. J. Appl. Phycol. 15 (2/3), 127–136.

Uchiyama, K., Naito, Y., Hasegawa, G., Nakamura, N., Takahashi, J., Yoshikawa, T., 2002. Astaxanthin protects beta-cells against glucose toxicity in diabetic db/db mice. Redox Rep. 7 (5), 290–293.

Vaquero, I., Vázquez, M., Ruiz-Domínguez, M.C., Vílchez, C., 2014. Enhanced production of a lutein-rich acidic environment microalga. J. Appl. Microbiol. 116 (4), 839–850.

Ventura, S.P.M., Nobre, B.P., Ertekin, F., Hayes, M., Vieira, F., 2017. Extraction of value-added compounds from microalgae. Microalgae-Based Biofuels and Bioproducts 461–483.

Vo, T.T., Lee, C., Han, S.I., Kim, J.Y., Kim, S., Choi, Y.E., 2016. Effect of the ethylene precursor, 1-aminocyclopropane-1-carboxylic acid on different growth stages of Haematococcus pluvialis. Bioresour. Technol. 220, 85–93.

Wan, M., Hou, D., Li, Y., Fan, J., Huang, J., et al., 2014a. The effective photoinduction of Haematococcus pluvialis for accumulating astaxanthin with attached cultivation. Bioresour. Technol. 163, 26–32.

Wan, M., Zhang, J., Hou, D., Fan, J., Li, Y., et al., 2014b. The effect of temperature on cell growth and astaxanthin accumulation of Haematococcus pluvialis during a light-dark cyclic cultivation. Bioresour. Technol. 167, 276–283.

Wang, J., Han, D., Sommerfeld, M.R., Lu, C., Hu, Q., 2013a. Effect of initial biomass density on growth and astaxanthin production of Haematococcus pluvialis in an outdoor photobioreactor. J. Appl. Phycol. 25 (1), 253–260.

Wang, J., Sommerfeld, M.R., Lu, C., Hu, Q., 2013b. Combined effect of initial biomass density and nitrogen concentration on growth and astaxanthin production of Haematococcus pluvialis (Chlorophyta) in outdoor cultivation. ALGAE 28 (2), 193–202.

Wei, D., Chen, F., Chen, G., Zhang, X., Liu, L., Zhang, H., 2008. Enhanced production of lutein in heterotrophic Chlorella protothecoides by oxidative stress. Sci. China Ser. C Life Sci. 51 (12), 1088–1093.

Wells, M.L., Potin, P., Craigie, J.S., Raven, J.A., Merchant, S.S., et al., 2017. Algae as nutritional and functional food sources: revisiting our understanding. J. Appl. Phycol. 29, 949–982.

Yoo, J.J., Choi, S.P., Kim, B.W., Sim, S.J., 2012. Optimal design of scalable photo-bioreactor for phototropic culturing of Haematococcus pluvialis. Bioproc. Biosyst. Eng. 35 (1–2), 309–315.

Yoshida, H., Yanai, H., Ito, K., Tomono, Y., Koikeda, T., et al., 2010. Administration of natural astaxanthin increases serum HDL-cholesterol and adiponectin in subjects with mild hyperlipidemia. Atherosclerosis 209 (2), 520–523.

Zhang, L., Wang, H., 2015. Multiple mechanisms of anti-cancer effects exerted by astaxanthin. Mar. Drugs 13 (7), 4310–4330.

Zhang, B.Y., Geng, Y.H., Li, Z.K., Hu, H.J., Li, Y.G., 2009. Production of astaxanthin from Haematococcus in open pond by two-stage growth one-step process. Aquaculture 295 (3–4), 275–281.

Zhang, W., Wang, J.J., Liu, T., 2014. Attached cultivation of Haematococcus pluvialis for astaxanthin production. Bioresour. Technol. 158C, 329–335.

Zhao, L., Sweet, B.V., 2008. Lutein and zeaxanthin for macular degeneration. Am. J. Heal. Pharm. 65 (13), 1232–1238.

Zhu, L., 2015. Biorefinery as a promising approach to promote microalgae industry: an innovative framework. Renew. Sustain. Energy Rev. 41, 1376–1384.

Zhu, X., Chen, Y., Chen, Q., Yang, H., Xie, X., 2018. Astaxanthin promotes Nrf2/ARE signaling to alleviate renal fibronectin and collagen IV accumulation in diabetic rats. J. Diabetes Res. 2018, 1–7.

Zou, H.-F., Zhang, Y.-Q., Wei, W., Chen, H.-W., Song, Q.-X., et al., 2013. The transcription factor AtDOF4.2 regulates shoot branching and seed coat formation in Arabidopsis. Biochem. J. 449 (2), 373–388.

CHAPTER

8

Biotechnological exploitation of cyanobacteria and microalgae for bioactive compounds

Shailendra Kumar Singh[a], Rupali Kaur[a], Ajay Bansal[b], Suman Kapur[c], Shanthy Sundaram[a]

Centre of Biotechnology, University of Allahabad, Prayagraj, India[a]; Department of Chemical Engineering, Dr B R Ambedkar National Institute of Technology, Jalandhar, India[b]; Birla Institute of Technology & Science, Pilani, Hyderabad Campus, Telangana, India[c]

Chapter outline

Biotechnological Production of Bioactive Compounds. https://doi.org/10.1016/B978-0-444-64323-0.00008-4

1. Introduction

More than 70% of Earth's surface is covered with water, in which algae is one of the most dominant group of photosynthetic living organisms, not just in aquatic part but also on terrestrial region (Mimouni et al., 2012). Being the primary producers of our planet, algae can rapidly produce biomass from the consumption of inexpensive solar energy, atmospheric CO_2 and common water nutrients in a wide range of environmental conditions (Singh et al., 2012). Thus, they have an incredible impact on the sustainability of Earth's ecosystem (Singh et al., 2016a). It is estimated that more than 200,000 and several million species of microalgae exist (Norton et al., 1996), but only a limited number, of around 30,000 have been studied and documented well (Shalaby, 2011). Thus, this unexplored group of photosynthetic microorganisms provide a virtually untapped source of high value products.

Due to the bulk of the natural oil made by microalgae is in the form of triglycerides, which is the right type of oil for producing biofuel, currently microalgal research are the exclusive focus in the algae-to-biofuel arena (Singh et al., 2013). However, in the last decades, increasing attention has been paid to the potential of growing these tiny microorganisms commercially as they can synthesize numerous bioactive molecules that can be harnessed for commercial use. The first use of microalgae by humans, dates back 2000 years to the Chinese, who used *Nostoc* to survive during famine (Choopani et al., 2016). In general, bioactive compounds are phytochemicals, found in foods of humans, animals, or plants and capable of modulating metabolic processes of the organism consuming it (Guaadaoui et al., 2014). They exhibit beneficial effects such as antioxidant activity (Violeta et al., 2017), inhibition or induction of enzymes (Gbonjubola et al., 2016), inhibition of receptor activities (Yun et al., 2010), and induction and inhibition of gene expression (Iwata et al., 2017). Thus, bioactive compounds comprise a very diverse class of compounds (carotenoids, polyphenols, phytosterols, tocopherols, and organosulfur compounds) with varied chemical assemblies (hydrophilic or lipophilic), range of concentrations both in foods and in the human body, specific to species or ubiquitously occurrence in nature, mode of action, specificity of bioactive molecules, and effectiveness against free radicals. Bioactive compounds of microalgae comprise a vital market in which compounds carotenoids (e.g. β-carotene, astaxanthin etc.), polyunsaturated fatty acid (PUFA) (e.g. docosahexaenoic acid (DHA) and eicosapentaenoic acid (EPA)) and polysaccharides (e.g. β-glucan) dominate. The most widely used microalgal class and their dominating species in commercial production includes Cyanophyceae (blue-green algae) e.g. *Oscillatoria, Nostoc, Arthrospira (Spirulina)*), Chlorophyceae (green algae) e.g. *Chlamydomonas, Dunaliella salina, Chlorella,* Prymnesiophyceae e.g. *Isochrysis galbana,* and Bacillariophyceae (including diatoms) e.g. *Thalassiosira weissflogii* and *Cyclotella cryptica*.

The present article will provide a brief overview of algae-derived bioactive compounds with potential benefits to human health. This chapter covers the differences between cyanobacteria and microalgae at the cellular level and discuss the biological activities of the bioactive compounds with a focus on human nutrition and health.

Another section will also be devoted to novel technologies concerning the commercial cultivation of microalgae and extraction of bioactive compound from algal biomass.

2. Algae as a source of bioactive compounds

Algae represent one of the oldest (evolve 3.8 billion years ago) (Encarnação et al., 2015) and most diversified photosynthetic organisms (over 2,80,000 reported species) in freshwater and marine systems (Norton et al., 1996; Jha and Zi-Rong, 2004), convert light energy from the Sun through photosynthesis into chemical energy. Based on size, algae could be categorized into two broad groups; microalgae and macroalgae. Thus, photosynthetic microalgae belong to the subgroup of algae, comprise several thousand species. However, in its broadest definition, microalgae include prokaryotic cyanobacteria, green algae, diatoms, and many other eukaryotic groups (Norton et al., 1996). They are classified as (1) cyanobacteria (BGA), (2) rhodophytes, (rhodophytes), (3) chlorophytes (chlorophytes), and 4) chomophytes (all others). Microalgae contain up to 50%–70 % protein (up to 50% in meat, and 15%–17% in wheat), 30% lipids, over 40% glycerol, up to 8%–14% carotene and a high concentration of vitamins B_1, B_2, B_3, B_6, B_{12}, E, K, D, etc., compared with other plants or animals (Priyadarshani and Rath, 2012).

The second group macroalgae is a collective term used for large aquatic photosynthetic plants like seaweeds and other benthic (attached to the bottom) marine algae that can be generally seen without the aid of a microscope (Rindi et al., 2012; Kim and Chojnacka, 2015). While limited advantage of natural populations microalgae has been used for thousands of years by indigenous populations (*Nostoc* in Asia and *Spirulina* in Africa and North America for sustenance), it is only now the biotechnological potential of algae have realized (Olaizola, 2003; Choopani et al., 2016).

Algae have remarkable abilities to endure in range of extreme environmental conditions such as high-low salinity, heat-cold, desiccation, osmotic stress, high-low temperature, anaerobiosis, nitrogen deficiency, photooxidation, UV–visible irradiation etc (Rai and Gaur, 2012; Holzinger and Karsten, 2013). However, extreme conditions trigger an imbalance between the production of free radicals and the ability of the cells to detoxify their harmful effects through neutralization by antioxidants. In order to cope with oxidative stress, algal cells produce a variety of useful bioactive compounds which helps to protect algae from harmful effects of free radicals (Priyadarshani and Rath, 2012). As microalgae are tiny organism, possess a much simpler reproductive cycle than terrestrial plants and their simple growth needs make them good candidates to grow easily in bioreactors, considered as natural factories to produce specific bioactive compounds at an industrial scale (Olaizola, 2003). Thus, various bioactive constituents can be extracted from microalgae to get antioxidative, anti-inflammatory, antimicrobial, antiviral, antimalarial, anticancer, antihypertensive, immune modulator products that find an application in health food, cosmetics, feed and food supplements, pharmaceuticals and therapeutic industries.

3. Algae: cellular structure

Taxonomically, earlier algae were classified with plants as they are photosynthetic autotrophs possess chlorophyll and chloroplasts and superficially appear like plants. However, study of their cell structure under electron microscope has revealed one major fact that blue-green algae have prokaryotic type of cell like that of bacteria and hence they are more related to them than to other algae with which they were traditionally grouped. Thus, algae have been placed in Kingdom Protista along with protozoa. All other algae have eukaryotic type of cell (see Fig. 8.1 and Table 8.1). Algae are widely distributed in nature wherever there is plenty of water and sunshine. They even inhabit harsh habitats. Although simple in structure, lacking differentiation, algae exhibit great diversity in size and appearance. Their size ranges from simple microscopic to giant thallus extending several meters in length as in kelps.

3.1 Prokaryotic microalgae: cyanobacteria

Cyanobacteria are prokaryotic oxygenic phototrophs, belong to the kingdom Monera and division Cyanophyta. They are among the most primitive forms of life and found in almost every conceivable habitat on earth. There are an estimated 150 genera, containing approximately 2000 species of cyanobacteria (Hitzfeld et al., 2000). Their cellular structure is simple prokaryote and performs photosynthesis, resembling plants but lack plant cell walls resembling primitive bacteria. As like prokaryotic life form, they do not have membrane bound nuclei or any internal membrane system. Like Gram-negative bacteria, cyanobacterial cell membrane has an

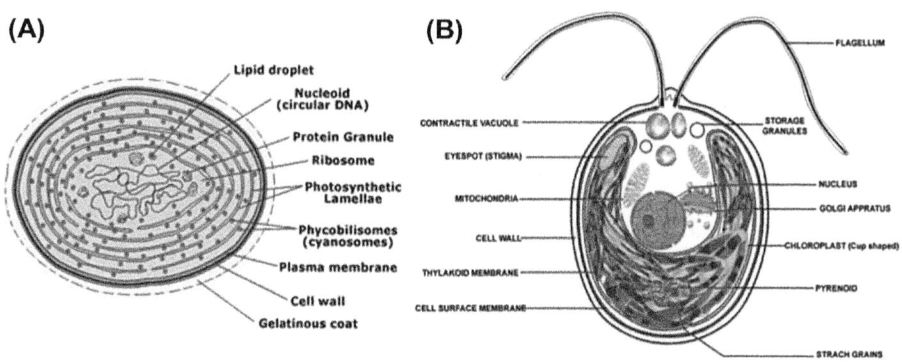

FIG. 8.1

Cellular Structure of Algae (A) Cyanobacteria (B) Eukaryotic microalgae.

Table 8.1 Difference between cyanobacteria and eukaryotic microalgae (Singh et al., 2014).

	Cyanobacteria	Green algae
Habitat	Any photosynthetic bacteria, which often form colonies in the form of filaments, spheres or sheets and occur in diverse environments	Refer to any green color algae found in freshwater habitat.
Nucleus	Prokaryotes; nuclear materials, deoxyribo-nucleic acid (DNA), is dispersed to some degree throughout the cell.	Eukaryotes; DNA is delimited from the remainder of the protoplasm by a perforated nuclear membrane.
Membrane-bound organelles	Do not consist of membrane-bound organelles; photosynthetic lamellae occur freely in the cytoplasm.	Consist of membrane-bound organelles like plastids, endoplasmic reticulum (ER), mitochondria and Golgi apparatus.
Chloroplasts	Do not contain chloroplasts, so they are homogenous colored, appear pale green blue, slightly brown-green etc.	Contain one or more chloroplast per cell Chlorophyta have green color.
Pigments	Contain Chl a, phycoerythrin and phycocyanin as photosynthetic pigments	Contain Chl a, Chl b, carotenoids, and xanthophyll as photosynthetic pigments
Mode of nutrition	Photoautotrophs or heterotrophs	Photoautotrophs
Nitrogen fixation	Contains heterocysts organelle for nitrogen fixation	Not involved in nitrogen fixation
Storage of nutrients	Show an efficient storage of nutrients. In most cases, the Iodine test for starch is positive.	Have less ability to store nutrients, About starch test: not all eukaryotic algae accumulate starch (e.g., Xanthophyceae (yellow-green algae) have green chloroplasts but do not have starch.
Mobility	Cannot swim but, they can change their buoyancy, changing the depth in water	Can swim through water
Reproduction	• Asexual reproduction methods include cell division and the formation of a plate, seprating the two cells. • Does not sexually reproduce	• Asexual reproduction methods include fission, budding, fragmentation or zoospores formation. • Sexually reproduce by the formation of gametes
Examples	*Nostoc, Anabaena, Oscillatoria* and *Spirulina*	*Chlamydomonus, Ulva, Spirogyra, Chlorella* and green seaweeds

outer peptidoglycan layer, which is a polymer composed of two sugar derivatives; N-acetylglucosamine (NAG) and N-acetylmuramic acid (NAM), and several different amino acids (Singh et al., 2014). These also resemble animals in having complex sugars like glycogen on their cell membrane. The peripheral protoplasm of cyanobacteria is primarily made of thylakoids with water-soluble chromo proteins phycobilisomes (phycobiliproteins) and glycogen granules. They exist in diverse morphologies including unicellular forms exist as single cells, suspended or benthic, or aggregates and filamentous forms may be thin or thick, single trichome or bundles either with or without a sheath (Castenholz et al., 2001; Abed et al., 2009). All cyanobacteria carry out oxygenic photosynthesis. However, under anoxic conditions and during the dark, some cyanobacterial species can switch to the typical bacterial anoxygenic photosynthesis using sulfide as electron donor (Stal, 1995; Lau et al., 2015). Some cyanobacteria form heterocysts and could fix atmospheric nitrogen (Singh et al., 2011). Certain genera of cyanobacteria could produce diverse structural classes of secondary metabolites. Thus, during last decades, some cyanobacterial genera have gained much attention as a promising group of diverse taxa of organisms (e.g. *Aphanizomenon flos-aquae, Lyngbya, Spirulina platensis, S. maxima, Spirulina fusiformis, Oscillatoria, Symploca, Calothrix, Leptolyngbya, Dichothrix, Nostoc, Geitlerinema, Schizothrix, Aphanothece, Blennothrix, Synechocystis*) with new and rich source of bioactive compounds (Nagarajan et al., 2012). Major part of cyanobacterial secondary metabolites are peptides or possess peptidic substructures, synthesized by non-ribosomal peptide synthetase (NRPS) or NRPS/PKS (polyketide synthase) hybrid pathways. To date, some 600 cyanobacterial peptides have been reported (Gkelis et al., 2015). Most of the bioactive compounds isolated from cyanobacteria tend to be lipopeptides, i.e. an amino acid derived fragment linked to a fatty acid derived portion. Nagle et al. (1996) analyzed of 424 marine cyanobacterial natural products and showed that 40.2% are lipopeptides (cyclic or linear), 5.6% are of pure amino acid, 4.2% are fatty acids, 4.2% macrolides and 9.4% are amides. However, the biosynthetic mechanism for many of the cyanobacterial peptides is yet unrevealed.

3.2 Eukaryotic microalgae

Depending on the species, their sizes can range from a few micrometers (μm) to a few hundreds of micrometers (mm). However, all eukaryotic algae share some basic features such as; a thin, rigid cellulosic cell wall, typical eukaryotic nucleus with pores, chromatin, nucleolus, karyolymph etc., lamellar, discoid or tubular cristae shaped mitochondria and chloroplasts with enclosed thylakoids as a place of photosynthesis (Andersen, 1992). Some species like model microalga *Chlamydomonas reinhardtii* may have one or more flagella for cell motility and a dense, proteinaceous area called pyrenoid, usually associated with synthesis and storage of the starch (Nguyen et al., 2013). Cell wall is mainly composed of hydroxyproline-rich glycoprotein and non-cellulosic polysaccharides, instead of cellulose. The central nucleus is enclosed in chloroplast, contains single large pyrenoid at the posterior end of the chloroplast. Pyrenoids are RuBisCO rich micro-compartments which act as CO_2 fixation center in microalgae, analogous role to carboxysomes of cyanobacteria (Singh et al., 2014). It also helps to formation and deposition of starch. Other organelles like, golgi stacks and endoplasmic reticulum has also been visualized inside the cell.

4. Microalgal bioactive compounds

In the last decades, a considerable attention has been increased toward the biotechnology of microalgae because of their ability to produce a variety of bioactive products (Singh et al., 2016b; Sharma and Sharma, 2017). Both, primary and secondary metabolism of these microorganisms has been demonstrated to play a key role in the production of special chemicals. Primary metabolites are those directly involved in normal growth development, reproduction, cell division, or metabolism. However, secondary metabolites production requires additional energy (ATP, NADPH) and C and N allocation and they serve no obvious function in the cell for their primary needs (Encarnação et al., 2015). Their production in the cell appear to be associated with the exposure of specific environmental stress conditions (Singh et al., 2018). Thus, secondary metabolites showed a wide variation in their chemical structures such as acyclic peptides, linear decadepsipeptides, linear lipopeptides, linear alkynoic lipopeptides, cyclic depsipeptides, cyclic undecapeptides, cyclic hexapeptides, lipophilic cyclic peptides, paracyclophanes, alkaloids, sesterterpenes, swinholides, fatty acid amides, brominated fatty acids, glycolipids, polyphenolic ethers, macrolactones, glycosidic macrolides and iminotetrasaccharides (Nagarajan et al., 2012) (see Fig. 8.2 and Table 8.2).

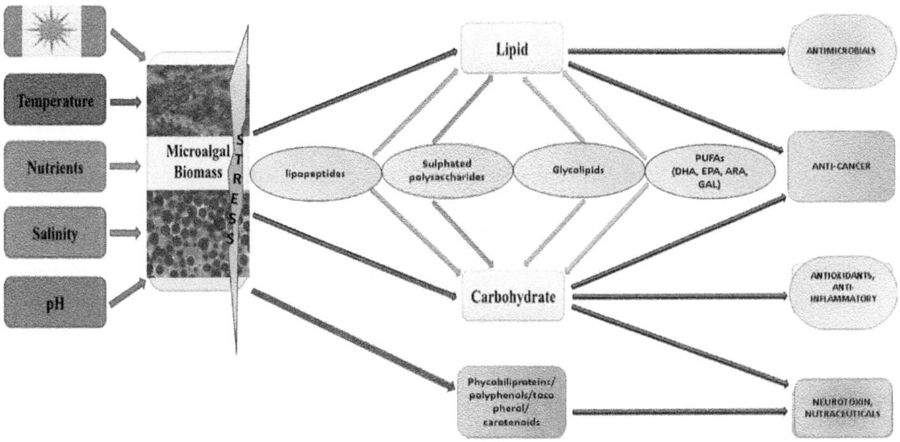

FIG. 8.2

Production of bioactive compounds in algae under stress conditions.

Table 8.2 Important bioactive compounds and their derivatives (Singh et al., 2014).

S. No.	Bioactive compounds	Derivatives
1.	*Pigments/Carotenoids*	β-carotene, astaxanthin, lutein, zeaxanthin, canthaxanthin, chlorophyll, phycocyanin, phycoerythrin, fucoxanthin
2.	*Polyunsaturated fatty acids (PUFAs)*	DHA (Docosahexaenoic Acid), EPA(Eicosapentaeneoic acid), ARA(Arachidonic acid), GAL(C18:3)
3.	*Vitamins*	A, B_1, B_6, B_{12}, C, E, biotin, riboflavin, nicotinic acid, pantothenate, folic acid
4.	*Antioxidants*	Catalases, polyphenols, superoxide dismutase, tocopherols
5.	*Other*	Antimicrobial, antifungal, antiviral agents, toxins, amino acids, proteins, sterols, MAAs for light protection.

These naturally produced secondary metabolites have been shown to possess one or more of anti-bacterial (Najdenski et al., 2013; Sibi, 2015; Syed et al., 2015), anti-fungal (Danyal et al., 2013; Najdenski et al., 2013; Shishido et al., 2015), anti-viral (Yim et al., 2004; Lee et al., 2006), anti-cancer (Quintana et al., 2014; Iwasaki et al., 2015), algicide (Volk and Furkert, 2006; Yingying et al., 2008), immunosuppressive agents (Koehn et al. 1992), sodium-channel blockers (Shimizu, 2003; Taniguchi et al., 2019) and activators (Choi et al., 2010), Ca++ Activator (Li et al., 2001), and Protease inhibitors (Matthew et al., 2009; Gunasekera, et. al., 2009). Thus, microalgae have demonstrated a vast biotechnological application with clear health benefits and just barely explored potential. Considering the diverse potential and applications, an over-view of various bioactive compounds including toxins, vitamins, enzymes and pharmaceuticals uses are presented here.

4.1 Antioxidant compounds

Antioxidants are compounds that delay autoxidation by inhibiting formation of free radicals or by interrupting propagation of the harmful free radicals (Lobo et al., 2010). Recent decades have witnessed a growing attention in antioxidant compounds (see Table 8.3) because such compounds can play favourable roles in food preservation and in human health (Wilson et al., 2017). Currently, synthetic antioxidants are used in many products manufactured by the food and pharmaceutical industry but due to their toxicity danger posed to the safety of humans (Cushen et al., 2012). Therefore, there is a need to search for new natural inhibitors of the oxidation process, which will not cause side effects. Microalgae can produce a wide range of secondary metabolites with antioxidant actions, for example, tocopherol, carotenoids (e.g. β-carotene, chlorophyll *a*, lycopene, lutein), phycobiliproteins

Table 8.3 Important antioxidant substance from algal strains.

Algal strains	Antioxidants	References
Chlamydocapsa nivalis	Phenolic antioxidants	Duval et al. (1999)
Botryococcus braunii	β-Carotene, BHT	Borowitzka (1992)
Chlorella pyrenoidosa	Vitamins E	Becker (2004)
Chlorella spp.	Vitamins C	Running et al. (2002)
Chlorella vulgaris	Vitamins C and E Lutein (carotenoid)	Borowitzka (1992) Cha et al. (2008)
Scenedesmus obliquus	Vitamins C and E	Borowitzka (1992)
Scenedesmus quadricauda	Vitamins C and E	Becker (2004)
Chlamydocapsa spp.	Lutein, canthaxanthin and astaxanthin (carotenoids)	Leya et al. (2009)
Dunaliella salina	β-carotene	Mendes et al. (2003); Plaza et al. (2009)
Haematococcus pluvialis	Astaxanthin, cantaxanthin Lutein	Plaza et al. (2009), Nobre et al. (2006)
Chlorella vulgaris	Cantaxanthin, astaxanthin	Plaza et al. (2009), Mendes et al. (1995)
Chlorella pyrenoidosaL	Lutein, violaxanthin	Plaza et al. (2009); Cha et al. (2008)

(e.g. phycocyanin, allophycocyanin, phycoerythrin), polyphenols (e.g. phlorotannins, catechins, flavonoids, tannins, lignans, and mycosporine-like amino acids (MAA) (Nagarajan et al., 2012). These compounds have shown potent antioxidant properties i.e. quench the excitation energy of singlet oxygen radicals (SOR) into their complex ringed chain (Blokhina et al., 2003). The radical scavenging lessens tissue decay by reducing the damage to molecules such as DNA, proteins, and membrane lipids (Lobo et al., 2010). Furthermore, the processes brought about by SOR have also been linked to the aging process and to the pathogenesis of several diseases, viz., cancer, cardiovascular diseases, atherosclerosis, rheumatoid arthritis, muscular dystrophy, cataracts, and several neurological disorders (Guedes et al., 2011b). For instance, C-phycocyanin (CPC) was evaluated as an antioxidant *in vitro* as it was able to scavenge alkoxyl, hydroxyl and peroxy radicals and inhibited microsomal lipids peroxidation *in vitro* (Romay et al., 2003). CPC can constitute up to 15% of the dry weight of a blue-green algae harvest and contribute to the antioxidant, anti-inflammatory, neuroprotective and hepato-protective effects (Jensen et al., 2015). Lutein, a carotenoid vitamin, has also been recommended for the prevention of cancer and diseases related to retinal degeneration (Mozaffarieh et al., 2003). Nakashima et al. (2009) studies indicated that the decline of cognitive ability in Alzheimer's disease may be caused by the persistent oxidative stress in the brain. They showed in transgenic mice fed with *Chlorella* sp. containing carotenoids such as β-carotene and lutein, progression in cognitive impairment could be prevented to a significant extent. β-carotene serves as an essential nutrient and also has high demand in the market as a natural food coloring agent, as an additive to cosmetics and also as a health food (Bogacz-Radomska and Harasym, 2018). Another imperative pigment, astaxanthin, produced from *Haematococcus pluvialis* is unique among carotenoids for being able to span the lipid bilayers and provide superior antioxidant defense (Shah et al., 2016).

Astaxanthin's antioxidant capacities are exceptional and it is much greater than tocopherols and β-carotene when tested via *in vitro* lipid peroxidation and radical scavenging models as well as *in vivo* vitamin E-deficient rat model (Kurashige et al., 1990). Several studies have demonstrated that microalgae, such as *Chlorella vulgaris* and *Scenedesmus quadricauda* are able to presumably produce sulfated polysaccharides that function as protection against microcystin oxidative stress (Mohamed, 2008). The production of natural antioxidant compounds from microalgae in larger scale after bioprocess optimization is urged, even though photoautotrophic biotechnology is far from being fully mastered.

4.2 Anti-inflammatory compounds

Feeding of algae has been verified to promote immunity and to protect against oxidative and inflammation stress (see Table 8.4), implicated in several health conditions such as cardiovascular diseases (Ku et al., 2013), colitis (Coskun et al., 2011), atherosclerosis (Mao et al., 2005), arthritis (Rasool et al., 2006) and allergic rhinitis (Selmi et al., 2011) in animal and human studies. Besides its free radical scavenging effect, C-Phycocyanin, a 35 kDa water-soluble orange carotenoid proteins, also have anti-inflammatory activities, inhibiting the formation of leucotriene, an inflammatory metabolite of arachidonic acid (Romay et al., 2003). CPC also acts as a selective inhibitor of cytochrome oxidase 2, and has hepatoprotective (Vadiraja et al., 1998) and anti-inflammatory effect (Shih et al., 2009). Thus, CPC could be potentially useful for treatment of LPS-related acute lung injury by inhibiting inflammatory responses and apoptosis in lung tissues (Leung et al., 2013). Another anti-inflammatory compound, sulphated polysaccharides was extracted from *Porphyridium* sp., composed of 10 sugars (primarily xylose, glucose, and galactose) in addition to glycoproteins and inorganic sulfate, *Chlorella stigmatophora* and *Phaeodactylum tricornutum*, which attained 25%–30% of that exhibited by the reference compound indomethacin (Matsui et al., 2003). In the former case, the inhibited the spreading of immune cell recruitment toward inflammatory stimuli *in vivo*. Further studies revealed an anti-inflammatory capability in crude extracts of active components were found to be water soluble (Talhouk et al., 2007). *A. flos-aquae* and *Spirulina* contain significant amounts of ώ -3—α-linolenic acid and ώ-6-γ-linolenic acid (Fan and Chapkin, 1998), respectively which inhibit the formation and decrease the level of inflammatory metabolite postaglandins and arachidonate compounds. Thus, compounds have a lowering action on blood pressure and plays an important role in lipid metabolism.

Table 8.4 Vital anti-inflammatory substance from algae.

Microalga source	Bioactive compound	Mechanism of action	Reference
Haematococcus pluvialis, H. lacustris, Scenedesmus sp, C. zofingiensis, C. vulgaris, Chlorococcum sp.	Astaxanthin ($C_{40}H_{52}O_4$, 3,3′-dihydroxy-β,β′-carotene-4,4′-dione) member of xanthophyll family of carotenoids	Inhibition of intracellular ROS accumulation, decrease of H_2O_2-induced NF-κB activation, expression of inducible NOS in primary macrophages, suppression of serum levels of NO	Ma and Chen (2001); Del Campo et al. (2004); Qin et al. (2008)
Porphyridium sp.	Sulphated polysaccharides	Inhibition of (polymorphonuclear leukocytes) PMN migration toward chemoattracting molecules, partial blocking of adhesion to endothelial cells	Ginzberg et al. (2000)
Gyrodinium impudicum strain KG03	p-KG03, a highly sulfated exopolysaccharide	Immunostimulatory effects	Yim et al. (2005); Bae et al. (2006)
Chlorella stigmatophora; Phaeodactylum tricornutum	Hydrosoluble components, Sterols	Immunosuppressant effects	Guzman et al. (2003); Guzman et al. (2001)

4.3 Antimicrobial compounds

In order to out-compete other microorganisms, microalgae have shown a broad spectrum of antimicrobial activity. This section has summarised important new advances in the discovery of novel bioactive antimicrobial compounds.

4.3.1 Antibacterial compounds

Antibiotic sensitivity of multidrug-resistant bacterial strains has caused an increased concern in healthcare institutions worldwide as they are not susceptible to most conventional antibiotics (Exner et al., 2017). Thus, to overcome the resistance problem, it is crucial to discover novel antibacterial compounds that follow distinct biochemical mechanisms of action. As summarized in Table 8.5, many studies have shown microalgae could produce secondary metabolites with antibacterial activity (Catarina et al., 2011; de Jesus Raposo et al., 2014; Dussault et al., 2016; Nowruzi

Table 8.5 Important anti-bacterial products from algae.

Microalga source	Bioactive compound	Target bacteria	Reference
Cyanobacteria			
Fischerella ambigua PTCC 1635	Parsiguine (cyclic polymer)	*Staphylococcus epidermidis PTCC 1114*	Ghasemi et al. (2004)
Nodularia harveyana Nostoc insulare	Norharmane (9H-pyrido (3,4-b) indole) and 4,40-dihydroxybipheny	*Escherichia coli ATCC 8739, Pseudomonas aeruginosa ATCC 9027, Staphylococcus aureus ATCC 6538, Bacillus subtilis ATCC 6633, Bacillus cereus ATCC 14579*	Volk and Furkert (2006)
Cyanobacterium sp. Cyanobium sp. Gloeotrichia sp. Myxosarcina sp. Phormidium autumnale	Aeruginosin, cyanopeptolin, microcystin, and cylindrospermopsin	*B. cereus, E. coli, S. typhimurium, P. aeruginosa, S. pasteuri, M. luteus, Enterococcus faecalis, S. aureus*	Silva-Stenico et al. (2013)
Skeletonema costatum	Extrametabolites	*Listeria monocytogenes*	Terekhova et al. (2009)
Microcystis aeruginosa	Kawaguchipeptin B (cyclic undecapeptide)	*Staphylococcus aureus*	Ishida et al. (1997)
Green Microalgae			
Chlorella vulgaris, Chlorella pyrenoidosa	Chlorellin	*Bacillus subtilis, S. aureus, Streptococcus pyogenes, Escherichia coli, Pseudomonas aeruginosa*	Pratt et al. (1944)
Chlorella vulgaris and Dunaliella salina	Algal extracts	*Streptococcus mutans*	Jafari et al. (2018)
Chlorella vulgaris	Phenolic compounds	*E. coli, Klebsiella sp., Bacillus sp., Pseudomonas sp.*	Syed et al. (2015)
C. ellipsoidea *C. protothecoides* *C. pyrenoidosa*	Saturated and unsaturated fatty acids	*Propionibacterium acnes*	Sibi (2015)
Chlorococcum HS-101	Alpha-linolenic polyunsaturated fatty acid	*B. subtilis, B. cereus, S. aureus, MRSA, Enterobacter aerogenes*	Ohta et al. (1995)

Chlorococcum humicola	Pigments (carotenoid, chlorophyll)	*B. subtilis, S. aureus, E. coli, P. aeruginosa, Salmonella typhimurium, Klebsiella pneumoniae, Vibrio cholerae*	Bhagavathy et al. (2011b)
Dunaliella primolecta	α-linolenic polyunsaturated fatty acid	*B. cereus, B. subtilis, S. aureus, MRSA, E. aerogenes*	Ohta et al. (1995)
D. salina	Indolic derivative, polyunsaturated fatty acids, β ionone and neophytadiene	*S. aureus, E. coli, P. aeruginosa*	Herrero et al. (2006) Mendiola et al. (2008) Pane et al. (2015)
Haematococcus pluvialis	Short-chain fatty acid	*S. aureus, E. coli*	Santoyo et al. (2009)
Scenedesmus obliquus	Long-chain fatty acid	*S. aureus, E. coli, P. aeruginosa, Salmonella sp*	Guedes et al. (2011a)
Tetraselmis suecica	Fatty acid	*Proteus sp., S. pyogenes*	Bai and Krishnakumar (2013)
Red microalgae			
Porphyridium aerugineum, P. cruentum	Phycobiliprotein	*S. aureus, S. pyogenes* *Salmonella typhimurium*	Najdenski et al. (2013)
Rhodella reticulata	Extracellular sulfated polysaccharides	*S. aureus, S. pyogenes, B. cereus, S. typhimurium*	Najdenski et al. (2013)
Diatoms			
Chaetoceros muelleri	Unsaturated fatty acid	*B. subtilis, S. aureus, E. coli*	Mendiola et al. (2007)
Phaeodactylum tricornutum	Fatty acid	*B. cereus, S. aureus, S. epidermidis, MRSA*	Desbois et al. (2008); Desbois et al. (2009)
Haslea karadagensis	Marennine-like pigment	*Polaribacter irgensii, Pseudoalteromonas elyakowii, V. aestuarianus*	Gastineau et al. (2012)
H. ostrearia	Marennine	*P. irgensii, P. elyakowii, V. coralliilyticus, V. tubiashii, V. aestuarianus*	Gastineau et al. (2014) Falaise et al. (2016)
Skeletonema costatum	Unsaturated, saturated long chain fatty acids	*Vibrio* spp.	Naviner et al. (1999)

et al., 2017; Habibi et al., 2018). For instance, Refs. Desbois et al. (2009) reported that activity of cell lysates of *P. tricornutum* was very effective against both Gram-positive (G^+) and Gram-negative (G^-) bacteria, even at micromole levels (Desbois et al., 2009). Furthermore, lysates have very potent antibacterial activity against the methicillin-resistant *Staphylococcus aureus* (MRSA), a bacterium causing infections that kill thousands of people per year and which can be highly resistant to conventional antibiotics. The major components of cell lysate were attributed to three different unsaturated fatty acids: the polyunsaturated fatty acid (PUFA) eicosapentaenoic acid (EPA), the monounsaturated fatty acid palmitoleic acid (PA) and the relatively unusual polyunsaturated fatty acid hexadecatrienoic acid (HTA) (Desbois et al., 2008, 2009, 2010). Likewise, PUFAs from other microalgal strains, *Chlorococcum* strain HS-101 and *Dunaliella primolecta* have also demonstrated antibacterial activity against the MRSA.

Furthermore, short chain fatty acids from *H. pluvialis* (Rodríguez-Meizoso et al., 2010; Santoyo et al., 2009) and long chain fatty acids from *Scenedesmus obliquus* (Guedes et al., 2011a) present antibacterial activity against *Escherichia coli* and *S. aureus*. Catarina et al. (2011) studied the effect of intra- and extracellular extracts of thirteen cyanobacteria and twelve microalgae for their activity against food-borne pathogens such as *Salmonella sp., Pseudomonas aeruginosa, E. coli* and *S. aureus*. The study showed that intracellular extracts were more powerful than their extracellular counterparts of algal cells. The extracts with antibacterial features were found to possess high contents of long chain fatty acids (Exner et al., 2017). The exact mechanism of the released bactericidal fatty acids is not yet fully elucidated, but it seems that they may act on multiple cellular targets against pathogenic bacteria, even though cell membranes are the most probable ones (Desbois et al., 2009). The microalgal bioactive fatty acids may inhibit the fatty acid synthesis in bacteria and damage the bacterial membrane, leading to leakage of cellular protoplast nutrient as well as decrease cellular respiration (Dussault et al., 2016). Thus, studies have shown that gram positive bacteria are more susceptible to microalgae bioactive compounds than gram negative bacteria. For instance, the phycobiliproteins and exopolysaccharides from the red microalgae *Porphyridium aerugineum* and *Rhodella reticulata* respectively, were active against the gram-positive bacteria *S. aureus* and *Streptoccocus pyogenes* but presented no effect against the gram-negative bacteria *E. coli* and *P. aeruginosa* (Najdenski et al., 2013). The diatom *P. tricornutum* did not demonstrate antibacterial effect against these 2-g negative bacteria either, whereas a good antibacterial activity against the gram-positive MRSA was observed (Desbois et al., 2009). Thus, the difference in sensitivities between bacteria may be due to their complex membrane permeability, making the penetration and the bactericidal action of the compound more difficult.

4.3.2 Antifungal compounds

As a matter of fact, fungi have been considered as harmful human pathogens since the 1970s, especially in developing and rural regions where diagnostics and treatment are difficult to distribute (Jung et al., 2018). Due to inefficient access and expensive medicines, mortality induced by life-threatening invasive fungal infections have been increased. Furthermore, growing use of antifungal agents in recent years has led to the development of antifungal resistance, a major clinical challenge to clinicians due to the limited antifungal agents (Ghannoum and Rice, 1999; Wiederhold, 2017). Thus, there is a need for novel drugs and several studies were recently conducted to find fungicide activity from natural marine products against human pathogenic fungi (El Amraoui et al., 2014; Cheung et al., 2014) including antifungal agents from microalgae (de Cano et al., 1990; Shishido et al., 2015). Shishido et al. (2015) detected potent antifungal compounds macrolide scytophycin and glycolipopeptide hassallidin in *Anabaena* sp. and *Nostoc* sp. Interestingly, all the strains shown to produce antifungal compounds in his study belong to Nostocales or Stigonematales cyanobacterial orders. Soltani et al. (2005) examined antifungal activity of soil cyanobacteria. They reported that six cyanobacterial isolates were inhibited the growth of *Candida kefyr* ATCC 1140, and one cyanobacterium species inhibited the growth of *Candida albicans* ATCC 14053. Kellam et al. (1988) also indicated that marine microalgae showed more potential in the search for new antifungal agents than freshwater species. As illustrated in Table 8.6, several antifungal compounds from various microalgae have been characterized. Polysaccharides with high molecular weight were identified in the diatom *Chaetoceros lauderi* (Gueho et al., 1977). They presented a large spectrum of activity against dermatophytes, molds and phytofungi, but no activity was detected against the yeasts tested (Viso et al., 1987). Nagai et al. (1993) also showed that dinoflagellate *Gambierdiscus toxicus* secrets Gambieric acids which have antifungal activity against several dermatophytes and molds but no activity against yeasts like Candida albicans or Saccharomyces cerevisiae. In contrary, diatom *Thalassiothrix frauenfeldii* was meanwhile active against yeasts and molds but not against dermathophytes (Walter and Mahesh, 2000). Other compounds such as pigments like β-carotene, chlorophyll-a and chlorophyll-b from *Chlorococcum humicola* (Bhagavathy et al., 2011a, 2011b), or phycobiliproteins from *P. aerugineum* have also demonstrated antifungal activities (Najdenski et al., 2013). The polyene-polyhydroxy metabolites amphidinols were extracted from the dinoflagellate *Amphidinium klebsii* (Satake et al., 1991; Houdai et al., 2004). Additionally, some cyanobacterial strains are known to produced effective antifungal compounds include Scytophycin from *S. pseudohofmanni* (Burja et al., 2001); Tolytoxin (Patterson and Bolis, 1997) and Phytoalexin (Patterson and Carmeli, 1992) from *S. ocellatum*; Toyocamycin from *Tolypothrix tenuis* (Banker and Carmeli, 1998); Hapalindole from *Hapalosiphon fontinalis* (Burja et al., 2001); Tjipanazoles from *T. tjipanasensis* (Bonjouklian et al.,1991) and Fisherellin from *Fischerella muscicola* (Dahms et al., 2006). These few results illustrate that the search for novel antifungal compounds from microalgae has not been greatly developed so far, although an increasing number of fungi display drug resistance phenomenon.

Table 8.6 Important anti-fungal products from algae.

Microalga source	Bioactive compound	Target fungus	Reference
Amphidinium sp.	Karatungiols (C_{69}-linear polyol)	*Aspergillus niger*	Washida et al. (2006)
Goniodoma pseudogoniaulax	Goniodomin A (polyether macrolide)	*Mortierella ramannianus and Candida albicans*	Murakami et al. (1988)
Lyngbya majuscula	Lyngbyabellin A, B	*Candida albicans*	Milligan et al. (2000)
	Tanikolide	*Candida albicans*	Singh et al. (1999)
Nostoc strain ATCC 53789	Cryptophycin	*Armillaria sp., Penicillium expansum, Rosellinia sp., and Sclerotinia sclerotiorum*	Biondi et al. (2004)
Pithophora oedogonium	Ethanol extract	*Penicillium viridicatum1101, Fusariumsolini1127*	Danyal et al. (2013)
Gloeocapsa sp.	*Exopolysaccharides*	*Candida albicans*	Najdenski et al. (2013)
Haematococcus pluvialis	*Butanoic acid and methyl lactate*	*Candida albicans*	Santoyo et al. (2009)
Fischerella ambigua PTCC 1635	Parsiguine (cyclic polymer)	*Candida krusei ATCC 44507*	Ghasemi et al. (2004)
Nostoc insulare	Norharmane (9H-pyrido (3,4-b) indole) and 4,40-dihydroxybipheny	*Candida albicans ATCC 10231*	Volk and Furkert, 2006
Nostoc sp., Anabaena sp., Scytonema sp.	Hassallidin Scytophycin	*Candida albicans, Aspergillus flavus*	Shishido et al. (2015)
Nannochloris sp	Aponin	*Ptychodiscus brevis*	Halvorson et al. (1984)

4.3.3 Algicidal compounds

Some of cyanobacterial strains can produce algicidal compounds to constrain growth of surrounding algal population (see Table 8.7). Yingying et al. (2008) showed that bioactive compounds of *I. galbana* may inhibit growth of eight species of microalgae either from interference on chlorophyll and protein synthesis or from changes in membrane permeability. Thus, algicidal may leads to obstruct photosynthesis by dissociation of phycobilin assemblages on the thylakoid membranes that lead to leakage across the cell wall (Wu et al., 1998). Several other metabolites such as Nostocyclamide, Nostocine A, Nostocarboline, etc., have also been

Table 8.7 Important algicidal products from algae.

Microalga source	Bioactive compound	Target algal strain	Reference
Peridinium bipes	Water-soluble extract	*Microcystis aeruginosa*	Wu et al. (1998)
Isochrysis galbana	$C_{22}H_{38}O_7$ from cell-free filtrates of cultural medium	*Dunaliella salina, Platymonas elliptica, Chlorella vulgaris, Chaetoceros muelleri, Chlorella gracilis, Nitzschia closterium,*	Yingying et al. (2008)
Nostoc spongiaeforme	Nostocine A	*Chlamydomonas reinhardtii* *Chlorella pyrenoidosa*	Hirata et al. (1996)
Nostoc linckia CALU 892	Cyanobacterin LU-1	*Synechococcus sp. R-2 (PCC 7942)*	Gromov et al. (1991) Flores & Wolk (1986)
Scytonema hofmanni	Cyanobactericin	*Nitzschia pusilla*	Abarzua et al. (1999)
Gomphosphaeria aponina *Nannochloris sp.*	Aponin	*Chattonella subsala* *Karenia brevis*	Halvorson and Martin (1981)
Nodularia harveyana Nostoc insulare	Norharmane (9H-pyrido (3,4-b) indole) and 4,40-dihydroxybipheny	*Arthrospira (Spirulina), Chroococcus minutus, Nostoc insulare and Synechocystis aquatilis*	Volk and Furkert (2006)

described from different species of *Nostoc* that are related with algacidal function (Nowruzi et al., 2017).

4.3.4 Antiviral compounds

Viral pathogens are the leading cause of human diseases and mortality worldwide. Generally, virus infected the host into three steps: Step 1, adsorption and invasion into host cells; Step 2, eclipse phase; and Step 3, maturity and release of viral progeny. Many researchers have shown that microalgae could inactivates the initial viral function right after Step 1. Treatments to block the entry of virus or its replication directly are difficult to design, as they can have adverse effects on the infected host cells (Kitazato et al., 2007). Thus, treatment against diseases caused by viruses are limited, and resistance to these available treatments demonstrate the need for new medicines (Yasuhara-Bell and Lu, 2010). Microalgae have consequently received more attention as they are easier to culture and possess relevant antiviral metabolites. A few selected examples are listed in Table 8.8. Antiviral compounds extracted from microalgae are mainly polysaccharides. Sulphated polysaccharides of microalgae appear to be a particularly attractive option for the viral therapy.

Table 8.8 Important antiviral products from algae.

Microalga source	Active compound	Mechanism of action	Target virus	Reference
Navicula directa	Polysaccharide	Inhibition of hyaluronidase	Herpes Simplex Viruseses (HSV1, 2); Influenza A virus	Lee et al. (2006)
Gyrodinium impudicum KG03	p-KG03 exopolysaccharide	Inhibition (or slowing down) of cytopathic effect	Encephalomyocarditis (EMCV) virus	Yim et al. (2004)
Dunaliella primolecta	Pheophorbide α-, β-like compounds	Inhibition of cytopathic effect	HSV1	Ohta et al. (1998)
Chlorella autotrophica, Ellipsoidon sp.	Sulfated polysaccharides	Replication inhibition *in vitro* of *C. autotrophica*: range 47.4–67.4%; *Ellipsoidon sp.*: up to 44%	Haemorrhagic Septicaemia Virus (VHSV) and African Swine Fever Virus (ASFV)	Fabregas et al. (1999)
Cryptomonads	Allophycocyanin	Inhibition of cytopathic effect, delay in synthesis of viral RNA	Enterovirus 71	Shih et al. (2003)
Oscillatoria raoi	Acetylated sulfoglyco lipids	Inhibition of reverse transcriptase of HIV-1	Human Immunodeficiency Virus (HIV-1)	Reshef et al. (1997)
Spirulina platensis	Calcium spirulan (Ca-SP), sulfated polysaccharide	Inhibition of cytopathic effect and syncytium formation induced by HIV-1	HIV-1 and HSV-1	Hayashi et al. (1996a)
Scytonema varium	Scytovirin	Anticytopathic activity	HIV-1	Bokesch et al. (2003)
Nostoc ellipsosporum	Cyanovirin-N (CV–N)	High-affinity interactions with the viral surface envelope glycoproteins	HIV	Esser et al. (1999)

The sulphated polysaccharide isolated from the cyanobacteria *S. platensis*, named spirulan, has demonstrated potent antiviral activity against the herpes simplex virus type 1 (HSV-1) and against the human immunodeficiency virus type 1 (HIV-1) (Hayashi et al., 1996a). A spirulan-like molecule isolated from *Arthrospira platensis* has also been indicated to possess antiviral activities against these two viruses, with absence of cytotoxic effects (Rechter et al., 2006).

Due to their antiviral spectrum against, e.g., HSV and HIV-1 viruses and pleiotropic mode of action, sulphated exopolysaccharides offer unique advantages for e.g. they are less likely to promote the development of resistant mutants when compared with other compounds that attack only one target throughout the viral life cycle. Furthermore, some microalgal bioactive compounds have been claimed to interfere with Stage I of some enveloped viruses. For instance, the anti-HSV activity of the routinely applied antiviral compound acyclovir® is expressed at Step 2, but the anti-HSV factor from *Dunaliella* sp. extracts inactivates the initial viral function right after Step 1. (Elion et al., 1977; Schaeffer et al., 1978).

Another unique virucidal protein, cyanovirin-N (CV—N) extracted from *Nostoc ellipsosporum* is irreversibly inactivates diverse primary strains of HIV-1 and blocks cell-to-cell transmission of HIV infection (Burja et al., 2001). The compounds isolated from *Lyngbya lagerhaimanii* and *Phormidium tenue* has also been shown to have anti-HIV activity (Schaeffer and Krylov, 2000). Some sulphated polysaccharides present a broad antiviral spectrum against enveloped viruses. Naviculan, extracted from the diatom *Navicula directa*, or A_1 and A_2 extracted from *Cochlodinium polykrikoides* demonstrated to potent antiviral activity against several enveloped viruses such as HIV-1, HSV-1 or influenza virus type A (IFV-A) (Hasui et al., 1995; Yim et al., 2004). The efficiency and the low toxicity of some of the microalgal compounds tested attest their advantageous use as antiviral agents. They could help to control viral diseases occurring in humans, but also in animal species with economic value (Mori and Boyd, 2001).

4.4 Antitumoral compounds

Microalgae are significant source for cytotoxic bioactive metabolites, having anticancer properties with applications in cancer chemotherapy. There is a well-established evidence confirming that microalgal carotenoids possess potent cancer chemo preventive features. In this regard red ketocarotenoid astaxanthin produced by microalgae has attracted considerable interest because of its outstanding antitumoral activity. Kang et al. (2001) reported that astaxanthin protects rat liver from damage induced by carbon tetrachloride, via inhibiting lipid peroxidation and stimulating its cell antioxidant system. Additionally, dietary astaxanthin displays antitumoral effects in the post-initiation phase of carcinogen-induced colon and oral cancer models (Palozza et al., 2009). Suppressive effects thereof have also been reported in transplantable tumor cells, including methylcholanthrene-induced fibrosarcoma and murine mammary tumor cells (Verreault et al., 2011). It has been further suggested that astaxanthin attenuates liver metastasis induced by stress in mice, thus promoting the immune response via inhibition of lipid peroxidation (Kurihara et al., 2002). The effects of astaxanthin on human breast cancer cells have been studied, in which case astaxanthin inhibited the proliferation of an MCF-7 cell line although less effectively than β-carotene and lycopene (Li et al., 2002). In addition to astaxanthin, more than 1,000 natural acetylenic carotenoids have been isolated from a wide variety of sources, including microalgal species (Raposo et al., 2015).

In addition to pigments, a sterol compound, campesterol, has been reported in *Tetraselmis suecica*, to be antiangiogenic (Fábregas et al., 1997). Borophycin, a boron containing metabolite of *Nostoc linckia* and *Nostoc spongiaeforme,* have exhibited potent cytotoxicity against human epidermoid carcinoma and human colorectal adenocarcinoma cell lines (Burja et al., 2001). Furthermore, cryptophycin which is known as a potent fungicide, produced from *Nostoc* sp. ATCC 53789 and *Nostoc* sp. GSV 224, have also shown potent cytotoxicity against human tumor cell lines (Trimurtulu et al., 1994). It also shows good activity against a broad-spectrum drug-sensitive and drug-resistant murine and human solid tumor (Burja et al., 2001). The mechanism of cytotoxicity of the cryptophycins is tubulin-interaction, with a disruption of tubulin-dynamics, resulting in apoptosis of tumor cells (Panda et al., 1998).

Another cyanobacterial cytotoxic compound, curacin A, is a major lipid component of tropical marine species *Lyngbya majuscula.* Curacin A shows strong cytotoxicity toward L1210 leukemic cell lines and inhibits tubulin polymerization by binding to it (Burja et al., 2002). Furthermore, biological activities of *L. majuscula* are very diverse and shares approximately 30% of all the natural products isolated from marine cyanobacteria. Muniaraj et al. (2018) performed *in silico* molecular docking between the cyanobacterial bioactive compounds and lung cancer causing receptor. They showed that Lyngbyastatin from *Lyngbya majuscula* have highest docking score against human lung cancer cell line (A549) using its methanolic extract. Additional examples of antitumoral compounds synthesized by algae are provided in Table 8.9.

4.5 Additional bioactive compounds

In the previous sections, significance of microalgae as diverse sources of bioactive compounds has been discussed with health and commercial values. However, the enormous potential of microalgae is still hold for discovery of new bioactive molecules (see Table 8.10). A lipophilic neurotoxin compounds, brevitoxins are produced by *Ptychodiscus brevis* which depolarizes the open voltage gated sodium (Na^+) ion channel in the cell wall. Thus, brevitoxins leads to uncontrolled Na^+ influx into the cell (Baden, 1983) and consequently, hyperexcited to nervous and muscular cells. The biosynthesis of natural sunscreen compounds, mycosporines and mycosporine like amino acids (MAAs) have also been reported in cyanobacteria such as *Lyngbya* sp. (Rastogi and Incharoensakdi, 2014), and *Nostoc commune* (Nazifi et al., 2013). Researchers have shown ability of MAAs to prevent UV– induced damage *in vivo* in mice (de la Coba et al., 2009) and fibroblast cells in human (Oyamada et al., 2008).

Table 8.9 Important antitumoral products from algae.

Microalga source	Active compound	Mechanism of action	Target virus	Reference
Haematococcus pluvialis	Astaxanthin	Expression decrease of cyclin D1, increase of p53 and some cyclin kinase inhibitors (p21WAF-1/CIP-1, p27)	Colon cancer cell lines	Palozza et al. (2009)
Gymnodinium sp.	Extracellular polysaccharide GA3P, D-galactan sulfate associated with L-(+)-lactic acid	Catalytic inhibition of DNA topoisomerase (topo I, topo II)	Human myeloid leukemia K562 cells	Umemura et al. (2003)
Lyngbya majuscula	Curacin A	Microtubulin assembly inhibitors	Colon, Renal, And Breast-Cancer-Derived Cell	Gerwick et al. (1994)
Oscillatoria nigroviridis	Almiramides D	Antitumour	Human tumor cell lines (A549, MCF-7, HeLa, PC3 and MDA-MB231)	Quintana et al. (2014)
Lyngbya sp.	Jahanyne	Antitumor	HeLa cells	Iwasaki et al. (2015)
L. bouillonii	Apratoxins F and G H-460	Antitumor	Human lung cancer cells	Tidgewell et al. (2010)
Phormidium spp.	Caylobolide B	Antitumor	HT29 colorectal adenocarcinoma and HeLa cervical carcinoma cells	Salvador et al. (2010)
L. majuscula	Cocosamides A and B	Antitumor	MCF7 and HT-29 cells	Gunasekera et al. (2011)
L. crossbyana	Crossbyanol A	Antitumor	H-460 human lung cancer cells	Choi et al. (2010)

Table 8.10 Important bioactive compounds from algae.

Microalga source	Active compound	Application	Reference
Lyngbya majuscula	Lyngbyatoxins and Kalkitoxin	Na^+ channel blocker	Shimizu (2003)
	Palmyralide A		Taniguchi et al. (2019)
	Jamaicamide A-C		Edwards et al. (2004)
	Tanikolide dimer		Gutiérrez et al. (2009)
	Hoiamide A		Pereira et al. (2009)
	Antillatoxin B	Na^+ channel activator	Nogle et al. (2001)
	Crossbyanol A		Choi et al. (2010)
	Antillatoxin	Ca^{++} Activator	Li et al. (2001)
Arthrospira	γ-Linolenic acid (GLA)	Infant formulas for full-term infants, Nutritional supplements	Kovaa et al. (2013); Qawasmi et al. (2013)
Porphyridium	Arachidonic acid (AA)	Infant formulas for full-term/preterm infants Nutritional supplements	
Nannochloropsis, Phaeodactylum, Nitzschia	Eicosapentaenoic acid (EPA)	Nutritional supplements, Aquaculture	
Crypthecodinium, Schizochytrium	Docosahexaenoic acid (DHA)	Infant formulas for full-term/preterm infants, Nutritional supplements, Aquaculture	
Anabaena circinalis	Anabaeno peptilides	protease inhibitor Chymotrypsin	Harada (2004)
Dichothrix utahensis	Molassamide	Protease inhibitors Elastase and chymotrypsin	Gunasekera et al. (2009)
L. confervoides	Tiglicamides A-C	Protease inhibitor Porcine pancreatic elastase	Matthew et al. (2009)
Lyngbya cryptovaginata	Geosmin	Odorous metabolites	Jüttner and Watson (2007)
Phormidium favosum	2- Methyliso- borneol	Odorous metabolites	Sugiura et al. (1998)

Some of the cyanobacteria constitute potential sources for large-scale production of vitamins. *Spirulina* sp is claimed to be the richest whole-food source of vitamin B_{12} and tocopherol (vitamin E) with 20 g of *Spirulina* sp fulfilling the daily body requirements of vitamin B_{12} and 70% of B_1, 50% of vitamin B_2 and 12% of B_3 (Watanabe, and Hall, 1996). *Aphanizomenon. flosaquae, Anabaena hassali, Microcystis pulverana, Nostoc punctiforme* and *Phormidium bijugatum* contain the B-complex vitamins, nicotinic acid and pentothene (Sharma et al., 2011; Raja et al., 2016).

5. Microalgal cell factories for bioactive compounds

The production of bioactive compounds from microalgal biomass consists of many steps such as; Upstream processing (preparation of microalgae for cultivation), Cultivation (growth in photobioreactors under optimized conditions), Downstream processing (cell harvesting by centrifugation, extraction and isolation of bioactive compounds), Formulation, Preservation, and Conditioning.

5.1 Microalgae cultivation

Microalgae are obligate photoautotrophs that require light for growth (Zaslavskaia et al., 2001). However, several microalgae are also capable of heterotrophic growth having organic substrate as single carbon source (Perez-Garcia and Bashan, 2015). Here we summarised the different ways of cultivation of microalgae.

5.1.1 Photoautotrophic cultivation system

There are three main ways to cultivate microalgae: (a) open pond system; (b) closed photobioreactor with natural sunlight; and (c) closed photobioreactor with artificial illumination (Chen et al., 2011). Mass cultivation of microalgae takes place in an open pond where production systems are located outdoors and identical to the natural growth environment, rely on inexpensive natural light for illumination. The most common technical designs for open pond systems are cultivations driven by paddle wheels and usually operating at water depths 15–20 cm (Rogers et al., 2014). At this water depth, biomass concentration up to 1000 mg. L^{-1} and productivities up to 60–100 mg. $L^{-1}.d^{-1}$ (10–25 g $m^{-2}.d^{-1}$) are possible. However, it is indeed difficult to control temperature, light distribution, and CO_2 concentration therein and to prevent microbial contamination that would reduce the throughput rate and compromise the quality of the final product(s). Thus, production of high value bioactive products from microalgae for biotechnological application appears to be feasible only in closed algal photobioreactors where vessels have a high surface to volume ratio and processing conditions could be reproduced (Posten, 2009).

A variety of closed photobioreactor configurations have accordingly been proposed to suit the characteristics of distinct microalgal strains. Their designs include horizontal or helical tubular system (glass, plastic, bags), cascade reactors, alveolar flattened plate system, vertical flat panels, or bubble columns, ultra-thin immobilized configurations etc (Ting et al., 2017). Among all proposed designs, vertical arrangement of horizontally running tubes and plates system seems to be preferred due to proper light distribution and appropriate medium flow. Important cultivation parameters need to be maintained for optimum production includes: species efficient light incidence into photobioreactor lumen, light path, layer thickness, O_2 release from the total system volume, CO_2/O_2 balance, salinity, nutrients, pH, temperature and turbulence (Zittelli et al., 2013).

5.1.2 Heterotrophic cultivation system

Heterotrophic cultivation systems are a cost-effective and viable solution to deprived biomass, high production cost and low productivity of photoautotrophic cultivation systems. As sugar and organic acids are used as the sole carbon source in heterotrophic cultivation which eliminates the requirement of light, resulting increase in the cell density and productivity. Several industrial algal strains have been successfully cultivated under heterotrophic culture conditions. For examples, heterotrophic *Nitzschia laevis* to produce eicosapentaenoic acid and docosahexaenoic acid (Wen and Chen, 2000), *Chlorella protothecoides* to produce biodiesel (Gao et al., 2010) and *Galdieria sulphuraria* to produce C-phycocyanin (Graverholt and Eriksen, 2007). The heterotrophic mode can further be modified to fed-batch (Izzati and Mardlijah, 2015), chemostat culture (Grima et al., 1993), which further reduces the cost of bioactive compounds recovery. Heterotrophic production requires an organism that can divide in the dark and can grow on inexpensive and sterilizable media, can adapt quickly to new environments and can withstand hydrodynamic stresses in the fermenter and peripheral environments. Other factors influencing the production of EPA are culture age, carbon source, nitrogen source, C/N ratio, C/P ratio and environmental factors like temperature (low temperature favors high PUFA content), pH (a pH range of 6.0–8.8 with optimum of 7.6) and salinity (Nielsen, 1999).

5.2 Extraction of algae-derived bioactive compounds

There are different techniques which can be used to extract biologically active compounds from cultivated microalgal biomass. However, choice of the appropriate method has a significant influence on the yield and activity of the extracted compounds. Thus, optimal extraction process conditions like proper solvent (e.g., water and organic solvents), pH, temperature, pressure etc. should be based on the nature of extracted bioactive compound. Conventional extraction techniques include: extraction in Soxhlet apparatus, Solid–Liquid Extraction (SLE) and Liquid–Liquid Extraction (LLE) (Ibañez et al., 2011). In these classic extraction approaches, organic solvents (i.e., petroleum ether, hexane, cyclohexane, isooctane, toluene, benzene, diethyl ether, dichloromethane, isopropanol, chloroform, acetone, methanol, ethanol etc.) are used. However, traditional methods have some disadvantages such as use of high volume of toxic solvents, difficult medium separation and long extraction times. Therefore, in the recent years, solvent-free methods of extraction such as supercritical fluid extraction (SFE), microwave-assisted extraction (MAE), ultrasound-assisted extraction (UAE), enzyme-assisted extraction (EAE), and pressurized liquid extraction (PLE) have been proposed (Michalak et al., 2015). These techniques are characterized by a higher extraction yield, a shorter process time, and finally they are more environmentally friendly than traditional extraction methods. The main advantage of novel extraction techniques is the extraction of biologically active compounds without loss of their activity. Among extraction techniques, SFE has the highest potential of all the new "green" techniques to be applied in the industrial scale. SFE is an economically efficient (recyclable used CO_2 as a solvent) and eco-friendly (solvent free extraction) technique, characterized by fast extraction rate, and high product yield (Herrero et al., 2010).

6. Concluding remarks and future directions

The manifold range of products such as fatty acids, antioxidants, proteins and even antibiotics accessible from diverse cyanobacterial and microalgal species, clearly demonstrates a wide range of applications and importance of these versatile microbes as cellular factories in biotechnology. If done in a proper and effective way, concepts for efficient cultivation of microalgae for products needed by mankind can exploit for various biotechnological applications such as high-value products, including nutraceuticals, and bioactive molecules. Furthermore, there is great potential to subsidize combating climate change, or even diminish it to a certain extent by CO_2 sequestering. Realizing needed enhancements of each process step in microalgal cultivation and extraction, substantial progress toward a cost-efficient algal-based technology for bioactive compounds is possible. Until today, especially the lacking synopsis of skills in process design and well-grounded understanding of intracellular activities hampers the broad industrial employment of these powerful phototrophic cell-factories. There is a clear and almost unrevealed potential in the development of innovative nutraceutical and pharmaceutical industries based on cultivation of microalgae and cyanobacteria and their exploitation in the production of bioactive substances.

References

Abarzua, S., Jakubowski, S., Eckert, S., Fuchs, P., 1999. Biotechnological investigation for the prevention of marine biofouling II. Blue-green algae as potential producers of biogenic agents for the growth inhibition of microfouling organisms. Bot. Mar. 42, 459–465.

Abed, R.M.M., Dobretsov, S., Sudesh, K., 2009. Applications of cyanobacteria in biotechnology. J. Appl. Microbiol. 106, 1–12.

Andersen, R.A., 1992. Diversity of eukaryotic algae. Biodivers. Conserv. 1, 267–292.

Baden, D.G., 1983. Marine food-borne dinoflagellate toxins. In: International Review of Cytology, vol. 82. Elsevier, pp. 99–150.

Bae, S.-Y., Yim, J.H., Lee, H.K., Pyo, S., 2006. Activation of murine peritoneal macrophages by sulfated exopolysaccharide from marine microalga *Gyrodinium impudicum* (strain KG03).: involvement of the NF-kB and JNK pathway. Int. Immunopharmacol. 6, 473–484.

Bai, V.D.M., Krishnakumar, S., 2013. Evaluation of antimicrobial metabolites from marine microalgae *Tetraselmis suecica* using gas chromatography: mass spectrometry (GC-MS). analysis. Int. J. Pharm. Pharm. Sci. 5, 17–23.

Banker, R., Carmeli, S., Tenuecyclamides, A.-D., 1998. Cyclic Hexapeptides from the Cyanobacterium *Nostoc spongiaeforme* var. tenue. Journal of natural products 61, 1248–1251.

Becker, W., 2004. 18 microalgae in human and animal nutrition. In: Handbook of Microalgal Culture: Biotechnology and Applied Phycology, p. 312.

Bhagavathy, S., Sumathi, P., Madhushree, M., 2011a. Antimutagenic assay of carotenoids from green algae *Chlorococcum humicola* using *Salmonella typhimurium* TA98, TA100 and TA102. Asian Pac. J. Trop. Dis. 1, 308–316.

Bhagavathy, S., Sumathi, P., Bell, I.J.S., 2011b. Green algae *Chlorococcum humicola*-a new source of bioactive compounds with antimicrobial activity. Asian Pac. J. Trop. Biomed. 1, S1–S7.

Biondi, N., Piccardi, R., Margheri, M.C., Rodolfi, L., Smith, G.D., Tredici, M.R., 2004. Evaluation of *Nostoc* strain ATCC 53789 as a potential source of natural pesticides. Appl. Environ. Microbiol. 70, 3313–3320.

Blokhina, O., Virolainen, E., Fagerstedt, K.V., 2003. Antioxidants, oxidative damage and oxygen deprivation stress: a review. Ann. Bot. 91, 179–194.

Bogacz-Radomska, L., Harasym, J., 2018. β-carotene properties and production methods. Food Qual Saf. 2, 69–74.

Bokesch, H.R., O'Keefe, B.R., McKee, T.C., Pannell, L.K., Patterson, G.M.L., Gardella, R.S., Sowder, R.C., Turpin, J., Watson, K., Buckheit, R.W., 2003. A potent novel anti-HIV protein from the cultured cyanobacterium *Scytonema varium*. Biochemistry 42, 2578–2584.

Borowitzka, M.A., 1992. Algal biotechnology products and processes: matching science and economics. J. Appl. Phycol. 4, 267–279.

Bonjouklian, R., Smitka, T.A., Doolin, L.E., Molloy, R.M., Debono, M., Shaffer, S.A., Moore, R.E., Stewart, J.B., Patterson, G.M.L., 1991. Tjipanazoles, new antifungal agents from the blue-green alga *Tolypothrix tjipanasensis*. Tetrahedron 47, 7739–7750.

Burja, A.M., Banaigs, B., Abou-Mansour, E., Burgess, J.G., Wright, P.C., 2001. Marine cyanobacteria: a prolific source of natural products. Tetrahedron 57, 9347–9377.

Burja, A.M., Abou-Mansour, E., Banaigs, B., Payri, C., Burgess, J.G., Wright, P.C., 2002. Culture of the marine cyanobacterium, *Lyngbya majuscula* (Oscillatoriaceae)., for bioprocess intensified production of cyclic and linear lipopeptides. J. Microbiol. Methods 48, 207–219.

Castenholz, R.W., Wilmotte, A., Herdman, M., Rippka, R., Waterbury, J.B., Iteman, I., Hoffmann, L., 2001. Phylum BX. cyanobacteria. In: Bergey's Manual of Systematic Bacteriology. Springer, pp. 473–599.

Catarina, G.,A., Barbosa, C.R., Amaro, H.M., Pereira, C.u.I., Xavier Malcata, F., 2011. Microalgal and cyanobacterial cell extracts for use as natural antibacterial additives against food pathogens. Int. J. Food Sci. Technol. 46, 862–870.

Cha, K.H., Koo, S.Y., Lee, D.-U., 2008. Antiproliferative effects of carotenoids extracted from *Chlorella ellipsoidea* and *Chlorella vulgaris* on human colon cancer cells. J. Agric. Food Chem. 56, 10521–10526.

Chen, C.-Y., Yeh, K.-L., Aisyah, R., Lee, D.-J., Chang, J.-S., 2011. Cultivation, photobioreactor design and harvesting of microalgae for biodiesel production: a critical review. Bioresour. Technol. 102, 71–81.

Cheung, R.C.F., Wong, J.H., Pan, W.L., Chan, Y.S., Yin, C.M., Dan, X.L., Wang, H.X., Fang, E.F., Lam, S.K., Ngai, P.H.K., 2014. Antifungal and antiviral products of marine organisms. Appl. Microbiol. Biotechnol. 98, 3475–3494.

Choi, H., Engene, N., Smith, J.E., Preskitt, L.B., Gerwick, W.H., 2010. Crossbyanols A-D, toxic brominated polyphenyl ethers from the Hawaiian bloom-forming Cyanobacterium *Leptolyngbya crossbyana*. J. Nat. Prod. 73, 517–522.

Choopani, A., Poursoltan, M., Fazilati, M., Latifi, A.M., Salavati, H., 2016. Spirulina: a source of gamma-linoleic acid and its applications. J. Appl. Biotechnol. Rep. 3, 483–488.

Coskun, Z.K., Kerem, M., Gurbuz, N., Omeroglu, S., Pasaoglu, H., Demirtas, C., Lortlar, N., Salman, B., Pasaoglu, O.T., Turgut, H.B., 2011. The study of biochemical and histopathological effects of *Spirulina* in rats with TNBS-induced colitis. Bratisl. Lek. Listy 112, 235–243.

Cushen, M., Kerry, J., Morris, M., Cruz-Romero, M., Cummins, E., 2012. Nanotechnologies in the food industry : recent developments, risks and regulation. Trends Food Sci. Technol. 24, 30–46.

Dahms, H.-U., Ying, X., Pfeiffer, C., 2006. Antifouling potential of cyanobacteria: a mini-review. Biofouling 22, 317–327.

Danyal, A., Mubeen, U., Malik, K.A., 2013. Investigating two native algal species to determine antibiotic susceptibility against some pathogens. Curr. Res. J. Biol. Sci. 5, 70–74.

de Cano, M.M.S., de Mulé, M.C., de Caire, G.Z., de Halperin, D.R., 1990. Inhibition of *Candida albicans* and *Staphylococcus aureus* by phenolic compounds from the terrestrial cyanobacterium *Nostoc muscorum*. J. Appl. Phycol. 2, 79–81.

de Jesus Raposo, M.F., de Morais, A.M.M.B., de Morais, R.M.S.C., 2014. Influence of sulphate on the composition and antibacterial and antiviral properties of the exopolysaccharide from *Porphyridium cruentum*. Life Sciences 101, 56–63.

De la Coba, F., Aguilera, J., Figueroa, F.L., De Ga¡lvez, M.V., Herrera, E., 2009. Antioxidant activity of mycosporine-like amino acids isolated from three red macroalgae and one marine lichen. J. Appl. Phycol. 21, 161–169.

Del Campo, J.A., Rodriguez, H., Moreno, J., Vargas, M.A., Rivas, J., Guerrero, M.G., 2004. Accumulation of astaxanthin and lutein in *Chlorella zofingiensis* (Chlorophyta). Appl. Microbiol. Biotechnol. 64, 848–854.

Desbois, A.P., Lebl, T., Yan, L., Smith, V.J., 2008. Isolation and structural characterisation of two antibacterial free fatty acids from the marine diatom, *Phaeodactylum tricornutum*. Appl. Microbiol. Biotechnol. 81, 755–764.

Desbois, A.P., Mearns-Spragg, A., Smith, V.J., 2009. A fatty acid from the diatom *Phaeodactylum tricornutum* is antibacterial against diverse bacteria including multi-resistant *Staphylococcus aureus* (MRSA). Mar. Biotechnol. 11, 45–52.

Desbois, A.P., Walton, M., Smith, V.J., 2010. Differential antibacterial activities of fusiform and oval morphotypes of *Phaeodactylum tricornutum (Bacillariophyceae)*. J. Mar. Biol. Assoc. U. K. 90, 769–774.

Dussault, D., Vu, K.D., Vansach, T., Horgen, F.D., Lacroix, M., 2016. Antimicrobial effects of marine algal extracts and cyanobacterial pure compounds against five foodborne pathogens. Food Chem. 199, 114–118.

Duval, B., Shetty, K., Thomas, W.H., 1999. Phenolic compounds and antioxidant properties in the snow alga Chlamydomonas nivalis after exposure to UV light. J. Appl. Phycol. 11, 559.

Edwards, D.J., Marquez, B.L., Nogle, L.M., McPhail, K., Goeger, D.E., Roberts, M.A., Gerwick, W.H., 2004. Structure and biosynthesis of the jamaicamides, new mixed polyketide-peptide neurotoxins from the marine cyanobacterium *Lyngbya majuscula*. Chem. Biol. 11, 817–833.

El Amraoui, B., El Amraoui, M., Cohen, N., Fassouane, A., 2014. Anti-Candida and anti-Cryptococcus antifungal produced by marine microorganisms. J. Med. Mycol. 24, e149–e153.

Elion, G.B., Furman, P.A., Fyfe, J.A., De Miranda, P., Beauchamp, L., Schaeffer, H.J., 1977. Selectivity of action of an antiherpetic agent, 9-(2-hydroxyethoxymethyl). guanine. Proc. Natl. Acad. Sci. Unit. States Am. 74, 5716–5720.

Encarnação, T., Pais, A.A.C.C., Campos, M.G., Burrows, H.D., 2015. Cyanobacteria and microalgae: a renewable source of bioactive compounds and other chemicals. Sci. Prog. 98, 145–168.

Esser, M.T., Mori, T., Mondor, I., Sattentau, Q.J., Dey, B., Berger, E.A., Boyd, M.R., Lifson, J.D., 1999. Cyanovirin-N binds to gp120 to interfere with CD4-dependent human immunodeficiency virus type 1 virion binding, fusion, and infectivity but does not affect the CD4 binding site on gp120 or soluble CD4-induced conformational changes in gp120. J. Virol. 73, 4360–4371.

Exner, M., Bhattacharya, S., Christiansen, B., Gebel, J., Goroncy-Bermes, P., Hartemann, P., Heeg, P., Ilschner, C., Kramer, A., Larson, E., 2017. Antibiotic resistance: what is so special about multidrug-resistant Gram-negative bacteria? GMS hyg. Infect. Control 12.

Fábregas, J., Arán, J., Morales, E.D., Lamela, T., Otero, A., 1997. Modification of sterol concentration in marine microalgae. Phytochemistry 46, 1189–1191.

Fabregas, J., Garcaa, D., Fernandez-Alonso, M., Rocha, A.I., Gomez-Puertas, P., Escribano, J.M., Otero, A., Coll, J.M., 1999. In vitro inhibition of the replication of haemorrhagic septicaemia virus (VHSV). and African swine fever virus (ASFV). by extracts from marine microalgae. Antivir. Res. 44, 67–73.

Falaise, C., Franaois, C., Travers, M.-A., Morga, B., Haure, J., Tremblay, R., Turcotte, F., Pasetto, P., Gastineau, R., Hardivillier, Y., 2016. Antimicrobial compounds from eukaryotic microalgae against human pathogens and diseases in aquaculture. Mar. Drugs 14, 159.

Fan, Y.-Y., Chapkin, R.S., 1998. Importance of dietary γ-linolenic acid in human health and nutrition. J. Nutr. 128, 1411–1414.

Flores, E., Wolk, C.P., 1986. Production, by filamentous, nitrogen-fixing cyanobacteria, of a bacteriocin and of other antibiotics that kill related strains. Arch. Microbiol. 145, 215–219.

Gao, C., Zhai, Y., Ding, Y., Wu, Q., 2010. Application of sweet sorghum for biodiesel production by heterotrophic microalga *Chlorella protothecoides*. Appl. Energy 87, 756–761.

Gastineau, R., Davidovich, N., Bardeau, J.-F., Caruso, A., Leignel, V., Hardivillier, Y., Jacquette, B., Davidovich, O.I., Rinca, Y., Gaudin, P., 2012. *Haslea karadagensis* (Bacillariophyta).: a second blue diatom, recorded from the Black Sea and producing a novel blue pigment. Eur. J. Phycol. 47, 469–479.

Gastineau, R., Davidovich, N., Hansen, G., Rines, J., Wulff, A., Kaczmarska, I., Ehrman, J., Hermann, D., Maumus, F., Hardivillier, Y., 2014. Haslea ostrearia-like diatoms: biodiversity out of the blue. In: Advances in Botanical Research, vol. 71. Elsevier, pp. 441–465.

Gbonjubola, V.A., Koorbanally, N.A., Islam, M.S., 2016. Inhibition of key enzymes linked to type 2 diabetes by compounds isolated from *Aframomum melegueta* fruit AU - mohammed, Aminu. Pharmaceut. Biol. 55, 1010–1016.

Gerwick, W.H., Proteau, P.J., Nagle, D.G., Hamel, E., Blokhin, A., Slate, D.L., 1994. Structure of curacin A, a novel antimitotic, antiproliferative and brine shrimp toxic natural product from the marine cyanobacterium *Lyngbya majuscula*. J. Org. Chem. 59, 1243–1245.

Ghannoum, M.A., Rice, L.B., 1999. Antifungal agents: mode of action, mechanisms of resistance, and correlation of these mechanisms with bacterial resistance. Clin. Microbiol. Rev. 12, 501–517.

Ghasemi, Y., Yazdi, M.T., Shafiee, A., Amini, M., Shokravi, S., Zarrini, G., 2004. Parsiguine, a novel antimicrobial substance from *Fischerella ambigua*. Pharmaceut. Biol. 42, 318–322.

Ginzberg, A., Cohen, M., Sod-Moriah, U.A., Shany, S., Rosenshtrauch, A., Arad, S.M., 2000. Chickens fed with biomass of the red microalga *Porphyridium* sp. have reduced blood cholesterol level and modified fatty acid composition in egg yolk. J. Appl. Phycol. 12, 325–330.

Gkelis, S., Lanaras, T., Sivonen, K., 2015. Cyanobacterial toxic and bioactive peptides in freshwater bodies of Greece: concentrations, occurrence patterns, and implications for human health. Mar. Drugs 13, 6319–6335.

Graverholt, O.S., Eriksen, N.T., 2007. Heterotrophic high-cell-density fed-batch and continuous-flow cultures of *Galdieria sulphuraria* and production of phycocyanin. Appl. Microbiol. Biotechnol. 77, 69–75.

Grima, E.M., Parez, J.A.S.n., Camacho, F.G., SÃ¡nchez, J.L.G., Alonso, D.L., 1993. n-3 PUFA productivity in chemostat cultures of microalgae. Appl. Microbiol. Biotechnol. 38, 599–605.

Gromov, B.V., Vepritskiy, A.A., Titova, N.N., Mamkayeva, K.A., Alexandrova, O.V., 1991. Production of the antibiotic cyanobacterin LU-1 by *Nostoc linckia* CALU 892 (cyanobacterium). J. Appl. Phycol. 3, 55–59.

Guaadaoui, A., Benaicha, S., Elmajdoub, N., Bellaoui, M., Hamal, A., 2014. What is a bioactive compound? A combined definition for a preliminary consensus. Int. J. Nutr. Food Sci. 3, 174–179.

Guedes, A.C., Amaro, H.M., Pereira, R.D., Malcata, F.X., 2011a. Effects of temperature and pH on growth and antioxidant content of the microalga *Scenedesmus obliquus*. Biotechnol. Prog. 27, 1218–1224.

Guedes, A.C., Amaro, H.M., Malcata, F.X., 2011b. Microalgae as sources of high added-value compounds - a brief review of recent work. Biotechnol. Prog. 27, 597–613.

Gueho, E., Pesando, D., Barelli, M., 1977. Proprietes antifongiques d'une diatomee *Chaetoceros lauderi* ralfs CC. Mycopathologia 60, 105–107.

Gunasekera, S.P., Miller, M.W., Kwan, J.C., Luesch, H., Paul, V.J., 2009. Molassamide, a depsipeptide serine protease inhibitor from the marine cyanobacterium Dichothrix utahensis. Journal of natural products 73, 459–462.

Gunasekera, S.P., Owle, C.S., Montaser, R., Luesch, H., Paul, V.J., 2011. Malyngamide 3 and cocosamides A and B from the marine cyanobacterium *Lyngbya majuscula* from Cocos Lagoon, Guam. J. Nat. Prod. 74, 871–876.

Gutiérrez, M., Andrianasolo, E.H., Shin, W.K., Goeger, D.E., Yokochi, A., Schemies, J., Jung, M., France, D., Cornell-Kennon, S., Lee, E., 2009. Structural and synthetic investigations of tanikolide dimer, a SIRT2 selective inhibitor, and tanikolide seco-acid from the Madagascar marine cyanobacterium *Lyngbya majuscula*. J. Org. Chem. 74, 5267–5275.

Guzman, S., Gato, A., Calleja, J.M., 2001. Antiinflammatory, analgesic and free radical scavenging activities of the marine microalgaeChlorella stigmatophora andPhaeodactylum tricornutum. Phytother Res. 15, 224–230.

Guzman, S., Gato, A., Lamela, M., Freirea Garabal, M., Calleja, J.M., 2003. Anti-inflammatory and immunomodulatory activities of polysaccharide from *Chlorella stigmatophora* and *Phaeodactylum tricornutum*. Phytother Res. 17, 665–670.

Habibi, Z., Imanpour Namin, J., Ramezanpour, Z., 2018. Evaluation of antimicrobial activities of microalgae *Scenedesmus dimorphus* extracts against bacterial strains. Casp. J. Environ. Sci. 16, 25–36.

Harada, K.-i., 2004. Production of secondary metabolites by freshwater cyanobacteria. Chemical and Pharmaceutical bulletin 52, 889–899.

Halvorson, M.J., Martin, D.F., 1981. Cytolysis of Chattonella subsalsa by aponin. J. Environ. Sci., Health Part A 16, 373–379.

Halvorson, M.J., Testrake, D., Martin, D.F., 1984. Effect of aponin, a substance from a green alga *Nannochloris* species, on the spore germination of two fungi. Microbios 41, 105–113.

Hasui, M., Matsuda, M., Okutani, K., Shigeta, S., 1995. In vitro antiviral activities of sulfated polysaccharides from a marine microalga (*Cochlodinium polykrikoides*). against human immunodeficiency virus and other enveloped viruses. Int. J. Biol. Macromol. 17, 293–297.

Hayashi, K., Hayashi, T., Kojima, I., 1996a. A natural sulfated polysaccharide, calcium spirulan, isolated from Spirulina platensis: in vitro and ex vivo evaluation of anti-herpes simplex virus and anti-human immunodeficiency virus activities. AIDS Res. Hum. Retrovir. 12, 1463–1471.

Hayashi, T., Hayashi, K., Maeda, M., Kojima, I., 1996b. Calcium spirulan, an inhibitor of enveloped virus replication, from a blue-green alga *Spirulina platensis*. J. Nat. Prod. 59, 83–87.

Herrero, M., Jaime, L., Martan-A lvarez, P.J., Cifuentes, A., Ibaj, E., 2006. Optimization of the extraction of antioxidants from *Dunaliella salina* microalga by pressurized liquids. J. Agric. Food Chem. 54, 5597–5603.

Herrero, M., Mendiola, J.A., Cifuentes, A., Iba, E., 2010. Supercritical fluid extraction: recent advances and applications. J. Chromatogr. A 1217, 2495–2511.

Hirata, K., Takashina, J., Nakagami, H., Ueyama, S., Murakami, K., Kanamori, T., Miyamoto, K., 1996. Growth inhibition of various organisms by a violet pigment, nostocine A, produced by *Nostoc spongiaeforme*. Biosci. Biotechnol. Biochem. 60, 1905–1906.

Hitzfeld, B.C., Hager, S.J., Dietrich, D.R., 2000. Cyanobacterial toxins: removal during drinking water treatment, and human risk assessment. Environ. Health Perspect. 108, 113–122.

Holzinger, A., Karsten, U., 2013. Desiccation stress and tolerance in green algae: consequences for ultrastructure, physiological and molecular mechanisms. Front. Plant Sci. 4, 327.

Houdai, T., Matsuoka, S., Matsumori, N., Murata, M., 2004. Membrane-permeabilizing activities of amphidinol 3, polyene-polyhydroxy antifungal from a marine dinoflagellate. Biochim. Biophys. Acta Biomembr. 1667, 91–100.

Ibañez, E., Herrero, M., Mendiola, J.A., Castro-Puyana, M., 2011. Extraction and characterization of bioactive compounds with health benefits from marine resources: macro and micro algae, cyanobacteria, and invertebrates. In: Marine Bioactive Compounds. Springer, pp. 55–98.

Ishida, K., Matsuda, H., Murakami, M., Yamaguchi, K., 1997. Kawaguchipeptin B, an antibacterial cyclic undecapeptide from the cyanobacterium *Microcystis aeruginosa*. J. Nat. Prod. 60, 724–726.

Iwasaki, A., Ohno, O., Sumimoto, S., Ogawa, H., Nguyen, K.A., Suenaga, K., 2015. Jahanyne, an apoptosis-inducing lipopeptide from the marine cyanobacterium *Lyngbya* sp. Org. Lett. 17, 652–655.

Iwata, M., Sawada, R., Iwata, H., Kotera, M., Yamanishi, Y., 2017. Elucidating the modes of action for bioactive compounds in a cell-specific manner by large-scale chemically-induced transcriptomics. Sci. Rep. 7, 40164.

Izzati, N., Mardlijah, M., 2015. Optimal feeding strategy on microalgae growth in fed-batch bioreactor model. Int. J. Appl. Math. Comput. Sci. 1, 1–5.

Jafari, S., Mobasher, M.A., Najafipour, S., Ghasemi, Y., Mohkam, M., Ebrahimi, M.A., Mobasher, N., 2018. Antibacterial potential of *Chlorella vulgaris* and *Dunaliella salina* extracts against *Streptococcus mutans*. Jundishapur J. Nat. Pharm. Prod. 13, e13226.

Jensen, G.S., Attridge, V.L., Beaman, J.L., Guthrie, J., Ehmann, A., Benson, K.F., 2015. Antioxidant and anti-inflammatory properties of an aqueous cyanophyta extract derived from arthrospira platensis: contribution to bioactivities by the non-phycocyanin aqueous fraction. J. Med. Food 18, 535–541.

Jha, R., Zi-Rong, X., 2004. Biomedical compounds from marine organisms. Mar. Drugs 2, 123–146.

Jung, E.H., Meyers, D.J., Bosch, J., Casadevall, A., 2018. Novel antifungal compounds discovered in medicines for malaria Venturea™ malaria box. mSphere 3, e00517–e00537.

Jüttner, F., Watson, S.B., 2007. Biochemical and ecological control of geosmin and 2-methylisoborneol in source waters. Appl. Environ. Microbiol. 73, 4395–4406.

Kang, J.O., Kim, S.J., Kim, H., 2001. Effect of astaxanthin on the hepatotoxicity, lipid peroxidation and antioxidative enzymes in the liver of CCl. Methods Find. Exp. Clin. Pharmacol. 23, 79–84.

Kellam, S.J., Cannell, R.J.P., Owsianka, A.M., Walker, J.M., 1988. Results of a large-scale screening programme to detect antifungal activity from marine and freshwater microalgae in laboratory culture. Br. Phycol. J. 23, 45–47.

Kim, S.K., Chojnacka, K., 2015. Marine Algae Extracts, 2 Volume Set: Processes, Products, and Applications. John Wiley , Sons, p. 784.

Kitazato, K., Wang, Y., Kobayashi, N., 2007. Viral infectious disease and natural products with antiviral activity. Drug Discov Ther 1, 14–22.

Koehn, F.E., Longley, R.E., Reed, J.K., 1992. Microcolins A and B, new immunosuppressive peptides from the blue-green alga Lyngbya majuscula. Journal of natural products 55, 613–619.

Kovaa, D.J., Simeunovia, J.B., Babia, O.B., Mia¡an, A., Milovanovia, I.L., 2013. Algae in food and feed. Food Feed Res 40, 21–31.

Ku, C.S., Yang, Y., Park, Y., Lee, J., 2013. Health benefits of blue-green algae: prevention of cardiovascular disease and nonalcoholic fatty liver disease. J. Med. Food 16, 103–111.

Kurashige, M., Okimasu, E., Inoue, M., Utsumi, K., 1990. Inhibition of oxidative injury of biological membranes by astaxanthin. Physiol. Chem. Phys. Med. NMR 22, 27–38.

Kurihara, H., Koda, H., Asami, S., Kiso, Y., Tanaka, T., 2002. Contribution of the antioxidative property of astaxanthin to its protective effect on the promotion of cancer metastasis in mice treated with restraint stress. Life Sciences 70, 2509–2520.

Lau, N.-S., Matsui, M., Abdullah, A.A.-A., 2015. Cyanobacteria: Photoautotrophic Microbial Factories for the Sustainable Synthesis of Industrial Products.

Lee, J.-B., Hayashi, K., Hirata, M., Kuroda, E., Suzuki, E., Kubo, Y., Hayashi, T., 2006. Antiviral sulfated polysaccharide from *Navicula directa*, a diatom collected from deep-sea water in Toyama Bay. Biol. Pharm. Bull. 29, 2135–2139.

Leung, P.-o., Lee, H.-H., Kung, Y.-C., Tsai, M.-F., Chou, T.-C., 2013. Therapeutic effect of C-phycocyanin extracted from blue green algae in a rat model of acute lung injury induced by lipopolysaccharide. Evid. Based Complement Altern. Med. 2013.

Leya, T., Rahn, A., Latz, C., Remias, D., 2009. Response of arctic snow and permafrost algae to high light and nitrogen stress by changes in pigment composition and applied aspects for biotechnology. FEMS Microbiol. Ecol. 67, 432–443.

Li, W.I., Berman, F.W., Okino, T., Yokokawa, F., Shioiri, T., Gerwick, W.H., Murray, T.F., 2001. Antillatoxin is a marine cyanobacterial toxin that potently activates voltage-gated sodium channels. Proc. Natl. Acad. Sci. Unit. States Am. 98, 7599–7604.

Li, Z., Wang, Y., Mo, B., 2002. The effects of carotenoids on the proliferation of human breast cancer cell and gene expression of bcl-2. Zhonghua yu fang yi xue za zhi 36, 254–257.

Lobo, V., Patil, A., Phatak, A., Chandra, N., 2010. Free radicals, antioxidants and functional foods: impact on human health. Phcog. Rev. 4, 118.

Ma, R.Y.-N., Chen, F., 2001. Enhanced production of free trans-astaxanthin by oxidative stress in the cultures of the green microalga *Chlorococcum* sp. Process Biochem. 36, 1175–1179.

Mao, T.K., Water, J.V.d., Gershwin, M.E., 2005. Effects of a *Spirulina*-based dietary supplement on cytokine production from allergic rhinitis patients. J. Med. Food 8, 27–30.

Matsui, M.S., Muizzuddin, N., Arad, S., Marenus, K., 2003. Sulfated polysaccharides from red microalgae have antiinflammatory properties *in vitro* and *in vivo*. Appl. Biochem. Biotechnol. 104, 13–22.

Matthew, S., Paul, V.J., Luesch, H., 2009. Tiglicamides A–C, cyclodepsipeptides from the marine cyanobacterium *Lyngbya confervoides*. Phytochemistry 70, 2058–2063.

Mendes, R.L., Coelho, J.P., Fernandes, H.L., Marrucho, I.J., Cabral, J.M.S., Novais, J.l.M., Palavra, A.a.F., 1995. Applications of supercritical CO_2 extraction to microalgae and plants. J. Chem. Technol. Biotechnol. 62, 53–59.

Mendes, R.L., Nobre, B.P., Cardoso, M.T., Pereira, A.P., Palavra, A.F., 2003. Supercritical carbon dioxide extraction of compounds with pharmaceutical importance from microalgae. Inorganica Chimica Acta 356, 328–334.

Mendiola, J.A., Torres, C.F., Tor, A., Mart lvarez, P.J., Santoyo, S., Arredondo, B.O., Se¡ns, F.J., Cifuentes, A., Ibez, E., 2007. Use of supercritical CO_2 to obtain extracts with antimicrobial activity from *Chaetoceros muelleri* microalga. A correlation with their lipidic content. Eur. Food Res. Technol. 224, 505–510.

Mendiola, J.A., Santoyo, S., Cifuentes, A., Reglero, G., Ibanez, E., SeÃ±orÃ¡ns, F.J., 2008. Antimicrobial activity of sub-and supercritical CO_2 extracts of the green alga *Dunaliella salina*. J. Food Prot. 71, 2138–2143.

Michalak, I., Dmytryk, A., Wieczorek, P.P., RÃ³j, E., Åska, B., Garka, B., Messyasz, B., Lipok, J., Mikulewicz, M., Wilk, R., 2015. Supercritical algal extracts: a source of biologically active compounds from nature. J. Chem. 2015.

Milligan, K.E., Marquez, B.L., Williamson, R.T., Gerwick, W.H., 2000. Lyngbyabellin B, a toxic and antifungal secondary metabolite from the marine cyanobacterium *Lyngbya majuscula*. J. Nat. Prod. 63, 1440–1443.

Mimouni, V., Ulmann, L., Pasquet, V., Mathieu, M., Picot, L., Bougaran, G.I., Cadoret, J.-P., Morant-Manceau, A., Schoefs, B., 2012. The potential of microalgae for the production of bioactive molecules of pharmaceutical interest. Curr. Pharmaceut. Biotechnol. 13, 2733–2750.

Mohamed, Z.A., 2008. Polysaccharides as a protective response against microcystin-induced oxidative stress in *Chlorella vulgaris* and *Scenedesmus quadricauda* and their possible significance in the aquatic ecosystem. Ecotoxicology 17, 504.

Mori, T., Boyd, M.R., 2001. Cyanovirin-N, a potent human immunodeficiency virus-inactivating protein, blocks both CD4-dependent and CD4-independent binding of soluble gp120 (sgp120). to target cells, inhibits sCD4-induced binding of sgp120 to cell-associated CXCR4, and dissociates bound sgp120 from target cells. Antimicrob. Agents Chemother. 45, 664–672.

Mozaffarieh, M., Sacu, S., Wedrich, A., 2003. The role of the carotenoids, lutein and zeaxanthin, in protecting against age-related macular degeneration: a review based on controversial evidence. Nutr. J. 2, 20.

Muniaraj, S., Subramanian, V., Srinivasan, P., Palani, M., 2018. In silico and in vitro studies on *Lyngbya majuscula* using against lung cancer cell line (A549). Phcog. J. 10.

Murakami, M., Makabe, K., Yamaguchi, K., Konosu, S., Wälchli, M.R., 1988. Goniodomin A, a novel polyether macrolide from the dinoflagellate *Goniodoma pseudogoniaulax*. Tetrahedron Lett. 29, 1149–1152.

Nagai, H., Mikami, Y., Yazawa, K., Gonoi, T., Yasumoto, T., 1993. Biological activities of novel polyether antifungals, gambieric acids A and B from a marine dinoflagellate *Gambierdiscus toxicus*. J. Antibiot. 46, 520–522.

Nagarajan, M., Maruthanayagam, V., Sundararaman, M., 2012. A review of pharmacological and toxicological potentials of marine cyanobacterial metabolites. J. Appl. Toxicol. 32, 153–185.

Nagle, D.G., Paul, V.J., Roberts, M.A., 1996. Ypaoamide, a new broadly acting feeding deterrent from the marine cyanobacterium Lyngbya majuscula. Tetrahedron Lett. 37, 6263–6266.

Najdenski, H.M., Gigova, L.G., Iliev, I.I., Pilarski, P.S., LukavskÃ½, J., Tsvetkova, I.V., Ninova, M.S., Kussovski, V.K., 2013. Antibacterial and antifungal activities of selected microalgae and cyanobacteria. Int. J. Food Sci. Technol. 48, 1533–1540.

Nakashima, Y., Ohsawa, I., Konishi, F., Hasegawa, T., Kumamoto, S., Suzuki, Y., Ohta, S., 2009. Preventive effects of Chlorella on cognitive decline in age-dependent dementia model mice. Neurosci. Lett. 464, 193–198.

Naviner, M., Berga, J.P., Durand, P., Le Bris, H., 1999. Antibacterial activity of the marine diatom *Skeletonema costatum* against aquacultural pathogens. Aquaculture 174, 15–24.

Nazifi, E., Wada, N., Yamaba, M., Asano, T., Nishiuchi, T., Matsugo, S., Sakamoto, T., 2013. Glycosylated porphyra-334 and palythine-threonine from the terrestrial cyanobacterium *Nostoc commune*. Mar. Drugs 11, 3124–3154.

Nguyen, H.M., Cuina, S., Beyly-Adriano, A., Lageret, B., Billon, E., Auroy, P., Beisson, F., Peltier, G., Li-Beisson, Y., 2013. The green microalga *Chlamydomonas reinhardtii* has a single ώ-3 fatty acid desaturase which localizes to the chloroplast and impacts both plastidic and extraplastidic membrane lipids. Plant Physiol. https://doi.org/10.1104/pp.113.223941.

Nielsen, J., 1999. Fermentation, monitoring, design and optimization. In: The Encyclopedia of Bioprocess Technology: Fermentation, Biocatalysis, and Bioseparation. Wiley-Interscience, New York, pp. 1147–1156.

Nobre, B., Marcelo, F., Passos, R., Beirao, L., Palavra, A.n., Gouveia, L., Mendes, R., 2006. Supercritical carbon dioxide extraction of astaxanthin and other carotenoids from the microalga *Haematococcus pluvialis*. Eur. Food Res. Technol. 223, 787–790.

Nogle, L.M., Okino, T., Gerwick, W.H., 2001. Antillatoxin B, a neurotoxic lipopeptide from the marine cyanobacterium *Lyngbya majuscula*. J. Nat. Prod. 64, 983–985.

Norton, T.A., Melkonian, M., Andersen, R.A., 1996. Algal biodiversity. Phycologia 35, 308–326.

Nowruzi, B., Haghighat, S., Fahimi, H., Mohammadi, E., 2017. *Nostoc* cyanobacteria species: a new and rich source of novel bioactive compounds with pharmaceutical potential. J. Pharm. Health Serv. Res. 9, 5–12.

Ohta, S., Shiomi, Y., Kawashima, A., Aozasa, O., Nakao, T., Nagate, T., Kitamura, K., Miyata, H., 1995. Antibiotic effect of linolenic acid from *Chlorococcum* strain HS-101 and *Dunaliella primolecta* on methicillin-resistant *Staphylococcus aureus*. J. Appl. Phycol. 7, 121–127.

Ohta, S., Ono, F., Shiomi, Y., Nakao, T., Aozasa, O., Nagate, T., Kitamura, K., Yamaguchi, S., Nishi, M., Miyata, H., 1998. Anti-herpes simplex virus substances produced by the marine green alga, *Dunaliella primolecta*. J. Appl. Phycol. 10, 349–356.

Olaizola, M., 2003. Commercial development of microalgal biotechnology: from the test tube to the marketplace. Biomol. Eng. 20, 459–466.

Oyamada, C., Kaneniwa, M., Ebitani, K., Murata, M., Ishihara, K., 2008. Mycosporine-like amino acids extracted from scallop (Patinopecten yessoensis). ovaries: UV protection and growth stimulation activities on human cells. Mar. Biotechnol. 10, 141–150.

Patterson, G.M.L., Carmeli, S., 1992. Biological effects of tolytoxin (6-hydroxy-7-O-methyl-scytophycin b), a potent bioactive metabolite from cyanobacteria. Archives of microbiology 157, 406–410.

Patterson, G.M.L., Bolis, C.M., 1997. Fungal cell-wall polysaccharides elicit an antifungal secondary metabolite (Phytoalexin) in the cyanobacterium Scytonema ocellatum. Journal of phycology 33, 54–60.

Palozza, P., Torelli, C., Boninsegna, A., Simone, R., Catalano, A., Mele, M.C., Picci, N., 2009. Growth-inhibitory effects of the astaxanthin-rich alga Haematococcus pluvialis in human colon cancer cells. Cancer Lett. 283, 108–117.

Panda, D., DeLuca, K., Williams, D., Jordan, M.A., Wilson, L., 1998. Antiproliferative mechanism of action of cryptophycin-52: kinetic stabilization of microtubule dynamics by high-affinity binding to microtubule ends. Proc. Natl. Acad. Sci. Unit. States Am. 95, 9313–9318.

Pane, G., Cacciola, G., Giacco, E., Mariottini, G.L., Coppo, E., 2015. Assessment of the antimicrobial activity of algae extracts on bacteria responsible of external otitis. Mar. Drugs 13, 6440–6452.

Pereira, A., Cao, Z., Murray, T.F., Gerwick, W.H., 2009. Hoiamide a, a sodium channel activator of unusual architecture from a consortium of two Papua New Guinea cyanobacteria. Chem. Biol. 16, 893–906.

Perez-Garcia, O., Bashan, Y., 2015. Microalgal heterotrophic and mixotrophic culturing for bio-refining: from metabolic routes to techno-economics. In: Algal Biorefineries. Springer, pp. 61–131.

Plaza, M., Herrero, M., Cifuentes, A., Ibanez, E., 2009. Innovative natural functional ing ents from microalgae. J. Agric. Food Chem. 57, 7159–7170.

Posten, C., 2009. Design principles of photo-bioreactors for cultivation of microalgae. Eng. Life Sci. 9, 165–177.

Pratt, R., Daniels, T.C., Eiler, J.J., Gunnison, J.B., Kumler, W.D., Oneto, J.F., Strait, L.A., Spoehr, H.A., Hardin, G.J., Milner, H.W., 1944. Chlorellin, an antibacterial substance from *Chlorella*. Science 351–352.

Priyadarshani, I., Rath, B., 2012. Commercial and industrial applications of micro algae: a review. J. Algal Biomass Utiln. 3, 89–100.

Qawasmi, A., Landeros-Weisenberger, A., Bloch, M.H., 2013. Meta-analysis of LCPUFA supplementation of infant formula and visual acuity. Pediatrics 131, e262–e272.

Qin, S., Liu, G.-X., Hu, Z.-Y., 2008. The accumulation and metabolism of astaxanthin in *Scenedesmus obliquus* (Chlorophyceae). Process Biochem. 43, 795–802.

Quintana, J., Bayona, L.M., Castellanos, L., Puyana, M., Camargo, P., Aristiza¡bal, F., Edwards, C., Tabudravu, J.N., Jaspars, M., Ramos, F.A., 2014. Almiramide D, cytotoxic peptide from the marine cyanobacterium *Oscillatoria nigroviridis*. Bioorg. Med. Chem. 22, 6789–6795.

Rai, L.C., Gaur, J.P., 2012. Algal Adaptation to Environmental Stresses: Physiological, Biochemical and Molecular Mechanisms. Springer Science , Business Media.

Raja, R., Hemaiswarya, S., Ganesan, V., Carvalho, I.S., 2016. Recent developments in therapeutic applications of Cyanobacteria. Crit. Rev. Microbiol. 42, 394–405.

Raposo, M., de Morais, A., de Morais, R., 2015. Carotenoids from marine microalgae: a valuable natural source for the prevention of chronic diseases. Mar. Drugs 13, 5128–5155.

Rasool, M., Sabina, E.P., Lavanya, B., 2006. Anti-inflammatory effect of Spirulina fusiformis on adjuvant-induced arthritis in mice. Biol. Pharm. Bull. 29, 2483–2487.

Rastogi, R.P., Incharoensakdi, A., 2014. Characterization of UV-screening compounds, mycosporine-like amino acids, and scytonemin in the cyanobacterium *Lyngbya* sp. CU2555. FEMS Microbiol. Ecol. 87, 244–256.

Rechter, S., König, T., Auerochs, S., Thulke, S., Walter, H., Darnenburg, H., Walter, C., Marschall, M., 2006. Antiviral activity of *Arthrospira*-derived spirulan-like substances. Antivir. Res. 72, 197–206.

Reshef, V., Mizrachi, E., Maretzki, T., Silberstein, C., Loya, S., Hizi, A., Carmeli, S., 1997. New acylated sulfoglycolipids and digalactolipids and related known glycolipids from cyanobacteria with a potential to inhibit the reverse transcriptase of HIV-1. J. Nat. Prod. 60, 1251–1260.

Rindi, F., Soler-Vila, A., Guiry, M.D., 2012. Taxonomy of marine macroalgae used as sources of bioactive compounds. In: Marine Bioactive Compounds. Springer, pp. 1–53.

Rodríguez-Meizoso, I., Jaime, L., Santoyo, S., Señoráns, F.J., Cifuentes, A., Ibáñez, E., 2010. Subcritical water extraction and characterization of bioactive compounds from *Haematococcus pluvialis* microalga. J. Pharm. Biomed. Anal. 51, 456–463.

Rogers, J.N., Rosenberg, J.N., Guzman, B.J., Oh, V.H., Mimbela, L.E., Ghassemi, A., Betenbaugh, M.J., Oyler, G.A., Donohue, M.D., 2014. A critical analysis of paddlewheel-driven raceway ponds for algal biofuel production at commercial scales. Algal Res. 4, 76–88.

Romay, C.H., Gonzalez, R., Ledon, N., Remirez, D., Rimbau, V., 2003. C-phycocyanin: a biliprotein with antioxidant, anti-inflammatory and neuroprotective effects. Curr. Protein Pept. Sci. 4, 207–216.

Running, J.A., Severson, D.K., Schneider, K.J., 2002. Extracellular production of L-ascorbic acid by *Chlorella protothecoides*, Prototheca species, and mutants of *P. moriformis* during aerobic culturing at low pH. J. Ind. Microbiol. Biotechnol. 29, 93–98.

Salvador, L.A., Paul, V.J., Luesch, H., 2010. Caylobolide B, a macrolactone from symplostatin 1-producing marine cyanobacteria *Phormidium* spp. from Florida. J. Nat. Prod. 73, 1606–1609.

Santoyo, S., Rodraguez-Meizoso, I., Cifuentes, A., Jaime, L., Reina, G.G.-B., Seaorans, F.J., Ibaez, E., 2009. Green processes based on the extraction with pressurized fluids to obtain potent antimicrobials from *Haematococcus pluvialis* microalgae. LWT-Food Sci. Technol. 42, 1213–1218.

Satake, M., Murata, M., Yasumoto, T., Fujita, T., Naoki, H., 1991. Amphidinol, a polyhydroxy-polyene antifungal agent with an unprecedented structure, from a marine dinoflagellate, *Amphidinium klebsii*. J. Am. Chem. Soc. 113, 9859–9861.

Schaeffer, D.J., Krylov, V.S., 2000. Anti-HIV activity of extracts and compounds from algae and cyanobacteria. Ecotoxicol. Environ. Saf. 45, 208–227.

Schaeffer, H.J., Beauchamp, L., de Miranda, P., Elion, G.B., Bauer, D.J., Collins, P., 1978. 9-(2-hydroxyethoxymethyl). guanine activity against viruses of the herpes group. Nature 272, 583.

Selmi, C., Leung, P.S.C., Fischer, L., German, B., Yang, C.-Y., Kenny, T.P., Cysewski, G.R., Gershwin, M.E., 2011. The effects of *Spirulina* on anemia and immune function in senior citizens. Cell. Mol. Immunol. 8, 248.

Shah, M., Mahfuzur, R., Liang, Y., Cheng, J.J., Daroch, M., 2016. Astaxanthin-producing green microalga *Haematococcus pluvialis*: from single cell to high value commercial products. Front. Plant Sci. 7, 531.

Shalaby, E.A., 2011. Algal biomass and biodiesel production. In: Biodiesel-Feedstocks and Processing Technologies. InTech.

Sharma, P., Sharma, N., 2017. Industrial and biotechnological applications of algae: a review. J. Adv. Plant Biol. 1, 01.

Sharma, N.K., Tiwari, S.P., Tripathi, K., Rai, A.K., 2011. Sustainability and cyanobacteria (blue-green algae).: facts and challenges. J. Appl. Phycol. 23, 1059–1081.

Shih, S.R., Tsai, K.N., Li, Y.S., Chueh, C.C., Chan, E.C., 2003. Inhibition of enterovirus 71-induced apoptosis by allophycocyanin isolated from a blue green alga *Spirulina platensis*. J. Med. Virol. 70, 119–125.

Shih, C.-M., Cheng, S.-N., Wong, C.-S., Kuo, Y.-L., Chou, T.-C., 2009. Antiinflammatory and antihyperalgesic activity of C-phycocyanin. Anesth. Analg. 108, 1303–1310.

Shimizu, Y., 2003. Microalgal metabolites. Curr. Opin. Microbiol. 6, 236–243.

Shishido, T.K., Humisto, A., Jokela, J., Liu, L., Wahlsten, M., Tamrakar, A., Fewer, D.P., Permi, P., Andreote, A.P.D., Fiore, M.F., 2015. Antifungal compounds from cyanobacteria. Mar. Drugs 13, 2124–2140.

Sibi, G., 2015. Inhibition of lipase and inflammatory mediators by *Chlorella* lipid extracts for antiacne treatment. J. Adv. Pharm. Technol. Res. 6, 7.

Silva-Stenico, M.E., Kaneno, R., Albani Zambuzi, F., Vaz, M.G., Alvarenga, D.O., Fatima Fiore, M., 2013. Natural products from cyanobacteria with antimicrobial and antitumor activity. Curr. Pharmaceut. Biotechnol. 14, 820–828.

Singh, I.P., Milligan, K.E., Gerwick, W.H., 1999. Tanikolide, a toxic and antifungal lactone from the marine cyanobacterium *Lyngbya majuscula*. J. Nat. Prod. 62, 1333–1335.

Singh, K.S., Bansal, A., Jha, M.K., Dey, A., 2011. Comparative studies on uptake of wastewater nutrients by immobilized cells of *Chlorella minutissima* and dairy waste isolated algae. Indian Chem. Eng. 53, 211–219.

Singh, S.K., Bansal, A., Jha, M.K., Dey, A., 2012. An integrated approach to remove Cr (VI). using immobilized *Chlorella minutissima* grown in nutrient rich sewage wastewater. Bioresour. Technol. 104, 257–265.

Singh, S.K., Bansal, A., Jha, M.K., Jain, R., 2013. Production of biodiesel from wastewater grown *Chlorella minutissima*. Indian J. Chem. Technol. 20, 341–345.

Singh, S.K., Sundaram, S., Kishor, K., 2014. Photosynthetic Microorganisms: Mechanism for Carbon Concentration. Springer.

Singh, S.K., Sundaram, S., Sinha, S., Rahman, M.A., Kapur, S., 2016a. Recent advances in CO_2 uptake and fixation mechanism of cyanobacteria and microalgae. Crit. Rev. Environ. Sci. Technol. 46, 1297–1323.

Singh, S.K., Rahman, A., Dixit, K., Nath, A., Sundaram, S., 2016b. Evaluation of promising algal strains for sustainable exploitation coupled with CO_2 fixation. Environ. Technol. 37, 613–622.

Singh, S.K., Major, S.R., Cai, H., Chen, F., Hill, R.T., Li, Y., 2018. Draft genome sequences of *Cloacibacterium normanense* IMET F, a microalgal growth-promoting bacterium, and *Aeromonas jandaei I*MET J, a microalgal growth-inhibiting bacterium. Genome Announc. 6, e00503–00518.

Soltani, N., Khavari-Nejad, R.A., Tabatabaei Yazdi, M., Shokravi, S., Fernandez-Valiente, E., 2005. Screening of soil cyanobacteria for antifungal and antibacterial activity. Pharmaceut. Biol. 43, 455–459.

Stal, L.J., 1995. Physiological ecology of cyanobacteria in microbial mats and other communities. New Phytol. 131, 1–32.

Sugiura, N., Iwami, N., Inamor, Y., Nishimura, O., Sudo, R., 1998. Significance of attached cyanobacteria relevant to the occurrence of musty odor in Lake Kasumigaura. Water Res 32, 3549–3554.

Syed, S., Arasu, A., Ponnuswamy, I., 2015. The uses of *Chlorella vulgaris* as antimicrobial agent and as a diet: the presence of bio-active compounds which caters the vitamins, minerals in general. Int. J. Biosci. Biotechnol. 7, 185–190.

Talhouk, R.S., Karam, C., Fostok, S., El-Jouni, W., Barbour, E.K., 2007. Anti-inflammatory bioactivities in plant extracts. J. Med. Food 10, 1–10.

Taniguchi, M., Nunnery, J.K., Engene, N., Esquenazi, E., Byrum, T., Dorrestein, P.C., Gerwick, W.H., 2009. Palmyramide A, a cyclic depsipeptide from a Palmyra Atoll collection of the marine cyanobacterium *Lyngbya majuscula*. J. Nat. Prod. 73, 393–398.

Terekhova, V.E., Aizdaicher, N.A., Buzoleva, L.S., Somov, G.P., 2009. Influence of extrametabolites of marine microalgae on the reproduction of the bacterium *Listeria monocytogenes*. Russ. J. Mar. Biol. 35, 355–358.

Tidgewell, K., Engene, N., Byrum, T., Media, J., Doi, T., Valeriote, F.A., Gerwick, W.H., 2010. Evolved diversification of a modular natural product pathway: apratoxins F and G, two cytotoxic cyclic depsipeptides from a Palmyra collection of *Lyngbya bouillonii*. Chembiochem 11, 1458–1466.

Ting, H., Haifeng, L., Shanshan, M., Zhang, Y., Zhidan, L., Na, D., 2017. Progress in microalgae cultivation photobioreactors and applications in wastewater treatment: a review. Int. J. Agric. Biol. Eng. 10, 1–29.

Trimurtulu, G., Ohtani, I., Patterson, G.M.L., Moore, R.E., Corbett, T.H., Valeriote, F.A., Demchik, L., 1994. Total structures of cryptophycins, potent antitumor depsipeptides from the blue-green alga *Nostoc* sp. strain GSV 224. J. Am. Chem. Soc. 116, 4729–4737.

Umemura, K., Yanase, K., Suzuki, M., Okutani, K., Yamori, T., Andoh, T., 2003. Inhibition of DNA topoisomerases I and II, and growth inhibition of human cancer cell lines by a marine microalgal polysaccharide. Biochem. Pharmacol. 66, 481–487.

Vadiraja, B.B., Gaikwad, N.W., Madyastha, K.M., 1998. Hepatoprotective effect of C-Phycocyanin: protection for carbon tetrachloride andR-(+).-pulegone-mediated hepatotoxicty in rats. Biochem. Biophys. Res. Commun. 249, 428–431.

Verreault, M., Strutt, D., Masin, D., Anantha, M., Yung, A., Kozlowski, P., Waterhouse, D., Bally, M.B., Yapp, D.T., 2011. Vascular normalization in orthotopic glioblastoma following intravenous treatment with lipid-based nanoparticulate formulations of irinotecan (Irinophore C™)., doxorubicin (Caelyx®). or vincristine. BMC Cancer 11, 124.

Violeta, N., Trandafir, I., Cosmulescu, S., 2017. Bioactive compounds, antioxidant activity and nutritional quality of different culinary aromatic herbs. Not. Bot. Horti Agrobot. Cluj-Napoca 45, 179–184.

Viso, A.C., Pesando, D., Baby, C., 1987. Antibacterial and antifungal properties of some marine diatoms in culture. Bot. Mar. 30, 41–46.

Volk, R.-B., Furkert, F.H., 2006. Antialgal, antibacterial and antifungal activity of two metabolites produced and excreted by cyanobacteria during growth. Microbiol. Res. 161, 180–186.

Walter, C.S., Mahesh, R., 2000. Antibacterial and antifungal activities of some marine diatoms in culture. Indian J. Mar. Sci. 29, 238–242.

Washida, K., Koyama, T., Yamada, K., Kita, M., Uemura, D., 2006. Karatungiols A and B, two novel antimicrobial polyol compounds, from the symbiotic marine dinoflagellate *Amphidinium* sp. Tetrahedron Lett. 47, 2521–2525.

Watanabe, Y., Hall, D.O., 1996. Photosynthetic production of the filamentous cyanobacteriumSpirulina platensis in a cone-shaped helical tubular photobioreactor. Appl. Microbiol. Biotechnol. 44, 693–698.

Wen, Z.-Y., Chen, F., 2000. Production potential of eicosapentaenoic acid by the diatom *Nitzschia laevis*. Biotechnol. Lett. 22, 727–733.

Wiederhold, N.P., 2017. Antifungal resistance: current trends and future strategies to combat. Infect. Drug Resist. 10, 249.

Wilson, D., Nash, P., Buttar, H., Griffiths, K., Singh, R., De Meester, F., Horiuchi, R., Takahashi, T., 2017. The role of food antioxidants, benefits of functional foods, and influence of feeding habits on the health of the older person: an overview. Antioxidants 6, 81.

Wu, J.-T., Kuo-Huang, L.-L., Lee, J., 1998. Algicidal effect of peridinium bipes on Microcystis aeruginosa. Curr. Microbiol. 37, 257–261.

Yasuhara-Bell, J., Lu, Y., 2010. Marine compounds and their antiviral activities. Antivir. Res. 86, 231–240.

Yim, J.H., Kim, S.J., Ahn, S.H., Lee, C.K., Rhie, K.T., Lee, H.K., 2004. Antiviral effects of sulfated exopolysaccharide from the marine microalga Gyrodinium impudicum strain KG03. Mar. Biotechnol. 6, 17–25.

Yim, J.H., Son, E., Pyo, S., Lee, H.K., 2005. Novel sulfated polysaccharide derived from red-tide microalga *Gyrodinium impudicum* strain KG03 with immunostimulating activity in vivo. Mar. Biotechnol. 7, 331–338.

Yingying, S., Changhai, W., Jing, C., 2008. Growth inhibition of the eight species of microalgae by growth inhibitor from the culture of *Isochrysis galbana* and its isolation and identification. J. Appl. Phycol. 20, 315–321.

Yun, K.-J., Yang, G., Feng, Z., Nenkep, V.N., Xavier, S.-N., Leutou, A.S., Kim, G.-D., Cho, H.-Y., Choi, H.-D., Son, B.-W., 2010. Screening on receptor tyrosine kinase inhibitory activity of marine algae-derived symbiotic microorganisms. Korean J. Pharmacogn. 41, 43–47.

Zaslavskaia, L.A., Lippmeier, J.C., Shih, C., Ehrhardt, D., Grossman, A.R., Apt, K.E., 2001. Trophic conversion of an obligate photoautotrophic organism through metabolic engineering. Science 292, 2073–2075.

Zittelli, G.C., Biondi, N., Rodolfi, L., Tredici, M.R., 2013. Photobioreactors for mass production of microalgae. Handbook of microalgal culture: Applied Phycology and Biotechnology 2, 225–266.

CHAPTER

Biotechnological production of sweeteners

9

Andrés Felipe Hernández-Pérez[a], Fanny Machado Jofre[a], Sarah de Souza Queiroz[a], Priscila Vaz de Arruda[b], Anuj K. Chandel[a], Maria das Graças de Almeida Felipe[a]

Departamento de Biotecnologia, Escola de Engenharia de Lorena, Universidade de São Paulo, Lorena, SP, Brasil[a]; Universidade Tecnológica Federal do Paraná, Câmpus Toledo, PR, Brasil[b]

Chapter outline

Biotechnological Production of Bioactive Compounds. https://doi.org/10.1016/B978-0-444-64323-0.00009-6

1. Introduction

While sugar consumption has an important role on human health, in recent years, intake of sugar per capita has been limited by human due to increased health awareness. There has been a surge in zero-calorie sweeteners in last 10–15 years. During this time period, a large population has shifted from conventional sugar to low-calorie or zero-calorie sweeteners (Philippe et al., 2014). However, this change is primarily due to modifications in life styles, less physical activities, automatic work culture and increased dependency on machines which caused less calorie burning eventually causing diabetes and other associated health issues. Despite the increasing demand of sweeteners, there are some striking limiting factors in growth of sweeteners such as characteristic taste qualities, long term health concerns, low production yields, high cost and sustainable practices in agriculture (Philippe et al., 2014). Among the sweeteners, plant-derived sweeteners such as stevia and xylitol, which is obtained from lignocellulosic biomass, have gained significant attention.

In general, sweeteners can be classified into six major groups: Artificial sweeteners; Modified sugars; Natural caloric sweeteners; Natural zero calorie sweeteners; Sugars; and Sugar alcohols (Misra, 2016). Artificial sweeteners are non-nutritive sweeteners but very sweet compounds. Usually, these sweeteners have no role in tooth decay without having any effect on blood glucose or calories. Modified sugars are cariogenic and have high glycemic index. Inverted sugar and high fructose corn syrup are the commonly used modified sugars in confectionaries and processed food. Natural calorie sweeteners are derived primarily from sugar beets, sugarcane, maple trees etc and some nutrients are also added. Natural zero-caloric sweeteners constitute major fraction of sweeteners as they tend to have very low calories and less harmless to teeth. Thaumatin, glycyrrhizic acid, brazzein, rebaudioside A, stevioside are some examples of these sweeteners. These sweeteners have zero glycemic index and are harmless to teeth, but like artificial sweeteners. Sugars are conventional carbohydrates (monosaccharides, disaccharides) naturally occurring in fruits, vegetables, milk and cereals and have high glycemic index. Sugar alcohols or polyols are naturally available in vegetables, plants and cereals. They are safe and are used in sugar-free candies, cookies, and chewing gums. Xylitol, erythritol, sorbitol, etc are the common sugar alcohols (Silva and Chandel, 2012). Artificial sweeteners having non-nutritive value, are generally considered as sugar substitutes for human consumption spanning various applications. The role of non-nutritive artificial sweeteners on human health and environment is not yet well described (Praveena et al., 2019).

Currently the major artificial sweeteners and sugar-alcohols are commercially produced by chemical methods (O'Donnell, 2012; DuBois and Prakash, 2012; de Cock, 2012; Deis and Kearsley, 2012; Sentko and Willibald-Ettle, 2012; Zacharis, 2012a, 2012b). Biotechnological production of sweeteners is a viable and sustainable solution to cater the growing demand. Both natural plant extraction, lignocellulosic biomass and microbial production based all sweeteners can be produced via biotechnological production routes. However, there are gross technological and scale up challenges in sweeteners production via biotechnological routes (Park et al., 2016; Chakraborty and Das, 2019). For instance, recovery of hemicellulosic sugars from biomass with utmost yields and minimum inhibitors is an essential parameter for the commercial xylitol production. Later, the role of modern biotechnological routes employing metabolic engineering to develop designer microbial strains may be suitable for the commercially production of economic sweeteners. Future efforts for increasing microbial production of natural sweeteners will profit from novel methods such as CRISPR/Cas9 and from combination of rational strain engineering, adaptive laboratory evolution and high-throughput screening approaches (Schallmey et al., 2014; Eggeling et al., 2015; Shalem et al., 2015).

This chapter presents the classification of sweeteners, critically appraises the commercial outlook and demand, and their effects on human health. Furthermore, biotechnological production of the most important sugar alcohols is reviewed.

2. Classification of sweeteners

Under the name of sweeteners, the Food and Agricultural Organization includes products used for sweetening that are either derived from sugar crops, cereals, fruits, or milk, or produced by insects (Popkin and Nielsen, 2003). Many types of sweeteners are available and it is possible to categorize them into six groups: Artificial Sweeteners; Modified Sugars; Natural Caloric Sweeteners; Natural Zero Calorie Sweeteners; Sugars and Sugar Alcohols (Misra, 2016). Independent of the category, all of them require to exhibit a sucrose-like taste quality with properties such as demonstrated non-toxicity and lack of any offensive odor. This chapter is focused on the sweeteners produced by microbiological pathway, but first we will show the main characteristics of each category.

2.1 Artificial sweeteners

Artificial Sweeteners or low-calorie sweetener are a class of highly sweet compounds with low or not contribution of calories (Swithers, 2013). This type of sweeteners is also referred as non-nutritive sweeteners, high-intensity sweeteners, and non-caloric sweeteners, since it provides little or no calories or carbohydrates and do not increase blood sugar. Because of the last characteristic, they could be used in beverages and in foods to lower calorie content while maintaining palatability (Sylvetsky et al., 2011). In addition to these characteristics, this kind of sweeteners has the advantage of not promoting tooth decay and offer a sweet taste without increasing blood glucose or calories.

According to the European Parliament and the Council of the European Union, Regulation No 1333/2008, all food additives are identified by an E number and must be always stated on the packaging in the ingredient lists. All food additives prior their usage must be authorized by particular legislations such as European Union and Food and Drug Administration (Regulation no 1333/2008, 2008). Among the main artificial sweeteners approved for use as food additives by the US Food and Drug Administration (FDA, 2018) are acesulfame-potassium (E950), advantame, aspartame (E951), neotame (E961), saccharin (E954) and sucralose (E955). Table 9.1 shows the principal characteristics of this kind of sweeteners.

2.2 Modified sugars

According to Ref. Misra (2016), these are the most harmful of all sugars, because they tend to have a high glycemic index and they are cariogenic. The main modified sugars include caramel, golden syrup, high fructose corn syrup, refined syrup and inverted sugar. These are typically sugars produced by converting starch using enzymes and are often used in cooking or in processed foods.

2.3 Natural caloric sweeteners

These are the oldest known sweeteners and include high fructose corn syrup as prime example because it is the sweetener used in all U.S. soft drinks. There are two major sugar crops: sugar beets and sugar cane. Sugar and syrups are also produced from the sap of certain species of maple trees, from sweet sorghum when cultivated explicitly for making syrup, and from sugar palm (Popkin and Nielsen, 2003). They contain sugar but also other nutritive qualities and tend to have a lower glycemic index than sugar, but they still need to be taken in moderation as they can be detrimental to health in large quantities, as well as harmful to teeth. The main natural caloric sweeteners include honey, maple syrup, coconut palm sugar and sorghum syrup (Misra, 2016).

2.4 Natural zero calorie sweeteners

Global demand for naturally sourced, zero-calorie sweeteners has increased significantly over the last decade as consumers have become increasingly health conscious, since this kind of sweetener is a better alternative to artificial sweeteners. Relatively few sweet-tasting plant-derived natural products have been commercially launched to date, but those numbers are rapidly increasing. These include thaumatin, glycyrrhizic acid, monatin, brazzein, rebaudioside A, stevioside and mogroside V (DuBois and Prakash, 2012). *Stevia rebaudiana* and *Siraitia grosvenorii* or Luo Han Guo (LHG) are two plant species, which contain natural sweeteners, as rebaudioside A, stevioside and mogroside V, respectively, and these compounds have become increasingly interesting targets for commercial production mainly by biotechnological bioconversion (Philippe et al., 2014). These sweeteners have zero glycemic index and are harmless to teeth, but like artificial sweeteners they can have an aftertaste (Misra, 2016). The main natural zero calorie sweeteners includes: Luo Han Guo, stevia, thaumatin, pentadin, monellin and brazzein.

Table 9.1 The most important artificial sweeteners and their characteristics.

Artificial sweeteners	Synthesis	Sweetness powder	Molecular formula and weight	Brand names	References
Acesulfame-k (E950)	Artificial chemical process	120–200 times sweeter than sucrose	$C_4H_4KNO_4S$ 201.24 g/mol	Sunett®, Sunette, Sweet One®, Ace-K	Rymon Lipinski (1991); Nabors (2002); Godshall (2007), FDA (2018)
Advantame	Artificial chemical process, from aspartame and vanillin	Cooking, baking and food additive 20.000–37.000 times sweeter than sucrose	$C_{24}H_{30}N_2O_7$ 476.52 g/mol	Not yet branded. Available to the commercial and food ingredients market.	Otabe et al. (2011); O'Donnell (2012) FDA (2018)
Aspartame (E951)	Artificial chemical synthesis	Sweetener in soft drinks, cookies, chewing gum and just about any diet product. 180–200 times sweeter than sucrose	$C_{14}H_{18}N_2O_5$ 294.3 g/mol	Equal®, NutraSweet®, Candere, AminoSweet, NatraTaste, Sugar Twin®	Ager et al. (1998); Arora et al. (2009); Chattopadhyay et al. (2014), FDA (2018)
Neotame (E961)	Chemoenzymatic and Hydrogenation synthesis process	Chewing gum, Lemon tea, Yogurt, Soft drinks and Cake 7000–13.000 times sweeter than sucrose	$C_{20}H_{30}N_2O_5$ 378.469 g/mol	Newtame®	Nofre and Tinti (2000), Prakash and Zhao (2001), Prakash (2007), FDA (2018)
Saccharin (E954)	Artificial chemical synthesis	Medicine products, beverages, processed food and sugar substitute 200–700 times sweeter than sucrose	$C_7H_5NO_3S$ 183.18 g/mol	Sweet and Low®, Sweet Twin®, Sweet'N Low®, Necta Sweet®	Rosenman (1978) Misra (2016) FDA (2018)
Sucralose (E955)	Artificial chemical process	Beverages, Dairy products, Confectionery, Baked products, Pharmaceuticals 450–650 times sweeter than sucrose	$C_{12}H_{19}Cl_3O_8$ 397.64 g/mol	Splenda®	Chattopadhyay et al. (2014), FDA (2018)

2.5 Sugars

Sugars are carbohydrates that contain 4 calories per gram and are found naturally in many foods including fruit, vegetables, cereals and milk. This category includes a wide variety of monosaccharides (glucose, fructose, galactose and tagatose) and disaccharides (sucrose, lactose, maltose, isomaltulose, and trehalose), which exist either in a crystallized state as sugar or in thick liquid form as syrups (Misra, 2016). Monosaccharides require the least effort by the body to break down, meaning they are available for energy more quickly than disaccharides. They can be harmful to teeth and tend to have a high glycemic index (Aller et al., 2011).

2.6 Sugar alcohols

Sugar alcohols (polyols or polyhydric alcohols) are carbohydrates that occur naturally in small amounts in vegetables, plants and cereals, they are low digestible which are obtained by substituting and aldehyde group with a hydroxyl one (Livesey, 2003; Shankar et al., 2013; Misra, 2016). Another name of sugar alcohol is alditols, because of the most of them are produced from their corresponding aldose sugars (Mäkinen, 2011). This category of sweetener is generally recognized as safe for food additives and do not promote tooth decay or cause a sudden increase in blood glucose. They are used primarily to sweeten sugar-free candies, cookies, and chewing gums (FDA, 2018). Others common use is in the food industry as additives to give color, to sweeten or to help in food preservation (Regulation no 1333/2008, 2008).

Sugar alcohols include hydrogenated monosaccharides, such as xylitol (E967), erythritol (E968), sorbitol (E420) and mannitol (E421); and hydrogenated disaccharides, such as isomalt (E953), maltitol (E965) and lactitol (E966) (Regulation no1333/2008, 2008). Some other sugar alcohols used outside of the US and EU are arabitol, dulcitol, galactitol, iditol, ribitol and threitol (Misra, 2016). Table 9.2 shows the principal characteristics of this kind of sweetener.

3. Commercial outlook and demand of sweeteners

A wide variety of food products contained sweeteners aiming to reduce number of calories, particularly to address obesity, which is a major risk factor for cardiovascular diseases, diabetes, musculoskeletal disorders, and some types of cancer (Tahergorabi et al., 2016). A large number of studies to improve the biosynthesis of different sweeteners have been reported, employing biotechnological techniques as substrates with low-cost, as molasses, crude glycerol or hemicellulosic and cellulosic hydrolysates, fermentation conditions coupled with metabolic engineering, which may result in the reduction of production costs and finally the retail price (Rzechonek et al., 2018).

Table 9.2 The most important sugar alcohol sweeteners and their characteristics.

Artificial sweeteners	Synthesis	Sweetness powder	Molecular formula and weight	Brand names	References
Xylitol (E967)	Catalytic hydrogenation synthesis process (commercial) and fermentation or enzymatic processes	Almost equal (0.95) sweetening potency compared with sucrose (=1.0)	$C_5H_{12}O_5$ 152,15 g/mol	Polysweet®, Xylosweet®, Xyla®	Zacharis (2012b), Ecogal (2018), Kallscheuer (2018)
Erythritol (E968)	Fermentation processes by osmophilic yeast	0.6–0.7 sweetening potency compared with sucrose, usually it need to be mixed with other sweeteners to achieve the required texture and sweetness level.	$C_4H_{10}O_4$ 122.12 g/mol	C*Eridex®, ZSweet®, Zerose®	de Cock (2012); Grembecka (2015); Ecogal (2018)
Sorbitol (E420)	Catalytic hydrogenation process (commercial process) and fermentation processes	0.55 sweetening potency compared with sucrose (=1.0)	$C_6H_{14}O_6$ 182,172 g/mol	It is not usually sold on its own, but is added by manufacturers to various products.	Deis and Kearsley (2012), Ecogal (2018), Kallscheuer (2018)
Mannitol (E421)	Catalytic hydrogenation process (commercial process), Enzymatic catalysis and fermentation process	0.50 sweetening potency compared with sucrose (=1.0)	$C_6H_{14}O_6$ 182,172 g/mol	Mannidex®	Deis and Kearsley (2012), Bhatt et al. (2013), Ecogal (2018), Kallscheuer (2018), Zhang et al. (2018)
Isomalt (E953)	Chemical reduction of saccharose into isomaltulose	0.45–0.60 sweetening potency compared with sucrose (=1.0)	$C_{12}H_{24}O_{11}$ 344.324 g/mol	ClearCut™, DiabetiSweet,	Haghighatian et al. (2008) Sentko and Willibald-Ettle (2012) Ecogal (2018)
Lactitol (E966)	Catalytic hydrogenation process	0.4 sweetening potency compared with sucrose (=1.0)	$C_{12}H_{24}O_{11}$ 344.3124 g/mol	Lacty®	Ecogal (2018) Zacharis (2012a)
Maltitol (E965)	Catalytic hydrogenation synthesis process	0.9 sweetening potency compared with sucrose (=1.0)	$C_{12}H_{24}O_{11}$ 344,31 g/mol	SweetPearl®, Maltisweet®	Kearsley and Deis (2012), Ecogal (2018)

According to a research done in 2013 by Helgi Library and published in 2018, with 157 countries about "Which country eats the most sugar and sweeteners?", China ranked the highest in sugar and sweetener consumption per capita with 149.45 kg followed by India and USA. Average sugar and sweetener consumption per capita reached 24.0 kg in 2013 in the World according to Faostat. This is 0.95% more than in the previous year and 7.87% more than 10 years ago (Helgi Analytics calculation, 2013). According to Ref. Business Wire (2018), United States and Canada are among the largest consumers. The consumption of high-intensity sweeteners is expected to remain steady in the near future, mainly because its largest application, i.e. diet beverage market, is already matured. However, for the western countries, the demand of sweeteners is expected to increase slowly compared to that of the emerging countries because of the saturated market.

In 2012 the global market for sugar and sweeteners was US$ 77.5 billion, from which the sugar alone accounted for US$ 65 billon, whereas the markets for natural sweeteners and stevia accounted for $1.5 billion and $1.7 billion, respectively (Saini et al., 2016). In 2016, acesulfame-k held the lion's share (40%), followed by aspartate (30%). However, due to the growing concerns about the aspartame effects on health, European Food Safety Authority (EFSA) predicted a decline in the aspartame market to U$360 million in 2017 from US$ 480 million in 2013. According to a study by Transparency Market Research (TMR), the global market for natural sweeteners is likely to achieve US$ 39 million in terms of revenue by 2026 and a CAGR (Compound Annual Growth Rate) of 4.5% between 2017 and 2026. In terms of application, dairy and frozen food products are among the fastest growing segments of the global food sweetener market, registering a CAGR of 3.46% during the forecast period. Confectionery is the third largest application segment of the global food sweeteners market, accounting for a bulk of the total sugar demand (TMR, 2018).

Global increase in low calories food additives demand from end-use industries such as processed foods, ready-to-drink (RTD) beverages, soft and energy drinks, and confectioneries is expected to drive the non-sugar sweeteners demand over the next years. The growing utilization of natural and sugar alcohols sweeteners, including stevia, thaumatin, xylitol and sorbitol, in the diet beverages segment is expected to drive the sweeteners demand in the next decade (Grand View Reserch, 2015). Demand for caloric sweeteners — sugar and high fructose corn syrup - is expected to grow 1.5 % until 2027, reaching 213 Mt (OECD-FAO, 2018). This low growth rate results from the slowdown in global population growth and stagnation of the per capita consumption in developed countries and certain developing countries (Brazil, Egypt, Mexico, Paraguay, South Africa, Turkey), where per capita consumption has reached levels that raise health concerns. In countries with lower consumption levels, particularly in Asia and Africa, population growth and urbanization are expected to sustain the increment in sugar consumption, driven by increased consumption of sweetened beverages and prepared food products (OECD-FAO, 2018).

The market of sweeteners is highly fragmented with a large number of key players across the global value chain. Key industry players include Archer Daniels

Midland (ADM), Cargill, Celanese Corporation, Ajinomoto Co. Inc., DuPont – Danisco, GLG Life Tech, Ingredion, Kerry Group, Pure Circle, Nutrinova Inc., Mitsui & Co., and Zydus Wellness Ltd. Other prominent players operating in the global market include Nutrasweet Company, Roquette, Naturex, Hermes sweetener Ltd., Merisant Worldwide Inc., and Imperial Sugar Company (Grand View Reserch, 2015; TMR, 2018).

According to Ref. Philippe et al. (2014), few chemical entities that possess sweet or sweetness-enhancing properties have high commercial potential to promote the development of a process for their production. These authors suggested that high commercial potential for sweeteners depends on the following factors: availability in scale, taste quality, safety, stability, solubility, cost and patentability. The taste quality is one of the most important factors, since consumers conventionally prefer the taste of sucrose to artificial sweeteners (DuBois and Prakash, 2012). Furthermore, sweeteners must be safe to the consumer, not toxic or mutagenic, and be stable at varying pH, temperature and light conditions, without undesirable reactions and formation of unsafe derived compounds (Philippe et al., 2014).

4. Health effects of sweeteners and regulations for consumptions

Consumer awareness has led to a steady use of artificial low-calorie sweeteners. Due to this growing demand, studies on the effects of these products on the human and animal health are increasing. Furthermore, the fact that these molecules can be potential contaminants of the environment should be evaluated for their presence in effluents. Health effects of the main sweeteners, including sugar alcohol that can be produced by biotechnological routes, are presented.

4.1 Xylitol

Xylitol has a low glycemic index and low caloric content (3 kcal/g) compared to sucrose, being used in the food, pharmaceutical and dental industries (Livesey, 2003). It is a non-toxic compound certified as GRAS (Generally Regarded as Safe) by Food and Drug Administration (FDA), and it is recommended that the consumption does not exceed 60 g/day to avoid the risk of laxative effect (Mussatto and Roberto, 2002). One of the advantages of this compound is its chemical and microbiological stability, which guarantees its performance even in low concentrations. It has insulin-independent metabolism and does not cause large fluctuations in blood glucose levels. Some beneficial effects of xylitol use on health are the anticariogenic action (Gargouri et al., 2018), treatment and prevention of acute otitis and pulmonary infection, which are to the fact that xylitol is toxic to the bacteria that cause these infections; as well as prevention of osteoporosis since its consumption can stimulate the increase in the absorption of calcium, and therapeutic agent in the treatment of patients with hemolytic anemia (Mussatto and Roberto, 2002).

4.2 Erythritol

Among the polyols that can be used as sweeteners, erythritol is highlighted by its low caloric value, almost zero (0–0.2 $kcalg^{-1}$); cooling effect and classification as GRAS by the FDA. Like all polyols, can provide tooth protection and provide significant support in oral reducing dental plaque (Mäkinen et al., 2005; Rzechonek et al., 2018). Erythritol has anti-oxidant properties because it is a free radical scavenger and it can act while circulating throughout the body before it is excreted in the urine, since it is not metabolized (Bornet et al., 1996). Moreover, this sweetener does not induce changes on insulin levels in blood, making it appropriate for diabetic patients (Moon et al., 2010; Rzechonek et al., 2018). On the other hand, erythritol has approximately 70% of the relative sweetness of sucrose, which is lower than other polyols, such as xylitol or sorbitol (Carocho et al., 2017), and it has a relatively high price (Rzechonek et al., 2018).

4.3 Sorbitol

Sorbitol has applications not only as a sweetener, but also food stabiliser, humectant, anticariogenic, texturizer and softener (Cazetta et al., 2005). This polyol has a relative sweetness of around 60% of sucrose and it has a 20-fold higher solubility in water than mannitol (Silveira and Jonas, 2002; Carocho et al., 2017). Sorbitol is a low-calorie polyol, with 2.5 kcal/g, and it was established by the FDA as a Generally Regarded As Safe (GRAS) compound. The daily intake of sorbitol may exceed 50g, but above this value there is a risk of having a laxative effect (Lee, 2015). Sorbitol has a lower glycemic response in the blood compared to sucrose and can be an alternative sweetener for patients with diabetes, as well as inhibiting the growth of cariogenic species in the oral mucosa, preventing cavities (Livesey, 2003). Salivation is stimulated after ingestion of this sweetener, which might be used for treatment for the elderly who suffer from oral problems caused by insufficient salivation (Kim et al., 2015). Despite their beneficial effects on health, there are studies, still in rat trials, which have shown that intake of 0.15–150 mg kg^{-1} day^{-1} of sorbitol interfered in the milk composition of lactating rats and induced significant metabolic changes after 14 days of exposure to this polyol, leading to weight gain in adult rats (Cardoso et al., 2016).

4.4 Mannitol

Mannitol is a 6-carbon polyol, isomer of sorbitol, and widely found in nature (Wisselink et al., 2002). This compound is considered as GRAS, included on the chemical composition of various functional foods, and used in fine chemicals and pharmaceutical industries (Zhang et al., 2018). It has 50%–70% of the sweetness of sucrose (Dai et al., 2017). This sweetener has an energy value of 1.5 kcal/g and it is metabolized without major interferences in blood glucose levels, independent of insulin, which makes it possible for consumption by patients with diabetes (Dai et al., 2017). It is recommended that the daily intake does not exceed 20g, otherwise it may cause a laxative effect (Nabors and Gelardi, 2001). Like other polyols, mannitol acts as a low-energy carbohydrate that repairs and prevents the growth of cariogenic microorganisms (Livesey, 2003), and exhibits synergistic antioxidant effect as shown in some studies (Belda et al., 2005; André and Villain, 2017).

4.5 Sucralose

Sucralose is classified as a high-potency sweetener, which is made from chemical modifications of the sucrose molecule that include a five-step process that selectively substitutes three atoms of chlorine for three hydroxyl groups in the sucrose molecule (Roberts et al., 2000). Sucralose is non-nutritive, non-cariogenic and non-caloric and is poorly absorbed in humans and the established acceptable daily intake is 15 $mgkg^{-1}$ day^{-1} (Baird et al., 2000). It is approved for use in food products in most countries around the world and one of the reasons for this great use is due to the fact that it has excellent stability (FAO/WHO, 1989). Its numerous applications in food systems include like beverages, processed fruit and fruit spreads, milk products, frozen desserts and salad dressings are found from this sweetener (Finn and Lord, 2000). Studies of carcinogenicity, reproductive toxicology, neurotoxicity and genetic toxicology concluded that there was no evidence that sucralose was not safe for human consumption (Grice and Goldsmith, 2000).

4.6 Aspartame

The high-intensity sweetener aspartame (L-α-aspartyl-Lphenylalanine methyl ester) is one of the most used artificial sweeteners by the population and the established acceptable daily intake is 40 $mgkg^{-1}$ body weight (Lange et al., 2012). Based on the Joint FAO/WHO Expert Committee on Food Additives (JECFA), aspartame is considered nontoxic and not carcinogenic, based on both, animal and human studies, at dosages up to 4000 mg kg^{-1} day^{-1}. However, several studies have alleged that its consumption has been linked to behavioral and cognitive problems (Anton et al., 2010). The danger is based on the fact that consumption of aspartame, unlike dietary protein, can elevate the levels of phenylalanine and aspartic acid in the brain (Haighton et al., 2019). Studies were done to evaluate safety of aspartame use by potentially sensitive subpopulations (i.e., individuals heterozygous for the rare genetic disease phenylketonuria and individuals with Parkinson's disease, dizziness, depression, liver disease, and renal disease) (Renwick and Nordmann, 2007). Due to this, aspartame consumption needs to be approached with caution due to the possible effects on health. It is important to note that diabetics are advised to consume artificial sweeteners in minimal quantities, since new evidence suggests that long-term use can be harmful.

5. Biotechnological production of sweeteners

5.1 Xylitol

Current commercial production of xylitol occurs by a chemical route based on the catalyzed hydrogenation of purified xylose using nickel as catalyst, high temperatures and pressures. The chemical process allows converting only 50%–60% of xylan in biomass into xylitol, reaching efficiency up to 98% from pure xylose (Arcaño et al., 2018). This process is considered costly, energy-consuming and non-eco-friendly, due to the complexity of the xylose purification step, the high cost of the catalyst, the need for high pressure and temperature (Cheng et al., 2014; Dasgupta et al., 2017; Arcaño et al., 2018). As an alternative to this process, biotechnological production has advantages such as it is a selective biological process, not requiring the successive steps of xylose purification, and mild requirement of temperature and pressure (Hou-Rui, 2012; Canilha et al., 2013). Biotechnological production of xylitol consists mainly in the use of specific enzymes or microorganisms, which perform xylose to xylitol conversion. Xylose is the main constituent monomer of the hemicellulosic fraction of several xylan-rich lignocellulosic biomasses, including hardwood biomasses and agro-industrial byproducts (Felipe, 2004).

Lignocellulosic biomasses are submitted to a pre-treatment process to release the fermentable sugars in their monomer form. Considering xylitol production, dilute-acid hydrolysis is one the most used pre-treatments since it promotes preferentially hydrolysis of the hemicellulosic fraction (Chandel et al., 2012). Conventionally, dilute-acid hydrolysis is carried out using sulfuric acid 0.5%–1.5% (w/w) and temperatures above 121–160°C, which allows sugar recovery above 70% up to > 95%. The challenge of the pretreatment step is to maximize xylose recovery with minimum formation of compounds that can be potentially toxic to the microorganisms, by disturbing its metabolism and structure. The most important inhibitors are phenolic compounds, derived from partial degradation of lignin; furfural and 5-hydroxymethyl-furfural, formed by dehydration at high temperature and prolonged time of pentoses and hexoses, respectively; acetic acid formed from de-acetylation of hemicellulose; as well as trace metal ions (iron, chromium, nickel, zinc) from the material of the hydrolysis reactor and/or soil (Felipe, 2004; Jönsson et al., 2013; Rao et al., 2016).

Detoxification procedures previous to fermentation step are necessary to remove toxic compounds to minimal concentrations that do not interfere with the fermentative performance of the microorganisms (Rao et al., 2016). Several detoxification methods have been studies, such as pH alteration, combination of pH alteration and adsorption with activated charcoal, adsorption with activated charcoal followed by ionic exchange resins, and treatment only with ion exchange resins, biopolymers, nanofiltration membranes, reverse osmosis and biodetoxification (Mpabanga et al., 2012; Rao et al., 2016; Silva-Fernandes et al., 2017). Among them, adsorption by activated charcoal is one of the most used detoxification methods, because of its efficiency, low cost and properties, such as high chemical affinity for organic compounds, especially phenolics, and large surface area per unit mass.

After detoxification of the xylose-rich hemicellulosic hydrolysate, it will be used as culture medium for microbial fermentation. Yeasts are considered the best xylitol producers, such as *Candida guilliermondii, Candida tropicalis, Candida maltosa, Kluyveromyces marxianus, Debaromyces hansenii,* among others. Xylitol production is the result of the incomplete metabolism of xylose in these microorganisms, which is promoted by the fermentation conditions, mainly oxygen availability. Fig. 9.1 shows a model of the yeast metabolism involved in xylose, glucose, arabinose, fructose and glycerol catabolism and production of xylitol, erythritol and arabitol. In the case of xylitol, a xylose reductase (XR) dependent of NAD(P)H reduces this pentose into xylitol, which can be excreted or oxidized to xylulose by a xylitol dehydrogenase (XDH) dependent of NAD^+, and then xylulose is integrated on the non-oxidative phase of the Pentoses Phosphate Pathway (PPP) and the central carbon metabolism of carbon for cofactors regeneration, formation of intermediates, energy and cellular biomass (Flores et al., 2000; Granström et al., 2007).

Xylitol production depends on the high activity of the XR and the low activity of XDH, which are influenced by different fermentation conditions. The main factor is the restriction of oxygen availability in microaerobic atmosphere, since the redox imbalance of $NADH/NAD^+$ reduces the XDH activity under this condition

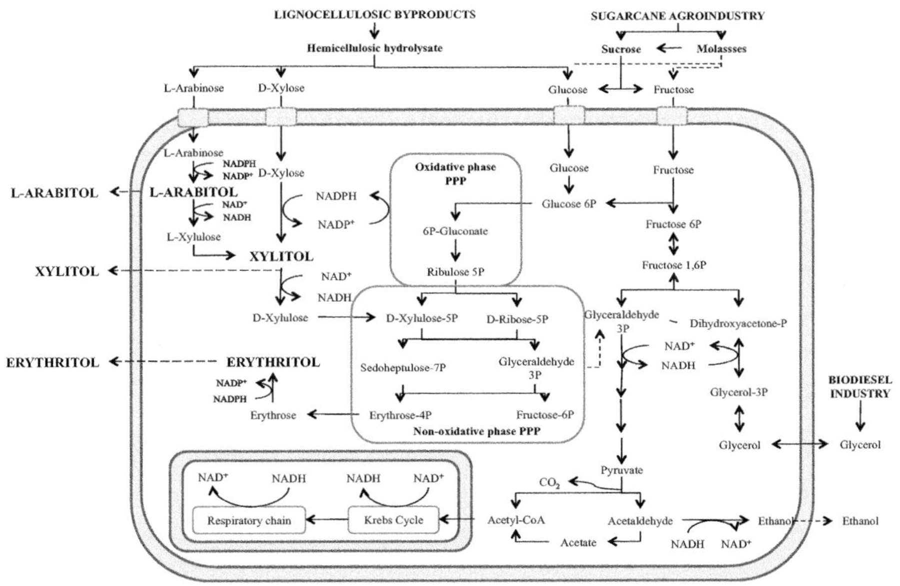

FIG. 9.1

Model of yeast metabolism involved on the production of xylitol, erythritol and arabitol.

Based on Flores, C.L., Rodríguez, C., Petit, T., Gancedo, C. 2000. Carbohydrate and energy-yielding metabolism in non-conventional yeasts. FEMS Microbiol. Rev. 24, 507–529., Granström, T.B., Izumori, K., Leisola, M., 2007. A rare sugar xylitol. Part II: biotechnological production and future applications of xylitol. Appl. Microbiol. Biotechnol. 74, 277–281., Kordowska-Wiater, M. 2015. Production of arabitol by yeasts: current status and future prospects. J. Appl. Microbiol. 119, 303–314 and Rzechonek, D.A., Dobrowolski, A. Rymowicz W., Mirończuk, A.M. 2018. Recent advances in biological production of erythritol. Curr. Rev. Biotechnol. 38 (4), 620–633.

(Granström et al., 2007). Xylitol accumulation reduces the carbon flux through the central metabolic pathways, particularly through PPP, which is the main NADPH-producing pathway (Kim et al., 1999; Granström et al., 2007). Consequently, regeneration of NADPH becomes an important challenge in xylitol production. Other important bottleneck for efficient bioconversion of hemicellulosic hydrolysates into xylitol is the sequential use of glucose and xylose is one the main that need to be addressed. This fact has been attributed to competition for transport systems between these sugars, and glucose inhibition of the enzymatic machinery for xylose assimilation (Zhang et al., 2015; Hou et al., 2017). Besides of these challenges, studies have been focused on formulation of the fermentation media regarding high concentration of xylose and nutritional supplementation; utilization of free or immobilized cells in various reactor types and operation modes such as batch, fed batch or continuous; scale-up of the whole process; and metabolic engineering of the microorganisms. Recent results on xylitol production from lignocellulosic biomasses are summarized in Table 9.3.

Xylitol recovery after fermentation is complex and expensive, due to the low xylitol concentration and composition of the fermented broth, regarding compounds not-assimilated by the microorganisms from the hemicellulosic hydrolysates or nutrients supplemented (Sampaio et al., 2006; Aliakbarian et al., 2012). Different methods have been investigated for xylitol separation, including clarification with activated charcoal, chromatographic methods, membrane separation, adsorption, concentration and crystallization (Aliakbarian et al., 2012; Martínez et al., 2015). After clarification, fermented broth must be concentrated to achieve a xylitol concentration at least as high as 750 gL^{-1} to favor the crystallization (Canilha et al., 2008; Martínez et al., 2008). Crystallization can be performed under different temperatures, with addition of solvent, such as ethanol, and xylitol crystals to start seeding process (Sampaio et al., 2006; Canilha et al., 2008; Martínez et al., 2008).

5.2 Erythritol

Erythritol can be chemically produced from dialdehyde starch using high temperatures and nickel as catalyst, however this process is not industrially employed because of its low efficiency (Moon et al., 2010; Carly and Fickers, 2018). Erythritol is currently produced by fermentative processes at industrial scale, using filamentous fungi and yeasts such as *Moniliella pollinis*, *Trichosporonoides megachiliensis* and *Yarrowia lipolytica*, and mainly from glucose as substrate, derived from hydrolyzed corn or wheat (Rzechonek et al., 2018). After fermentation, erythritol is recovered by membrane filtration of the fermented broth, concentration, ion exchange chromatography, treatment with activated carbon and crystallization (Moon et al., 2010; Rakicka et al., 2016; Rzechonek et al., 2018). Similar to other polyols, the cost of the purification procedures makes the bioprocess expensive.

Erythritol is produced by bacteria, filamentous fungi and yeasts. Bacteria produced this polyol during alternative NADPH regeneration through the phosphoketolase pathway under anaerobic conditions (Moon et al., 2010), as in the case of the

Table 9.3 Recent results on xylitol production from lignocellulosic biomasses.

Hydrolysate source	Yeast	Xylitol production			References
		Titer (gL^{-1})	Yp/s (gg^{-1})	Qp ($gL^{-1}h^{-1}$)	
Sugarcane bagasse Dilute-acid hydrolysis (Xylose 60 gL^{-1})	*C. guilliermondii* FTI 20037	41.8	0.66	0.29	Arruda et al. (2017)
Sugarcane straw Dilute-acid hydrolysis (Xylose 57 gL^{-1})	*C. guilliermondii* FTI 20037	36.1	0.65	0.75	Hernández-Pérez et al. (2016)
Corncob Dilute-acid hydrolysis (Xylose 54 gL^{-1})	*Saccharomyces cerevisiae* overexpressing *GRE3* (endogenous aldose reductase) and *SUT1* (xylose transporter)	47	0.87	0.32	Kogje and Ghosalkar (2017)
Corncob Dilute-acid hydrolysis (Xylose 160 gL^{-1})	*C. maltose* Xu316 (Adaptative evolution)	120	0.81	2.50	Jiang et al. (2016)
Rice straw Liquid hot water (Xylose 59.3 gL^{-1})	*S. cerevisiae* YPH499 expressing cytosolic XR, along with ß-glucosidase, xylosidase and xylanase displayed on cell surface	37.9	0.63	0.39	Guirimand et al. (2016)
Waste xylose mother liquor Liquid residue from xylose purification during xylitol chemical production (Xylose 127 gL^{-1})	*C. tropicalis* X828 (co-fermentation with *Bacillus subtilis*)	95	0.73	0.86	Wang et al. (2016)

heterofermentative *Oenococcus oeni* (former *Leuconostoc oenos)* (Veiga-da-Cunha et al., 1993) and other lactic acid bacteria (Tyler et al., 2016). Yeasts and filamentous fungi produce erythritol via PPP using mainly glucose as substrate (Fig. 9.1), despite other compounds can be also employed, such as fructose, sucrose, xylose or glycerol. PPP consists in an oxidative phase, in which NADPH and ribulose-5-phosphate are produced, and a non-oxidative phase with erythrose-4-phosphate as final product. Erythrose-4-phosphate is dephosphorylated and then reduced to erythritol by an Erythrose Reductase (ER) dependent of NAD(P)H (Moon et al., 2010; Lee et al., 2010; Rzechonek et al., 2018). During this process other polyols can be produced as byproducts, such as mannitol, arabitol, ribitol and glycerol, which affect the erythritol yield and purification (Lin et al., 2010; Yang et al., 2014).

Erythritol production by yeasts has been associated primarily to hyperosmotic stress, particularly during stationary phase (Kobayashi et al., 2013; Carly and Fickers, 2018). Fermentation conditions should be carefully established, particularly regarding oxygen availability, to avoid glycerol formation in detriment of erythritol production, since glycerol is the main osmoprotectant in yeasts (Regnat et al., 2018) High sugar (ranging from 200 to 400 gL^{-1}) and/or salt concentrations promote and, in some cases, enhance production of this polyol and reduction of byproducts formation (Tomaszewska et al., 2012; Yang et al., 2014; Carly and Fickers, 2018). However, Tomaszewska et al. (2012) found that these conditions can also reduce the productivity or the process because of an extension of the lag phase. Other factors related to erythritol production are (a) oxidative stress, as found in *Moniliella* (Kobayashi et al., 2015b); (b) pH, as observed for *Y. lipolytica*, which produced more erythritol and less byproducts from glycerol at pH 3 (Rymowicz et al., 2009), but it achieved a higher productivity at pH 5.5 when grown in glucose (Ghezelbash et al., 2012); (c) nutritional supplementation of the fermentation media, mainly regarding nitrogen source, such as yeast extract (Tomaszewska et al., 2014b), inorganic ions, such as Cu, Mn and Zn (Lee et al., 2000; Tomaszwkska et al., 2014a) and vitamins, specifically thiamine (Tomaszewska et al., 2014b) and inositol (Lee et al., 2001).

Recent researches aiming to reduction of production costs are focused on utilization of low-cost substrates, increase of the productivity and reduction of byproducts formation. Regarding utilization of low-cost substrates, crude glycerol, which is a byproduct of the biodiesel industry, was already used for erythritol production by the yeasts *Y. lipolytica* (Rakicka et al., 2016) and *Moniliella megachiliensis* (Kobayashi et al., 2015a). According to Refs. Rakicka et al. (2017) and Mirończuk et al. (2015), an interesting feature of glycerol fermentation is the reduction of byproducts formation compared with the bioprocesses that employ sugars as substrates, fact that consequently favors erythritol purification. Furthermore, Rakicka et al. (2016) reported that erythritol production was favored by the presence of NaCl and other mineral salts in crude glycerol. Xylose can be used also as carbon source for erythritol production, as demonstrated by Ref. Guo et al. (2016) with the yeast *Aureobasidium pullulans*. After UV mutagenesis and optimization of fermentation medium, these authors obtained a final erythritol titer of 26.35 gL^{-1} from corncob hemicellulosic hydrolysate, corresponding to a productivity of 0.18 $gL^{-1}h^{-1}$ and a yield of 0.12 gg^{-1} (Guo et al., 2016).

An approach to increase the bioprocess productivity has been the careful control of the osmotic pressure during the fermentation, which is related not only with the concentration of the carbon source, but also with the mode of addition of the substrate or salts. In this regard, Rakicka et al. (2017) investigated a continuous system of two sequential chemostats with different dilution rate, in which the osmotic pressure derived from the utilization of high concentration of crude glycerol in the first stage was alleviated in the second one. This fermentation system resulted in a final erythritol titer of 199.4 gL^{-1}, yield of 0.6 gg^{-1} and productivity 0.8 $gL^{-1}h^{-1}$. Tomaszweska and Rywińska (2016) studied the gradual addition of salt, achieving the highest concentration of erythritol with NaCl 75 gL^{-1} but after 215 h of fermentation. Nonetheless, addition of salt resulted also in the necessity of desalination of the fermented broth prior to erythritol recovery (Rakicka et al., 2016).

Regarding metabolic engineering for improvement of erythritol production, approaches have been focused on increasing the carbon flux through the PPP by overexpression of important genes of this pathway, such as glucose-6-phosphate dehydrogenase (G6PDH), 6-phosphogluconate dehydrogenase (6GPDH), transaldolase, transketolase and ER (Carly et al., 2017; Mirończuk et al., 2017; Cheng et al., 2018); reducing the erythritol assimilation by deletion of the gene encoding the erythrulose kinase (Carly et al., 2017); and improving the uptake of substrates, such glycerol, by overexpression of the genes encoding glycerol kinase (Mirończuk et al., 2016), and sucrose, by heterologous expression of invertase gene from *Saccharomyces cerevisiae* in *Yarrowia lipolytca* (Rakicka et al., 2017). For instance, Ref. Cheng et al. (2018) demonstrated that overexpression genes encoding for ER, G6PDF and 6GPDH led to increases of 23.5% and 50% in erythritol yield (0.63 gg^{-1}) and productivity (2.4 $gL^{-1}h^{-1}$), respectively, by *Y. lipolytica,* compared to the wild strain.

5.3 Arabitol

D-arabitol is a five-carbon polyol, stereoisomer of xylitol, with sweetness similar to sucrose but with a lower caloric content (0.2 kcal g^{-1}) and employed in food industries (Kordowska-Wiater, 2015; Loman et al., 2018). This pentitol can be used also as building block for production of several chemicals, such as arabinoic and xylonic acids, xylitol, propylene, ethylene glycol, among others (Koganti and Ju, 2013; Yoshikawa et al., 2014a; Kordowska-Wiater, 2015). Arabitol, along with xylitol, is considered as one of the most important high added value chemicals that can be directly produced from sugars derived from lignocellulosic biomass (Werpy and Petersen, 2004). Currently, this polyol is industrially produced by the catalytic reduction of arabinonic and lyxonic acids, which requires high temperature and expensive catalyst (Kordowska-Wiater, 2015).

Arabitol can be produced by biochemical conversion of arabinose (Fonseca et al., 2007), xylose (Jagtap and Rao, 2018), glucose (Qi et al., 2015), or crude glycerol (Yoshikawa et al., 2014a). Among these substrates, glucose and glycerol could be more promising, since glucose is easily and preferably consumed by the arabitol-producing microorganisms, as well as it may be abundant in biorefineries, as can be also the case for glycerol (Koganti et al., 2011; Guo et al., 2019). Furthermore, various studies have shown that when mixture of sugars were used as cultivation media, as in the case of hemicellulosic hydrolysates, sugars were consumed sequentially as follows: glucose > xylose > arabinose (McMillan and Boynton, 1994; Fonseca et al., 2007; Loman et al., 2018).

D-arabitol production from glucose in yeasts can follow two alternative biochemical pathways, which can be distinguished after glucose-6-phosphate enters in the PPP, as shown in Fig. 9.1. In some microorganisms, glucose-6-phosphate can be converted in D-ribulose-5-phosphate, which is dephosporylated and then reduced to D-arabitol by a NADPH-dependent D-arabitol dehydrogenase (Kordowska-Wiater, 2015). The other pathway consists in the formation of D-xylulose-5-phosphate pathway from glucose-6-phosphate, pentose that is dephosphorylated and then reduced to D-arabitol by a NADH-dependent D-arabitol dehydrogenase (Zhang et al., 2014; Kordowska-Wiater, 2015). In the case of glycerol, it is expected glucose-6-phosphate to be formed from this polyol through gluconeogenic pathway and then D-arabitol produced by one of the routes mentioned (Kordowska-Wiater, 2015).

L-arabitol can be produced from L-arabinose through the reduction of this pentose by a NADPH-dependent L-arabinose reductase (Kordowska-Wiater, 2015). Accumulation of this polyol is promoted by limited oxygen availability because of a NADH/NAD^+ imbalance that inhibits the oxidation of L-arabitol to L-xylulose by a NAD-dependent L-arabitol-4-dehydrogenase, and then its integration to the central metabolic pathways (Kordowska-Wiater, 2015). In the case of D-arabitol production from xylose, this pentose is firsly reduced to xylitol by a NAD(P)H-dependent reductase, which is further oxidized to D-xylulose by a NAD^+-dependent xylitol dehydrogenase and finally this pentose is reduced to D-arabitol by a NADH-dependent arabitol dehydrogenase (Jagtap and Rao, 2018).

D-arabitol production have been mostly demonstrated in studies with osmotolerant yeasts using different substrates, as was interestingly reviewed by Refs. Kordowska-Wiater (2015). According to this author, fermentation conditions include temperature between 28 and 45°C, pH between 3.6 and 7.0, substrate concentrations of 20–600 gL^{-1} for glucose, 20–100 gL^{-1} for arabinose and 100–350 gL^{-1} for glycerol, and different oxygen conditions depending on the microorganism and the substrate. Guo et al. (2019) found a maximum titer of 72.69 gL^{-1} from optimized conditions of 200 gL^{-1} initial glucose, 5% initial inoculum, 30°C, pH 5.0, 200 rpm, 96 h in Erlenmeyer flasks. A further increment to 76.32 gL^{-1} was achieved by optimization of media composition regarding yeast extract (10 gL^{-1}), $(NH_4)_2SO_4$ (2 gL^{-1}) and peptone (7.5 gL^{-1}).

Koganti and Ju (2013) investigated D-arabitol production from raw glycerol by *Debaryomyces hansenii* depending on the N:P ratio, dissolved oxygen (DO) glycerol to maintain its concentration near to 100 gL^{-1}. The highest D-arabitol production, corresponding to 40 gL^{-1}, productivity of 0.33 $gL^{-1}h^{-1}$ and efficiency of 55%, was achieved with N:P = 9, 30°C, DO of 5% air saturation and pH 3.5. Yoshikawa et al. (2014a) conducted a similar study, in which *Candida quercitrusa* was selected by its arabitol-producing ability from raw glycerol and the culture conditions were optimized. The highest arabitol titer, 85.1 gL^{-1} after 10 days corresponding to a yield of 0.40 gg^{-1}, was achieved in jar fermenter, 28°C and with a medium containing 250 gL^{-1} glycerol, 6 gL^{-1} yeast extract and 2 gL^{-1} $CaCl_2$.

Jagtap and Rao (2018) studied D-arabitol production from xylose by the oleaginous yeast *Rhodosporidium toruloides* in a nitrogen-rich medium. These authors achieved a maximum arabitol production of 49 gL^{-1} from 150 gL^{-1} of xylose after 168 h cultivation in 250 mL Erlenmeyer flasks with 50 mL of medium, at 30 C and 250 rpm. Mixture of enzymatic hydrolysates of soybean flour and hulls was used for arabitol production by *D. hansenii* in the work of Loman et al. (2018). These authors studied the effect of the C:N ratio, mineral nutrient supplementation, DO and pH control first in Erlenmeyer flasks and then scaled up to 2.5 L fermentor. Arabitol production in the fermenter was 43 gL^{-1} from 80 gL^{-1} of total sugars in 48 h, in a medium with a C:N ratio of 28.5, supplemented with K_2HPO_4 (2.4 gL^{-1}), DO maintained in 5% and without pH control (Loman et al., 2018).

Regarding arabitol separation and purification, the fermented broth must be treated with activated charcoal for clarification, desalinized in ion-exchange resins and then concentrated by vacuum evaporation to approximately 70% (w/v) of arabitol (Mingguo et al., 2011; Kordowska-Wiater, 2015). Arabitol crystallization can be achieved by slow cooling from 70 to 4 C, centrifugation, ethanol 95% wash and dry, resulting in powdery crystals of approximately 98% of purity (Mingguo et al., 2011; Kordowska-Wiater, 2015). Arabitol may not be the sole product of the yeast metabolism depending on the fermentation process and the microorganisms and frequently mixtures of polyols are found in the fermented broths, whose separation is complicated and compromises the purification procedures (Mingguo et al., 2011; Kordowska-Wiater, 2015). In this regard, Mingguo et al. (2011) suggested an innovative approach to eliminate other polyols from the fermented broth, by using bacteria from the genus *Bacillus* able to consume xylitol, sorbitol, mannitol, but not arabitol.

5.4 Sorbitol

Sorbitol is industrially produced from glucose syrup (50% w/v) or glucose and fructose mixtures, by catalytic hydrogenation using Raney nickel as catalyst, temperature between 120 and 150°C and pressure of 70 bar (Silveira and Jonas, 2002). For sorbitol recovery, firstly the catalyst is separated by precipitation and filtration, next sorbitol is purified by ion exchange chromatography and activated charcoal treatment, and concentrated by vacuum evaporation until 70% solution, which is the most common commercial product (Silveira and Jonas, 2002). According to these authors, chemical process was preferred to the biotechnological process because of its lower cost.

Biotechnological production of sorbitol has been studied by using mainly *Zymomonas mobilis* and recombinant strains of heterofermentative lactic acid bacteria. *Z. mobilis* produces sorbitol, along with ethanol or gluconic acid, when glucose and fructose are used as carbon sources under anaerobic conditions (Silveira and Jonas, 2002; Liu et al., 2010; Park et al., 2016). This process is based on the activity of the periplasmic constitutive enzyme Glucose-Fructose oxidoreductase (GFOR), which uses NADPH and $NADP^+$ to reduce fructose to sorbitol and to oxidize glucose to gluconic acid, respectively (Zachariou and Scopes, 1986; Liu et al., 2010). An interesting feature of this bioprocess is that NADPH and $NADP^+$ are sufficiently regenerated and recycled (Liu et al., 2010). GFOR activity is probably related to the regulation of osmotic stress by production or sorbitol, which acts as a compatible solute (Loos et al., 1994; Liu et al., 2010). Furthermore, Silveira et al. (1999) indicated that utilization of high concentrations of substrate favored sorbitol production because inhibition of the normal metabolism of the microorganism and utilization of substrates preferentially by GFOR.

Studies on sorbitol production by *Z. mobilis* have been mostly focused on the utilization of permeabilized cells in order to inhibit ethanol formation without affecting the GFOR activity. According to the review elaborated by Refs. Silveira and Jonas (2002), this strategy has allowed to obtain high efficiencies (more than 95%) and productivities (more than 1.5 $gL^{-1}h^{-1}$). Despite the advantages of immobilization regarding the operation of the process and reutilization of the biocatalysts, it has also disadvantages related to the stability of the immobilization matrix, loss of the enzyme activity and mass transfer limitation (Silveira and Jonas, 2002; Liu et al., 2010). Liu et al. (2010) reported an interesting approach to selectively inhibit enzymes of the ethanol-producing pathway Entner-Duodoroff by addition of divalent ions, mainly Zn^{2+}, in a recombinant strain of *Z. mobilis* with overexpression of GFOR. By using this approach, after optimization of culture conditions (glucose concentration 160 gL^{-1} and pH 6.0) and without immobilization, sorbitol yield was almost 100% and ethanol production was highly reduced (Liu et al., 2010).

Engineered strains of heterofermentative lactic acid bacteria are able to produce sorbitol by reversing the catabolic pathway of this polyol (Ladero et al., 2007; De Boeck et al., 2010; Hatti-Kaul et al., 2018). To do so, strategies studied have been the overexpression of sorbitol-6-phosphate dehydrogenase (Stl6PDH), disruption of mannitol phosphate dehydrogenase (M1PDH) and lactate dehydrogenase (LDH) for reducing by-products formation and glycolytic flux, improvement of redox balance, and inactivation of sorbitol transport and catabolic system to avoid sorbitol utilization (Ladero et al., 2007; De Boeck et al., 2010; Park et al., 2016). Ladero et al. (2007) engineered a strain of *Lactobacillus plantarum* for sorbitol production from glucose or maltose, by reversing the sorbitol catabolic pathway through Stl6PDH overexpression and LDH disruption. These authors found that 61%–65% of the carbon flux from fructose-6-phosphate was redirected to sorbitol formation using resting cells and controlled pH. De Boeck studied sorbitol production from glucose and lactose by a recombinant strain of *Lactobacillus casei*, in which Stl6PDH was overexpressed, LDH and M1PDH were disrupted and the phosphoenolpyruvate

(PTS) sorbitol-specific phosphotrasnferase system was inactivated. Implementation of these strategies led to sorbitol production without mannitol formation and without assimilation of the sorbitol produced, even after glucose exhaustion.

Relative high cost of the substrates, particularly fructose, is one of the major bottlenecks for scale-up of the biotechnological production of sorbitol. Studies have been done for sorbitol production from byproducts, such as molasses (Cazetta et al., 2005), whey permeate (Ladero et al., 2007), corn steep liquor (Silveira et al., 2001). Other bottleneck is the separation and purification of sorbitol, since procedures that have been used in laboratory scale, such as basic anion exchange resin (Chun and Rogers, 1988), selective precipitation with organic solvents (Silveira et al., 1994) and electrodialysis (Ferraz et al., 2000), may be expensive and inefficient for an industrial scale (Silveira and Jonas, 2002).

5.5 Mannitol

Nowadays, mannitol is produced concomitantly with sorbitol, by the chemical hydrogenation of mixture of glucose and fructose, using raney nickel as catalyst, high temperatures (120–160 C), and obtaining solutions of approximately 25% of mannitol, which is further recovery by ion-exclusion chromatography and crystallization (Dai et al., 2017). Biological production of mannitol is advantageous compared to the chemical production, since the latter requires pure substrate and has high costs and low efficiency (Zhang et al., 2018). Mannitol production is highly common in nature, mainly by bacteria, yeasts and filamentous fungi, in which can be used as carbon and energy source and as osmoprotectant (Dai et al., 2017). In the case of yeasts, some species studied are *Candida magnolia*, *S. cerevisiae* and *Torulaspora delbruickii* (Lee et al., 2003; Zhang et al., 2018). Lee et al. (2003) studied mannitol production by *C. magnolia* using fructose (150 gL^{-1}) as substrate, obtaining a final titer of 67 gL^{-1}, corresponding to a productivity of 0.81 $gL^{-1}h^{-1}$ and yield of 0.45 gg^{-1}.

The most commonly studied mannitol-producing bacteria are mainly lactic acid bacteria, which are industrially interesting due to the fact that these are safe organisms for employment in the food sector. Heterofermentative and homofermentative lactic acid bacteria have different metabolic pathways for mannitol production. In heterofermentative bacteria and under aerobic conditions, fructose is partially reduced to mannitol by a NADH-dependent mannitol dehydrogenase, while the other portion of fructose is further metabolized via Phosphoketolase pathway to produce lactic acid, CO_2, and acetic acid or ethanol (Wisselink et al., 2002; Zhang et al., 2018). Mannitol production is favored by limitation of the oxygen availability in detriment of ethanol formation, as well as by avoiding pH decrease because of the acid production (Zhang et al., 2018). In the case of homofermentative species, the mannitol production pathway begins with the reduction of fructose-6P through the enzyme mannitol-1P dehydrogenase, obtaining mannitol-1P, which is then dephosphorylated to mannitol by the enzyme mannitol phosphatase (Dai et al., 2017). Nevertheless, only recombinant homofermentative strains with deficiency in LDH are able to produce mannitol from fructose or glucose (Dai et al., 2017; Zhang et al., 2018).

One of the most important factor that affected mannitol production costs is the substrate, since utilization of fructose represents a high cost, therefore, various alternative low-cost substrates have been investigated. Meng et al. (2017) reported mannitol production of 68.5 gL^{-1}, yield of 0.34 gg^{-1} and productivity of 0.95 $gL^{-1}h^{-1}$ by *Candida parapsilosis* from glucose (200 gL^{-1}), which is cheaper than fructose. Yoshikawa et al. (2014b) studied mannitol production by *Candida azyma* from crude glycerol (300 gL^{-1}) and obtained a final titer of 50.8 gL^{-1}, yield of 0.3 gg^{-1} and productivity of 0.3 $gL^{-1}h^{-1}$. Saha (2006) used mixtures (1:1) of sugarcane molasses and fructose syrup as carbon sources and nutrients for *Lactobacillus intermedius*. After 16 h of fermentation, these authors observed a final titer of 104.8 gL¯1, yield of 0.87 gg^{-1} and productivity of 4.76 $gL^{-1}h^{-1}$; however ethanol and lactic acid were also produced. Similarly, Saha and Racine (2010) reported a final mannitol titer of 124 gL^{-1} and a productivity of 5.2 $gL^{-1}h^{-1}$ by *L. intermedius* from 250 gL^{-1} of fructose and 4.5 gL^{-1} of glucose and after pH optimization.

Recovery of mannitol can occur by cooling (4°C) the fermentative medium which contains 180 gL^{-1} or more of this polyol (Saha and Nakamura, 2003). Techniques of electrodialysis followed by crystallization, filtration and concentration by evaporation were reported as efficient methods for recovery and purification of mannitol (Soetaert et al., 1999; Itoh et al., 1992).

6. Conclusion and future directions

Sweeteners play a big role in human food with adding enjoyment, palatability and nutrition. Artificial sweeteners, if consumed, within the rules and regulations are normally safe for health. Because of diabetes menace, obesity concerns artificial sweeteners are being widely consumed by human. Therefore, in last few decades, sweeteners have gained huge demand in primarily food and pharmaceutical sectors. Among the sweeteners, xylitol is the most common and thus got maximum attention for its production to fulfill the growing demand. Biotechnological production of sweeteners harnessing the lignocellulosic biomass, and plants could be a sustainable platform for the commercial demand. Recent developments in extraction and purification of plant-based extracts, lignocellulosic biomass processing, modern genetic engineering tools to develop designer microorganisms have a profound role in sustainable production of sweeteners at large scale in the biorefineries under the sustainability regime. Advances in fermentation methods, nutrients formulation, economic carbon and nitrogen sources will definitely set the pace of artificial sweeteners large scale production to cater the burgeoning demand in various industrial sectors. Next 5-years research in artificial sweeteners will be based on omics-based approaches to develop designer super bugs for the maximum sugars utilization and simultaneously yielding high titers of artificial sweeteners after microbial fermentation. Using the agroindustrial residues, artificial sweetners can be produced in integrated biorefineries (sugarcane and corn processing facilities) in order to cut down the production costs under lignocellulose biorefinery regime.

Acknowledgments

AKC is grateful to the CAPES-Brazil for the financial assistance through visiting professor and researcher program (Processo USP no 15.1.1118.1.0). This work was supported by the São Paulo Research Foundation (FAPESP) (process 2016/22179-0 and scholarship 2016/05971-2) and CNPq (Conselho Nacional de Desenvolvimento Científico e Tecnológico, Brazil).

References

TMR – Transparency Market Research, 2018 TMR – Transparency Market Research. 2018. Natural Sweeteners Market (Product Types - High Intensity, Low Intensity; End User - Food and Beverages, Pharmaceuticals, Direct Sales, Other End Users; Application - Bakery Goods, Sweet Spreads, Confectionery and Chewing Gums, Beverages, Dairy Products, Others) - Global Industry Analysis, Size, Share, Growth, Trends and Forecast 2017 – 2026. Available at: https://www.transparencymarketresearch.com/natural-sweeteners-market.html.

WHO Food Additive Series: 24, 1989. Toxicological evaluation of certain food additives and contaminants. In: The 33rd Meeting of the Joint FAO/WHO Expert Committee on Food Additives, Geneva, Switzerland, pp. 1–45.

Ager, D.J., Pantaleone, D.P., Henderson, S.A., Katritzky, A.R., Prakash, I., Walters, D.E., 1998. Commercial, synthetic nonnutritive sweeteners. Angew. Chem. Int. Ed. 37, 1802–1817.

Akinterinwa, O., Khankal, R., Cirino, P.C., 2012. Metabolic engineering for bioproduction of sugar alcohols. Curr. Opin. Biotechnol. 19, 461–467.

Aliakbarian, B., Faveri, D., Perego, P., Converti, A., 2012. D-Xylitol. An assessment on xylitol recovery methods. In: Silva, S.S., Chandel, A.K. (Eds.), D-xylitol. Fermentative Production, Application and Commercialization. Springer-Verlag, Germany, pp. 229–244.

Aller, E.E.J.G., Abete, I., Astrup, A., Martinez, J.A., van Baak, M.A., 2011. Starches, sugars and obesity. Nutrients 3, 341–369.

André, P., Villain, F., 2017. Free-radical scavenging properties of mannitol and its role as a constituent of hyaluronic acid fillers: a literature review. Int. J. Cosmet. Sci. 39, 355–360.

Anton, S.D., Martin, C.K., Han, H., Coulon, S., Cefalu, W.T., Geiselman, P., Williamson, D.A., 2010. Effects of stevia, aspartame, and sucrose on food intake, satiety, and postprandial glucose and insulin levels. Appetite 55, 37–43.

Arcaño, Y.D., García, O.D.V., Mandelli, D., Carvalho, W.A., Pontes, L.A.M., 2018. Xylitol: a review on the progress and challenges of its production by chemical route. Catal. Today. Available at: https://doi.org/10.1016/j.cattod.2018.07.060.

Arora, S., Singh, V.P., Sharma, V., Wadhwa, B.K., George, V., Singh, A.K., Sharma, G.S., 2009. Analysis of sucralose and its storage stability in barfi. J. Food Sci. Technol. 46, 114–117.

Arruda, P.V., Santos, J.C., Rodrigues, R.C.L.B., Silva, D.D.V., Yamakawam, C.K., Rocha, G.J.M., Nolasco Junior, J., Pradella, J.G.C., Rosell, C.E.V., Felipe, M.G.A., 2017. Scale up of xylitol production from sugarcane bagasse hemicellulosic hydrolysate by *Candida guilliermondii* FTI 20037. J. Ind. Eng. Chem. 47, 297–302.

Auerbach, M.H., Locke, G., Hendrick, M.E., 2001. Alitame. In: O'Brien-Nabors, L. (Ed.), Alternative Sweeteners: Third Edition, Revised and Expanded. CRC Press, New York, pp. 31–40.

Baird, I.M., Shephard, N.W., Merritt, R.J., Hildick-Smith, P., 2000. Repeated dose study of sucralose tolerance in human subjects. Food Chem. Toxicol. 38, 123–130.

Belda, J.I., Artola, A., García-Manzanares, M.D., Ferrer, C., Haroun, H.E., Hassanein, A., Baeyens, V., Munoz, G., Alió, J.L., 2005. Hyaluronic acid combined with mannitol to improve protection against free-radical endothelial damage: experimental model. J. Cataract Refract. Surg. 31, 1213–1218.

Bhatt, S.M., Mohan, A., Srivastava, S.K., 2013. Challenges in enzymatic route of mannitol production. ISRN Biotechnol 1–13. Avaliable from: https://doi.org/10.5402/2013/914187.

Bornet, F.R.J., Blayo, A., Dauchy, F., Slama, G., 1996. Plasma and urine kinetics of erythritol after oral ingestion by healthy humans. Regul. Toxicol. Pharmacol. 24, 280–286.

Bozell, J.J., Petersen, G.R., 2010. Technology development for the production of biobased products from biorefinery carbohydrates—the US Department of Energy's "Top 10" revisited. Green Chem. 12, 539–554.

Business Wire, 2018. The Global Food Sweetener Market Growth, Trends & Forecast 2018–2023. Available at: https://www.businesswire.com/news/home/20180516005625/en/Global-Food-Sweetener-Market-Growth-Trends-Forecast.

Canilha, L., Carvalho, W., Giulietti, M., Felipe, M.G.A., Silva, J.B.A., 2008. Clarification of a wheat straw-derived medium with ion-exchange resins for xylitol crystallization. J. Chem. Technol. Biotechnol. 83, 715–721.

Canilha, L., Rodrigues, R.C.L.B., Antunes, F.A.F., Chandel, A.K., Milessi, T.S.S., Felipe, M.G.A., Silva, S.S., 2013. Bioconversion of hemicellulose from sugarcane biomass into sustainable products. In: Chandel, A.K. (Ed.), Sustainable Degradation of Lignocellulosic Biomass – Techniques, Applications and Commercialization. Intech Open Access, United Kingdom, pp. 15–45.

Cardoso, F.S., Araújo-Lima, C.F., Aiub, C.A.F., Felzenszwalb, I., 2016. Exposure to sorbitol during lactation causes metabolic alterations and genotoxic effects in rat offspring. Toxicol. Lett. 260, 36–45.

Carly, F., Fickers, P., 2018. Erythritol production by yeasts: a snapchot of current knowledge. Yeast 35, 455–463.

Carly, F., Gamboa-Melendez, H., Vandermies, M., Damblon, C., Nicaud, J.M., Fickers, P., 2017. Identification and characterization of EYK1, a key gene for erythritol catabolism in *Yarrowia lipolytica*. Appl. Microbiol. Biotechnol. 101 (17), 6587–6596.

Carocho, M., Morales, P., Ferreira, I.C.F.R., 2017. Sweeteners as food additives in the XXI century: a review of what is known, and what is to come. Food Chem. Toxicol. 107, 302–317.

Cazetta, M.L., Celligoi, M.A.P.C., Buzato, J.B., Scarmino, I.S., Silva, R.S.F., 2005. Optimization study for sorbitol production by *Zymomonas mobilis* in sugar cane molasses. Process Biochem. 40, 747–751.

Chakraborty, R., Das, A., 2019. Artificial sweeteners. Ency. Food Chem. 1, 30–34.

Chandel, A.K., Silva, S.S., Carvalho, W., Singh, O.V., 2012. Sugarcane bagasse and leaves: foreseeable biomass of biofuel and bioproducts. J. Chem. Technol. Biotechnol. 87, 11–20.

Chattopadhyay, S., Raychaudhuri, U., Chakraborty, R., 2014. Artificial sweeteners—a review. J. Food Sci. Technol. 51, 611–621.

Cheng, K.K., Wu, J., Lin, Z.N., Zhang, J.A., 2014. Aerobic and sequential anaerobic fermentation to produce xylitol and ethanol using non-detoxified acid pretreated corncob. Biotechnol. Biofuels 7, 166.

Cheng, H., Wang, S., Bilal, M., Ge, X., Zhang, C., Fickers, P., Cheng, H., 2018. Identification, characterization of two NADPH-dependent erythrose reductases in the yeast *Yarrowia lipolytica* and improvement of erythritol productivity using metabolic engineering. Microb. Cell Factories 17, 133.

Chun, U.H., Rogers, P.L., 1988. The simultaneous production of sorbitol and gluconic acid by *Zymomonas mobilis*. Appl. Microbiol. Biotechnol. 29, 19–24.

Dai, Y.W., Meng, Q., Mu, W., Zhang, T., 2017. Recent advances in theapplications and biotechnological production of mannitol. J. Funct.Foods. 36, 404–409.

Dasgupta, D., Bandhu, S., Adhikari, D.K., Ghosh, D., 2017. Challenges and prospects of xylitol production with whole cell bio-catalysis: a review. Microbiol. Res. 197, 9–21.

De Boeck, R., Sarmiento-Rubiano, L.A., Nadal, I., Monedero, V., Pérez-Martínez, G., Yebra, M.J., 2010. Sorbitol production from lactose by engineered *Lactobacillus casei* deficient in sorbitol transport system and mannitol-1-phosphate dehydrogenase. Appl. Microbiol. Biot. 85, 1915–1922.

De Cock, P., 2012. Erythritol. In: ÓDonnell, K., Kearsley, M.W. (Eds.), Sweeteners and Sugar Alternatives in Food Technology. Wiley-Blackwell, pp. 213–241.

Deis, R.C., Kearsley, M.W., 2012. Sorbitol and mannitol. In: ÓDonnell, K., Kearsley, M.W. (Eds.), Sweeteners and Sugar Alternatives in Food Technology. Wiley-Blackwell, pp. 331–346.

DuBois, G.E., Prakash, I., 2012. Non-caloric sweeteners, sweetness modulators, and sweetener enhancers. Annu. Rev. Food Sci. Technol. 3, 353–380.

Ecogal, 2018. The Sweet Names of Sugar. Available at: http://ecogal.tv/the-sweet-names-of-sugar/.

Eggeling, L., Bott, M., Marienhagen, J., 2015. Novel screening methods—biosensors. Curr. Opin. Biotechnol. 35, 30–36.

Felipe, M.G.A., 2004. Biotechnological production of xylitol from lignocellulosic materials. In: Saha, B., Hayashi, K. (Eds.), Lignocellulose Biodegradation. American Chemical Society Publications, Oxford University Press, USA, pp. 300–315.

Ferraz, H.C., Borges, C.P., Alves, T.L.M., 2000. Produção de sorbitol e ácido glicônico por células permeabilizadas e imobilizadas de *Zymomonas mobilis* com separação simultânea dos produtos por eletrodiálise. In: Proceedings of the 13th National Symposium on Fermentation. Terezópolis, Brazil.

Finn, J.P., Lord, G.H., 2000. Neurotoxicity studies on sucralose and its hydrolysis products with special reference to histopathologic and ultrastructure changes. Food Chem. Toxicol. 38, 7–17.

Flores, C.L., Rodríguez, C., Petit, T., Gancedo, C., 2000. Carbohydrate and energy-yielding metabolism in non-conventional yeasts. FEMS Microbiol. Rev. 24, 507–529.

Fonseca, C., Romão, R., Souza, H.R., Hahn-Hägerdal, B., Spencer-Martins, I., 2007. L-Arabinose transport and catabolism in yeasts. FEBS J. 274, 3589–3600.

Gargouri, W., Zmantar, T., Kammoun, R., Kechaou, N., Ghoul-Mazgar, S., 2018. Coupling xylitol with remineralizing agents improves tooth protection against demineralization but reduces antibiofilm effect. Microb. Pathog. 123, 177–182.

Ghezelbash, G.R., Nahvi, I., Rabbani, M., 2012. Study of polyols production by *Yarrowia lipolytica* in batch culture and optimization of growth condition for maximum production. Jundishapur J. Microbiol. 5, 546–549.

Glória, M.B.A., 2003. Sweeteners - others. In: Caballero, B. (Ed.), Encyclopedia of Food Sciences and Nutrition, second ed. Academic Press, pp. 5695–5702.

Godshall, M.A., 2007. The expanding world of nutritive and NonNutritive sweeteners. Sugar J. 69, 12–20.

Grand View Reserch, 2015. Non Sugar Sweeteners Market Analysis, Market Size, Application Analysis, Regional Outlook, Competitive Strategies, and Forecasts, 2015 to 2022. Available at: https://www.grandviewresearch.com/industry-analysis/non-sugar-sweeteners-market.

Granström, T.B., Izumori, K., Leisola, M., 2007. A rare sugar xylitol. Part II: biotechnological production and future applications of xylitol. Appl. Microbiol. Biotechnol. 74, 277–281.

Grembecka, M., 2015. Sugar alcohols – their role in the modern world of sweeteners - a review. Eur. Food Res. Technol. 241, 1–14.

Grice, H.C., Goldsmith, L.A., 2000. Sucralose - an overview of the toxicity data. Food Chem. Toxicol. 38, 1–6.

Guirimand, G., Sasaki, K., Inokuma, K., Bamba, T., Hasunuma, T., Kondo, A., 2016. Cell surface engineering of *Saccharomyces cerevisiae* combined with membrane separation technology for xylitol
production from rice straw hydrolysate. Appl. Microbiol. Biotechnol. 100, 3477–3487.

Guo, J., Li, J., Chen, Y., Guo, X., Xiao, D., 2016. Improving erythritol production of *Aureobasidium pullulans* from xylose by mutagenesis and medium optimization. Appl. Biochem. Biotechnol. 180, 717–727.

Guo, Q., Zabed, H., Zhang, H., Wang, X., Yun, J., Zhang, G., Yang, M., Sun, W., Qi, X., 2019. Optimization of fermentation medium for a newly isolated yeast strain (*Zygosaccharomyces rouxii* JM-C46) and evaluation of factors affecting biosynthesis of D-arabitol. LWT – Food Sci. Tecnhol. 99, 319–327.

Haghighatian, M., Mofid, M.R., Nekouei, M.K., Yaghmaei, P., Tafreshi, A.H., 2008. Isomalt production by cloning, purifying and expressing of the MDH gene from *Pseudomonas fluorescens* DSM 50106 in different strains of *E. coli*. Pakistan J. Biol. Sci. 11 (16), 2001–2006.

Haighton, L., Roberts, A., Walters, B., Lynch, B., 2019. Systematic review and evaluation of aspartame carcinogenicity bioassays using quality criteria. Regul. Toxicol. Pharmacol. 103, 332–344.

Hatti-Kaul, R., Chen, L., Dishisha, T., Enshasy, H.E., 2018. Lactic acid bacteria: from starter cultures to producers of chemicals. FEMS Microbiol. Lett. 365, 1–20.

Helgi Analytics Calculation. 2013. Food & Drink: Which Country Eats the Most Sugar & Sweeteners? Sugar & Sweetener Consumption (Total) (kt), Available at: https://www.helgilibrary.com/charts/which-country-eats-the-most-sugar-sweeteners/.

Hernández-Pérez, A.F., Costa, I.A.L., Silva, D.D.V., Dussán, K.J., Villela, T.R., Canettieri, E.V., Carvalho Jr., J.A., Soares-Neto, T.G., Felipe, M.G.A., 2016. Biochemical conversion of sugarcane straw hemicellulosic hydrolyzate supplemented with co-substrates for xylitol production. Bioresour. Technol. 200, 1085–1088.

Hou, J., Qiu, C., Shen, Y., Li, H., Bao, X., 2017. Engineering of *Saccharomyces cerevisiae* for the efficient co-utilization of glucose and xylose. FEMS Yeast Res. 17 (4), 1–11.

Hou-Rui, Z., 2012. Key drivers influencing the large scale production of xylitol. In: Silva, S.S., Chandel, A.K. (Eds.), D-xylitol. Fermentative Production, Application and Commercialization. Springer-Verlag, Germany, pp. 267–290.

Itoh, Y., Tanaka, A., Araya, H., Ogasawara, K., Inaba, H., Sakamoto, Y., Koga, J., 1992. *Lactobacillus* sp. B001 and method of producing mannitol, acetic acid and lactic acid. European Patent 486024.

Jagtap, S.S., Rao, C.V., 2018. Production of D-arabitol from D-xylose by the oleaginous yeast *Rhodosporidium toruloides* IFO0880. Appl. Microbiol. Biotechnol. 102, 143–151.

Jeya, M., Lee, K.M., Tiwari, M.K., Kim, J.S., Gunasekaran, P., Kim, S.Y., Kim, I.W., Lee, J.K., 2009. Isolation of a novel high erythritol-producing *Pseudozyma tsukubaensis* and scale-up of erythritol fermentation to industrial level. Appl. Microbiol. Biotechnol. 83, 225–231.

Jiang, X., He, P., Qi, X., Lin, Y., Zhang, Y., Wang, Q., 2016. High-efficient xylitol production by evolved *Candida maltosa* adapted to corncob hemicellulosic hydrolysate. J. Chem. Technol. Biotechnol. 91, 2994–2999.

Jönsson, L.J., Alriksson, B., Nilvebrant, N.O., 2013. Bioconversion of lignocellulose: inhibitors and detoxification. Biotechnol. Biofuels 6, 16.

Kallscheuer, N., 2018. Engineered microorganisms for the production of food additives approved by the European Union: a systematic analysis. Front. Microbiol. 9, 1746.

Kearsley, M.W., Deis, R.C., 2012. Maltitol powder. In: ÓDonnell, K., Kearsley, M.W. (Eds.), Sweeteners and Sugar Alternatives in Food Technology. Wiley-Blackwell, pp. 295–308.

Kim, J., Ryu, Y., Seo, J., 1999. Analysis and optimization of a two-substrate fermentation for xylitol production using *Candida tropicalis*. J. Ind. Microbiol. Biotechnol. 22, 181–186.

Kim, B.S., Chang, J.Y., Kim, Y.Y., Kho, H.S., 2015. The effects of xylitol and sorbitol on lysozyme- and peroxidase-related enzymatic and candidacidal activities. Arch. Oral Biol. 60, 998–1006.

Kobayashi, Y., Yoshida, J., Iwata, H., Koyama, Y., Kato, J., Ogihara, J., Kasumi, T., 2013. Gene expression and function involved in polyol biosynthesis of *Trichosporonoides megachiliensis* under hyper-osmotic stress. J. Biosci. Bioeng. 115, 645–650.

Kobayashi, Y., Iwata, H., Mizushima, D., Ogihara, J., Kasumi, T., 2015a. Erythritol production by *Moniliella megachiliensis* using nonrefined glycerol waste as carbon source. Lett. Appl. Microbiol. 60, 475–480.

Kobayashi, Y., Iwata, H., Yoshida, J., Ogihara, J., Kato, J., Kasumi, T., 2015b. Metabolic correlation between polyol and energy-storing carbohydrate under osmotic and oxidative stress condition in *Moniliella megachiliensis*. J. Biosci. Bioeng. 120, 405–410.

Koganti, S., Ju, L.K., 2013. *Debaryomyces hansenii* fermentation for arabitol production. Biochem. Eng. J. 79, 112–119.

Koganti, S., Kuo, T.M., Kurtzman, C.P., Smith, N., Ju, L.K., 2011. Production of arabitol from glycerol: strain screening and study of factors affecting production yield. Appl. Microbiol. Biotechnol. 90, 257–267.

Kogje, A., Ghosalkar, A., 2017. Xylitol production by genetically modified industrial strain of *Saccharomyces cerevisiae* using glycerol as co-substrate. J. Ind. Microbiol. Biotechnol. 44, 961–971.

Kordowska-Wiater, M., 2015. Production of arabitol by yeasts: current status and future prospects. J. Appl. Microbiol. 119, 303–314.

Ladero, V., Ramos, A., Wiersma, A., Goffin, P., Schanck, A., Kleerebezem, M., Hugenholtz, J., Smid, E.J., Hols, P., 2007. High-level production of the low-calorie sugar sorbitol by *Lactobacillus plantarum* through metabolic engineering. Appl. Environ. Microbiol. 73, 1864–1872.

Lange, F.T., Scheurer, M., Brauch, H.J., 2012. Artificial sweeteners - a recently recognized class of emerging environmental contaminants: a review. Anal. Bioanal. Chem. 403, 2503–2518.

Lee, J., 2015. Sorbitol, *Rubus* fruit, and misconception. Food Chem. 166, 616–622.

Lee, J.K., Ha, S.J., Kim, S.Y., Oh, D.K., 2001. Increased erythritol production in *Torula* sp. with inositol and phytic acid. Biotechnol. Lett. 23, 497–500.

Lee, J.K., Song, J.Y., Kim, S.Y., 2003. Controlling substrate concentration in fed-batch *Candida magnoliae* culture increases mannitol production. Biotechnol. Prog. 19, 768–775.

Lee, J.K., Ha, S.J., Kim, S.Y., Oh, D.K., 2000. Increased erythritol production in *Torula* sp. by Mn2+ and Cu2+. Biotechnol. Lett. 22, 983–986.

Lee, D.H., Lee, Y.J., Ryu, Y.W., Seo, J.H., 2010. Molecular cloning and biochemical characterization of a novel erythrose reductase from *Candida magnoliae* JH110. Microb. Cell Factories 9, 43.

Lin, S.J., Wen, C.Y., Wang, P.M., Huang, J.C., Wei, C.L., Chang, J.W., Chu, W.S., 2010. High-level production of erythritol by mutants of osmophilic *Moniliella* sp. Process Biochem. 45, 973–979.

Liu, C., Dong, H., Zhong, J., Ryu, D.D.Y., Bao, J., 2010. Sorbitol production using recombinant *Zymomonas mobilis* strain. J. Biotechnol. 148, 105–112.

Livesey, G., 2003. Health potential of polyols as sugar replacers, with emphasis on low-glycaemic properties. Nutr. Res. Rev. 16, 163–191.

Loman, A.A., Islam, S.M.M., Ju, L.K., 2018. Production of arabitol from enzymatic hydrolysate of soybean flour by *Debaryomyces hansenii* fermentation. Appl. Microbiol. Biotechnol. 102, 641–653.

Loos, H., Krämer, R., Sahm, H., Sprenger, G.A., 1994. Sorbitol promotes growth of *Zymomonas mobilis* in environments with high concentrations of sugar: evidence for a physiological function of glucose-fructose oxidoreductase in osmoprotection. J. Bacteriol. 176, 7688–7693.

Mäkinen, K.K., 2011. Sugar alcohol sweeteners as alternatives to sugar with special consideration of xylitol. Med. Princ. Pract. 20, 303–320.

Mäkinen, K.K., Saag, M., Isotupa, K.P., Olak, J., Nommela, R., Soderling, E., Makinen, P.L., 2005. Similarity of the effects of erythritol and xylitol on some risk factors of dental caries. Caries Res. 39, 207–215.

Martínez, E.A., Giulietti, M., Silva, J.B.D.A., Derenzo, S., 2008. Kinetics of the xylitol crystallization in hydro-alcoholic solution. Chem. Eng. Process 47, 2157–2162.

Martínez, E.A., Canettieri, E.V., Bispo, J.A., Giulietti, M., Silva, J.B.A., Converti, A., 2015. Strategies for xylitol purification and crystallization: a review. Separ. Sci. Technol. 50, 2087–2098.

McMillan, J.D., Boynton, B.L., 1994. Arabinose utilization by xylose-fermenting yeasts and fungi. Appl. Biochem. Biotechnol. 45/46, 569–584.

Meng, Q., Zhang, T., Wei, W.T., Mu, W.M., Miao, M., 2017. Production of mannitol from a high concentration of glucose by *Candida parapsilosis* SK26.001. Appl. Biochem. Biotechnol. 181, 391–406.

Mingguo, J., Wang, B., Yang, L., Lin, S., Cheng, H., 2011. Microbiological purification of L-arabitol from xylitol mother liquor. J. Microbiol. Biotechnol. 21, 43–49.

Mironczuk, A.M., Rakicka, M., Biegalska, A., Rymowicz, W., Dobrowolski, A., 2015. A two-stage fermentation process of erythritol production by yeast *Y. lipolytica* from molasses and glycerol. Bioresour. Technol. 198, 445–455.

Mirończuk, A.M., Rzechonek, D.A., Biegalska, A., Rakicka, M., Dobrowolski, A., 2016. A novel strain of *Yarrowia lipolytica* as a platform for value-added product synthesis from glycerol. Biotechnol. Biofuels 9, 180.

Mirończuk, A.M., Biegalska, A., Dobrowolski, A., 2017. Functional overexpression of genes involved in erythritol synthesis in the yeast *Yarrowia lipolytica*. Biotechnol. Biofuels 10, 77.

Misra, S., 2016. Sugars and starches. In: Reversing Diabetes, the High 5 Way. Educreation Publishing, New Delhi, pp. 157–167.

Moon, H.J., Jeya, M., Kim, I.W., Lee, J.K., 2010. Biotechnological production of erythritol and its applications. Appl. Microbiol. Biotechnol. 86, 1017–1025.

Mpabanga, T.P., Chandel, A.K., Silva, S.S., Singh, O.V., 2012. Detoxification strategies applied to lignocellulosic hydrolysates for improved xylitol production. In: Silva, S.S., Chandel, A.K. (Eds.), D-xylitol. Fermentative Production, Application and Commercialization. Springer-Verlag, Germany, pp. 63–82.

Mussatto, S., Roberto, I.C., 2002. Xilitol: edulcorante com efeitos benéficos para a saúde humana. Rev. Bras. Ciencias Farm. 38, 401–413.

Nabors, L.O., 2002. Sweet choices: sugar replacements for foods and beverages. Food Technol. 56, 28–32.

Nabors, L.O., Gelardi, R., 2001. Alternative sweeteners: an overview. In: Nabors, L.O. (Ed.), Alternative Sweeteners, 4ed. CRC Press, USA, pp. 1–10.

Nofre, C., Tinti, J.M., 2000. Neotame: discovery, properties, utility. Food Chem. 69 (3), 245–257.

OECD/FAO, 2018. "OECD-FAO Agricultural Outlook", OECD Agriculture statistics (database). https://doi.org/10.1787/agr-outl-data-en.

Otabe, T., Fujieda, T., Masuyama, K., Ubukata, C., Lee, C., 2011. Advantame – an overview of the toxicity data. Food Chem. Toxicol. 49, S2–S7.

O'Donnell, K., 2012. Aspartame, neotame and advantame. In: ÓDonnell, K., Kearsley, M.W. (Eds.), Sweeteners and Sugar Alternatives in Food Technology. Wiley-Blackwell, pp. 117–136.

Park, Y.C., Oh, E.J., Jo, J.H., Jin, Y.S., Seo, J.H., 2016. Recent advances in biological production of sugar alcohols. Curr. Opin. Biotechnol. 37, 105–113.

Philippe, R.N., Mey, M.D., Anderson, J., Ajilkumar, P.K., 2014. Biotechnological production of natural zero-calorie sweeteners. Curr. Opin. Biotechnol. 26, 155–161.

Popkin, B.M., Nielsen, S.J., 2003. The sweetening of the world's diet. Obes. Res. 11, 1325–1332.

Prakash, I., 2007. Synthesis of N-[N-(3,3-dimethylbutyl)-L-α-aspartyl]-Lphenylalanine 1-methyl ester using 3,3-dimethylbutyraldehyde precursors. Ip.com US Patent 7288670. Available at: http://ip.com/patent/US7288670.

Prakash, I., Zhao, R.Y., 2001. Chemoenzymatic synthesis of neotame, Pub No. WO/2001/085977. Available at: http://www.wipo.int/pctdb/en/wo.jsp?wo=2001085977.

Praveena, S.M., Cheema, M.S., Guo, H.R., 2019. Non-nutritive artificial sweeteners as an emerging contaminant in environment: a global review and risks perspectives. Ecotoxicol. Environ. Saf. 170, 699–707.

Qi, X., Luo, Y., Wang, X., Zhu, J., Lin, J., Zhang, H., Chen, F., Sun, W., 2015. Enhanced D-arabitol production by *Zygosaccharomyces rouxii* JM-C46: isolation of strains and process of repeated-batch fermentation. J. Ind. Microbiol. Biotechnol. 42, 807–812.

Rakicka, M., Rukowicz, B., Rywińska, A., Lazar, Z., Rymowicz, W., 2016. Technology of efficient continuous erythritol production from glycerol. J. Clean. Prod. 136, 905–913.

Rakicka, M., Rywińska, A., Lazar, Z., Rymowicz, W., 2017. Two-stage continuous culture – technology boosting erythritol production. J. Clean. Prod. 168, 420–427.

Rao, L.V., Goli, J.K., Gentela, J., Koti, S., 2016. Bioconversion of lignocellulosic biomass to xylitol: an overview. Bioresour. Technol. 213, 299–310.

Regnat, K., Mach, R.L., Mach-Aigner, A.R., 2018. Erythritol as sweetener – wherefrom and whereto? Appl. Microbiol. Biotechnol. 102, 578–595.

Regulation (EC) No. 1333/2008 of the European Parliament and of the Council of 16 December 2008 on food additives. Available at: https://eur-lex.europa.eu/LexUriServ/LexUriServ.do?uri=OJ:L:2008:354:0016:0033:en:PDF.

Renwick, A.G., Nordmann, H., 2007. First European conference on aspartame: putting safety and benefits into perspective. Synopsis of presentations and conclusions. Food Chem. Toxicol. 45, 1308–1313.

Roberts, A., Renwick, A.G., Sims, J., Snodin, D.J., 2000. Sucralose metabolism and pharmacokinetics in man. Food Chem. Toxicol. 38, 31–41.

Rosenman, K., 1978. Benefits of saccharin: a review. Environ. Res. 15 (1), 70–81.

Rymon Lipinski, G.W., 1991. Sensory Properties of Acesulfame-K. Marcel Dekker, New York.

Rymowicz, W., Rywinska, A., Marcinkiewicz, M., 2009. High-yield production of erythritol from raw glycerol in fed-batch cultures of *Yarrowia lipolytica*. Biotechnol. Lett. 31, 377–380.

Rzechonek, D.A., Dobrowolski, A., Rymowicz, W., Mirończuk, A.M., 2018. Recent advances in biological production of erythritol. Curr. Rev. Biotechnol. 38 (4), 620–633.

Saha, B.C., 2006. A low-cost medium for mannitol production by *Lactobacillus intermedius* NRRL B-3693. Appl. Microbiol. Biotechnol. 72, 676–680.

Saha, B.C., Nakamura, L.K., 2003. Production of mannitol and lactic acid by fermentation with *Lactobacillus intermedius* NRRL B-3693. Biotechnol. Bioeng. 82, 864–871.

Saha, B.C., Racine, F.M., 2010. Effects of pH and corn steep liquor variability on mannitol production by *Lactobacillus intermedius* NRRL B-3693. Appl. Microbiol. Biotechnol. 87, 553–560.

Saini, R.K., Kumar, S.S., Priyanka, P.S., Kamireddy, K., Giridhar, P., 2016. Alternative sweeteners: current scenario and future innovations for value addition. In: Bica, J.L., Maróstica Jr., M.R., Pastore, G.M. (Eds.), Biotechnological Production of Natural Ingredients for Food Industry. Benthan Science, pp. 21–59.

Sampaio, F.C., Passos, F.M.L., Passos, F.J.V., Faveri, D., Perego, P., Converti, A., 2006. Xylitol crystallization from culture media fermented by yeasts. Chem. Eng. Process 45, 1041–1046.

Schallmey, M., Frunzke, J., Eggeling, L., Marienhagen, J., 2014. Looking for the pick of the bunch: high-throughput screening of producing microorganisms with biosensors. Curr. Opin. Biotechnol. 26, 148–154.

Sentko, A., Willibald-Ettle, I., 2012. Isomalt. In: ÓDonnell, K., Kearsley, M.W. (Eds.), Sweeteners and Sugar Alternatives in Food Technology. Wiley-Blackwell, pp. 242–274.

Shalem, O., Sanjana, N.E., Zhang, F., 2015. High-throughput functional genomics using CRISPR–Cas9. Nat. Rev. Genet. 16, 299.

Shankar, P., Ahuja, S., Sriram, K., 2013. Non-nutritive sweeteners: review and update. Nutrition 29, 1293–1299.

Silva, S.S., Chandel, A.K., 2012. D-xylitol. Fermentative Production, Application and Commercialization. Springer-Verlag, Germany, pp. 229–244.

Silva-Fernandes, T., Santos, J.C., Hasmann, F., Rodrigues, R.C.L.B., Izario Filho, H.J., Felipe, M.G.A., 2017. Biodegradable alternative for removing toxic compounds from sugarcane bagasse hemicellulosic hydrolysates for valorization in biorefineries. Bioresour. Technol. 243, 384–392.

Silveira, M.M., Wisbeck, E., Lemmel, C., Erzinger, G., Costa, J.P.L., Bertasso, M., Jonas, R., 1999. Bioconversion of glucose and fructose to sorbitol and gluconic acid by untreated cells of *Zymomonas mobilis*. J. Biotechnol. 75, 99–103.

Silveira, M.M., Jonas, R., 2002. The biotechnological production of sorbitol. Appl. Microbiol. Biotechnol. 59, 400–408.

Silveira, M.M., Costa, J.P.L, Jonas, R. 1994. Processo de produção e recuperação de sorbitol e ácido glucônico ou gluconato. Brazilian Patent PI 9.403.981–0.

Silveira, M.M., Wisbeck, E., Hoch, I., Jonas, R., 2001. Production of glucose-fructose oxidoreductase and ethanol by *Zymomonas mobilis* ATCC 29191 in medium containing corn steep liquor as a source of vitamins. Appl. Microbiol. Biotechnol. 55, 442–445.

Soetaert, W., Vanhooren, P.T., Vandamme, E.J., 1999. The production of D-mannitol by fermentation. Methods Biotechnol. 10, 261–275.

Swithers, S.E., 2013. Artificial sweeteners produce the counterintuitive effect of inducing metabolic derangements. Trends Endocrinol. Metabol. 24 (9), 431–441.

Sylvetsky, A., Rother, K.I., Brown, R., 2011. Artificial sweetener use among children: epidemiology, recommendations, metabolic outcomes, and future directions. Pediatr. Clin. 58 (6), 1467–1480.

Tahergorabi, Z., Khazaei, M., Moodi, M., Chamani, E., 2016. From obesity to cancer: a review on proposed mechanisms. Cell Biochem. Funct. 34, 533–545.

Tomaszewska, L., Rywińska, A., 2016. Erythritol biosynthesis from glycerol by *Yarrowia lipolytica* yeast: effect of osmotic pressure. Chem. Pap. 70, 272–283.

Tomaszewska, L., Rywinska, A., Gładkowski, W., 2012. Production of erythritol and mannitol by *Yarrowia lipolytica* yeast in media containing glycerol. J. Ind. Microbiol. Biotechnol. 39, 1333–1343.

Tomaszewska, L., Rymowicz, W., Rywinska, A., 2014a. Mineral supplementation increases erythrose reductase activity in erythritol biosynthesis from glycerol by *Yarrowia lipolytica*. Appl. Biochem. Biotechnol. 172, 3069–3078.

Tomaszewska, L., Rywinska, A., Rymowicz, W., 2014b. High selectivity of erythritol production from glycerol by *Yarrowia lipolytica*. Biomass Bioenergy 64, 309–320.

Tyler, C.A., Kopit, L., Doyle, C., Yu, A.O., Hugenholtz, J., Marco, M.,L., 2016. Polyol production during heterofermentative growth of the plant isolate *Lactobacillus florum* 2F. J. Appl. Microbiol. 120, 1336–1345.

United States Food and Drug Administration, 2018. Additional information about high-intensity sweeteners permitted for use in food in the United States. Available at: https://www.fda.gov/Food/IngredientsPackagingLabeling/FoodAdditivesIngredients/ucm397725.htm.

Veiga-Da-Cunha, M., Santos, H., van Schaftingen, E., 1993. Pathway and regulation of erythritol formation in *Leuconostoc oenos*. J. Bacteriol. 175, 3941–3948.

Wang, H., Li, L., Zhang, L., An, J., Cheng, H., Deng, Z., 2016. Xylitol production from waste xylosemother liquor containing miscellaneous sugars and inhibitors: one-pot biotransformation by *Candida tropicalis* and recombinant *Bacillus subtilis*. Microb. Cell Factories 15, 82.

Werpy, T., Petersen, G., 2004. Top value added chemicals from biomass: volume I—results of screening for potential candidates from sugars and synthesis gas. National Renewable Energy Lab., Golden, CO (US), p. 76.

Wisselink, H.W., Weusthuis, R.A., Eggink, G., Hugenholtz, J., Grobben, G.J., 2002. Mannitol production by lactic acid bacteria: a review. Int. Dairy J. 12, 151–161.

Yang, L.B., Zhan, X.B., Zheng, Z.Y., Wu, J.R., Gao, M.J., Lin, C.C., 2014. A novel osmotic pressure control fed-batch fermentation strategy for improvement of erythritol production by *Yarrowia lipolytica* from glycerol. Bioresour. Technol. 151, 120–127.

Yoshikawa, J., Habe, H., Morita, T., Fukuoka, T., Imura, T., Iwabuchi, H., Uemura, S., Tamura, T., Kitamoto, D., 2014a. Production of d-arabitol from raw glycerol by *Candida quercitrusa*. Appl. Microbiol. Biotechnol. 98 (7), 2947–2953.

Yoshikawa, J., Habe, H., Morita, T., Fukuoka, T., Imura, T., Iwabuchi, H., Uemura, S., Tamura, T., Kitamoto, D., 2014b. Production of mannitol from raw glycerol by *Candida azyma*. J. Biosci. Bioeng. 117 (6), 725–729.

Zachariou, M., Scopes, R.K., 1986. Glucose-fructose oxidoreductase, a new enzyme isolated from *Zymomonas mobilis* that is responsible for sorbitol production. J. Bacteriol. 167, 863–869.

Zacharis, C., 2012a. Lactitol. In: ÓDonnell, K., Kearsley, M.W. (Eds.), Sweeteners and Sugar Alternatives in Food Technology. Wiley-Blackwell, pp. 275–293.

Zacharis, C., 2012b. Xylitol. In: ÓDonnell, K., Kearsley, M.W. (Eds.), Sweeteners and Sugar Alternatives in Food Technology. Wiley-Blackwell, 347–282.

Zhang, G., Lin, Y., He, P., Li, L., Wang, Q., Ma, Y., 2014. Characterization of the sugar alcohol-producing yeast *Pichia anomala*. J. Ind. Microbiol. Biotechnol. 41, 41–48.

Zhang, G.C., Liu, J.J., Kong, I.I., Kwak, S., Jin, Y.S., 2015. Combining C6 and C5 sugar metabolism for enhancing microbial bioconversion. Curr. Opin. Chem. Biol. 29, 49–57.

Zhang, M., Gu, L., Cheng, C., Ma, J., Xin, F., Liu, J., Wu, H., Jiang, M., 2018. Recent advances in microbial production of mannitol: utilization of low-cost substrates, strain development and regulation strategies. World J. Microbiol. Biotechnol. 34, 41.

CHAPTER

Microbial production of omega-3 polyunsaturated fatty acids

10

Madan L. Verma[a,b], Kaushal Kishor[c], Deepka Sharma[d], Sanjeev Kumar[e], Krishan D. Sharma[f]

Department of Biotechnology, Dr Y. S. Parmar University of Horticulture and Forestry, Neri Campus, Himachal Pradesh, India[a]; Centre for Chemistry and Biotechnology, Deakin University, Geelong Campus, VIC, Australia[b]; Technology Research and Advisory, Aranca Pvt Ltd, Mumbai, India[c]; Department of Biotechnology, Dr Y. S. Parmar University of Horticulture and Forestry, Himachal Pradesh, India[d]; Department of Basic Sciences, Dr Y. S. Parmar University of Horticulture and Forestry, Hamirpur, Himachal Pradesh, India[e]; Department of Food Science and Technology, Dr Y. S. Parmar University of Horticulture and Forestry, Solan, Himachal Pradesh, India[f]

Chapter outline

Biotechnological Production of Bioactive Compounds. https://doi.org/10.1016/B978-0-444-64323-0.00010-2

1. Introduction

Omega-3 fatty acids are polyunsaturated fatty acids (PUFAs), recognized as essential fatty acids because they are significant for good health and growth of higher eukaryotes (Khan et al., 2017). Nutritionally, eicosapentaenoic acid (EPA, 20:5) and docosahexaenoic acid (DHA, 22:6) are on the whole important omega-3 fatty acids belonging to this group of bioactive admixture compounds. The term omega in omega-3 fatty acid, is named after the terminal carbon atom farthest from the functional carboxylic acid group. For example, further classification of polyunsaturated fatty acid omega-3 fatty acid is based on first unsaturation site relative to the omega end of that fatty acid (Insel et al., 2013). Thus, an omega-3 fatty acid like α-linolenic acid (ALA), which has three carbon-carbon double bonds. There are four major omega-3 fatty acids that are synthesized from precursor fatty acids or ingested in foods and then used by the body (Tocher et al., 2019). The omega-3 fatty acids are alpha-linolenic acid (ALA), eicosapentaenoic acid (EPA), docosapentaenoic acid (DPA), and docosahexaenoic acid (DHA).

Polyunsaturated fatty acids (PUFA), the omega-3 fatty acids (FA) docosahexaenoic acid (DHA) and eicosapentaenoic acid (EPA) is essential for development of eye, brains in infants as well as role in prevention and control of several cancer, coronary heart disease, hypertension, depression, type-2 diabetes mellitus, atherosclerosis, and thrombosis (Adams et al., 1996; Balk et al., 2004; Schacky and Harris, 2007; Tocher et al., 2019; Martínez Andrade et al., 2018). At present most of these fatty acids are derived or obtained from cold water fish such as salmon and tuna. However, fish derived oil has several issues ranging from its unattractive odor, presence of hazardous materials, stability, purity and objections from vegetarian. Furthermore, globally increasing demand of these lipids in food, nutraceutical and pharmaceuticals cannot be alone ensured by fish sources. Therefore, in order to meet these demands alternative production systems are needed (Tanaka et al., 2017).

Fish generally obtain them most of PUFA from zooplankton which feeds the photosynthetic microalgae and heterotrophic microorganisms. These lipids are generally originating from microbial resources such as Thraustochytrids (*Aurantiochytrium*, *Schizochytrium*) and *Crypthecodinium cohnii* which are potential DHA producers while strains such as *Phaeodactylum tricornutum* and *Nannochloropsis* have shown EPA production capability (Xie et al., 2017; Chua and Schenk, 2017; Steinrücken et al., 2017). Thraustochytrids is a disputed microbe in term of their classification between algae and fungi. *Thraustochytrids* generally accumulate up to 50% of their dry weight as lipids. DHA may be present up to 25% in these lipids (Puri, 2017). The lipids production employing these strains can serve a potential route due to their fast growth, ability to grow on cheap medium and non-arable land. The culturing these strains on commercial scale will ensure product uniformity and will be acceptable by vegetarians. The mass culturing of these organism will also valuable bioproducts such as food additives, vitamins, pigments, animal feed, pharmaceutical compounds, cosmetics, and biofuel (Allemann and Allen, 2018; Islam et al., 2018; Chen et al., 2018).

The present chapter discusses the bioprocessing of oleaginous microalgae for omega-3 fatty acid production. Different sources of omega-3 fatty acids are comprehensively discussed. Various methods of lipid extraction from microalgae available with respect to omega-3 fatty acid production is critically reviewed. All areas of lipid extraction methodologies including solvent extraction procedures, mechanical approaches, and solvent-free procedures apart from some of the latest extraction technologies is discussed.

2. Biosynthesis mechanism of omega-3 fatty acids

EPA and DHA are highly unsaturated fatty acids synthesized from alpha−linolenic acid (Lenihan-Geels et al., 2013). It involves three endoplasmic reticulum fatty acid elongation enzyme systems, and two different desaturases that involves a peroxisomal β-oxidation step as the final step in DHA synthesis (Ratledge, 2004). The two different desaturases involved in biosynthesis mechanism are Δ^5-eicosatrienoyl-CoA desaturase (D5D) and Δ^6-oleoyl(linolenoyl)-CoA desaturase (D6D). The D5D desaturases enzyme catalyses the desaturation of C5 to C6 bond in fatty acids. And the D6D desaturases enzyme catalyses the desaturation of the C6 to C7 bond in fatty acids (Adarme-Vega et al., 2012; Ofosu et al., 2017a) as shown in Fig. 10.1.

3. Sources of omega-3-fatty acids

Naturally, ALA (an essential fatty acid) is found mainly in plant oils such as flaxseed, soybean, and canola oils whereas DHA and EPA are found in fish and other seafood. One could get adequate amounts of omega-3s by eating a variety of foods, including: Fish and other seafood (especially cold-water fatty fish, such as salmon, mackerel, tuna, herring, and sardines (Gunstone, 1996; Whitehead, 1985), some nuts and seeds (such as flaxseed, Brussels Sprouts, chia seeds, and walnuts), plant oils (such as flaxseed oil, soybean oil, canola oil, perilla oil, hemp seed). The omega-3 fatty acid converts from ALA to EPA and finally to DHA from DPA in the body. EPA and DHA referred as "miracle food of 21st century" (Swinbanks, 1993; Bourre, 2007; Amiri-Jami and Griffiths, 2010) are the two-primary omega-3 PUFAs that serve as bioactive lipids (Khan et al., 2017).

Omega-3 fatty acids have proven to be very essential for human health due to their multiple health benefits (Islam et al., 2018; Tocher et al., 2019). These essential fatty acids (EFAs) need to be uptake through diet because they are unable to be produced by the human body. Omega-3 oil is primarily taken in food diet from various types of fish oil and microalgal oil (Fig. 10.2). Rough, scaly skin and a red, swollen, itchy rash can cause because of a deficiency of omega-3 fatty acids. Omega-3 fatty acids are vital integral components of the plasma membrane of the basic unit of the life. DHA concentration are quite high in brain, and eye. Omega-3s also provide

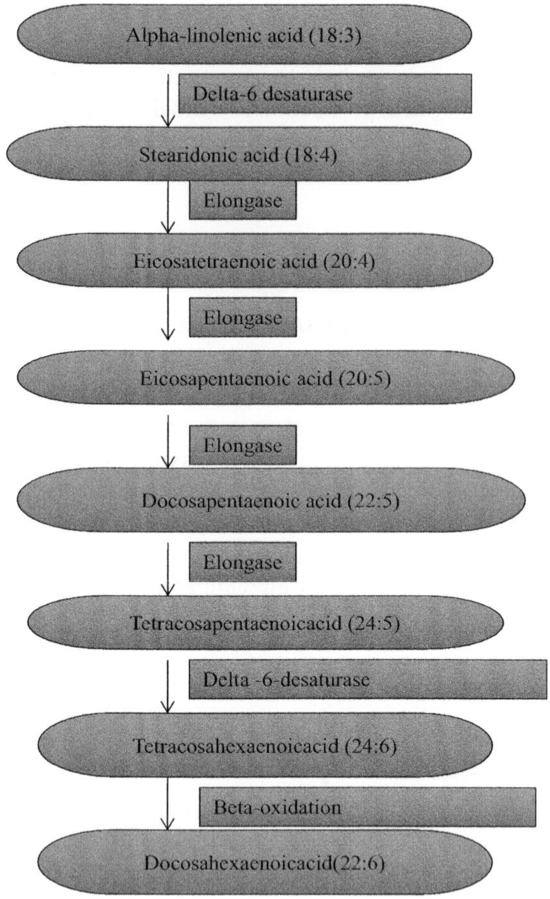

FIG. 10.1

Flow chart showing conversion of alpha-linolinic acid (ALA) to docosahexaenoic acid (DHA).

calories to give your body energy and have many functions in your heart, blood vessels, lungs, immune system, and endocrine system (the network of hormone-producing glands) (Martínez Andrade et al., 2018). Nowadays, omega-3s are found naturally in some foods and are added to some fortified food such as certain brands of eggs, yogurt, juices, milk, soy beverages, and infant formulas (Winwood, 2013; Puri, 2017; Ofosu et al., 2017).

Fish oils show unfavourable effects in inflammatory bowel disease, can cause cancers and cardiovascular complications. Thus, it necessitates the healthy and sustainable alternative sources of polyunsaturated fatty acid. Also due to the (1) rapid decline in fish species and number (2) mercury contamination (3) unpleasant smell and taste of fish (4) stability of fish oil (5) complicated process of purification an

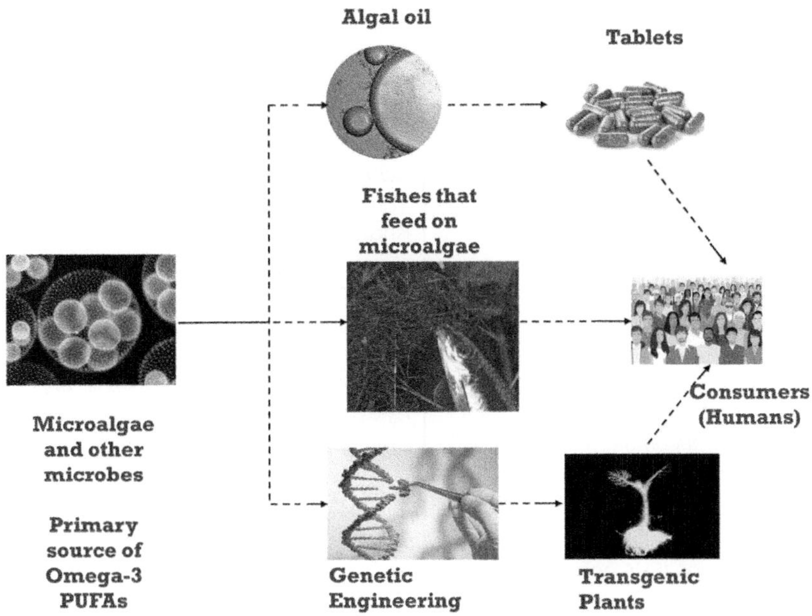

FIG. 10.2

Sustainable sources of polyunsaturated omega-3 fatty acids.

alternative source of these fatty acids is required (Mahaffey et al., 2008; Bourdon et al., 2010; Lenihan-Geels et al., 2013). Other sources of omega-3 fatty acids include fungi, marine algae and marine bacteria (Amiri-Jami and Griffiths, 2010).

3.1 Microflora

The prime source of the omega-3 fatty acids are microalgae and oleaginous organisms (Table 10.1).

3.1.1 Bacteria

Yazawa (1996) demonstrated *Shewanella putrefaciens* as a good source of EPA. In another work Yazawa et al. (1988a,b) described two strains *Pneumatophorus japonicas, Alteromonas putrefaciens* as another bacterial source of EPA. Santos-Merino et al. (2018) engineered the fatty acid synthesis pathway in *Synechococcus elongatus* PCC 7942 for improved omega-3 fatty acid production.

3.1.2 Microalgae

Massive number of algal species produce omega-3 fatty acids. However, their quantity is variable with respect to production of EPA and DHA. Some of algae and algae like microorganisms used for production of EPA and DHA are from families *Thraustochytriaceae* and *Crypthecodiniaceae*. *Thraustochytrium*, *Schizochytrium*

Table 10.1 List of microbial sources of omega-3 fatty acids.

Source of omega-3 fatty acids	References
Microalgae	
Thraustochytrium (EPA, DHA)	Scott et al. (2011)
Nannochloropsis (EPA, DHA)	Hu and Gao (2003), Patil et al. (2007), Pal et al. (2011), and Van Wagenen et al. (2012)
Pinguiococcus pyrenoidosus (EPA, DHA)	Sang et al. (2012)
Chlorella minutissima (EPA)	Yongmanitchai and Ward (1991) and Khozin-Goldberg et al. (2002)
Pavlova spp (EPA, DHA)	Hu et al. (2008), Carvalho et al., 2016, Guihéneuf et al. (2009)
Schizochytrium (DHA)	
Ulkenia (DHA)	Doughman et al. (2007), Chen et al. (2007), Borowitzka (2013), and Klok et al. (2014)
*Cryptocodinium cohnii (*DHA)	Kyle (2001)
Phaeodactylum tricornutum (EPA)	Wen and Chen (2003), Ji et al. (2015)
Nannochloropsis (EPA)	
Nitzchia (EPA)	
Monodus (EPA)	
Isochrysis galbana (EPA and DHA)	Guihéneuf et al. (2009)
Pavloba lutheria (EPA and DHA)	
Phaeodactylum tricornutum (EPA and DHA)	Hamilton et al. (2016)
Fungi	
Thraustochytrium aureum (EPA, DHA)	Ward and Singh (2005)
Mortierella (EPA)	Sakayu et al. (1988), Jareonkitmongkol et al. (1993), Jermsuntiea et al. (2011)
Pythium (EPA)	Athalye et al. (2009)
Trichoderma spp.	Liang et al. (2012)
Aspegillus niger (DHA and EPA)	Gayathri et al. (2010)
Transgenics	
Plants	Wu et al. (2005a,b)
Mustard	Kinney et al. (2004)
Soybean (EPA)	Cheng et al. (2010)
Brassica carinata (EPA)	Petrie et al. (2010)
Nicotiana benthamiana (EPA)	Petrie et al. (2010)
Camelina sativa	
Yeast	
Yarrowia lipolytica (EPA)	Xie et al. (2015)
Bacteria	
Shewanella putrefaciens (EPA)	Yazawa (1996)
Pneumatophorus japonicus (EPA)	Yazawa et al. (1988)
Alteromonas putrefaciens(EPA)	Yazawa et al. (1988)

and *Ulkenia* genera from family Thraustochytriaceae and Cryptocodinium genus from Crypthecodiniaceae family are rich source of DHA (Borowitzka, 2013; Klok et al., 2014). *Schizochytrium* sp. produces higher DHA amounts as compared to lower amounts of EPA (Doughman et al., 2007). Chen et al., 2007 had isolated a marine protist, *Schizochytrium mangrovei*, which can produce a very high yield of DHA. Sijtsma and de Swaaf (2004) reported current use of *Schizochytrium* sp. in food processing and their product thereof. In another study, thraustochytrid *Thraustochytrium* sp. exhibited higher levels of DHA synthesis that goes up to 35% total fatty acids. Additional studies by Kyle (2001) analyzed *Cryptocodinium cohnii*, another high-DHA synthesizing microalgae, and such oils are also used in commercial products. Cost, extraction and purification methods are currently limiting the potential of using micro algal oils on a larger-scale (Adarme-Vega; 2012). Moreover, microalgae like *P. tricornutum*, *Nannochloropsis*, *Nitzchia*, and *Monodus* are good sources of eicosapentaenoic acid (EPA) (Wen and Chen, 2003; Ji et al., 2015; Chua and Schenk, 2017; Steinrücken et al., 2017; Chen et al., 2018).

3.1.3 Fungi

Species of lower fungi are also able to accumulate a high percentage of EPA in the lipid fraction (Ward and Singh, 2005). Several filamentous fungi belonging to the genus *Mortierella* were found to produce large amounts of EPA in their mycelia when grown at low temperature (Sakayu et al., 1988; Jareonkitmongkol et al., 1993; Jermsuntiea et al., 2011; Cordova and Alper, 2018). Gayathri et al. (2010) investigated on the potential production of DHA and EPA (omega-3-fatty acids) from soil borne fungi. Out of isolated ten fungal cultures researchers found that only two cultures *Tichoderma* spp. and *Aspegillus niger* were able to produce considerable amount of EPA and DHA.

3.1.4 Genetically modified plants and other microorganisms

Algal oils have the benefit of being produced in a carefully controlled environment, as well as being suitable for those following a vegetarian diet, and having excellent sustainability credentials (Winwood, 2013). As production of these oils through microalgae requires higher cost and investments. Therefore, it is a need to find the most economical ways to produce cheap as well as abundant sources for fulfilling oil demands. So, Ruiz-Lopez et al. (2012) emphasized on effectiveness of production of omega-3 fatty acids through genetic modification of plants.

Scientists are focused on shift in the metabolic pathway of higher plants to produce fair quantities of omega- 3 fatty acids. Barclay et al. (1994) bioengineered *Arabidopsis thaliana* by overexpression of three microbial enzymes ΔD9 elongase (*Isochrysis galbana*), ΔD8 desaturase (*Euglena gracilis*), and a D5 desaturase (*Mortierella alpina*) in the leaf tissues which enhance the synthesis of ARA (7%) and EPA (3%) of the total fatty acids. The entire DHA biosynthetic pathway was later reconstituted in oilseed crop *Brassica juncea* by Wu et al. (2005a,b) through stepwise metabolic engineering. Here, transgenic plants produced up to 25%ARA and 15% EPA, as well as up to 1.5% DHA in seeds. Many investigations have been carried out to insights the omega-3 biosynthesis pathway ranging from different host system, in particular from microalgae to oilseed plants such. Recently, Kajikawa et al. (2008) inserted some genes Δ6 Des, a Δ6 Elo, and a Δ5 Des obtained from *Marchantia polymorpha* in tobacco, which has led to the accumulation of 15.5% ARA and 4.9% EPA. In another report by Xie et al. (2015), E.I. DuPont Company had metabolically engineered yeast, *Yarrowia lipolytica,* which is used to commercially produce EPA in an industrial mass scale.

Amiri-Jami and Griffiths (2010) successfully cloned a unique 35 kb gene cluster capable of producing EPA and DHA from *Shewanella baltica* MAC1 strain found in fish intestine in *Escherichia coli.* Transgenic *E. coli* has produced higher EPA as compared to the marine bacterium *S. baltica* MAC1 by 6.1 times. Also, in another investigation, Amiri-Jami et al. (2014) expressed the *S. baltica* MAC1 EPA/DHA gene cluster into *Lactococcus lactis* subsp. *cremoris* MG1363 which yielded high levels of EPA and DHA.

4. Bioprocessing of omega-3 fatty acids production

The production of polyunsaturated fatty acids (PUFA) such as DHA and EPA can be increased in these organisms via three routes (Chen et al., 2016; Liu et al., 2018):

1. Employing mutant or genetically modified strain
2. Change in physicochemical conditions
3. Combining above two methods

The genetic engineering can be employed to change the expression of PUFA by activating or deactivating certain pathways on genetic level in these microorganisms. These may be accomplished by gene exchange, addition or deletion or mutation. While the changes in physicochemical conditions such as light, temperature, pH, salt concentration (salinity), agitation speed, optimum nutrients such as carbon, nitrogen, nutrients and vitamins. The strains of microorganisms employed for production of DHA/EPA rich oils are the results of intensive research in collection, isolation, screening procedures which must have ability of high growth rate, high content of DHA/EPA as percentage of total lipids, and tolerance to withstand low salinity conditions, pH and temperature of fermentation process (Qu et al., 2013). In this section, focuses have been laid on review of optimization of physicochemical parameter on production of DHA and EPA from algal stains.

4.1 Optimization of physicochemical condition on production of DHA

Nutrient imbalances in culture medium lead accumulation of lipid in oleaginous microbes (Chen et al., 2016; Liu et al., 2018). The depletion nitrogen coupled with high amount of carbon sources in medium favors the accumulation of lipid in these organisms. The lipid accumulation in these organisms is slow at the initial phases of growth curve (0–24 h) and significant production take place during after 24 h of inoculation until late stationary phase (96 h). After this phase cell enters into death phase.

4.1.1 Effect of carbon sources

Carbon sources such as including pentoses, hexoses, saccharides, sugar alcohol and glycosides, alcohols and organic acids are most important nutrient for growth of a microorganism. The sugar molecules such as glucose, fructose, glycerol, etc. are most common carbon sources. The sugars are converted in fatty acid via conversion in pyruvic acid, which is further converted to acetyl CoA via the citric acid cycle. The acetyl CoA acts as building block molecule for lipid production. Most of studies reports to have carbon substrate concentration in the culture medium in range of 5–70 g/L with tolerance up to 150 g/L. Bajpai et al. (1991) investigated effect of carbon source such as fructose, sucrose, lactose, Starch, glucose, maltose and linseed oil on DHA production by *Thraustochytrium aureum* and found a maximum yield of DHA of 511 mg/L in light-exposed cultures containing 2.5% starch.

Singh et al. (1996) investigated effect of carbon source on DHA production by *Thrausfochytrium* sp. in shake-flask cultures incubated at 25°C for 5 days. Glucose, starch and line seed oil resulted similar biomass and lipid yield. The maximum DHA was produced in glucose containing medium. Maltose and sucrose resulted poor growth and DHA yield. Nagano et al. (2009) studied effect of carbon sources such as D-glucose, D-fructose, D-mannose, D-galactose, D-xylose, D-ribose, L-arabinose, lactose, sucrose, maltose, soluble starch and glycerol at a final concentration of 3% on growth of A. limacinum. D-Glucose, D-fructose, sucrose, glycerol, galactose and D-mannose favored cell growth while D-xylose, D-ribose, L-arabinose, lactose, maltose and soluble starch showed little growth of the cells.

Li et al. (2015) studied production of docosahexaenoic acid by *Aurantiochytrium limacinum* SR21 using glucose and glycerol as the mixed carbon sources in both flask and fed-batch cultures. The productivity of DHA was 15.24% higher in mixed substrate condition as compared with glucose as sole carbon source in the fed-batch culture with maintenance of a constant air supply. Abad and Turon (2015) compared glucose, pure glycerol and crude glycerol (83%) on growth of *A. limacinum* with concentration of 10 g/L and reported similar net growth rates, biomass and DHA productivity. Nazir et al. (2018) studied optimization of growth, lipid and DHA production of *Aurantiochytrium* SW1 using response surface methodology. They reported optimum conditions as 70 g/L fructose, 250 rpm agitation speed and 10 g/L monosodium glutamate.

In another study by Yokochi et al. (1998) investigated culture conditions of *Schizochytrium limacinum* for the purpose of microbial docosahexaenoic acid (DHA) production. The glucose, fructose, sachharose, lactose, maltose, starch, glycerol, oleic acid, or linseed oil was used as a carbon source. The glucose, fructose, glycerol, oleic acid, and linseed oil were suitable for high dry cell weight and total fatty acids while disaccharide and polysaccharide were not effective for cell growth. Also, glucose, fructose, and glycerol as carbon source resulted 32.5%, 30.9%, and

43.1% DHA content in fatty acid respectively. Zhu et al. (2008) studied growth of *S. limacinum* with carbon sources such as glucose, fructose, potato powder, starch and glycerol in order to identify a suitable carbon sources at a concentration of 30 g/L. The potato powder as carbon source yielded the highest cell biomass while employing glucose as carbon source reported maximum DHA production. Sahin et al. (2018) on growth of *Schizochytrium* sp. employing glucose, fructose, glycerol, ethanol as carbon sources and reported that glycerol has highest yield even with lower biomass production while addition of ethanol enhances the DHA production but yield was reduced due to decreased biomass production.

Gong et al. (2015) studied docosahexaenoic acid production by marine dinoflagellate *Crypthecodinium cohnii* using cheap by-product such as rapeseed meal hydrolysate and waste molasses as carbon feedstock in the batch fermentations. They achieved DHA yield up to 8.72 mg/L in a media composed of diluted RMH (7%) and 1–9% waste molasses. Hu et al. (2010) optimized production of 1,3- dihydroxyacetone (DHA) production by *Gluconobacter oxydans* in shake flasks and bubble column bioreactors. The 88.7 $\pm$ 3.2% conversion rate glycerol to DHA and 2.38 g/l/h of productivity was achieved in pulse glycerol feeding strategy. Wu et al. (2005) analyzed effect of carbon sources such as glucose, fructose, lactose, maltose, sucrose and starch on docosahexaenoic acid production; and all substrates promoted the cell growth and lipid production of *Schizochytrium* sp.

All substrates employed supported cell growth and lipid production. However, carbon sources such as lactose, maltose, sucrose and starch were found to be not suitable for DHA production. The most suitable substrate for docosahexaenoic acid production by *Schizochytrium* sp. was glucose (20 g/L) in 4-day incubation. Carvalho and Malcata (2005) studied photosynthetic *Pavlova lutheri* for eicosapentaenoic acid and docosahexaenoic acid production by using carbon dioxide as carbon source. Mass productivities of eicosapentaenoic acid and docosahexaenoic acid achieved in cultures supplied with 0.5% (v/v) CO_2, at a dilution rate of 0.297 units per hour.

4.1.2 Effect of nitrogen sources

The nitrogen is required in the initial phase fermentation where cell growth and development occur and amino acid and protein is synthesised. When nitrogen is depleted in the fermentation medium, organisms such as algae start production of fatty acids from the carbon source. Bajpai et al. (1991) investigated effect of nitrogen sources such as tryptone, peptone, malt extract, yeast extract, sodium glutamate on DHA Production by *T. aureum*. Among these highest amounts of DHA was produced using medium of sodium glutamate (269 mg/L) followed by yeast extract (247.7 mg/L). Singh et al. (1995) studied effect of nitrogen sources such as malt extract, tryptone, peptone, casamino acids and sodium glutamate on DHA production by *Thraustochyfrium* sp. ATCC 20692. Sodium glutamate (482 mg/L) and peptone (419 mg/L) were most efficient nitrogen source.

Yokochi et al. (1998) studied effect of inorganic nitrogen sources (urea, sodium nitrate, ammonium nitrate, ammonium acetate, ammonium sulfate) and organic nitrogen sources (corn steep liquor, yeast extract, tryptone, polypepton) on cell growth and DHA yield of *S. limacinum* SR21. The highest DHA was produced using corn steep liquor as nitrogen source followed by ammonium acetate and yeast extract. Wu et al. (2005a,b) investigated tryptone, peptone, yeast extract, monosodium glutamate, urea, sodium nitrate, ammonium chloride as nitrogen sources for production of docosahexaenoic acid production by *Schizochytrium* sp. S31. Among complex nitrogen sources, yeast extract was the most suitable for biomass, lipid and DHA production while, monosodium glutamate and ammonium chloride had similar potential for DHA production among the defined nitrogen sources. Jiang et al. (2001) analyzed effects of nitrogen source and vitamin B12 on docosahexaenoic acid production by *Crypthecodinium cohnii*. Among tryptone, yeast extract and corn steep liquor, tryptone concentration at 1 g/L resulted most suitable for cell growth and DHA production. Zhu et al. (2008) studied effects of culture conditions on growth and docosahexaenoic acid production from *S. limacinum* and found soybean cake hydrolysate (1.8 g/L). Wang et al. (2018) investigated nitrogen sources such as sodium glutamate, tryptone, peptone, yeast extract, peptone-yeast extract, ammonium sulfate, ammonium nitrate, and sodium nitrate for Improved production of docosahexaenoic acid in batch fermentation by *Schizochytrium* sp. and *Thraustochytriidae* sp. Yeast extract was best suited nitrogen source for *Schizochytrium* sp. while for *Thraustochytriidae* yeast extract, peptone yeast extract has similar performance for DHA production.

Sahin et al. (2018) suggested proteose peptone medium for docosahexaenoic acid production via *Schizochytrium* species. Ren et al. (2014) studied regulation of docosahexaenoic acid production in *Schizochytrium* sp. by addition of monosodium glutamate and ammonium sulfate. They found that monosodium glutamate accelerates the glucose consumption rate but lowers lipid accumulation while ammonium sulfate increases DHA content but extends the fermentation periods. Fidalgo et al. (1998) investigated *I. galbana* in nitrate, nitrite or urea medium and found urea as most suitable nitrogen source for DHA production (28.37 mg/g DW, 14.13% of total fatty acids).

Increase in DHA production under nitrogen starved condition have been reported by Jakobsen et al. (2008) for *Aurantiochytrium* sp. strain T66 grown in batch bioreactor cultures; *Schizochytrium* (Ganuza and Izquierdo, 2007) under batch and continuous culture.

4.1.3 Effect of pH

Bajpai et al. (1991) investigated the effect of initial pH on DHA production in the basal medium containing 2% glucose and reported DHA yield and content of biomass was optimal at an initial pH of 6.0 by *T. aureum* ATCC 34304. Singh et al. (1996) report optimal pH condition as 7.0 for *Thraustochytrium* sp. ATCC 20892. Yang et al. (2011) studied effects of pH and aeration on the production of docosahexaenoic acid by *T. aureum* in controlled batch fermenter cultures in glucose and maltose medium and indicated pH 5.5 most suitable for DHA production. Li and Ward (1994) studied effect of the initial pH of the growth medium on the production of DHA by *Thraustochytrium roseum* ATCC 28210 in basal medium was supplemented with 2.5% starch and reported initial pH 6.0 as most suitable condition. Jiang and Chen (2000) studied effects of medium glucose concentration and pH on docosahexaenoic acid content of heterotrophic *Crypthecodinium cohnii*. The highest DHA proportion of total fatty acids of 56.8% was observed at pH 7.2. Wu et al. (2005a,b) reported pH 7.0 for optimal production of docosahexaenoic acid production by *Schizochytrium* sp. S31. Zhu et al. (2008) reported initial pH 7.0 as optimal growth and docosahexaenoic acid production from *S. limacinum* OUC88. Wang et al. (2018) studied two different thraustochytrid strains, *Schizochytrium* sp. PKU#Mn4 and *Thraustochytriidae* sp. PKU#Mn16 and reported pH 6.47 as optimum condition for PKU#Mn4. Gao et al. (2013) studied DHA production in *Aurantiochytrium* species and find maximum DHA production at pH 6.0.

These studies indicate that optimal pH conditions of DHA production in *Thraustochytrium* sp., *Crypthecodinium cohnii*, *Schizochytrium* sp. are pH: 5.5–7; 7–8, 6–7, respectively.

4.1.4 Effect of temperature

Wang et al. (2018) studied two different thraustochytrid strains, *Schizochytrium* sp. PKU#Mn4 and *Thraustochytriidae* sp. PKU#Mn16 and suggest optimal temperature to obtain maximal DHA yield is 28°C for the two strains. They also conclude that lower temperature favor higher DHA yield but lower the cell biomass. Bajpai et al. (1991) reported optimal temperature for DHA production by *T. aureum* ATCC 34304 as 28°C. Singh et al. (1996) studied *Thraustochytrium* sp. ATCC 20892 and reported 25°C as optimal temperature for DHA production. Li and Ward (1994) reported maximum DHA production at 25°C in *T. roseum*. Zhu et al. (2008) studied and reported maximum cell growth and DHA production were obtained at 23°C from *S. limacinum* OUC88. Nakahara et al. (1996) reports the optimal temperature of 28°C for DHA production from *S. limacinum*. de Swaaf et al. (1999) reported the percentage of DHA at 27°C was 35.9% as compared to 40.4% at 30°C in *C. cohnii*. Gao et al. (2013) reports optimum temperature range between 20 and 28°C for production of DHA in *Aurantiochytrium* sp. SD116. Taoka et al. (2009) studied reports highest DHA production at culture temperature between 15 and 20°C for *Aurantiochytrium* sp. strain mh0186.

4.1.5 Effect of aeration on DHA production

Most of researchers have employed the aeration rate between 0.2 and 1.0 vvm while agitation rate was 50–400 rpm in their studies. Bailey et al. (2003), in patent US.6607900B2 applied higher level of dissolved oxygen present in fermentation medium during biomass density increasing stage in comparison with level of dissolved oxygen present in fermentation medium during said production stage. Chi et al. (2009) also proposed two-stage culture strategy for maximizing DHA Production in *S. limacinum* strain preferring high dissolved oxygen (DO) level for cell growth stage and low DO level for Lipid accumulation stage. Ren et al. (2010) also proposed stepwise aeration control strategy for efficient DHA production by *Schizochytrium* sp. they adopted 0.4 vvm for the first 24 h, then shifted to 0.6 vvm until 96 h, and then switched back to 0.4 vvm in end of the fermentation for high DHA production. Qu et al. (2011) studied docosahexaenoic acid production by *Schizochytrium* sp. using a two-stage oxygen supply control strategy based on oxygen transfer coefficient. During initial 40 h, KLa was controlled at 150 per hour for higher cell growth and then lowered to 88.5 per hour for DHA production. Chang et al. (2014) reported high kLa enhanced the conversion yield, DHA concentration and DHA productivity in *Schizochytrium* sp. S31.

4.1.6 Effect of salinity on DHA production

S. limacinum can be cultivated at all saline levels. Yokochi et al. (1998) reported 50% sea water concentration suitable for maximum cell growth of *S. limacinum* SR21. A salinity equivalent to 25% of natural sea water for DHA production was reported by Chatdumrong et al. (2007). *T. aureum* strain was not able to grow in zero-salinity-medium and as well as at very high saline conditions. The best salinity was half that of sea water (Iida et al., 1996). A similar salinity is also ideal for cultivating and lipid production by *C. cohni* (de Swaaf et al., 1999).

5. Extraction and quantification of microalgal omega-3 fatty acids

Microalgae offer a promising non-polluted resource for biotechnology and bioengineering of long chain-poly unsaturated fatty acid production (LC-PUFA) (Khozin-Goldberg et al., 2011). Microalgae produce an array of compounds to help in the adaptation and endurance of different environmental circumstances. Many marine microalgal strains have oil contents of flanked by 10%–50%, (w/w) and produce a high percentage of whole lipids (up to 30%–70% of dry weight) (Ward and Singh, 2005). The accumulation of fatty acids is directly correlated to microalgal growth stages, functioning as an energy store during adverse situations or cell division. Omega-3 is accumulated due to its high energy content, as well as the good flow properties vital for cellular functions (Tiez and Zeiger, 2010; Cohen et al., 2000). Up to now, the ω-3 fatty acid content of many microalgae strains has been studied.

Peltomaa et al. (2018) studied marine Cryptophytes as great sources of Omega 3 fatty acids. For lipid extraction, constituents were collected by centrifugation near the last part of the exponential growth stage. The obtained pellets were placed into −80°C until freeze-drying. The homogenized biomass samples were extracted using chloroform/methanol (2:1 v/v) and sonicated to maximize extraction, after which samples were vortexed and centrifuged. Toluene and sulfuric acid in methanol were used for the transesterification of fatty acid methyl esters (FAMEs). FAMEs were studied with a gas chromatograph equipped with mass detector (GC–MS), using helium as a carrier gas.

The quantity of omega-3 PUFAs (73.9%) was high in all of the studied strains, which is in conformity with previous studies on cryptophytes (Aboal et al., 2014). The highest all ω-3 PUFA proportions were found in *Storeatula major* (81.1%) and *Proteomonas sulcata* (80.3%). These two strains with *Teleaulax acuta* (79.6%), *Chroomonas mesostigmatica* (71.1%) also had the highest amount of Omega-3 PUFAs in proportion to dry weight. Besides this, *Guillardia theta* (65.4%) and *Hemiselmis* sp. (70.1%) contain 65.4 and 70.1% of all omega 3 fatty acids.

Oviyaasri et al. (2017) studied the isolation of polyunsaturated fatty acid from the marine alga sourced from *Tetraselmis* sp. The microalgae were grown in optimized media using marine water in the photo-incubator. The alga is lyophilized by using freeze drier followed by sonication or ultrasonic lysis to disrupt parts of the cell wall or the complete cell to release biological molecules. Then EPA content of selected alga is done by analysis of fatty acid methyl esters (FAMEs) using gas chromatography with cold on-column injection and flame ionization detection. The reaction condition of GC column using HP-88 capillary chromatography as a split/splitless injector, and detector Hewlett Packard EL-980 flame ionization detection (FID).

The sum of total lipids separated from *Tetraselmis* sp. was used as an implication of the decisive of the cell disruption method used (Halim et al., 2013; McMillan et al. 2013; Spiden et al., 2013; Lee et al. 2010). No significant differences were detected between the different cell-disruption methods. This indicates that the cell wall is dissolved by the solvents used, so it does not require to be ruined for optimum extraction. Though, when looking at the first separation only, bead thrashing led to more efficient extraction of lipid, although the difference was rather small (7.5%). During the second separation method, a smaller amount of lipids is extracted when bead thrashing is used as a cell-disruption method.

Abirami et al. (2016) studied the extraction of omega 3 fatty acids from *Nannochloropsis gaditana*. It is sufficiently rich in omega-3 LC-PUFA to serve as a potential alternative for fish oil. Ryckebosch et al. (2013) also showed that a blend of chloroform and methanol yield the highest extraction efficiency. Pre-treatment of algal biomass was originated to be a vital step to facilitate easier and faster lipid revival technique for fatty acid extraction. The methods had useful effects on the cell disruption of marine microalga which was subjected to acid hydrolysis. The suspension was added to cellulase enzyme, mixture was shaken and autoclaved followed by sonication. Lipid extraction by the method same to the Folch method (1957) i.e. dichloromethane/methanol in 2:1 (v/v) having butylated hydroxytoluene and NaCl. The organic phase containing lipids was dried and subjected to saponification. Further, the suspension was subjected to methylation and fatty acid methyl esters were then collected. Fatty acid methyl ester was charged on a thin layer of silica gel and evolved by ascending technique using three solvent systems, petroleum ether: ether (60:40) (v/v), hexane: ether (80:20) (v/v), toluene: ethyl acetate (90:10) (v/v) (Barma and Goswami, 2013; Deshpande et al., 2013; Chakraborty, 2010). After developing, the plates dehydrated at room temperature and placed in iodine cavity. Fatty acid methyl ester gave dark brown colored mark with iodine vapor. The colored spots are marked and the retardation factor (Rf) values of the spots are calculated.

Chromatography with standard methyl ester is conceded out and Rf values are correlated. The estimation of fatty acid by gas chromatography which is equipped with a capillary injector and a flame ionization detector. The omega 3 fatty acid in the samples were calculated by using EPA methyl ester (Eicosapentaenoic acid), DHA methyl ester (Docosahexaenoic acid) as external standards. EPA belongs to a group of fatty acids that are part of the phospholipids, which act as structural components in the cell wall. Concentration of these omega-3 fatty acids are variable under the special nutritional deficiency of nitrogen component. However, once nutritional deficiency is fulfilled, cell can synthesize higher dose of EPA.

Gupta et al. (2016) explored the omega-3 fatty acids producing thraustochytrids from Australian and Indian marine biodiversity. The fatty acids were extracted from dried biomass with solvent mixture containing a 2:1 ratio of chloroform to methanol. The upper layer was removed, dried over nitrogen gas and lipid content (% dry wt basis) was determined gravimetrically. For FAMEs, toluene was added to the tube followed by the addition of internal standard, methyl nonadecanoate (C19:0) and butylated hydroxytoluene (BHT). Fatty acid methyl esters (FAMEs) were extracted into hexane and concentrated using nitrogen gas. The samples were analyzed by a GC-FID system (Agilent Technologies, 6890 N, US). The GC was equipped with a capillary column. Fatty acids peaks were identified on comparison of retention time data with external standard. Peaks were quantified with Chemstation chromatography software. Chemotaxonomic similarity was studied using Multivariate software package. The presence of omega-3 fatty acids, particularly docosahexaenoic acid (DHA), in the fatty acid profile of microorganisms is a marker of thraustochytrids. Palmitic acid (C16:0) and DHA (C22:6n3) are the major fatty acids present in the thraustochytrid fatty acid profile, constituting 70%–90% of total fatty acids (TFAs), depending on strain and culture conditions. However, in the case of Indian thraustochytrids, oleic acid (C18:1) was also present in significant amounts (7.6%–30.6% of TFA), compared to trace amounts in the Australian thraustochytrids (0.4%–4.9% of TFA).

Dammak et al. (2016) studied the total lipid determination from dry biomass according to the method of Folch et al. (1957) as customized by Bligh and Dyer (1959) in microalga. The dry cells from cultures were extracted using chloroform/methanol/water (2/1/1 v/v) and agitated in orbital shaker. The extract was centrifuged and the organic phase was recovered, pellet was re-extracted in chloroform/methanol/water solution. Finally, the solvent phases were combined and vaporised to yield the lipid content that was calculated using the following equation:

$$\text{Lipid content } (\%) = WL/WA \times 100\%$$

Where WL (g) is the extracted algal lipids weight and WA (g) is the dry algal biomass). Gravimetric analysis of lipid content of Box-Behnken experiments were performed. The analysis of FAME was conducted using gas chromatography equipped with a Supelcowax 10 capillary column. The separation of lipids was performed by mono-dimensional High Performance-Thin Layer Chromatography (HP-TLC) using Silicagel 60 F254 plates as a stationary phase (Miao and Wu, 2006).

Similar to plants, many species of microalgae can accumulate polar and neutral lipids as energy and carbon storage (Vitova et al. 2015). HP-TLC studies showed that extracted lipids of *Tetraselmis* sp was primarily composed of TAG, 1,2 diacylglycerol, 1,3 diacylglycerol and FFA. In *Tetraselmis* sp. total fatty acid composition as determined by GC-FID analysis is 16.10.

Khan et al. (2015) studied the isolation of omega-3 polyunsaturated fatty acid from auto- and heterotrophic marine species. The alga is lyophilized by using freeze drier followed by ultrasonic assisted lysis to disrupt parts of the cell or the complete cell to release biological molecules. Then EPA content of selected alga is done by analysis of fatty acid methyl esters using gas chromatography. It is generally thought that photoautotrophic microalgae tend to produce higher levels of EPA than heterotrophs. *Nannochloropsis* sp., *P. tricornutum, Nitzschia laevis* and *Porphyridium cruentum* showed higher EPA levels in total fatty acids. However, low lipid contents of cells produce lower EPA amounts in the biomass (Table 10.2).

Several marine heterotroph microalgae are well thought-out as the most paramount alternative industrial sources of oils rich in DHA (Table 10.3), with approved use in human foods, particularly for application in infant formulas (Wynn and Ratledge, 2005; Raghukumar 2008), since they are measured to be non-pathogenic and nontoxigenic (Fedorova-Dahms et al. 2011).

Table 10.2 List of marine microalgae species characterized by EPA production by growing photoautotrophic mode.

Species	EPA content % of total fatty acids (%TFA)	EPA content % of biomass dry weight (DW)	References
Nannochloropsis sp.	38–39	2–3	Chaturvedi and Fujita (2006)
Phaeodactylum tricornutum	31	5	Meiser et al. (2004)
Nitzschia laevis	25–33	3–4	Cao et al. (2007)
Porphyridium cruentum	25	3	Wen et al. (2002)
Odontella aurita	26	NA	Guihéneuf et al. (2010)
Pavlova lutheri	22–29	NA	Guihéneuf et al. (2009)
Cyclotella cryptica	17–23	1	Pahl et al. (2010)
Cylindrotheca sp.	24–25	NA	Suman et al. (2012)

NA, *Not Available.*

Table 10.3 List of marine microalgae species characterized by DHA production by growing heterotopic mode.

Species	DHA content (% of total fatty acids (% TFA)	DHA content (% of biomass dry weight (% DW)	References
Schizochytrium mangrovei	31–41	12–21	Fan et al. (2001)
Schizochytrium limacinum	25–35	5–15	Ethier et al. (2011)
Thraustochytrium aureum	32–37	6–7	Taoka et al. (2011)
Thraustochytrium striatum	37	2	Fan et al. (2001)
Ulkenia sp.	10–23	5	Quilodran et al. (2010)
Aurantiochytrium sp.	40	18	Hong et al. (2011)
Crypthecodinium cohnii	53–57	5–6	Jiang and Chen (2000)

Abdo et al. (2015) used hexane-isopropanol extraction in microalgae. To investigate the potential production of omega 3-fatty acids from microalgae (*Chlamydomonas variabills*, *Chlorella vulgaris*, *Haematococcus pluvialis* and *Spirulina platensis*), Bligh and dyer (1959) method was used. The dried-up biomass of microalgae was crushed and blended with hexane isopropanol (3:2, v/v). The mixture was exposed to a magnetic stirrer and cell residue was taken apart by filtering. The filtrate was moved into a sorting out funnel and sufficient water was put on to induce biphasic layering. After settle up, the solvent mixture was separated into two distinctive phases, upper dark green hexane layer containing most of the extracted lipids and lower light green layer restrain most of the non-lipids. The lipid samples were methylated and were taken out with petroleum ether at 40–60°C. Ether extract was washed with distilled water and then dehydrated over anhydrous sodium sulfate, and finally filtered off. The GC/MS evaluation was accomplished using a Trace GC Ultra/ISQ Single Quadrupole MS, TG-5MS fused silica capillary column.

The result shows that *C. variabills* has a substantial concentration of omega 6-GLA (29%) and omega 3-EPA (6%). Besides this, *C. vulgaris* reveals a low and moderate concentrations of omega 6 (1 and 6%, respectively), with high amount of omega 3 (21%). *Haematococcus pluvialis* have higher concentration of omega 6 (15%) and a measurable concentration of (8,11-Eicosadienoic acid) omega 6 (2%). On the other hand, in *S. platensis,* omega 3 and omega 6 were observed (4.9% and 3.22%), while omega 6 was not reported. Tsurkan et al. (2015) found that sum of total PUFAs in green microalgae varied from 55.9 to 64.5%.

Ryckebosch et al. (2014) examined nutritional evaluation of microalgae oils rich in omega-3 long chain polyunsaturated fatty acids. Total lipids were separated from the microalgae according to the method formerly optimized by Ryckebosch et al. (2012). Biomass was extracted four times with chloroform/methanol (1:1): over again with and without the addition of water. The oil from all the four extractions was combined. The lipid content of the microalgal total lipid extracts and the fish oil sample was obtained by fractionation using silica solid phase extraction followed by gravimetric quantification as per the Ryckebosch et al. (2012) method.

To find out total fatty acid composition, the total lipid extracts, different lipid class samples and fish oil sample were methylated according to Ryckebosch et al. (2012). The FAMEs attained were extracted by gas chromatography with cold on-column injection and flame ionization detection. For quantification in mg fatty acid/g oil, an internal standard of fatty acid (C20:0, C19:0 or C20:1) was added to the oil before methylation.

In the total lipid extracts, the omega-3 polyunsaturated fatty acids PUFAs (ALA (α-linolenic acid), SDA (stearidonic acid), EPA, DPA (eicosapentaenoic acid, docosapentaenoic acid) and DHA (docosahexaenoic acid) were present at 0.3–92 mg/g oil; 0–43 mg/g oil; 3–193 mg/g oil; 0–2 mg/g oil and 0–46 mg/g oil. Total lipid extracts with high EPA content were obtained from *N. gaditana*, *Nannochloropsis oculata* and *Phaeodactylum*. Total lipids with high DHA content were separated from *Isochryris* and *Pavlova*. Total lipids with high SDA (Stearidonic acid) content were noticed in *Isochrysis* and *Rhodomonas*. *Pavlova* is the only microalga that gave a total lipid extract rich in DHA that also contained an extensive amount of EPA (Ryckebosch et al., 2012).

Breuer et al. (2013) described method to determine the content and composition of total fatty acids in microalgae. Fatty acids are one of the main components of microalgal biomass and make up between 5 and 50% of the cell dry weight. A microalga produces both saturated as well as highly unsaturated fatty acids. The latter include fatty acids with nutritional benefits (omega-3 fatty acids) like eicosapentaenoic acid and docosahexaenoic acid. There are two different methods for sample preparation. The first method is recommended when a limited amount of algae culture is available. The algal dry weight concentration (g/L) is determined as described by Breuer et al. (2012).

Then EPA content of selected alga is done by analysis of fatty acid methyl esters using gas chromatography with cold on-column injection and flame ionization detection. The fatty acid composition and content of *Scenedesmus obliquus* (Chlorophyceae) under both nitrogen replete and deplete conditions are studied. Fatty acid composition and content are highly affected by nitrogen starvation. In *S. obliquus,* C16:0 (palmitic acid) and C18:1 (oleic acid) are the two most abundant fatty acids. The fatty acid composition and content of *P. tricornutum* (diatom) under both nitrogen replete and deplete conditions are noticed. Similar to *S. obliquus*, the fatty acid content and composition are highly affected by nitrogen starvation. *P. tricornutum* also produces substantial amounts of highly unsaturated fatty acids such as C20:5 (eicosapentaenoic acid, EPA) as well as very long chain fatty acids (lignoceric acid, C24:0) that can be detected by this method.

Lohman et al. (2013) presented a method that allows for rapid and consistent extraction of lipids from a range of algae, followed by their characterization using gas chromatographic analysis. They feature a new method which uses microwave energy as a reliable, single-step cell disruption technique to extract lipids from live cultures of microalgae. After extractable lipid characterization by GC-FID, the same lipid extracts are transesterified into FAMEs. This method was tested on *P. tricornutum*, *Chlamydomonas reinhardtii*, and *C. vulgaris*. In the recent past, the Bligh and Dyer (1959) gravimetric method for quantifying lipid has been considered the standard for lipid extraction (Gouveia and Oliveira, 2009; Laurens et al. 2012; Teixeira, 2012). In recent times, researchers have demonstrated that cell lysis by mechanical means such as sonication and bead beating aids in lipid extraction (Zheng et al. 2011).

In *P. tricornutum,* lipids were extracted by the microwave extraction method. FAMEs composition was analyzed using gas chromatography—mass spectroscopy detection (GC—MS). Total FAME was 51.2% (w/w) and the sum of all extractable lipids was 31.5% (w/w) with TAG contributing the majority at 27.4% (w/w) or 87% of the extract. Results from direct in situ transesterification indicate that under these growth conditions this organism preferentially synthesized C16:1 and C16:0 lipid compounds. Total C16:1 and C16:0 FAMEs were 24.8% (w/w) and 14.7% (w/w) respectively.

C. vulgaris were sparged continuously with air amended with 5% CO_2. Once peak TAG accumulation was reached, as monitored by Nile Red fluorescence. The cultures were harvested for lipid extraction, total FAMEs determination and an estimate of FAMEs derived from extractable lipids. Interestingly, this freshwater green microalga preferentially synthesized unsaturated C18 fatty acids under the culture conditions employed here. Total FAME content for *C. vulgaris* was 33% (w/w) and the sum of all extractable lipids was 21.6% (w/w) with TAG contributing the majority at 17.4% (w/w) or 80.6% of the total extract.

C. reinhardtii was cultured as previously reported by Gardner et al. (2013) until peak TAG accumulation was achieved as monitored by Nile Red fluorescence. The cultures were harvested and lipid was extracted using the microwave extraction method. Total FAME content was determined to be 14.5% (w/w) and the sum of all extractable lipids was 8.5% (w/w) with TAG contributing the majority at 3.5% (w/w) or 41.2% of the extract and DAG accounting for 2.4% (w/w) or 28.2% of the extract.

Gupta et al. (2012) studied thraustochytrids (marine heterokonts) as a novel source of omega-3 oils and classified as oleaginous microorganisms due to their production of docosahexaenoic (DHA) and eicosapentaenoic (EPA) ω-3-fatty acids). Omega-3 fatty acids invention from microbes with special reference to thraustochytrids, has been achieving consideration in current years (Raghukumar 2008). Bligh and Dyer method (1959) is usually the first to be cited due in large part to its effortlessness in using common solvents such as chloroform, methanol and water mixtures. This technique has been modified by researchers to boost the substantial output of PUFAs from microbes by bringing together lipid extraction with trans-esterification followed by recovery using a hexane and chloroform mixture (Burja et al., 2007; Lewis et al. 2000). Direct saponification and direct trans-esterification, the acid Bligh and Dyer method link with trichloroacetic acid and the miniature Bligh and Dyer method were some of the customized extraction methods reported (Burja et al. 2007). Such methods yielded higher amounts of lipids from *Thraustochytrium* and *Schizochytrium* species. The main steps in the process involve extraction, the constructing of fatty acid methyl esters, GC optimisation and the usage of suitable internal and external standards for determining the concentration (Masood et al. 2005; Schreiner, 2005).

An improved procedure was derived with the adding up of acetyl chloride on a dry ice bath, followed by trans-esterification performed at room temperature, and an analysis of GC data using relative response factors. This method was established to be relevant in the revival of omega-3 fatty acids, DHA and EPA (Xu et al. 2010). A new method has also been developed by Jacobsen and co-workers for lipid separation with alteration of the previously described protocols. This modification involves heat treatment and protease digestion of freeze dried thraustochytrid cells. The use of heat and an enzyme in the lipid separation was a novel impression to be implemented in the lipid extraction procedure (Jakobsen et al., 2008). The quantification of fatty acids includes urea fractionation, thin layer chromatography and preparative scale gas chromatography followed by high performance liquid chromatography (HPLC) with some modifications such as reverse phase C18 column and silver nitrate HPLC.

A variety of investigative techniques have been reported for the quantification of fatty acids. These include urea fractionation; thin layer chromatography (TLC) and gas chromatography (Fuchs et al., 2011; Sowa and Subbaiah, 2004). This was followed by high performance liquid chromatography (HPLC) methods with some modifications such as reverse phase HPLC using C18 column and silver nitrate HPLC (Lima and Abdalla, 2002; Rao et al., 1995; Mehta et al., 1998; Rezanka and Votruba, 2002.). The percentage of DHA and EPA produced by different species of *Thraustochytrium* given in Table 10.4.

Table 10.4 Percentage of DHA and EPA produced by different *Thraustochytrium* and *Schizochytrium* species.

	% of total fatty acids (TFA)		
Strain	**DHA**	**EPA**	**References**
T. aureum	41–75	1.2–5.2	Bajpai et al. (1991)
T. roseum	48.3–58.2	NA	Singh and Ward (1997)
S. limacinum	6.0–43.1	<1	Yokochi et al. (1998)
S. mangrovei	28	NA	Bowles et al. (1999)
S. limacinum	34.9	1.6	Aki et al. (2003)

NA, *Not Available.*

Duong et al. (2012) find out that lipid determination in qualitative and quantitative analysis is vital for recognition of suitable strains for biodiesel production. Conventional method such as solvent extraction has been used by Bligh and Dyer (1959). Separation and profiling of lipid components require elaborate techniques includes thin layer chromatography (TLC), gas chromatography—mass spectroscopy (GC/MS). These methods are time-consuming for lipid extraction and analysis thus a rapid screening for lipid content in organisms or cells is necessary and important for high-throughput screening. Nile red (9-diethylamino-5-benzo[α] phenoxazinone), a lipophilic stain, maybe used for this purpose.

Mostly microalgae are the storehouse of higher lipids contents ranging from 10% to 30% on dry weight basis. Depending on the specific algae species and their cultivation conditions, however, microalgal lipid production may range widely from 2 to 75% (Mata et al. 2010) and in some intense cases, it can reach 70%—90% of dry weight (Chisti, 2007; Li et al. 2008). The percentages of polyunsaturated fatty acids in different microalgal species are as *Chaetoceros calcitrans* (8.7), *I. galbana* (17.0), *Nanaochloropsis* sp. (2.8), *P. tricornutum* (30.0), *P. cruentum* (17.1), *Skeletonema costatum* (5.1) and *Tetreselmis suecica* (20.9) (Chisti, 2007; Yan and Schenk, 2011)

Adarme-Vega et al. (2012) has given a promising approach toward sustainable omega-3 fatty acid production in microalgae. A typical extraction protocol is based on the method of Bligh and Dyer (1959), which uses a solvent mixture made of methanol/chloroform for the cell disruption and lipid extraction. The biomass is first de-watered either by filtration, dissolved air flotation, flocculation or sedimentation and then dried to form pellets or directly administrated to livestock (Reitan et al., 1997). The methods such as supercritical fluid extraction, winterization and fractional (molecular) distillation are used for the extraction and purification of PUFA from microalgae (Andrich et al. 2005; Herrero et al. 2006). A comparison shows that microalgae can reach much higher EPA and DHA contents and productivities compared with other possible sources (Table 10.5).

6. Extraction of omega-3 fatty acids from fungi

Fungi have the capability to produce vital omega-3 fatty acids, which is of great scientific significance, since these microorganisms are competent of developing under varied conditions, making their production a viable option (Czernichow et al. 2010). The great sources of omega-3 fatty acids are the microflora. The microflora has the eminences of being heterotrophic and able of producing high quantity of omega-3. The species of lower fungi are also able to gather a high percentage of EPA in the lipid. Mainly filamentous fungi, in solid and submerged bioprocesses to produce biomasses that have compounds with economically high and diversified yields, such as fatty acids (Zen et al., 2014). Agro-industrial wastes can be used as an alternative in culture media to develop fungi and produce lipids in biomass. The extracted lipids can be incorporated into animal feed and biodiesel production, among others (Zen et al. 2014). Scrutiny of literature indicates the production of PUFAs from marine

Table 10.5 Comparison of EPA and DHA fatty acid contents as percentage from total lipids in examples of fungi and microalgae.

Microorganisms	% EPA and/or DHA Production	References
Fungi		
Thraustochytrium aureum	62.9 EPA + DHA	Ward and Singh (2005)
Mortierella	20.0 EPA	Jareonkitmongkol et al. (1993)
Mortierella	13.0 EPA	Jareonkitmongkol et al. (1993)
Pythium	12.0 EPA	Athalye et al. (2009)
Pythium irregulare	8.2 EPA	Liang et al. (2012)
Algae		
Nannochloropsis sp.	26.7 EPA + DHA	Hu and Gao (2003)
Nannochloropsis oceanica	23.4 EPA	Patil et al. (2007)
Nannochloropsis salina	28 EPA	Van Wagenen et al. (2012)
Pinguiococcus pyrenoidosus	22.03 EPA + DHA	Sang et al. (2012)
Thraustochytrium sp.	45.1 EPA + DHA	Scott et al. (2011)
Chlorella minutissima	39.9 EPA	Yongmanitchai and Ward (1991)
Dunaliella salina	21.4 EPA	Bhosale et al. (2010)
Pavlova viridis	36.0 EPA + DHA	Hu et al. (2008)
Pavlova lutheri	27.7 EPA + DHA	Carvalho and Malcata (2005)
Pavlova lutheri	41.5 EPA + DHA	Guihéneuf et al. (2009)
Isocrysis galbana	28.0 EPA + DHA	Yago et al. (2011)

fungi (Deelai et al. 2015; Zen et al. 2014; Papanikolaou and Aggelis, 2002). Several fungi Trichoderma sp, Aspergillus niger and Mortierella alpine were set up to make large amounts of EPA in their mycelia.

Vadivelan and Venkateswaran (2014) investigated the production and enhancement of omega-3 fatty acid from *Mortierella* sp. It is the most overproducer microorganism for the industrial production of omega-3 PUFAs. Foremost members of the genus include *Mortierella alliacea*, *M. alpina*, *M. polycephala*, *Mortierella elongata*, and *M. spinosa*. But *M. alpina* has demonstrated an excellent source of lipids production. The separation of lipids from biomass was accomplished according to the method of Nisha et al. (2009). Evaluation of biomass production was studied by cells separartion by suction filtration. The dry biomass was ground to fine powder macerated and blended of chloroform/methanol (2:1), the mixtures filtered, sodium chloride solution was added. The chloroform solvent containing total fatty acid then evaporated and dried under nitrogen vacuum. Fatty acid methyl esters were prepared as described by Nisha and Venkateswaran (2011) and was used for gas chromatographic analysis. Lipids were analyzed by gas chromatography using RTX-2330 (fused silica) 30 m capillary column. Omega-3 fatty acid production by diverse strains of *Mortierella* sp. (*M. alpina, M. elongata, Mortierella horticola,* and *Mortierella exigus*) was detected. One of the isolates, *M. alpina* CFR-GV15, was found to be producing the maximum amount of omega-3 fatty acids. *M. alpina* CFR-GV15, the total fatty acid content significantly increased maximum of 54% AA, 5% EPA, and 4% DHA whereas the standard culture MTCC 6344 accumulated less amount of total fatty acid content 39% AA, 3% EPA, and 3% DHA. Similarly, *M. alpina* CFRGV15, *M. alpina* CFR-GVM15, and *M. horticola* CFR-GV10 was hyperproducer for AA, EPA, and DHA. However, *M. elongata* CFR-GV16 and CFR-GV17 and *M. exigus* CFR-GV11 had low yield.

Gayathri et al. (2010) analyzed the omega-3 fatty acids from soil microorganisms (*Trichoderma* sp and *Aspergillus niger*). The microorganisms used were *Trichoderma* sp. and *Aspergillus niger,* isolated from soil. The use of *Trichoderma* sp. is preferred since it produced substantial amounts of EPA and DHA. Both, *Trichoderma* sp and *Aspergillus niger* were grown in potato dextrose broth initially and later transferred to a medium containing oleic acid which is considered as a possible precursor for the synthesis of highly unsaturated fatty acids. Fungal mycelium of both species isolated from soil was crushed and subjected to saponification. To obtain a strong form of only omega-3 fatty acids, a base and alcohol mixture was used (David, 2005; Kang and Wang, 2005). Thus, obtained methyl esters of the fatty acid fractions were subjected to GC analysis using a fused silica capillary column. Of the two cultures, *Trichoderma* sp. produced 7.47 mg/g of DHA and 0.298 mg/g of EPA. Both oleic acids containing medium and potato dextrose broth inoculated with *Trichoderma* sp. revealed higher concentrations of DHA than EPA. This study has clearly shown that extensive amounts of DHA as compared to EPA were produced by *Trichoderma* sp. and to a lesser extent *Aspergillus* sp.

7. Conclusion and future directions

Microorganisms have the excellent capability to produce omega 3 fatty acids. Such polyunsaturated fatty acids have a vast significance to human due to their useful health benefits, such as lessening the cardiovascular problems and other chronic-degenerative diseases. Advances in media optimization and ease of bioreactor scale up study explore new venue for large biomass production that lead to higher level of omega-3 fatty acids production. Therefore, interest in the bioprospecting of microalgae and other oleaginous microorganisms for omega 3 fatty acids has been growing in the biotechnology industry. There is requirement of simple and economically valuable method for the extraction of omega-3 fatty acid since microalgal polyunsaturated fatty acid is considered safe for food processing application. Various physico-chemical methods have shown promising route in the higher yield extraction of omega-3 fatty acids. Thus, a market exists for eco-friendly extraction methods for the production of high-quality omega-3 fatty acids.

References

Abad, S., Turon, X., 2015. Biotechnological production of docosahexaenoic acid using *Aurantiochytrium limacinum*: carbon sources comparison and growth characterization. Mar. Drugs 13, 7275–7284.

Abdo, S.M., Ali, G.H., El-Baz, F.K., 2015. Potential production of omega fatty acids from microalgae. Int. J. Pharm. Sci. Rev. Res. 34, 210–215.

Abirami, S., Murugesan, S., Sivaswamy, N.S., 2016. Effect of various pretreatment methods prior to extraction of omega 3 fatty acids from *Nannochloropsis gaditana*. Int. J. Appl. Res. 2, 81–85.

Aboal, M., González-Silvera, D., Roldán, M., Hernández-Mariné, M., López-Jiménez, J.Á., Whitton, B.A., 2014. The freshwater alga *Chroothece richteriana* (Rhodophyta) as a potential source of lipids. Food Chem. 162, 143–148.

Adams, P.B., Lawson, S., Sanigorski, A., Sinclair, A.J., 1996. Arachidonic acid to eicosapentaenoic acid ratio in blood correlates positively with clinical symptoms of depression. Lipids 31, 157–161.

Adarme-Vega, T.C., Lim, D.K.Y., Timmins, M., Vernen, F., Li, Y., Schenk, P.M., 2012. Microalgal biofactories: a promising approach towards sustainable omega-3 fatty acid production. Microb. Cell Factories 11, 86–96.

Aki, T., Hachida, K., Yoshinaga, M., Katai, Y., Yamasaki, T., Kawamoto, S., et al., 2003. Thraustochytrid as a potential source of carotenoids. J. Am. Oil Chem. Soc. 80, 789–794.

Allemann, M.N., Allen, E.E., 2018. Characterization and application of marine microbial omega-3 polyunsaturated fatty acid synthesis. Methods Enzymol. 605, 3–32.

Amiri-Jami, M., Griffiths, M.W., 2010. Recombinant production of omega-3 fatty acids in *Escherichia coli* using a gene cluster isolated from *Shewanella baltica* MAC1. J. Appl. Microbiol. 109, 1897–1905.

Amiri-Jami, M., LaPointe, G., Griffiths, M.W., 2014. Engineering of EPA/DHA omega-3 fatty acid production by *Lactococcus lactis* subsp. cremoris MG1363. Appl. Microbiol. Biotechnol. 98, 3071–3080.

Andrich, G., Nesti, U., Venturi, F., Zinnai, A., Fiorentini, R., 2005. Supercritical fluid extraction of bioactive lipids from the microalga *Nannochloropsis* sp. Eur. J. Lipid Sci. Technol. 107, 381–386.

Athalye, S.K., Garcia, R.A., Wen, Z., 2009. Use of biodiesel-derived crude glycerol for producing eicosapentaenoic acid (EPA) by the fungus *Pythium irregulare*. J. Agric. Food Chem. 57, 2739–2744.

Bailey, R.B., DiMasi, D., Hansen, J.M., Mirrasoul, P.J., Ruecker, C.M., Veeder III, G.T., Kaneko, T., Barclay, W.R., 2003. OmegaTech Inc. Enhanced production of lipids containing polyenoic fatty acid by very high density cultures of eukaryotic microbes in fermentors, U.S. Patent 6,607,900.

Bajpai, P.K., Bajpai, P., Ward, O.P., 1991. Optimization of production of docosahexaenoic acid (DHA) by *Thraustochytrium aureum* ATCC 34304. J. Am. Oil Chem. Soc. 68, 509–514.

Balk, E., Chung, M., Lichtenstein, A., Chew, P., Kupelnick, B., Lawrence, A., DeVine, D., Lau, J., 2004. Effects of omega-3 fatty acids on cardiovascular risk factors and intermediate markers of cardiovascular disease. Evid. Rep. Technol. Assess. 93, 1–6.

Barclay, W.R., Meager, K.M., Abril, J.R., 1994. Heterotrophic production of long chain omega-3 fatty acids utilizing algae and algae-like microorganisms. J. Appl. Phycol. 6, 123–129.

Barma, K.S., Goswami, B.C., 2013. Identification and estimation of fatty acids in fresh water fish *Anabus testudineous*. Ann. Biol. Res. 4, 18–24.

Bhosale, R.A., Rajabhoj, M., Chaugule, B., 2010. *Dunaliella salina* Teod. as a prominent source of eicosapentaenoic acid. Int. J. Algae 12, 185–189.

Bligh, E.G., Dyer, W.J., 1959. A rapid method for total lipid extraction and purification. Can. J. Biochem. Physiol. 37, 911–917.

Borowitzka, M.A., 2013. High-value products from microalgae their development and commercialization. J. Appl. Phycol. 25, 743–756.

Bourdon, J., Bazinet, T., Arnason, T., Kimpe, L., Blais, J., White, P., 2010. Polychlorinated biphenyls (PCBs) contamination and aryl hydrocarbon receptor (AhR) agonist activity of omega-3 polyunsaturated fatty acid supplements: implications for daily intake of dioxins and PCBs. Food Chem. Toxicol. 48, 3093–3097.

Bourre, J.M., 2007. Dietary omega-3 fatty acids for women. Biomed. Pharmacother. 61, 105–112.

Bowles, R.D., Hunt, A.E., Bremer, G.B., Duchars, M.G., Eaton, R.A., 1999. Long-chain n-3 polyunsaturated fatty acid production by members of the marine protistan group the thraustochytrids: screening the isolates and optimisation of DHA production. J. Biotechnol. 70, 193–202.

Breuer, G., Lamers, P.P., Martens, D.E., Draaisma, R.B., Wijffels, R.H., 2012. The impact of nitrogen starvation on the dynamics of triacylglycerol accumulation in nine microalgae strains. Bioresour. Technol. 124, 217–226.

Breuer, G., Evers, W.A.C., Jeroen, H., de Vree, J.H., Kleinegris, D.M., Martens, D.E., Wijffels, R.H., Lamers, P.P., 2013. Analysis of fatty acid content and composition in microalgae. J. Vis. Exp. 80, 50628.

Burja, A.M., Armenta, R.E., Radianingtyas, H., Barrow, C.J., 2007. Evaluation of fatty acid extraction methods for *Thraustochytrium* sp. ONC-T18. J. Agric. Food Chem. 55, 4795–4801.

Cao, X.H., Li, S.Y., Wang, C.L., Lu, M.F., 2007. Potential use of the herbicide quizalofop-p-ethyl for eicosapentaenoic acid overproduction by the diatom *Nitzschia laevis*. Chin. J. Biotechnol. 23, 885–890.

Carvalho, A.P., Malcata, F.X., 2005. Optimization of ω-3 fatty acid production by microalgae: crossover effects of CO_2 and light intensity under batch and continuous cultivation modes. Mar. Biotechnol. 7, 381–388.

de Carvalho, Carla, C.C.R., 2016. Whole cell biocatalysts: essential workers from Nature to the industry. Microbiol Biotechnol. 10 (2), 250–263. https://doi.org/10.1111/1751-7915.12363.

Chakraborty, K., 2010. Preparation of eicosapentaenoic acid concentrates from sardine oil by *Bacillus circulans* lipase. Food Chem. 120, 433–442.

Chang, G., Wu, J., Jiang, C., Tian, G., Wu, Q., Chang, M., Wang, X., 2014. The relationship of oxygen uptake rate and kLa with rheological properties in high cell density cultivation of docosahexaenoic acid by *Schizochytrium* sp. S31. Bioresour. Technol. 152, 234–240.

Chatdumrong, W., Yongmanitchai, W., Limtong, S., Worawattanamateekul, W., 2007. Optimization of docosahexaenoic acid (DHA) production and improvement of astaxanthin content in a mutant *Schizochytrium limacinum* isolated from mangrove forest in Thailand. Nat. Sci. 41, 324–334.

Chaturvedi, R., Fujita, Y., 2006. Isolation of enhanced eicosapentaenoic acid producing mutants of *Nannochloropsis oculata* ST-6 using ethyl methane sulfonate induced mutagenesis techniques and their characterization at mRNA transcript level. Phycol. Res. 54, 208–219.

Chen, F., Fan, K.W., Jiang, Y., Faan, Y.W., 2007. Lipid characterization of mangrove thraustochytrid– *Schizochytrium mangrovei*. J. Agric. Food Chem. 55, 2906–2910.

Chen, W., Zhou, P., Zhu, Y., Xie, C., Ma, L., Wang, X., Bao, Z., Yu, L., 2016. Improvement in the docosahexaenoic acid production of *Schizochytrium* sp. S056 by replacement of sea salt. Bioproc. Biosyst. Eng. 39, 315–321.

Chen, C.Y., Nagarajan, D., Cheah, W.Y., 2018. Eicosapentaenoic acid production from *Nannochloropsis oceanica* CY2 using deep sea water in outdoor plastic-bag type photobioreactors. Bioresour. Technol. 253, 1–7.

Cheng, B., Wu, G., Vrinten, P., Falk, K., Bauer, J., Qiu, X., 2010. Towards the production of high levels of eicosapentaenoic acid in transgenic plants: the effects of different host species, genes and promoters. Transgenic Res. 19, 221–229.

Chi, Z., Liu, Y., Frear, C., Chen, S., 2009. Study of a two-stage growth of DHA-producing marine algae *Schizochytrium limacinum* SR21 with shifting dissolved oxygen level. Appl. Microbiol. Biotechnol. 81, 1141–1148.

Chisti, Y., 2007. Biodiesel from microalgae. Biotechnol. Adv. 25, 294–306.

Chua, E.T., Schenk, P.M., 2017. A biorefinery for Nannochloropsis: induction, harvesting, and extraction of EPA-rich oil and high-value protein. Bioresour. Technol. 244, 1416–1424.

Cohen, Z., Khozin-Goldberg, I., Adlerstein, D., Bigogno, C., 2000. The role of triacylglycerol as a reservoir of polyunsaturated fatty acids for the rapid production of chloroplastic lipids in certain microalgae. Biochem. Soc. Trans. 28, 740–744.

Cordova, L.T., Alper, H.S., 2018. Production of α-linolenic acid in *Yarrowia lipolytica* using low-temperature fermentation. Appl. Microbiol. Biotechnol. 102, 8809–8816.

Czernichow, S., Thomas, D., Bruckert, E., 2010. n-6 Fatty acids and cardiovascular health: a review of the evidence for dietary intake recommendations. Br. J. Nutr. 104, 788–796.

Dammak, M., Haase, S.M., Miladi, R., Amor, F.B., Barkallah, M., Gosset, D., Pichon, C., Huchzermeyer, B., Fendri, I., Denis, M., Abdelkafi, S., 2016. Enhanced lipid and biomass production by a newly isolated and identified marine microalga. Lipids Health Dis. 15, 209.

David, R., 2005. Method for Preparing Pure EPA and DHA. US Patent # 6846942.

de Swaaf, M.E., de Rijk, T.C., Eggink, G., Sijtsma, L., 1999. Optimisation of docosahexaenoic acid production in batch cultivations by *Crypthecodinium cohnii*. Prog. Ind. Microbiol. 35, 185–192.

Deelai, S., Satinee Suetrong, S., Damrianant, S., Unagul, P., Sakkayawong, N., 2015. Isolation and identification of native lower fungi for polyunsaturated fatty acid (PUFA) production in Thailand, and the effect of carbon and nitrogen sources on growth and production. Afr. J. Biotechnol. 14, 1449–1460.

Deshpande, S., Patil, T., Alone, S., Duragkar, N., 2013. Microbial conversion of plant based polyunsaturated fatty acid (pufa) to long chain pufa and its identification by gas chromatography. J. Biotechnol. Biomater. S13, 006.

Doughman, D., Krupanidhi, S., Sanjeeve, C., 2007. Omega-3 fatty acids for nutrition and medicine considering microalgae oil as a vegetarian source of EPA and DHA. Curr. Diabetes Rev. 3, 198–203.

Duong, V.T., Li, Y., Nowak, E., Schenk, P.M., 2012. Microalgae isolation and selection for prospective biodiesel production. Energies 5, 1835–1849.

Ethier, S., Woisard, K., Vaughan, D., Wen, Z., 2011. Continuous culture of the microalgae *Schizochytrium limacinum* on biodiesel-derived crude glycerol for producing docosahexaenoic acid. Bioresour. Technol. 102, 88–93.

Fan, K.W., Chen, F., Jones, E.B.G., Vrijmoed, L.L.P., 2001. Eicosapentaenoic and docosahexaenoic acids production by and okara-utilizing potential of thraustochytrids. J. Ind. Microbiol. Biotechnol. 27, 199–202.

Fedorova-Dahms, I., Marone, P.A., Bailey-Hall, E., Ryan, A.S., 2011. Safety evaluation of algal oil from *Schizochytrium* sp. Food Chem. Toxicol. 49, 70–77.

Fidalgo, J.P., Cid, A., Torres, E., Sukenik, A., Herrero, C., 1998. Effects of nitrogen source and growth phase on proximate biochemical composition, lipid classes and fatty acid profile of the marine microalga *Isochrysis galbana.* Aquaculture 166, 105–116.

Folch, J., Lees, M., Sloane-Stanley, G.H., 1957. A simple method for the isolation and purification of total lipids from animal tissues. J. Biol. Chem. 226, 497–509.

Fuchs, B., Suß, R., Teuber, K., Eibisch, M., Schiller, J., 2011. Lipid analysis by thin-layer chromatography-a review of the current state. J. Chromatogr. A 1218, 2754–2774.

Ganuza, E., Izquierdo, M.S., 2007. Lipid accumulation in *Schizochytrium* G13/2S produced in continuous culture. Appl. Microbiol. Biotechnol. 76, 985–990.

Gao, M., Song, X., Feng, Y., Li, W., Cui, Q., 2013. Isolation and characterization of *Aurantiochytrium* species: high docosahexaenoic acid (DHA) production by the newly isolated microalga, *Aurantiochytrium* sp. SD116. J. Oleo Sci. 62, 143–151.

Gardner, R.D., Lohman, E., Gerlach, R., Cooksey, K.E., Peyton, B.M., 2013. Comparison of CO(2) and bicarbonate as inorganic carbon sources for triacylglycerol and starch accumulation in *Chlamydomonas reinhardtii*. Biotechnol. Bioeng. 110, 87–96.

Gayathri, S., Prabhu, S., Mitra, A., Kumar, V., Shenbaga Sheela, J., Rajani, G., Sundar, N., Sivvaswamy, S.N., 2010. Soil microorganisms produce omega-3 fatty acids. Indian J. Sci. Technol. 3, 499–503.

Gong, Y., Liu, J., Jiang, M., Liang, Z., Jin, H., Hu, X., Wan, X., Hu, C., 2015. Improvement of omega-3 docosahexaenoic acid production by marine dinoflagellate *Crypthecodinium cohnii* using rapeseed meal hydrolysate and waste molasses as feedstock. PLoS One 10 e0125368.

Gouveia, L., Oliveira, A.C., 2009. Microalgae as a raw material for biofuels production. J. Ind. Microbiol. Biotechnol. 36, 269–274.

Guihéneuf, F., Mimouni, V., Ulmann, L., Tremblin, G., 2009. Combined effects of irradiance level and carbon source on fatty acid and lipid class composition in the microalga *Pavlova lutheri* commonly used in mariculture. J. Exp. Mar. Biol. Ecol. 369, 136–143.

Guihéneuf, F., Fouqueray, M., Mimouni, V., Ulmann, L., Jacquette, B., Tremblin, G., 2010. Effect of UV stress on the fatty acid and lipid class composition in two marine microalgae *Pavlova lutheri* (Pavlovophyceae) and *Odontella aurita* (Bacillariophyceae). J. Appl. Phycol. 22, 629–638.

Gunstone, F.D., 1996. Fatty Acid and Lipid Chemistry. Black Academic and Professional, London.

Gupta, A., Barrow, C.J., Puri, M., 2012. Omega-3 biotechnology: thraustochytrids as a novel source of omega-3 oils. Biotechnol. Adv. 30, 1733–1745.

Gupta, A., Singh, D., Byreddy, A.R., Thyagarajan, T., Sonkar, S.P., Mathur, A.S., Tuli, D.K., Barrow, C.J., Puri, M., 2016. Exploring omega-3 fatty acids, enzymes and biodiesel producing thraustochytrids from Australian and Indian marine biodiversity. Biotechnol. J. 11, 345–355.

Halim, R., Rupasinghe, T.W.T., Tull, D.L., Webley, P.A., 2013. Mechanical cell disruption for lipid extraction from microalgal biomass. Bioresour. Technol. 140, 53–63.

Hamilton, M.L., Powers, S., Napier, J.A., Sayanova, O., 2016. Heterotrophic production of omega-3 long-chain polyunsaturated fatty acids by trophically converted marine diatom *Phaeodactylum tricornutum*. Mar. Drugs 14 pii: E53.

Herrero, M., Cifuentes, A., Ibañez, E., 2006. Sub- and supercritical fluid extraction of functional ingredients from different natural sources: plants, food-byproducts, algae and microalgae: a review. Food Chem. 98, 136–148.

Hong, W.K., Rairakhwada, D., Seo, P.S., Park, S.Y., Hur, B.K., Kim, C.H., Seo, J.W., 2011. Production of lipids containing high Levels of docosahexaenoic acid by a newly isolated microalga, *Aurantiochytrium* sp KRS101. Appl. Biochem. Biotechnol. 164, 1468–1480.

Hu, H., Gao, K., 2003. Optimization of growth and fatty acid composition of a unicellular marine picoplankton, *Nannochloropsis* sp. with enriched carbon sources. Biotechnol. Lett. 25, 421–425.

Hu, C., Li, M., Li, J., Zhu, Q., Liu, Z., 2008. Variation of lipid and fatty acid compositions of the marine microalga *Pavlova viridis* (Prymnesiophyceae) under laboratory and outdoor culture conditions. World J. Microbiol. Biotechnol. 24, 1209–1214.

Hu, Z.C., Liu, Z.Q., Zheng, Y.G., Shen, Y.C., 2010. Production of 1, 3-dihydroxyacetone from glycerol by *Gluconobacter oxydans* ZJB09112. J. Microbiol. Biotechnol. 20, 340–345.

Iida, I., Nakahara, T., Yokochi, T., Kamisaka, Y., Yagi, H., Yamaoka, M., Suzuki, O., 1996. Improvement of docosahexaenoic acid production in a culture of *Thraustochytrium aureum* by medium optimization. J. Ferment. Bioeng. 81, 76–78.

Insel, P., Ross, D., McMohan, K., 2013. Discovering Nutrition. Jones and bartlett Publishers.

Islam, M.S., Castellucci, C., Fiorini, R., Greco, S., Gagliardi, R., Zannotti, A., Giannubilo, S.R., Ciavattini, A., Frega, N.G., Pacetti, D., Ciarmela, P., 2018. Omega-3 fatty acids modulate the lipid profile, membrane architecture, and gene expression of leiomyoma cells. J. Cell. Physiol. 233, 7143–7156.

Jakobsen, A.N., Aasen, I.M., Josefsen, K.D., Strom, A.R., 2008. Accumulation of DHA rich lipid in *thraustochytrid; Aurantiochytrium* sp. strain T66: effects of N and P starvation and O_2 limitation. Appl. Microbiol. Biotechnol. 80, 297–306.

Jareonkitmongkol, S., Shimizu, S., Yamada, H., 1993. Production of an eicosapentaenoic acid-containing oil by a Δ12 desaturase-defective mutant of *Mortierella alpina* 1S-4. J. Am. Oil Chem. Soc. 70, 119–123.

Jermsuntiea, W., Aki, T., Kawamoto, S., Ono, K., 2011. Metabolism and synthesis of lipids in the polyunsaturated fatty acid-producing fungus *Mortierella alliacea*. J. Oleo Sci. 60, 11.

Ji, X., Ren, L., Huang, H., 2015. Omega 3 biotechnology:a green and sustainable process for omega-3 fatty acids production. Frontiers Bioeng Biotechnol. 3, 1–2.

Jiang, Y., Chen, F., 2000. Effects of medium glucose concentration and pH on docosahexaenoic acid content of heterotrophic *Crypthecodinium cohnii*. Process Biochem. 35, 1205–1209.

Jiang, Y., Chen, F., Li, H.B., 2001. Effects of nitrogen source and vitamin B12 on docosahexaenoic acid production by *Crypthecodinium Cohnii*. In: Algae and Their Biotechnological Potential. Springer, Dordrecht, pp. 69–78.

Kajikawa, M., Matsui, K., Ochiai, M., Tanaka, Y., Kita, Y., Ishimoto, M., Kohzu, Y., Shoji, S.I., Yamato, K.T., Ohyama, K., Fukuzawa, H., 2008. Production of arachidonic and eicosapentaenoic acids in plants using bryophyte fatty acid Δ6-desaturase, Δ6-elongase, and Δ5-desaturase genes. Biosc. Biotech. Biochem. 72, 435–444.

Kang, J.X., Wang, J., 2005. A simplified method for analysis of poly unsaturated fatty acids. BMC Biochem. 6, 5.

Khan, M., Rahman, M.M., Zaman, S., Jahangir, T.A., Razu, M.H., 2015. Omega-3 polyunsaturated fatty acids from algae. In: Liu, J., Sun, Z., Gerken, H. (Eds.), Recent Advances in Microalgal Biotechnology. Omics Group Publisher, pp. 1–17.

Khan, W.A., Chun-Mei, H., Khan, N., Iqbal, A., Lyu, S., Shah, F., 2017. Bioengineered plants can be a useful source of Omega-3 fatty acids. BioMed Res. Int. 7348919, 9.

Khozin-Goldberg, I., Didi-Cohen, S., Shayakhmetova, I., Cohen, Z., 2002. Biosynthesis of eicosapentaenoic acid (EPA) in the fresh water eustigmatophyte *Monodus subterraneus* (Eustigmatophyceae). J. Phycol. 38, 745–756.

Khozin-Goldberg, I., Iskandarov, U., Cohen, Z., 2011. LC-PUFA from photosynthetic microalgae: occurrence, biosynthesis, and prospects in biotechnology. Appl. Microbiol. Biotechnol. 91, 905–915.

Kinney, A.J., Cahoon, E.B., Damude, H.G., Hitz, W.D., Kolar, C.W., Liu, Z., 2004. Production of Very Long Chain Polyunsaturated Fatty Acids in Oilseed Plants. Patent WO 2004, 71467, A2.

Klok, A.J., Lamers, P.P., Martens, D.E., Draaisma, R.B., Wijffels, R.H., 2014. Edible oils from microalgae: in sights in TAG accumulation. Trends Biotechnol. 32, 521–528.

Kyle, D., 2001. The large-scale production and use of a single-cell oil highly enriched in docosahexaenoic acid. ACS Symp. Ser. 788, 92–107.

Laurens, L., Dempster, T.A., Jones, H.D.T., Wolfrum, E.J., Van Wychen, S., McAllister, J.S.P., Rencenberger, M., Parchert, K.J., Gloe, L.M., 2012. Algal biomass constituent analysis: method uncertainties and investigation of the underlying measuring chemistries. Anal. Chem. 84, 1879–1887.

Lee, J.Y., Yoo, C., Jun, S.Y., Ahn, C.Y., Oh, H.M., 2010. Comparison of several methods for effective lipid extraction from microalgae. Bioresour. Technol. 101, S75–S77.

Lenihan-Geels, G., Bishop, K.S., Ferguson, L.R., 2013. Alternative sources of Omega-3 Fats: can we find a sustainable substitute for fish? Nutrients 5, 1301–1315.

Lewis, T., Nichols, P.D., McMeekin, T.A., 2000. Evaluation of extraction methods for recovery of fatty acids from lipid-producing microheterotrophs. J. Microbiol. Methods 43, 107–116.

Li, Z.Y., Ward, O.P., 1994. Production of docosahexaenoic acid by *Thraustochytrium roseum.* J. Ind. Microbiol. 13, 238–241.

Li, Q., Du, W., Liu, D., 2008. Perspectives of microbial oils for biodiesel production. Appl. Microbiol. Biotechnol. 80, 749–756.

Li, J., Liu, R., Chang, G., Li, X., Chang, M., Liu, Y., Jin, Q., Wang, X., 2015. A strategy for the highly efficient production of docosahexaenoic acid by *Aurantiochytrium limacinum* SR21 using glucose and glycerol as the mixed carbon sources. Bioresour. Technol. 177, 51–57.

Liang, Y., Zhao, X., Strait, M., Wen, Z., 2012. Use of dry-milling derived thin stillage for producing eicosapentaenoic acid (EPA) by the fungus *Pythium irregulare*. Bioresour. Technol. 111, 404–409.

Lima, E.S., Abdalla, D.S.P., 2002. High-performance liquid chromatography of fatty acids in biological samples. Anal. Chim. Acta 465, 81–91.

Liu, L., Wang, F., Yang, J., Li, X., Cui, J., Liu, J., Shi, M., Wang, K., Chen, L., Zhang, W., 2018. Nitrogen feeding strategies and metabolomic analysis to alleviate high-nitrogen inhibition on docosahexaenoic acid production in *Crypthecodinium cohnii*. J. Agric. Food Chem. 66, 10640–10650.

Lohman, E.J., Gardner, R.D., Halverson, L., Macur, R.E., Peyton, B.M., Gerlach, R., 2013. An efficient and scalable extraction and quantification method for algal derived biofuel. J. Microbiol. Methods 94, 235–244.

Mahaffey, K.R., Clickner, R.P., Jeffries, R.A., 2008. Methylmercury and omega-3 fatty acids: co-occurrence of dietary sources with emphasis on fish and shellfish. Environ. Res. 107, 20–29.

Martínez Andrade, K.A., Lauritano, C., Romano, G., Ianora, A., 2018. Marine microalgae with anti-cancer properties. Mar. Drugs 16, E165.

Masood, A., Stark, K.D., Salem, N., 2005. A simplified and efficient method for the analysis of fatty acid methyl esters suitable for large clinical studies. J. Lipid Res. 46, 2299–2305.

Mata, T.M., Martins, A.A., Caetano, N.S., 2010. Microalgae for biodiesel production and other applications: a review. Renew. Sustain. Energy Rev. 14, 217–232.

McMillan, J.R., Watson, I.A., Ali, M., Jaafar, W., 2013. Evaluation and comparison of algal cell disruption methods: microwave, waterbath, blender, ultrasonic and laser treatment. Appl. Energy 103, 128–134.

Mehta, A.A., Oeser, M., Carlson, M.G., 1998. Rapid quantitation of free fatty acids in human plasma by high-performance liquid chromatography. J. Chromatogr. B Biomed. Appl. 719, 9–23.

Meiser, A., Schmid-Staiger, U., Trosch, W., 2004. Optimization of eicosapentaenoic acid production by *Phaeodactylum tricornutum* in the flat panel airlift (FPA) reactor. J. Appl. Phycol. 16, 215–225.

Miao, X., Wu, Q., 2006. Biodiesel production from heterotrophic microalgal oil. Bioresour. Technol. 97, 841–846.

Nagano, N., Taoka, Y., Honda, D., Hayashi, M., 2009. Optimization of culture conditions for growth and docosahexaenoic acid production by a marine thraustochytrid, *Aurantiochytrium limacinum* mh0186. J. Oleo Sci. 58, 623–628.

Nakahara, T., Yokochi, T., Higashihara, T., Tanaka, S., Yaguchi, T., Honda, D., 1996. Production of docosahexaenoic and docosapentaenoic acids by *Schizochytrium* sp. isolated from Yap Islands. J. Am. Oil Chem. Soc. 73, 1421–1426.

Nazir, Y., Shuib, S., Kalil, M.S., Song, Y., Hamid, A.A., 2018. Optimization of culture conditions for enhanced growth, lipid and docosahexaenoic Acid (DHA) production of *Aurantiochytrium* SW1 by response surface methodology. Sci. Rep. 8, 8909.

Nisha, A., Venkateswaran, G., 2011. Effect of culture variables on mycelial arachidonic acid production by *Mortierella alpine*. Food Bioprocess Technol. 4, 232–240.

Nisha, A., Muthukumar, S.P., Venkateswaran, G., 2009. Safety evaluation of arachidonic acid rich *Mortierella alpina* biomass in albino rats-a subchronic study. Regul. Toxicol. Pharmacol. 53, 186–194.

Ofosu, F.K., Daliri, E.B.M., Lee, B.H., Yu, X., 2017. Current trends and future perspectives on omega-3 fatty acids. Res. J. Biol. 5, 14–20.

Ofosu, F.K., Daliri, E.B., Lee, B.H., Yu, X., 2017a. Current trends and future perspectives on omega-3 fatty acids. RRJOB 5 (2), 1–34.

Oviyaasri, M., Manjuladevi, M., Anitha, R., Narmatha, M., Kavitha, J., 2017. Extraction and isolation of polyunsaturated fatty acids from microalgae *Tetraselmis*.Sp. Int. J. Innovative Res. Sci. Eng. Technol. 6, 4810–4814.

Pahl, S.L., Lewis, D.M., Chen, F., King, K.D., 2010. Heterotrophic growth and nutritional aspects of the diatom *Cyclotella cryptica* (Bacillariophyceae): effect of some environmental factors. J. Biosci. Bioeng. 109, 235–239.

Pal, D., Khozin-Goldberg, I., Cohen, Z., Boussiba, S., 2011. The effect of light, salinity, and nitrogen availability on lipid production by *Nannochloropsis* sp. Appl. Microbiol. Biotechnol. 90, 1429–1441.

Papanikolaou, S., Aggelis, G., 2002. Lipid production by *Yarrowia lipolytica* growing on industrial glycerol in a single-stage continuous culture. Bioresour. Technol. 82, 43–49.

Patil, V., Källqvist, T., Olsen, E., Vogt, G., Gislerød, H.R., 2007. Fatty acid composition of 12 microalgae for possible use in aquaculture feed. Aquacult. Int. 15, 1–9.

Peltomaa, E., Johnson, M.D., Taipale, S.J., 2018. Marine cryptophytes are great sources of EPA and DHA. Mar. Drugs 16, 3.

Petrie, J.R., Shrestha, P., Mansour, M.P., Nichols, P.D., Liu, Q., Singh, S.P., 2010. Metabolic engineering of omega-3 long chain polyunsaturated fatty acids in plants using an acyl-CoA Δ6-desaturase with 3-preference from the marine microalga *Micromonas pusilla*. Metab. Eng. 12, 233–240.

Puri, M., 2017. Algal biotechnology for persuing omega-3 fatty acid (bioactive) production. Focus. Microbiology Australia 85–88.

Qu, L., Ji, X.J., Ren, L.J., Nie, Z.K., Feng, Y., Wu, W.J., Ouyang, P.K., Huang, H., 2011. Enhancement of docosahexaenoic acid production by *Schizochytrium* sp. using a two-stage oxygen supply control strategy based on oxygen transfer coefficient. Lett. Appl. Microbiol. 52, 22–27.

Qu, L., Ren, L.J., Sun, G.N., Ji, X.J., Nie, Z.K., Huang, H., 2013. Batch, fed-batch and repeated fed-batch fermentation processes of the marine thraustochytrid *Schizochytrium* sp. for producing docosahexaenoic acid. Bioproc. Biosyst. Eng. 36, 1905–1912.

Quilodran, B., Hinzpeter, I., Hormazabal, E., Quiroz, A., Shene, C., 2010. Docosahexaenoic acid (C22:6n-3, DHA) and astaxanthin production by Thraustochytriidae sp. AS4-A1 a native strain with high similitude to *Ulkenia* sp.: evaluation of liquid residues from food industry as nutrient sources. Enzym. Microb. Technol. 47, 24–30.

Raghukumar, S., 2008. Thraustochytrid marine protists: production of PUFAs and other emerging technologies. Mar. Biotechnol. 10, 631–640.

Rao, M.S., Hidajat, K., Ching, C.B., 1995. Reversed-phase HPLC: the separation method for the characterization and purification of long chain polyunsaturated fatty acids-a review. J. Chromatogr. Sci. 33, 9–21.

Ratledge, C., 2004. Fatty acid biosynthesis in microorganisms being used for single cell oil production. Biochimie 86, 807–815.

Reitan, K.I., Rainuzzo, J.R., Øie, G., Olsen, Y., 1997. A review of the nutritional effects of algae in marine fish larvae. Aquaculture 155, 207–221.

Ren, L.J., Ji, X.J., Huang, H., Qu, L., Feng, Y., Tong, Q.Q., Ouyang, P.K., 2010. Development of a stepwise aeration control strategy for efficient docosahexaenoic acid production by *Schizochytrium* sp. Appl. Microbiol. Biotechnol. 87, 1649–1656.

Ren, L.J., Sun, L.N., Zhuang, X.Y., Qu, L., Ji, X.J., Huang, H., 2014. Regulation of docosahexaenoic acid production by *Schizochytrium* sp.: effect of nitrogen addition. Bioproc. Biosyst. Eng. 37, 865–872.

Rezanka, T., Votruba, J., 2002. Chromatography of very long-chain fatty acids from animal and plant kingdoms. Anal. Chim. Acta 465, 273–297.

Ruiz-Lopez, N., Sayanova, O., Napier, J.A., Haslam, R.P., 2012. Metabolic engineering of the omega-3 long chain polyunsaturated fatty acid biosynthetic pathway into transgenic plants. J. Exp. Bot. 63, 2397–2410.

Ryckebosch, E., Muylaert, K., Foubert, I., 2012. Optimization of an analytical procedure for extraction of lipids from microalgae. J. Am. Oil Chem. Soc. 89, 189–198.

Ryckebosch, E., Bruneel, C., Termote-Verhalle, R., Lemahieu, C., Muylaert, K., Van Durme, J., Goiris, K., Foubert, I., 2013. Stability of omega-3 LC-PUFA-rich photoautotrophic microalgal oils compared to commercially available omega-3 LC-PUFA oils. J. Agric. Food Chem. 61, 10145–10155.

Ryckebosch, E., Bruneel, C., Termote-Verhalle, R., Muylaert, K., Foubert, I., 2014. Nutritional evaluation of omega- 3 LC-PUFA rich microalgae oils as an alternative for fish oil. Food Chem. 160, 393–400.

Sahin, D., Tas, E., Altindag, U.H., 2018. Enhancement of docosahexaenoic acid (DHA) production from *Schizochytrium* sp. S31 using different growth medium conditions. Amb. Express 8 (1), 7.

Sakayu, S., Shinmen, Y., Kawashima, H., Akimoto, K., Yamada, H., 1988. Fungal mycelia as a novel source of eicosapentaenoic acid: activation of enzyme(s) involved in eicosapentaenoic acid production at low temperature. Biochem. Biophys. Res. Commun. 150, 335–341.

Sang, M., Wang, M., Liu, J., Zhang, C., Li, A., 2012. Effects of temperature, salinity, light intensity, and pH on the eicosapentaenoic acid production of *Pinguiococcus pyrenoidosus*. J. Ocean Univ. China 11, 1–6.

Santos-Merino, M., Garcillán-Barcia, M.P., de la Cruz, F., 2018. Engineering the fatty acid synthesis pathway in *Synechococcus elongatus* PCC 7942 improves omega-3 fatty acid production. Biotechnol. Biofuels 11, 239.

Schacky, C., Harris, W.S., 2007. Cardiovascular benefits of omega-3 fatty acids. Cardiovasc. Res. 73, 310–315.

Schreiner, M., 2005. Quantification of long chain polyunsaturated fatty acids by gas chromatography: evaluation of factors affecting accuracy. J. Chromatogr. 1095, 126–130.

Scott, S.D., Armenta, R.E., Berryman, K.T., Norman, A.W., 2011. Use of raw glycerol to produce oil rich in polyunsaturated fatty acids by a Thraustochytrid. Enzym. Microb. Technol. 48, 267–272.

Sijtsma, L., de Swaaf, M., 2004. Biotechnological production and applications of the ω-3 polyunsaturated fatty acid docosahexaenoic acid. Appl. Microbiol. Biotechnol. 64, 146–153.

Singh, A., Ward, O.P., 1997. Microbial production of docosahexaenoic acid (DHA, C22:6). Adv. Appl. Microbiol. 45, 271–312.

Singh, A., Wilson, S., Ward, O.P., 1996. Docosahexaenoic acid (DHA) production by *Thraustochytrium* sp. ATCC 20892. World J. Microbiol. Biotechnol. 12 (1), 76–81.

Sowa, J.M., Subbaiah, P.V., 2004. Variable recoveries of fatty acids following the separation of lipids on commercial silica gel TLC plates: selective loss of unsaturated fatty acids on certain brands of plates. J. Chromatogr. B 813, 159–166.

Spiden, E.M., Yap, B.H.J., Hill, D.R.A., Kentish, S.E., Scales, P.J., Martin, G.J.O., 2013. Quantitative evaluation of the ease of rupture of industrially promising microalgae by high pressure homogenization. Bioresour. Technol. 140, 165–171.

Steinrücken, P., Erga, S.R., Mjøs, S.A., Kleivdal, H., Prestegard, S.K., 2017. Bioprospecting North Atlantic microalgae with fast growth and high polyunsaturated fatty acid (PUFA) content for microalgae-based technologies. Algal. Res. 26, 392–401.

Suman, K., Kiran, T., Devi, U.K., Sarma, N.S., 2012. Culture medium optimization and lipid profiling of *Cylindrotheca*, a lipid- and polyunsaturated fatty acid-rich pennate diatom and potential source of eicosapentaenoic acid. Bot. Mar. 55, 289–299.

Swinbanks, D., 1993. Fish-oil miracle additive brings benefits to some. Nature 364, 180.

Tanaka, T., Yabuuchi, T., Maeda, Y., Nojima, D., Matsumoto, M., Yoshino, T., 2017. Production of eicosapentaenoic acid by high cell density cultivation of the marine oleaginous diatom *Fistulifera solaris*. Bioresour. Technol. 245, 567–572.

Taoka, Y., Nagano, N., Okita, Y., Izumida, H., Sugimoto, S., Hayashi, M., 2009. Influences of culture temperature on the growth, lipid content and fatty acid composition of *Aurantiochytrium* sp. strain mh0186. Mar. Biotechnol. 11, 368–374.

Taoka, Y., Nagano, N., Okita, Y., Izumida, H., Sugimoto, S., Hayashi, M., 2011. Effect of Tween 80 on the growth, lipid accumulation and fatty acid composition of *Thraustochytrium aureum* ATCC 34304. J. Biosci. Bioeng. 111, 420–424.

Teixeira, R.E., 2012. Energy-efficient extraction of fuel and chemical feedstocks from algae. Green Chem. 14, 419–427.

Tiez, L., Zeiger, E., 2010. Plant Physiology, fifth ed. Sinauer Associates Inc, Publishers, Sunderland.

Tocher, D.R., Betancor, M.B., Sprague, M., Olsen, R.E., Napier, J.A., 2019. Omega-3 long-chain polyunsaturated fatty acids, EPA and DHA: bridging the gap between supply and demand. Nutrients 11, E89.

Tsurkan, Y., Karpenyuk, T., Guschina, I., Orazova, S., Goncharova, A., Beisembaeva, R., 2015. Identification of newly-isolated microorganisms containing valuable polyunsaturated fatty acids. J. Biotech. Res. 6, 14–20.

Vadivelan, G., Venkateswaran, G., 2014. Production and enhancement of omega-3 fatty acid from *Mortierella alpina* CFR-GV15: its food and therapeutic application. BioMed Res. Int. 657414, 9 pages.

Van Wagenen, J., Miller, T.W., Hobbs, S., Hook, P., Crowe, B., Huesemann, M., 2012. Effects of light and temperature on fatty acid production in *Nannochiloropsis salina*. Energies 5, 731–740.

Vitova, M., Bisova, K., Kawano, S., Zachleder, V., 2015. Accumulation of energy reserves in algae: from cell cycles to biotechnological applications. Biotechnol. Adv. 33, 1204–1218.

Wang, Q., Ye, H., Sen, B., Xie, Y., He, Y., Park, S., Wang, G., 2018. Improved production of docosahexaenoic acid in batch fermentation by newly-isolated strains of *Schizochytrium* sp. and *Thraustochytriidae* sp. through bioprocess optimization. Synthetic Systems Biotechnol. 3, 121–129.

Ward, O.P., Singh, A., 2005. Omega-3/6 fatty acids: alternative sources of production. Process Biochem. 40, 3627–3652.

Wen, Z.Y., Chen, F., 2003. Heterotrophic production of eicosapentaenoic acid by microalgae. Biotechnol. Adv. 21, 273–294.

Wen, Z.Y., Jiang, Y., Chen, F., 2002. High cell density culture of the diatom *Nitzschia laevis* for eicosapentaenoic acid production: fed-batch development. Process Biochem. 37, 1447–1453.

Whitehead, S., 1985. FAO species catalogue. In: Nations Undpfaaootu (Ed.), Clupeoid Fishes of the World, vol. 7. Rome: United Nations.

Winwood, R.J., 2013. Recent developments in the commercial production of DHA and EPA rich oils from micro-algae. Oilseeds Fats Crops Lipids 20, D604.

Wu, G., Truksa, M., Datla, N., Vrinten, P., Bauer, J., Zank, T., Cirpus, P., Heinz, E., Qiu, X., 2005a. Stepwise engineering to produce high yields of very long-chain polyunsaturated fatty acids in plants. Nat. Biotechnol. 23, 1013.

Wu, S.T., Yu, S.T., Lin, L.P., 2005b. Effect of culture conditions on docosahexaenoic acid production by *Schizochytrium* sp. S31. Process Biochem. 40 (9), 3103–3108.

Wynn, J.P., Ratledge, C., 2005. Oils from microorganisms. In: Shahidi, F. (Ed.), Bailey's Industrial Oil and Fat Products. John Wiley & Sons, Hoboken, NJ, USA.

Xie, D., Jackson, E.N., Zhu, Q., 2015. Sustainable source of omega-3 eicosapentaenoic acid from metabolically engineered *Yarrowia lipolytica*: from fundamental research to commercial production. Appl. Microbiol. Biotechnol. 99, 1599–1610.

Xie, D., Miller, E., Sharpe, P., Jackson, E., Zhu, Q., 2017. Omega-3 production by fermentation of *Yarrowia lipolytica*: from fed-batch to continuous. Biotechnol. Bioeng. 114, 798–812.

Xu, Z., Harvey, K., Pavlina, T., Dutot, G., Zaloga, G., Siddiqui, R., 2010. An improved method for determining medium- and long-chain FAMEs using gas chromatography. Lipids 45, 199–208.

Yago, T., Arakawa, H., Morinaga, T., Yoshie-Stark, Y., Yoshioka, M., 2011. Effect of wavelength of intermittent light on the growth and fatty acid profile of the haptophyte *Isochrysis galbana*. In: Ceccaldi, H.J., Dekeyser, I., Girault, M., Stora, G. (Eds.), Glob Chang: Mank-Mar Environ Interact. Springer, Dordrecht, pp. 43–45.

Yan, L., Schenk, P.M., 2011. Selection of Cultured Microalgae for Producing Omega-3 Biolipid Oil. Report for Queensland Sea Scallop Trading Pty Ltd.; The University of Queensland, Queensland, Australia, p. 36.

Yang, G., Yan, L., Gao, M., 2011. Effects of pH and aeration on the production of docosahexaenoic acid by *Thraustochytrium aureum* in controlled batch fermentor cultures. Adv. Mater. Res. 183, 50–54.

Yazawa, K., 1996. Production of eicosapentaenoic acid from marine bacteria. Lipids 31, S297–S300.

Yazawa, K., Araki, K., Okazaki, N., Watanabe, K., Ishikawa, C., Inoue, A., Numao, N., Kondo, K., 1988a. Production of eicosapentaenoic acid by marine bacteria. J. Biochem. 103, 5–7.

Yazawa, K., Araki, K., Watanabe, K., Ishikawa, C., Inoue, A., Kondo, K., Watabe, S., Hashimoto, K., 1988b. Eicosapentaenoic acid productivity of the bacteria isolated from fish intestines. Nippon Suisan Gakkaishi 54, 1835–1838.

Yokochi, T., Honda, D., Higashihara, T., Nakahara, T., 1998. Optimization of docosahexaenoic acid production by *Schizochytrium limacinum* SR21. Appl. Microbiol. Biotechnol. 49, 72–76.

Yongmanitchai, W., Ward, O.P., 1991. Growth of and omega-3 fatty acid production by *Phaeodactylum tricornutum* under different culture conditions. Appl. Environ. Microbiol. 57, 419–425.

Zen, C.K., Silva, K.P., Bertolin, T.E., Reinehr, C.O., Colla, L., 2014. Induçtion of synthesis of lipids and proteins by *Aspergillus niger*. CIATEC-UPF 6, 40–47.

Zheng, H., Yin, J., Gao, Z., Huang, H., Ji, X., Dou, C., 2011. Disruption of *Chlorella vulgaris* cells for the release of biodiesel-producing lipids: a comparison of grinding, ultrasonication, bead milling, enzymatic lysis, and microwaves. Appl. Biochem. Biotechnol. 164, 1215–1224.

Zhu, L., Zhang, X., Ren, X., Zhu, Q., 2008. Effects of culture conditions on growth and docosahexaenoic acid production from *Schizochytrium limacinum*. J. Ocean Univ. China 7, 83–88.

CHAPTER

Production of fungal and bacterial pigments and their applications

11

Salvador Sánchez-Muñoz[a], Gilda Mariano-Silva[a], Marcela O. Leite[a], Fabiana B. Mura[a], Madan L. Verma[b,c], Silvio S. da Silva[a], Anuj K. Chandel[a]

Department of Biotechnology, Engineering School of Lorena — University of São Paulo, Lorena, São Paulo, Brazil[a]; Department of Biotechnology, Dr Y. S. Parmar University of Horticulture and Forestry, Neri Campus, Himachal Pradesh, India[b]; Centre for Chemistry and Biotechnology, Deakin University, Geelong Campus, VIC, Australia[c]

Chapter outline

1. Introduction

The current global demand of pigments is approximately of 9.7 million tones (Ceresana, 2018). In 2014, around, 45% of the total demand in 2014 was related to the production of paints and varnishes and remaining demand is in various application sectors such as food/feed, pharmacy, cosmetics, etc. Colors are the most

Biotechnological Production of Bioactive Compounds. https://doi.org/10.1016/B978-0-444-64323-0.00011-4

pleasant and first parameter to be noticed about an article by the receptor (Tuli et al., 2015). Natural colors are gradually replacing synthetic ones, especially for use in food as additives, textiles, and cosmetics as they are biodegradable and nontoxic to human. On the other hand, synthetic compounds, have carcinogenic and teratogenic effects (Chiba et al., 2006; Subhasree et al., 2011; Mapari et al., 2005).

Because of healthier lifestyle, and the growing market for the natural food colorants, biopigments produced by microorganisms and plant sources are being investigated as readily available sources to replace chemically synthesized colorants (Dufosse et al., 2014; Rao et al., 2017; Saini et al., 2019). Cecchi et al. (2019) studied antioxidant properties of eumelanin (a natural biopigment from flora and fauna) in health, cosmetics and environmental sector. In 2011, global sale of natural colors was US$600 million, showing annual growth rate of ~7% (Tuli et al., 2015). The demand of dyes and organic pigments is expected to reach almost 10 million tons by 2017. For example, only carotenoids demand is estimated to be $1.4 billion in 2018 (Venil et al., 2014).

Bacteria and fungi produce thousands of various pigments molecules (Dufossé, 2016). Ingredients derived from microbial fermentation or extracted from algae are steadily gaining impact in the food industries. Thickening or gelling agents (e.g., polysaccharides such as xanthan, curdlan, gellan), flavor enhancers (yeast hydrolyzate, monosodium glutamate), polyunsaturated fatty acids, flavor compounds (gamma-decalactone, diacetyl, methyl-ketones), vitamins, essential amino acids, and acidulants (lactic acid, citric acid) are some examples illustrating this trend (Dufossé, 2016). Microorganisms produce a vast variety of biopigments such as carotenoids, melanins, flavins, quinones and more specifically monascins, violacein, phycocyanin or indigo (Dufossé et al., 2005).

Microorganisms such as Monascus, Paecilomyces, Serratia, Cordyceps, Streptomyces *Penicillium herquei* and *P. atrovenetum* secrete various kind of biopigments (Gunasekaran and Poorniammal, 2008). The potential of *Monascus purpureus* ATCC16436 was studied for the production of orange and red pigments under solid state fermentation employing corn cob and glycerol (Embaby et al., 2018). This study reported the maximal color value of 133.77 and 108.02 color value units/mL of orange and red pigments, respectively at 30 °C after 10 days using inoculum size (12×10^{11} spores/mL) and glycerol (2.17 M).

Fungal *Monascus* pigments, Arpink red™ (now Natural Red™) produced by *Penicillium oxalicum*, microalgal phycocyanin from *Arthrospira* (Spirulina) *platensis*, riboflavin from the mold fungus *Ashbya gossypii*, lycopene and β-carotene from the tropical mold *Blakeslea trispora*, β-carotene from the microalgae *Dunaliella salina*, and astaxanthin from the bacterium *Paracoccus carotinifaciens* and microalgae *Haematococcus pluvialis*, respectively (Dufosse et al., 2014). Recently, Saini et al. (2019) reviewed the biopigments production potential of cyanobacteria and their health benefits.

Microbial pigments have shown a potential in food applications because their production is natural and fast as compared with the animal or vegetable pigments (Carvalho et al., 2007). Type of bioreactors, composition of the nutrient medium (type of carbon and nitrogen sources and C/N ratio), ambient pH and cultural conditions have shown a great influence on biomass growth, and the production yield of secondary metabolites (Varmira et al., 2017; Pisareva and Kujumdzieva, 2010).

The first European success story in pigment production using a microorganism was β-carotene from the fungus Blakeslea. This natural source was launched in 1995 as a food Ingredients in Europe (Dufossé et al., 2005). European Union has approved the use of fungal caretenoids. This microbial production of pigments can cater round the year demand and supply of pigments in various applied sectors. (Mapari et al., 2010).

Microbial production of pigments using second-generation sugars from lignocellulosic sources can be a key factor to fulfill the commercial demand in sustainable manner. In this chapter, we discuss comprehensively about biopigments, their classification, microbial production and recovery of pigments.

1.1 Biopigments types, classification and properties

Colors are present in the nature. Microorganisms such as fungi, yeasts and bacteria produce a vast variety of biopigments such as carotenoids, melanins, flavins, quinones, monascins, violacein or indigo (Dufossé, 2006).

According to Delgado-Vargas et al. (2000), natural pigments can be classified as follows based on their structural characteristics:

- **Tetrapyrrole derivatives**: chlorophylls and heme colors.
- **Isoprenoid derivatives**: carotenoids and iridoids.
- **N-heterocyclic compounds different from tetrapyrroles**: pterins, purines, flavins, phenoxazines, phenazines and betalains.
- **Benzopyran derivatives (oxygenated heterocyclic substances)**: anthocyanins and alternative flavonoid pigments.
- **Quinones**: benzoquinone, naphthoquinone, anthraquinone.
- **Melanins**.

Tetrapyrrole derivatives have pyrrole rings in linear or cyclic arrays. Pyrrole is a colorless liquid substance, soluble in alcohol, ether, diluted acids and also in most organic chemicals (National Center for Biotechnology Information, 2018a). Chlorophyll structure contains a hydrocarbon tail connected to chlorin, an aromatic ring with tetrapyrrole. The magnesium ion in the center is bound to four pyrrole rings. Chlorophyll is a green pigment found in most algae, plants and cyanobacteria. Chen et al. (2012) reported that chlorophyll *a*, *d* and *f* are present in cyanobacteria. It is a flammable liquid with antitumor properties, antioxidant activities (Ferreira and Sant'Anna, 2017) and its derivatives have been found effective against dengue vectors (Azizullah et al., 2014).

Pyrrole

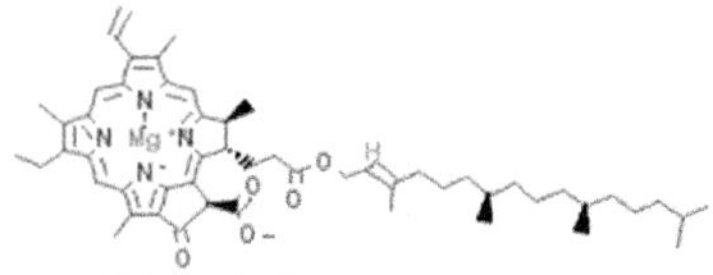

Chlorophylla

Sources: National Center for Biotechnology Information (NCBI). PubChem Compound Database; CID = 8027, https://pubchem.ncbi.nlm.nih.gov/compound/8027; National Center for Biotechnology Information. PubChem Compound Database; CID = 12085802, https://pubchem.ncbi.nlm.nih.gov/compound/12085802.

In general, carotenoids have eight isoprenoid units whose order is inverted at the molecule center. All carotenoids can be considered as lycopene ($C_{40}H_{56}$) derivatives by reactions adding or removing hydrogen, inserting oxygen, of cyclization, of double bond or methyl migration, of chain elongation and shortening (Delgado-Vargas et al., 2000). Lycopene is a solid red bright pigment, used to add color in food and to stabilize flavor formulation (NCBI, 2018c), with antioxidant, anti-inflammatory, and neuroprotective properties (Zhao et al., 2017). β-Carotene can be a red, brownish-red or purple-violet crystal or powder, depending on the solvent used for extraction and conditions of crystallization. Carotenoids are soluble in benzene, chloroform, carbon disulfide and moderately soluble in ether, petroleum ether and oils. β-Carotene is approved by U.S. Food and Drug Association to be used as a nutrient supplement and a as a source of vitamin A. Direct exposure of caretenoids can cause eye and skin irritation and harmful to aquatic life with long lasting effects (NCBI, 2018d).

In flavins, a pteridine and a benzene ring are condensed. Riboflavin, or vitamin B2, is the main molecule of this group, produced by all live cells of microorganisms and plants shortening (Delgado-Vargas et al., 2000). It is a solid orange-yellow pigment, soluble in water, sodium chloride solution and dilute alkali solution (National Center for Biotechnology Information, 2018e).

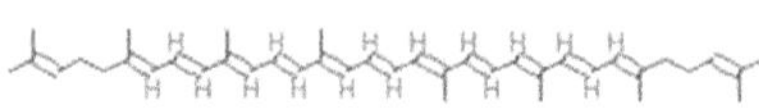

Lycopene

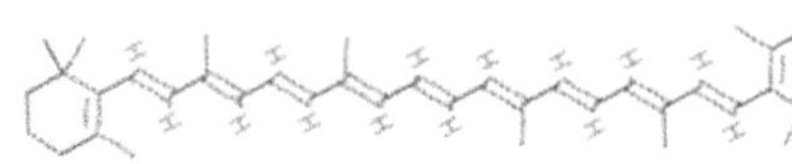

Betacarotene

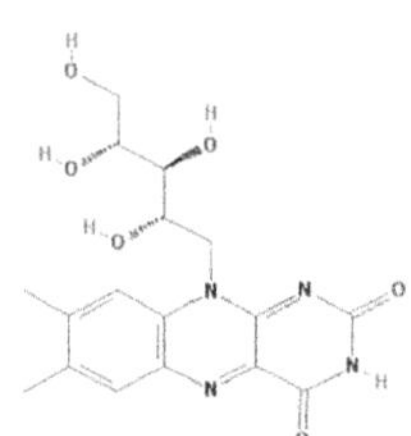

Riboflavin

Sources: National Center for Biotechnology Information. PubChem Compound Database; CID = 446925, https://pubchem.ncbi.nlm.nih.gov/compound/446925; National Center for Biotechnology Information. PubChem Compound Database; CID = 5280489, https://pubchem.ncbi.nlm.nih.gov/compound/5280489; National Center for Biotechnology Information. PubChem Compound Database; CID = 493570, https://pubchem.ncbi.nlm.nih.gov/compound/493570.

Chemically, betalains are compounds with structures based on the general formula shown:

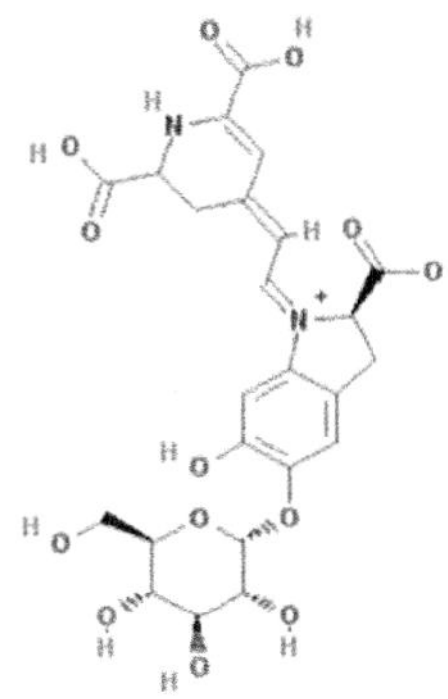

Betalain

Source: National Center for Biotechnology Information. PubChem Compound Database; CID = 56841626, https://pubchem.ncbi.nlm.nih.gov/compound/56841626.

Based on the structure, they can be divided into the yellow betaxanthins (480 nm) and red-purple betacyanins (540 nm) and their color is due to the resonating double bonds (NCBI, 2018f). Betalains present antioxidant, antiradical, antimicrobial and neuroprotective properties (Martins et al., 2017).

Among the flavonoids, the anthocyanins are the most relevant pigments. Chemically, they have 15 carbons with a chrome ring bearing a second aromatic ring B in position 2 (C6-C3-C6) and with sugar molecules bonded at different hydroxylated positions of the primary structure, which gives to them several possibilities of combinations and production of colors. They possess antimicrobial and antioxidant activities and can reduce the risk of cancer and heart disease shortening (Delgado-Vargas et al., 2000).

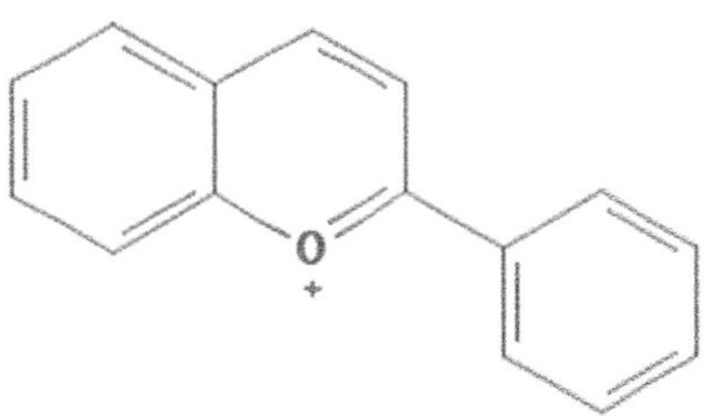

Primary structure anthocyanin

Source: National Center for Biotechnology Information. PubChem Compound Database; CID = 6780, https://pubchem.ncbi.nlm.nih.gov/compound/6780.

Quinones have the largest structural variation among other groups. The primary structure is a desaturated cyclic ketone, derived from an aromatic mono or polycyclic compound.

Menaquinones are found in bacteria and anthraquinones in fungi and lichens (Delgado-Vargas et al., 2000). Anthraquinones have an anthracene ring with a keto group on position 9, 10 and different functional groups such as −OH, −CH3, −OCH3, −CH2OH, −CHO, −COOH at various positions. They present antimicrobial, antiviral, antiparasitic and antioxidant properties (Fouillaud et al., 2018).

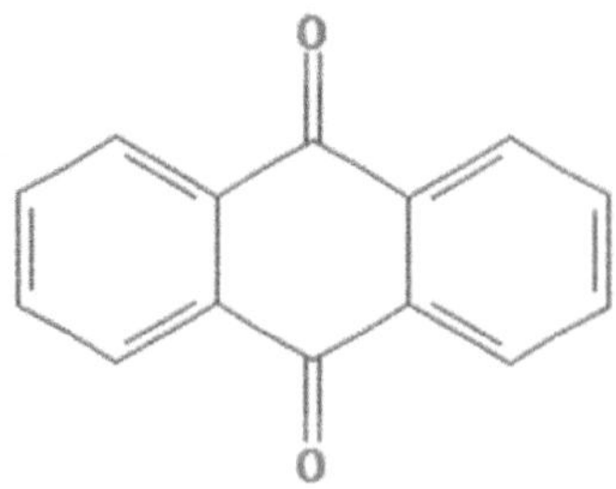

Anthraquinone

Source: National Center for Biotechnology Information. PubChem Compound Database; CID = 6325610, https://pubchem.ncbi.nlm.nih.gov/compound/6325610.

Melanins are nitrogenous polymeric compounds with the indole ring as monomer. In general, melanins are a mixture of macromolecules, characterized by molecular irregularity, photoconductivity and insolubility at most used solvents (Abbas et al., 2009). Allomelanins have been described in spores and fungi shortening (Delgado-Vargas et al., 2000).

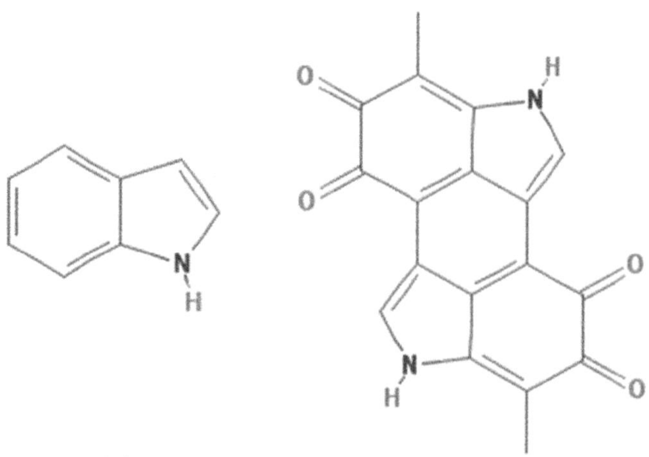

Indole *Melanin*

Sources: National Center for Biotechnology Information. PubChem Compound Database; CID = 8027, https://pubchem.ncbi.nlm.nih.gov/compound/8027; National Center for Biotechnology Information. PubChem Compound Database; CID = 12085802, https://pubchem.ncbi.nlm.nih.gov/compound/12085802.

2. Microorganisms: their role as assets in biopigments production

2.1 Physiological aspects of microbial biopigments production

Historically, mankind has been exploring a wide variety of plant species for biopigments (e.g henna, saffron, brazilwood, paprika and Marigold flowers) production and their early applications as natural pigments in cloth dying and food coloration. Production of biopigments from plants is not an ideal strategy, since it is limited by seasonal and geographic factors, which restricts their availability (Aberoumand, 2011; Mata-Gómez et al., 2014). Microbial production, on the other hand, offers several of advantages when compared to plants and, for this reason, recent research on biopigments production from bacteria, microalgae, fungi and yeasts have gained particular interests (Heer and Sharma, 2017). The use of microorganisms for the recovery of biopigments is considered more advantageous than extraction from plant not only due to its independence from atmospheric and geographical conditions, but also for the following reasons: (a) production feasibility (Heer and Sharma, 2017); (b) higher stability to heat, light exposure and pH variation and high solubility in water as compared to plant pigments (Nigam and Luke, 2016); (c) some kinds of pigments, like aryl carotenoids, are only produced by microorganisms (Dufossé, 2006); (d) using low-cost substrates, such as agro-industrial wastes, (Hernandez-Almanza et al., 2014; Venil et al., 2013); (e) possibilities of using modern genetic engineering tools to express biopigment production by industrially important species (e.g. *Escherichia coli* and *Saccharomyces cerevisiae*) (Venil et al., 2013) compared to plants, microbial diversity has enormous unexplored potential, expanding the possibility to find novel molecules (Nigam and Luke, 2016).

Biopigments are essential to metabolism of photosynthetic microorganisms, since these molecules use light and allow carbon fixation into organic molecules by photosynthesis. Chlorophyll and carotenoids are the main pigments involved in this process and carotenoids play the additional role of protecting cells against oxidative stress (Buzzini et al., 2007; Uenojo et al., 2007). Despite of biopigments relation to the photosynthetic metabolism, a wide variety of non-photosynthetic microorganisms synthesize these compounds. In this case, biopigment production is associated to protection against UV light and free radicals that are naturally derived from cell metabolism, such as 1O2, OH- and peroxides. Pigments inactivate free radicals and, therefore, protect cell structures and DNA from damages and consequently improve the organism survival chances (Buzzini et al., 2007; Hernandez-Almanza et al., 2014; Zoz et al., 2015). The reaction between a carotenoid, an antioxidant pigment, and a free radical is shown in Eqs. (11.1) and (11.2).

$$R + CAR(H) \rightarrow RH + {}^{*}CAR \tag{11.1}$$

$$1O2 + CAR \rightarrow 3O2 + {}^{*}CAR \tag{11.2}$$

R = Free radical
CAR = Carotenoid

The production of biopigment by microorganisms can also be a strategy for improving its ability to compete with other organisms for the production of natural resources for example antimicrobial activity of some natural pigments. *Monascus*, as an example, produces three major biopigments with recognized antimicrobial properties: ankaflavin, rubropunctatin and monascorubramin (Carvalho et al., 2014). Another important biopigment with antibiotic property is actinorhodin, which is synthesized by *Streptomyces* bacteria and is generally blue, but can assume different colors according to the medium pH (Charkoudian et al., 2010).

Is has been observed that biopigments produced by microorganisms in a response to environmental conditions are often associated with health benefits for humans. Prodigiosin, versatile red biopigment is synthesized by *Serratia* and other gram-negative bacteria. Prodigiosin is reported to have antibacterial, antiprotozoal, antifungal, cytotoxic and anti-inflammatory properties. Pigments from *Monascus* are also known for their pharmacological properties, such as the restriction of cholesterol synthesis, anti-cancer, anti-mutagenic, anti-obesity and antimicrobial activities. Additionally, antioxidant, anti-carcinogenic and immune response enhancement has also been described for biopigments produced by yeasts (Demain and Martens, 2017; Panesar et al., 2015). Melanin is another biopigment commonly found in microorganisms, plants and animals having antimicrobial, anti-inflammatory, antioxidant and immunogenic properties (El-Naggar and El-Ewasy, 2017; Rosas et al., 2000).

2.2 Microorganisms: an asset in biopigment production

The majority of biopigment producing microorganisms are bacteria, microalgae, yeasts and filamentous fungi. Studies involving pigment production are at early stages and an intense development of this research field is required to increase the industrial feasibility of biopigment producing processes. Bacteria cultivation offers a series of advantages that must be further explored, such as: fast and simple cultivation systems, continuous cultivation, high yields in fermentation processes, simple and easy biopigments extraction processes, easy genetic manipulation, propagation and strain selection (Venil et al., 2013).

Eukaryotic microorganisms, yeasts offer more advantages for biopigments production as compared to filamentous fungi. The unicellularity of yeasts and their elevated growth rates make them favorable organisms for biopigments production (Bhosale and Gadre, 2001). Biopigments produced by yeast species (e.g. *Rhodotorula* spp.) are harmless to human health while filamentous fungi derived biopigments have health concerns (Hernandez-Almanza et al., 2014). Nonetheless, biopigments from yeasts are intracellularly produced, which require cell disruption and extraction methodologies to obtain these products (Park et al., 2007).

Carotenoids are a remarkable class of biopigments with a global market of approximately US$1.4 billion in 2018 (Venil et al., 2014). There are more than 600 known structures of carotenoids which are generally produced by yeast, among which *Rhodotorula*, *Rhodosporidium*, *Sporobolomyces* and *Sporodiobolus* are well known. Studies aiming to characterize biopigment production by many yeasts species are known for more than 60 years, but very few conclusive data is available for carotenoids production because of difficulty in analysis and characterization (Buzzini et al., 2007).

Among carotenoids, beta-carotene is a noteworthy biopigment with industrial interest. An advantage presented by the carotenoids producing yeasts is that, usually, more than one carotenoid molecules are produced by yeasts. For example, *Rhodotorula* sp., is known to produce beta-carotene, torulene and torularhodin (Zoz et al., 2015). In recent years, there has been an increasing demand for natural pigments and beta-carotene are one of the molecules with available natural sources, however the cost of its production from microbes is estimated to be US$1000/kg, while the manufacture of its synthetic equivalent costs approximately US$500/kg. The main strategies for lowering the costs of microbial beta-carotene production involve the use of low cost substrates and the development of strain improvement techniques (Rao et al., 2017).

Similarly, to yeast, bacteria can produce carotenoids as a strategy of its protection against photo-oxidative damages. Canthaxanthin is a carotenoid widely used as food additive in animal feeding and its biosynthesis by microorganisms is well reported, being produced by bacteria from the genus *Diezia* and *Paracoccus*. Industrial production of canthaxanthin is from algae-*Haematococcus lacustris* and the main application of this biopigment is to improve the visual aspect of egg yolk, cheese, candies, meat, fish, beer and wine (Aberoumand, 2011).

Astaxanthin is another valuable carotenoid of great importance. It outstands as a highly effective antioxidant that is applicable in cosmetics, pharmaceutical products and food industry. Additionally, it is an important food additive for aquaculture (e.g. to give salmon flesh its characteristic orange color). This carotenoid, which market value ranges between US$ 2500 to US$7000/kg, is mainly produced and extracted by microalgae, but it can also be found in yeast species. *Phaffia rhodozyma* is a yeast largely studied for the production of this compound (0.35 mg/g of dry biomass) (Demain and Martens, 2017; Panis and Carreon, 2016).

Beside from carotenoids, another relevant biopigment that can be found in the microbial world is riboflavin. Riboflavin, or vitamin B12, is a molecule that can be found in a wide variety of organisms and plays a physiological role in photosynthesis and participates in several important metabolic processes, e.g. the electron transport chain, DNA damage repair and lipids metabolism. It is also known that a dietary deficiency of riboflavin can affect the health of eyes (Santos De Souza et al., 2005). This yellow-bright pigment is widely used in the food industry as an additive to beverages, instant desserts, ice cream dressings and similar products (Aberoumand, 2011). Daily consuming products such as cheese, pastas, sauces and cereals also have riboflavin (Nigam and Luke, 2016).

Riboflavin is a versatile molecule that has been applied not only in food products, but also in pharmaceutical applications. For many decades, this pigment was utilized, in association with phototherapy, for the treatment of neonatal jaundice. Additionally, it has been successfully employed on a migraine treatment. Ultimately, an effective usage for riboflavin is its combination with UV light to reduce levels of harmful pathogens in blood products (Rao et al., 2017).

The commercial production of riboflavin is mainly performed by three microorganisms: the ascomycetes *A. gossypii*, *Candida famata* (*Debaryozyma hansenii*) and the bacteria specie *Bacillus subtilis*. *A. gossypii* is the preferred microorganism for due to its high genetic stability and yield. Genetics and metabolic features of this microorganism has been extensively studied (Jeong et al., 2015; Ledesma-Amaro et al., 2014; Schwechheimer et al., 2018).

Another outstanding biopigments commonly found in nature are Melanins, a group of molecules that includes a wide range of polymeric substances (Gao and Garcia-Pichel, 2011). Melanins are most commonly found in brown or black coloration, although other colorations have also been described (El-Gamal et al., 2017; Langfelder et al., 2003). Melanin is also characterized by molecular irregularity and photoconductivity. Melanins are usually not soluble in the most solvents (Abbas et al., 2009). Regarding the production of melanin by *Aureobasidium pullulans*, Zheng et al. (2008) reported that approximately half amount of melanin are produced intracellularly and are stored in structures known as chlamydospores, while the other half is released into the extracellular medium as granules.

While, some biopigments are equally produced by bacteria, yeasts and filamentous fungi, others are markedly produced by bacteria. Biopigments such as pyocyanin, prodigiosin and violacein are normally produced by bacteria. *Pseudomonas aeruginosa* is the main known microorganism produces pyocyanin, a blue pigment whose antifungal and antibacterial property supported its employment as a biocontrol agent. On the other hand, prodigiosin is a red biopigment that has been isolated from a diversity of bacteria and presents unique pharmaceutical activities, such as antibacterial, antifungal, anti-malarial, immunosuppressant ant anticarcinogenic properties. Violacein, a purple compound, is also produced by a large number of bacterial species and is associated with numerous biological activities, such as enzyme modulation, anti-leishmanial, anti-viral, anti-ulcerogenic and anticancer characteristics (Rao et al., 2017).

2.3 Consolidated processes for microbial processes production

Considering large-scale or industrial processes that employ microorganisms for biopigment production, algae are the most used and competent organisms (e.g. *Dunaliella* and *Haematococcus*) to fulfill the industrial demand. However, there are fungal species with remarkable levels of pigments production that could be exploited at industrial levels. For example, *B. trispora* and *Phycomyces blakesleeanus*, can produce not only beta and gamma-carotene, but also phytoene, lycopene and organic acids (Ribeiro et al., 2011). Industrial scale production of biopigments is mainly limited not only by the high costs, but also by the concomitant production of toxins in some of these processes and by the resistance of the final product to extreme processes conditions, such as high temperature and extreme pHs (Rao et al., 2017).

B. trispora is a filamentous fungus whose beta-carotene production started to be explored at industrial levels in European countries in the years 2000. It was reported that beta-carotene production by this fungus can achieve 170 mg/g of dry biomass (Varzakakou and Roukas, 2010). This organism is an asset in biopigments production as, it is capable of producing high amounts of lycopene besides beta-carotene. Lycopene is another carotenoid of great dietary importance due to its contribution to reduce the oxidative stress. Studies show that *B. trispora* can produce lycopene of 15,000 μg/g of dry biomass, while for *E. coli*, for example, it reaches only 500 μg/g of dry biomass (Mantzouridou and Tsimidou, 2008). This yeast is also an asset for the industrial production of beta-carotene (Rao et al., 2017).

Another biopigment that conquered the industrial demand of biopigments is anthraquinone, copiously synthesized by *Penicillium* fungi. This pigment was authorized by the EU, in 2004, to be distributed in the market for two years (Dufossé et al., 2005). A remarkable fungus that has also been cultivated for the largescale production of biopigments is *Monascus* sp. This fungus can ferment red rice (also known as Ang-Khak) is widely consumed in Japan, Thailand, China. Solid-state fermentation system is employed for large scale biopigment production from *Monascus* sp., however red rice is not well-accepted in western countries due to the presence of toxins (citrinin) in the fermented rice slurry (Rao et al., 2017; Carvalho et al., 2014).

The filamentous fungus *Penicillium* has also been the protagonist of the industrial production of a pigment originally named Arpink RedTM, the first fungal-derived anthraquinone to be commercialized and manufactured by two companies from Czech Republic. The regulatory agency established limits for the presence of this biopigment for meat products, milk products, alcoholic and non-alcoholic drinks. This pigment i.e., Natural Red™ started to be manufactured by another company (Venil et al., 2014).

2.4 Prospecting microorganisms for biopigments and innovative applications of these metabolites

It is important to notice that, while the potential to discover new plant pigments is minimum, the colorful microbe universe is a huge unexploited field of knowledge (Nigam and Luke, 2016). However, prospecting microbes to explore their potential to produce novel pigments molecules is a work that requires a correct taxonomical approach. When a microorganism, especially a filamentous fungus, is misidentified, its ability to produce a toxin or to act as a human pathogen might not be considered, consequently the inadvertent usage of its biopigment or other metabolites as food or drugs additives can lead to severe health problems. On the other hand, the proper identification of a species allows researchers to elaborate biotechnological processes that leads to pigment production with the concomitant reduction of undesired metabolites formation (Mapari et al., 2005).

Prospecting new microorganisms is a costly and time-consuming approach, yet it is a way to achieve promising results. Recently, some studies have been focused on marine microbiota bioprospection. The unique conditions presented by marine environments favor the origin of exclusive molecules and studies have been showing that there are some biopigments produced only by marine fungi, e.g. a yellow pigment (anthracene-glycoside asperflavinribofuranoside) that is uniquely produced by *Microsporum* sp. (Li et al., 2006). Marine algae and cyanobacteria have been extensively investigated, while researches involving filamentous fungi are more recent and in future investigation will lead to search for potent and novel strains pigments producing microorganisms (Dewapriya and Kim, 2014).

Poorly explored environments characterization of an abundant biodiversity, such as the Amazon forest, are also a great potential source of new microorganisms and metabolites. A recent work that prospected fungi from amazon soil identified five species as biopigments producers and one of the identified pigments was sclerotiorin, a molecule with antimicrobial and antifungal activities and capability of inhibiting HIV-1 virus. It was observed that culture media composition, especially nitrogen sources, affected the production of this noteworthy metabolite (Celestino et al., 2014).

Advances in the field of biopigments production are driven by either the discovery of new microorganisms and novel metabolites or by finding new applications of biopigments. Considering the exploration of diversified properties of a known biopigment, antioxidant properties of biopigments by *P. aeruginosa*, *Acinetobacter mufti* and *Enterococcus hirae* presents unique properties (Masi et al., 2014). An innovative investigation was similarly conducted by (Manimala and Murugesan, 2014) showing antimicrobial and antioxidant properties of a carotenoid produced by *Sporobolomyces* yeasts. Melanin is another distinctive biopigment whose recent applications have gone further than its ability to interact with light: melanin is capable of binding to a variety of metals and this property has been used in the formulation of iron supplements to treat anemia and for the development of real-time metal detection sensors (Cordero and Casadevall, 2017).

2.5 Strain improvement methods for improved biopigments production

Currently, the production of most biopigments at industrial levels is still limited by processing issues at large scale operations. For instance, long fermentation time and low biopigments yield act as barriers for the economic feasibility of the biopigments production bioprocess. Effective approaches to overcome the drawbacks of microbial biopigments production (e.g. product yields) are the optimization of cultivation parameters and strain improvement. Strain improvement lies on mutation or others genetic and metabolic engineering tools to reshape a microbe genetic design in order to favor biochemical pathways responsible for the synthesis of a particular metabolite and increase its yield (Rao et al., 2017; Venil et al., 2014). Genetic manipulation of microorganisms can also lead to the development of "cell factories" employing the heterologous expression of metabolic pathways from known biopigment producers (Mapari et al., 2005).

There is a variety of studies in the literature that report the adoption of diverse mutation and genetic engineering procedures to enhance the efficiency of bioprocesses. Examples of the most frequently mutagenic agents used for biopigments production are presented below.

a) UV radiation: The effect of UV light as a mutagenic agent is based on the formation of pyrimidine dimers that impedes DNA polymerase activity (Rastogi et al., 2010). Response surface methodology was applied to increase total carotenoids production by *Micrococcus roseus* PTCC 1411 and UV radiation was used as a mutation agent with an irradiation distance of 30.45 cm and an irradiation time of 47.76 h. In this work, it was reported that optimum conditions were reached and total carotenoids concentration was of 7.2 mg/L (Yolmeh et al., 2017).
b) Gamma radiation: Astaxanthin production by *P. rhodozyma* was dramatically increased by a treatment with 5 kGy of gamma radiation: while the determined astaxanthin yield for the parental strain was of 196.6 μg/g dry biomass, the mutant strain produced 2562.5 μg/g dry biomass of this biopigment (Najafi et al., 2011).
c) Microwaves: The effect of microwaves is to promote polar molecules to vibrate and, as a result, the permeability of cell membrane and DNA structure can be modified. It is equally possible that microwaves stimulate the ionization of materials and a heat transfer to the cell. This mutation agent was efficiently employed to generate a *Serratia marcescens* with an increased prodigiosin yield (109% higher than the parental strain yield). It was also determined that purity level of prodigiosin obtained by this process was of 98% (Liu et al., 2013).
d) Atmospheric and room temperature plasma (ARTP): this mutation strategy acts by promoting several damages in plasmidial DNA and in oligonucleotides (Jin et al., 2018). ARTP was used to modify *B. trispora* strain to enhance the lycopene production and the mutated strain was capable of producing 55% more of this carotenoid than the wild-type strain. It also presented a smaller oxygen demand to reach maximum cell concentration, a characteristic favorable for industrial scale processes (Qiang et al., 2014).
e) Ethyl methanesulfonate (EMS): EMS, as others alkylating agents, transform guanine residues into O-6-ethylguanine, which promotes impairing and base changes. O-6-ethylguanine pairs with thymine instead of cytosine. In a study, where *S. marcescens* was mutated with EMS, prodigiosin production achieved 658.0 ± 46.0 mg/L, while the value registered for the parent strain was of 34.7 mg/L (El-Bialy and Abou El-Nour, 2015).
f) N-methyl-N-nitro-N-nitrosoguanidine (NTG): NTG is considered one of the most powerful mutagenic agents and it alters DNA by substituting GC by AT bases, causing mispairing in DNA synthesis (Harper and Lee, 2012). A mutant strain of *Rhodosporidium toruloides* was obtained by ARTP followed by NTG mutation procedures and it presented higher lipids and carotenoids yields when

compared to the parent strain. Carotenoids total production by the mutated strain was of 0.75 mg/g cell dry weight (Zhang et al., 2016).

A genetic tool that has been extensively employed for biopigments production in laboratory scale is heterologous expression, which grants either the expression of a metabolic route of pigment synthesis in a microorganism that originally did not produce any biopigment or the expression of transcriptional factors or specific enzymes form a biopigment producer microorganism to another. As an example, an engineered *E. coli* received genes derived from *Deinococcus wulumuqiensis* carotenogenic metabolic route with the purpose to obtain a high yield of lycopene. The resulting strain was able to produce 688 mg/L of carotenoids, which represented 2.2 higher production in comparison to the original strain (Xu et al., 2018).

Besides introducing pigment production metabolic pathways genes in microorganisms not capable of producing pigments, it is also possible to increase biopigment production rates in a species that is already a pigment producer. For instance, *Corynebacterium glutamicum*, a yellow-pigmented bacterium, had its sigma factors (contributors of promoter recognition and transcription initiation) manipulated to modify is carotenoids production: the sigma factor B (sigB) was deleted to favor the overexpression of sigA. As a result, lycopene yield increased eight times, decaprenoxanthin production augmented by a 2-fold factor and, additionally, the microorganism was able to synthesize non-native biopigments, such as beta-carotene and bisanhydrobacterioruberin (Taniguchi et al., 2017).

Riboflavin is another relevant biopigment whose production was successfully increased by genetic engineering tools. The synthesis of this metabolite is related to the purine pathway and it has been manipulated by heterologous expression. Dmytruk et al. (2011), for example, reported that riboflavin production by *C. famata* was increased four times by transferring a purinogenic enzyme and a transcriptional factor ortholog genes from *Debaryomyces hansenii* and overexpressing two genes that regulate this metabolic pathway.

As well as heterologous expression, adaptive laboratory evolution is also a powerful method for strain improvement. One of the researches that utilized adaptive evolution for biopigments was conducted by Reyes et al. (2014) and the target microorganism was *S. cerevisiae*. In the referred study, the yeast was engineered by using a shuttle vector to insert new genes in the strain and the carotenoids hiperproducer mutants were isolated. Further, the selected strains were treated with hydrogen peroxide for laboratory evolution experiments and RNA was extracted for subsequent analysis. The given procedure promoted an increase of carotenoids production by three times (from 6 mg/g dry cell weight to nearly 18 mg/g dry cell weight).

Other studies were based on metabolic engineering techniques. Buey et al. (2015) worked on gene disruption and overexpression in *Ashby gossypii* and, by augmenting the metabolic flux through the guanine pathway and it resulted in a 40% increase in riboflavin production. Similarly, Ledesma-Amaro et al. (2014) manipulated the same yeast and overexpressed the entire RIB genes sequence and the result was the generation of a strain with a riboflavin production capacity 5.4-fold higher than the wild-type strain.

Regarding filamentous fungi, the presence of a gene cluster that regulates secondary metabolites is well recognized and studies suggest that those genes are under expressed in laboratory or industrial scale conditions. To overexpress this gene set in *Monascus pilosus*, Lee et al. (2013) selected the laeA gene (which encodes a positive regulator of secondary metabolism) and substituted its promoter by alcA, a strong promoter from *Aspergillus nidulans*. The results were encouraging: the new developed strain was used for the production of *Monascus* fermented rice and, asides from the observation of higher pigment production and increased antioxidant activity of this product, it contained four times more monacolin K, a remarkable cholesterol-lowering metabolite.

The achievement of strain improvement can likewise be a product of the combination of different techniques. For instance, mutation with ARTP followed by a treatment with $NaNO_2$ and heterologous expression (overexpression of the HMG1 gene from *S. cerevisiae*) was applied on *Rhodotorula mucilaginosa* KC8 and the resulted strain was capable of producing 19.14 ± 0.09 mg/L of carotenoids, a concentration 121% higher than the obtained with the parent strain (Wang et al., 2017).

The effect of the association of genetic methods was also evaluated to achieve astaxanthin overproduction in *S. cerevisiae*. In this study, ARTP mutation was combined with the heterologous expression of genes involved in carotenoids synthesis from *Agrobacterium aurantiacum*. The resulted strain exhibited an astaxanthin production 1.78-fold greater than the parent strain and, additionally, astaxanthin proportion, among total carotenoids, was increased (Jin et al., 2018).

Strain improvement is not only important to obtain higher yields of biopigments, but also is a feasible method to reduce the production of undesired by-product. An example of this application regards *Monascus* pigments: it is known that the metabolic pathway that generates *Monascus* polyketides pigments also promotes the formation of citrinin, a hepatotoxin, and gene manipulation has been successfully applied to reduce the formation of this compound (Kang et al., 2014). Jia et al. (2010), for example, evaluated the effect of three different mutant agents (UV light, LiCl and diethyl sulfate) on the reduction of citrinin production and were able to produce a *Monascus* strain that did not synthesize any citrinin and presented slightly higher yields of yellow and red pigments.

3. Fermentation: methods and new formulations for biopigment production

3.1 Potential biomasses as a substrate for biopigment production

To economize the process on an industrial scale, there is a need to develop low-cost technology for the production of pigments that could replace synthetic ones (Panesar, 2014). Terán-Hilares et al. (2018), used sugarcane bagasse (SCB) hydrolysate based medium and found that the maximum red pigment production (18.71 AU 490 nm) was performed under the dark condition and the principal carbon sources (glucose and cellobiose) were metabolized. SCB enzymatic hydrolysate was demonstrated to be a promising carbon source for high thermal stability red pigment production by *Monascus ruber.* Likewise, Subhasree and colleagues (2011), investigated the feasibility of Jackfruit seed as a substrate for pigments production by *M. purpureus* in solid-state fermentation (SSF), when it was supplemented with several carbon sources like mannitol, lactose, starch and fructose and nitrogen sources like yeast extract, peptone, ammonium sulfate and ammonium nitrate. The maximum pigment production was found when the culture was supplemented with fructose (1.304 U 510 nm/g) as a carbon source and yeast extract (1.29 U510 nm/g) as the nitrogen source. Besides that, General et al. (2014), evaluated the feasibility and applicability of marine algal biomass *Saccharina* (Laminaria) *japonica* as a sole substrate for the production of pigments by *Talaromyces amestolkiae* GT11 in submerged fermentation. Results indicated that the fungus *T. amestolkiae* GT11 produced the highest amount of extracellular yellow (444.83 ± 22) and red (200.94 ± 12), and intracellular yellow (362.28 ± 34) and red (193.87 ± 10) pigments, utilizing 1% (w/v) of S. japonica powder at an initial pH of 5 and 30 °C. Another example is described by Panesar (2014), where was used kinnow (Citrus reticulata) peel waste as a substrate for the production of red pigment using *M. purpureus* MTCC 369 using submerged fermentation and it was reported a maximum pigment production when the media containing kinnow peel powder and pea pod where supplemented with magnesium sulfate (0.1%) with pH of 6.5, incubated at 35 °C after 9 days of incubation period.

3.2 Effect of carbon and nitrogen sources for the production of biopigments

Pombeiro et al. (2017) found that cell growth and fungi pigment production are affected by the composition of the culture medium used, by the nature and concentration of carbon and nitrogen sources, by pH, by the presence of oxygen and aeration, by light, among other factors. The way the conditions of culture are manipulated and the optimization of the composition of the medium can favor the production of pigments.

Gunasekaran and Poorniammal (2008), examined 10 carbon sources, glucose, fructose, mannose and sucrose were relatively favorable to the mycelial growth of *Penicillium* sp., although the pigment production was low. The maximum mycelial growth (5.50 g/L) was achieved in fructose medium, while the maximum pigment production (670 mg/L) was obtained in a soluble starch medium. Also, they made tests with peptone, peptone—yeast extract mixture, tryptone, and monosodium glutamate had a positive effect on pigment production, whereas soy peptone, beef extract, and potassium nitrate strongly inhibited red pigment synthesis. The best yield of the nitrogen sources tested was of peptone with a red pigment production of 880 mg/L.

Moreover, Cho et al. (2002), selected a high pigment-producing fungus *Paecilomyces sinclairii*, and 11 carbon sources were examined, glucose, fructose, mannose, sucrose, and maltose were relatively favorable to the mycelial growth of *P. sinclairii* although the pigment production was low, o maximum pigment production (467 mg/L) was obtained in a soluble starch medium. They show that organic nitrogen sources gave rise to higher mycelial growth in *P. sinclairii* compared with inorganic nitrogen sources. Meat peptone, casein peptone, peptone—yeast extract mixture and corn steep powder had a positive effect on pigment production, whereas soy peptone and malt extract strongly inhibited red pigment synthesis. Of all the nitrogen sources tested, meat peptone (428 mg/L) gave the highest yield for red pigment production.

In another work, five different cyanobacteria were exposed to different carbon sources by Sabarinathan et al. (2010), data revealed that sodium carbonate is the best-preferred carbon source followed by fructose, while ammonium carbonate is the least preferred carbon source for the phycocyanin pigment production. Among the cyanobacterial cultures tested *Westiellopsis*-ARM 48 produced maximum phycocyanin (55.53 μg mL-1) pigment. Sehrawat and colleagues also analyzed the effect of various nitrogen sources i.e., pea pod, taro leaves, soy okara, corn steep liquor, faba beans hull and peptone were examined with sweet potato peel powder, and it was found that addition of these have a significant effect, an increase in pigment production was observed. Among all the natural nitrogen sources tested, pea pod gave the highest color value (4.6 ± 0.06 CVU/g) and soy okara showed the minimum color value (2.7 ± 0.07 CVU/g). Almost similar results were obtained for peptone (4.8 ± 0.01 CVU/g) when compared to pea pod (4.6 CVU/g). Peapod was selected as best natural nitrogen source being waste residue and cheap as compared to peptone.

Sehrawat et al. (2017), also showed that carbon sources directly affect the growth of *Monascus* sp. Among the various carbon sources, i.e., wheat bran, rice bran, sweet potato peel powder, potato peel powder kinnow peel powder and fructose tested, they observed that sweet potato peel powder gave maximum biopigment production (3.3 ± 0.06 CVU/g). Potato peel powder (2.1 ± 0.09 CVU/g) gave minimum biopigment production after 15 days of incubation.

To improve astaxanthin productivity, Ni et al. (2007), analyzed the effects of nitrogen source on *P. rhodozyma* cultivation and astaxanthin production. Results showed nitrogen source could significantly affect *P. rhodozyma* cultivation with respect to carbon source utilization, growth rate, and astaxanthin accumulation, the experiment with different ratios of nitrogen sources, shown the optimal nitrogen mix was composed of 0.28 g/L $(NH_4)_2SO_4$, 0.49 g/L KNO_3 and 1.19 g/L beef extract, where the maximum amount of biomass and highest astaxanthin yield in terms of volume and in terms of biomass were 7.71 mg/L and 1.00 mg/g, respectively. One more example is the description of Kurbanoglu et al. (2015) for organic nitrogen sources such as peptones that contain different amino acids, which increase the production of prodigiosin, they utilized Ram Horn Peptone (RHP) (as a complete substrate and nitrogen source) in submerged culture for the production of prodigiosin by a local isolate, *S. marcescens* MO-1, to enhance prodigiosin production and reutilize this abundant animal waste. The concentration of 0.4% (w/v) RHP resulted in the greatest yield of prodigiosin (277.74 mg/L) after 48 h when mannitol was used as the sole source of carbon.

The effect of nitrogen source on biopigments production was also verified by Jiang et al. (2016). In this study, the authors evaluated melanin production by *Aureobasidium melanogenum* XJ5-1. Comparing results obtained in a PDA medium with or without 2 g/L of $(NH_4)_2SO_4$, it was observed that a large amount of melanin was synthesized in the absence of this nitrogen source and no melanin was production in the medium containing it.

3.3 Reactor configuration and fermentation modes

Solid-state fermentation systems could be advantageous due to its potential with several natural substrates, however, pigment production is conducted out commercially in submerged state fermentation. According to Zhang et al. (2013a), fermentation mode seems to be one of the most important factors which influence the yield for pigment production. They demonstrated the influence on pigment and citrinin production via submerged fermentation and solid-state fermentation from rice (*Oryza sativa* L.) by *M. purpureus* AS3.531, in which the pigment yield by submerged fermentation (7.93 U/g/g) could be comparable to that by solid state fermentation (6.63 U/g/g), however, the concentration of citrinin produced by SmF (19.02 μg/g) increased more than 100 times than that by SF (0.018 μg/g). Another comparative analysis was made by Sehrawat et al. (2017), where they observed that *M. purpureus* MTCC 369 in solid state fermentation showed a significant accumulation of biopigment (9.0 CVU/g) on the 9th day in comparison to submerged fermentation (5.1 CVU/g) on the 15th day.

In general, 75% of the fermenter volume is considered as maximum working volume as some headspace is necessary because of frothing during aeration and agitation and it has an effect on O_2 uptake. Effect of medium volume (2.0–4.5 L) on the production of an extracellular green pigment by growing *Bacillus cereus* $M^1$16 (MTCC 5521) in a complex medium was evaluated as follows pH-7.0 at 30 °C for 72 h in 5 L fermenter (submerged fermentation mode) by Banerjee et al. (2014).

Krairak et al. (2000), reported the maximum of pigment production in fed batch cultivation. They showed that the highest specific pigment production rate was achieved with a production rate (μ) of 0.02 h^{-1} in the feeding phase. However, among several fermentation experiments examined, the maximum pigment production in the final step was obtained at a μ of 0.01 h^{-1}; in each case the same amount of substrates was used. A maximum pigment concentration of about 1000 U/mL was obtained in fed-batch culture, which was twice as high as that obtained in batch culture. The superiority of fed-batch over batch operation arises from the extension of the pigment production phase for as long as possible.

Later, Varmira et al. (2017), used the following combination of the factors: aeration rate = 0.469 vvm, initial pH = 6.48, and light irradiation intensity = 1757.84

Lx to evaluate the carotenoids productions in an Airlift Photobioreactor (APBR) by *Rhodotorula rubra*, and their results showed a maximum performance in the APBR with biomass formation of 6.69 g/L, and carotenoids yield of 8.352 mg g/cell.

Some other examples are shown in Table 11.1, where the potential efficiency of different carbon and nitrogen sources, modes of operation and multiple species with potential application in biorefineries are summarized.

4. Biopigments recovery methods

Biopigment production by microbes may or may not involve the excretion of those metabolites to the culture medium. While some biopigments, such as yeasts carotenoids, are always produced intracellularly, other, like *Monascus* pigments, are produced extracellularly (Nigam and Luke, 2016; Park et al., 2007). In some cases (e.g. melanin), a pigment might be partially found in the inner of the cell and partially excreted to the medium. The predominance of one form or the other may depend on culture conditions (Langfelder et al., 2003). The recovery of intracellular biopigments involves essentially two steps: cell disruption and pigments extraction; the efficiency of this procedure lies on the preservation of biopigments from the process conditions, such as high temperatures and use of organic solvents (Nigam and Luke, 2016). An overview of different disruption techniques and extraction methodologies are presented in Table 11.2.

Cell disruption by mechanical methods, such as bead beater and biomass maceration, are traditionally the mostly used, nonetheless it is considered that chemical methods are more efficient since they are fast and efficient. Generally, chemical driven biopigments extractions are simple, low cost and fast procedures methods (Park et al., 2007). The choice for a proper extraction method for biopigments involves not only the assessment of the final yield in each procedure, but it is also a reflection about the extract composition. It is known that different extraction procedure will result in distinctive biopigments profile (Mapari et al., 2005). It is worth noting that intracellular biopigment extraction is still a challenge, since there are no standard and economic procedures for the extraction of biopigments (Park et al., 2007).

Another crucial point is the necessity to determine the molecular characteristics of the pigment in order to choose the best solvent or solvents combination for its extraction. Regarding solvent solubility, carotenoids, for instance, are stored in lipid vesicles and therefore can be extracted using various organic non-polar solvents (Park et al., 2007). Prodigiosin, diversely, is a pigment with higher hydrophobicity than many carotenoids and, for this reason, can be easily extracted with acetone (Sun et al., 2015).

Extracellular pigments from *Monascus* pigments, cell disruption is not required, but an extraction for the supernatant or from the interstices of microbial biomass may be executed. When obtaining pigments from *Monascus*, this procedure is usually performed with ethanol (Nimnoi and Lumyong, 2011). Extracellular melanin, on the other hand, can be precipitated with acetic acid or flocculated with alum (Jalmi et al., 2012).

Table 11.1 Biopigments production by different microorganisms on various carbon and nitrogen substrates under different cultivation conditions.

Strain	Mode/culture media	C and/or N source	Biopigment production	Reference
Monascus sp., LPB-31	SSF/different natural substrates with a 56% of moisture.	Rice Cassava Cassava flour	AU_{500nm}/g 210 119.6 98.1	Carvalho et al. (2007)
Penicillium sp.	SmF/basal 2% w/v of carbon or nitrogen source	**Carbon** Glucose | **Nitrogen** PP	AU_{500nm} 450 | AU_{500nm} 880	Gunasekaran and Poorniammal (2008)
P. resticulosum	SmF/synthetic media-Corn cob hydrolysate media	**Carbon** Xylose **Nitrogen** $NaNO_3$	Pigment yield (g/L) 0.55 ± 0.07 0.34 ± 0.04	Sopandi et al. (2013)
Monascus ruber	SmF/synthetic media-Glicose 10 g/L C:N ratio of 9	Media based in glicose Nitrogen sources: MGS	AU_{500nm}/g ~25	Said et al. (2014)
Monascus purpureus	SF/Jack fruit base media with a 60% of moisture.	Jackfruit seed supplemented with: Fructose **Nitrogen** Yeast extract	U_{510nm}/g 1.304, 1.29 | U_{410nm}/g 0.497, 0.725	Subhasree et al. (2011)
Monascus purpureus AS3.531	SF and SmF/SF: 40% (w/w) of initial total reducing sugar	Dried rice (*Oryza sativa L.*)	SF ($U_{420+505nm}/g$) ~200 | SmF (U/g) ~50	Zhang et al. (2013b)
Monascus M10	SmF/synthetic media	**Nitrogen** Glutamate	Total pigment (I + E) 4313 U/g	Zhang et al. (2013a)

Monascus pilosus C1	SmF/Chapec-Dox media was used with different carbon and nitrogen source	**Carbon** Glucose	OU_{400nm}/g 0.64 | OU_{500nm}/g 0.68	Pisareva and Kujumdzieva (2010)
Serratia marcescens	SmF/synthetic media	**Carbon** Succinic acid **Nitrogen** Yeast extract	Statistical responses 0.061 OD_{540nm} 0.035	Hardjito et al. (2002)
Serratia marcescens MO-1	SmF/synthetic media	**Carbon** Mannitol **Nitrogen** RHP	Prodigiosin (mg/L) 277.74 ± 7.6 277.74 ± 7.26	Kurbanoglu et al. (2015)
Westiellopsis-ARM 48	SmF/original carbon and nitrogen sources	**Carbon** Ammonium carbonate **Nitrogen** Ammonium sulfate	Phycocyanin (μg/m) 34.60 ± 0.4 35.21 ± 0.4	Sabarinathan et al. (2010)
Phaffia rhodozyma	SmF/synthetic media-Glucose 20 g/L $(NH_4)_2SO_4$	**Nitrogen** Carbamide	Astaxanthin (mg/L) 3.44	Ni et al. (2007)
Paecilomyces sinclairii	SmF/synthetic media	**Carbon** Glucose | **Nitrogen** PP-YE	mg/L 334 | mg/L 357	Cho et al. (2002)
Chromobacterium vaccinii CV5	SmF/synthetic media	**Carbon** Dextrose | **Nitrogen** $(NH4)_2SO_4$	Violacein (μg/m) ~1900 | Violacein (μg/m) ~350	Vishnu and Palaniswamy (2017)
Eremothecium ashbyii	SmF/synthetic media	**Carbon** Glucose **Nitrogen** Yeast extract	Riboflavin (mg/L) 656.14 ± 14.69 507.10 ± 19.11	Cheng et al. (2011)

SSF (solid state fermentation), SmF (Submerged fermentation), YE (Yeast extract), BE (Beef extract), MSG (monosodium glutamate, SP (Soya peptone), SS (Soluble starch), C (carbohydrate), P (protein), PP (peptone), Ext (extract), TP (tryptone), CSP (corn step powder), BP (Bacto Peptone), FP (fish peptone), RHP (ram horn peptone), CMC (carboxymethylcellulose), I (intracellular), E (extracellular).

Table 11.2 Commonly used techniques for cell disruption and recovery of intracellular biopigments.

Microorganism	Biopigment	Cell disruption method	Extraction solution	Product recovery	Reference
Sporidiobolus salmonicolor (CBS 2136)	Total carotenoids	Frozen biomass maceration (liquid N_2) and dimethylsulfoxide	Acetone/methanol (7:3, v/v)	913 μg/L	Valduga et al. (2009)
Phaffia rhodozyma NRRL Y- 17268	Total carotenoids	Frozen biomass maceration with diatomaceous earth followed by enzymatic lysis	20% (w/v) NaCl with petroleum ether	190.35 μg/g	Michelon et al. (2012)
Rhodotorula glutinis	Total carotenoids	Bead grinder	Acetone	125 mg/L	Aksu and Eren (2007)
Rhodobacter sphaeroides	Total carotenoids	Freeze-drying followed by HCl-assisted extraction procedure	Acetone	47,907 ± 25μg/g	Gu et al. (2008)
Phaffia rhodozyma MTCC 7536	Astaxanthin	Ultrasonication	β-glucanase digestion	435.71 ± 6.55 μg/g	Hasan et al. (2016)
Blakeslea trispora ATCC 14271	Beta-carotene	Bead-beater	Ethyl acetate	59.91 mg/g dry biomass	Jing et al. (2016)
Chromobacterium violaceum	Violacein	–	Methanol	0.82 g/L	Ahmad et al. (2012)
Lachnum singerianum YM-292		–	$NH_3 \cdot H_2O$ 0.5 mol/L	8.85 g/100 g dry biomass	Ye et al. (2011)

5. Potential applications

Pigments from fungi and bacteria not only have been used to add color in the food, textile, cosmetic and pharmaceutical industries, but also have other properties such as antioxidant, anticancer and antimicrobial activities.

The largest consumer of dyes and organic pigments are the textile industries and it is expected the growth of other sectors such as printing inks, paints, and coating agents (Rao et al., 2017). Alihosseini et al. (2008) isolated a strain of *Vibrio* sp. and dyed textile materials with its bright red pigment, obtaining good results with wool, nylon, acrylics and silk. Kumar et al. (2015) reported that red pigment prodigiosin from *Vibrio* sp. and *Serratia* sp. are suitable in textile industry for dyeing materials such as wool, nylon, acrylics, polyester, silk and cotton. Pigments from *Fusarium oxysporum, Trichoderma* sp. and *Alternaria* sp. apply to cellulosic fibers. Anthraquinones and pigments from *Trichoderma* sp. are suitable for dyeing wool fibers (Rao et al., 2017).

For any biopigment to be used as a food colorant, the production of mycotoxins by the species must be evaluated. In addition to the effect on humans and animals, these toxins are interactive with other secondary metabolites to produce synergistic effects. (Palencia et al., 2010). Nowadays, some fungal food grade pigments obtained by fermentation have been approved and include *Monascus* derived pigments in, Arpink red™ (now Natural Red™) produced *by P. oxalicum*, riboflavin from the mold fungus *A. gossypii*, lycopene and β-carotene from the tropical mold *B. trispora* (Dufossé, 2018).

Monascus sp. produce polyketide pigments, and the red and the yellow ones have more commercial value as food colorants. New food applications, like the coloration of processed meats (sausage, ham), tomato ketchup and marine products like fish paste and surimi were described (Dufossé, 2006). Despite being used traditionally in the South East Asia, it is still not approved in European Union (EU) and in the United States (US), mainly due to the mycotoxin citrinin. Kang et al. (2014) produced for the first time citrinin-free *Monascus* pigments at extremely low pH and Mapari et al. (2009) identified 10 strains belonging to four species: viz. *P. purpurogenum*, *P. aculeatum*, *P. funiculosum*, and *P. pinophilum* as potential pigment producers that produced Monascus-like pigments without known mycotoxins.

Fermented rice from *M. ruber* was used for coloring flavored milk, showed appealing color and was overall acceptable (Vidyalakshmi et al., 2009). Red yeast rice was effective against obesity-related inflammation and insulin resistance (Fujimoto et al., 2012).

Medina et al. (2018) reported, for the first time, synthesis of derived anthocyanin pigments by genera of *Metschnikowia* and *Hanseniaspora* yeasts. It showed the importance of non-*Saccharomyces* yeasts studies and yeast selection to benefit wine composition and improve wine quality. Riboflavin is a yellow food colorant with applications that include dressings, sherbet, beverages, instant desserts, ice creams, tablets and other products (Dufossé, 2006).

Aside from adding colors to clothes and foods, pigments have antimicrobial activities against various species of fungi and bacteria. *Duganella violaceinigra* str. NI28 crude violacein was studied using several multidrug resistant *Staphylococcus aureus*. Choi et al. (2015a) obtained that addition of 30 μM of violacein led to a 96% loss in the initial *S. aureus* population and the minimum inhibitory concentration was 1.8 μM. Besides, a study concluded that violacein works together with most commercial antibiotics and could be used as drug in combination with other antimicrobial agents, leading to significantly lower minimum inhibitory concentrations for *S. aureus, Klebsiella pneumonia* and *P. aeruginosa* (Subramaniam et al., 2014). Pigments produced by *M. purpureus* have been reported to inhibit the growth of fungi such as *Aspergillus, Trichoderma, Mucor, Penicillium* and *Fusarium* sp. and some bacteria such as *Bacillus, Pseudomonas, Escherichia* and *Streptomyces* sp. (Ferdes et al., 2009).

Łopusiewicz et al. (2018) studied the influence of gelatin coatings modified with fungal melanin on pork lard oxidative stability, containing 0.1%; 0.5%; 1% of fungal melanin after 7-, 14- and 21-days storage in controlled conditions. Lard with altered coatings reduced oxidative rancidity. Hence, modified coatings containing fungal melanin can be used effectively for the prevention of lard oxidation. Shanuja et al. (2018) studied the production, characterization, and application of a melanin precursor from *A. nidulans* sp. strain SG 28. Exhibiting pigment production with the yield of 0.700–0.800 g/L, with the SPF analysis, antioxidant property, and the UVB protection studies there is a potential use of pigment for skin care.

Melanins are able to dissipate more than 99% of the absorbed UV light. El-Naggar and El-Ewasy (2017) showed important applications of melanin, such as in plastic films, sunscreens and varnishes. It also has application in capillary and skin coloration and in sunglasses lenses (Silva et al., 2014). In addition, it can be used in industrial processes of cosmetics, pharmaceuticals, electronics and food industry, and it has biotechnological applications in the areas of biomedicine, materials science and nanotechnology. According to Pombeiro-Sponchiado et al. (2017), this pigment has interest from the biotechnological point of view, as it can be produced on a large scale with a low cost.

Anticancer activities of biopigments are also being studied in the literature. Prodigiosin induces apoptosis of cells from leukemia and other types of cancer (Campás et al., 2003; Díaz-Ruiz et al., 2001). Melanin suppresses 48–58% of proliferation of human bladder cancer cells in the concentration range of 0.032–0.5 mg/mL (Yakovlev et al., 2018). Violacein has been tested against various cancer cell lines and demonstrated low concentrations of cytotoxicity (Choi et al., 2015b). Various carotenoids, such as β-carotene, α-carotene, lycopene, lutein, zeaxanthin, β-cryptoxanthin, fucoxanthin, canthaxanthin and astaxanthin, have been proven to have anticancer activities (Tanaka et al., 2012).

Many studies involve agricultural subproducts as a carbon source to produce pigments. Ahmad et al. (2012) obtained violacein from *Chromobacterium violaceum* successfully using sugarcane bagasse, solid pineapple waste, molasses and brown sugar. Venil et al. (2017) used brown sugar and liquid pineapple waste to

Table 11.3 Biopigments from microorganisms and their applications.

Pigment	Bacteria	Application	Reference
Prodigiosin	*Vitrio* sp., *Serratia* sp., *Streptomyces* sp., *Pseudomonas* sp., *Pseudoalteromonas* sp.	Dyeing, Candles, Anticancer	Mehta and Shah (2015)
Violacein	*Chromobacterium violaceum, Janthinobacterium lividum, Duganella violaceinigra, Collimonas* sp., *Microbulbifer* sp., *Pseudoalteromonas*	Indicator of quorum sensing molecules, biofilm production, antibacterial, dyeing, anticancer	Venil et al. (2017), Choi et al. (2015b)
	Fungi/Yeast		
Melanin	*Auricularia auricula Nadsoniella nigra (fungi) Vibrio cholerae Shewanella colwelliana Alteromonas nigrifaciens Cellulophaga tyrosinoxydans(bacteria)*	Food colorant, anticancer	Zou and Ma (2018) Yakovlev et al. (2018) Soliev et al. (2011)
Anthraquinones	*Penicillium oxalicum* var. *Armeniaca, A. glaucus (fungi) Streptomyces* sp., *Photorhabdus temperata (bacteria)*	Dyeing, food colorant, antioxidant, anticancer	Kumar et al. (2015) Fouillaud et al. (2018)
Carotenoids	*Rhodotorula* sp. *Phycomyces blakesleeanus, Blakeslea trispora*	Precursor of vitamin A, anticancer, food colorant	Tanaka et al. (2012) Dufossé (2006)
Monascins	*Monascus* sp.	Food colorant, effective against obesity-related inflammation, insulin resistance	Fujimoto et al. (2012)

formulate red and violet pigments with application on plastic materials (polyvinyl butyral, ethyl acetate and methyl ethyl ketone) (Table 11.3).

6. Conclusion and future directions

Biopigments have a remarkable impact on the way we perceive objects around us. These molecules are widely used in the pharmaceutical, food, textile and paper and ink industries. Despite of the prevalence of chemically synthesized pigments for broad applications, the demand for natural pigments (i.e., biopigments) has been increasing. Compared to synthetic pigments, biopigments, mainly produced by plants and microorganisms, are less harmful to the environment, and human health, non-toxic and are also associated with beneficial biological activities.

Researches have demonstrated that biopigments production by microorganisms is a feasible strategy to recover these valuable metabolites, however some restriction factors are still existing (e.g., low yields, production of toxins and downstream processing costs). To meet the growing market demand for biopigments, it is necessary to establish efficient extraction and purification methods, optimal cultivation conditions and also to develop improved microbial strains. Regarding production costs, the use of agro-industrial wastes (lignocellulosic biomass) as carbon and nitrogen sources is a promising strategy and, considering the recent progress made in this field, the incorporation of the bioproduction of pigments into biorefineries is possible in near future. In next 5 years, microbial production of biopigments employing agro-industrial residues under biorefinery platform underpin the demand of biopigment in food, pharmaceutical, nutraceutical and textile industries. Modern biotechnological interventions in biorefinery will play a pivotal role in large scale biopigments production and warrant the successful commercialization eventually uplifting the bioeconomy.

Acknowledgments

AKC is grateful to the CAPES-Brazil for the financial assistance through visiting professor and researcher program (Processo USP nº 15.1.1118.1.0). Students fellowships were financed in part by the Coordenação de Aperfeiçoamento de Pessoal de Nível Superior - Brasil (CAPES) - Finance Code 001″

References

Abbas, M., D'Amico, F., Morresi, L., Pinto, N., Ficcadenti, M., Natali, R., et al., 2009. Structural, electrical, electronic and optical properties of melanin films. Eur. Phys. J. 28 (3), 285–291. https://doi.org/10.1140/epje/i2008-10437-9.

Aberoumand, A., 2011. A Review article on edible pigments properties and sources as natural biocolorants in foodstuff and food industry. World J. Dairy Food Sci. 6 (22), 71–78.

Ahmad, W.A., Yusof, N.Z., Nordin, N., Zakaria, Z.A., Rezali, M.F., 2012. Production and characterization of violacein by locally isolated *Chromobacterium violaceum* grown in agricultural wastes. Appl. Biochem. Biotechnol. 167 (5), 1220–1234. https://doi.org/10.1007/s12010-012-9553-7.

Aksu, Z., Eren, A.T., 2007. Production of carotenoids by the isolated yeast of *Rhodotorula glutinis*. Biochem. Eng. J. 35 (2), 107–113. https://doi.org/10.1016/j.bej.2007.01.004.

Alihosseini, F., Ju, K.S., Lango, J., Hammock, B.D., Sun, G., 2008. Antibacterial colorants: characterization of prodiginines and their applications on textile materials. Biotechnol. Prog. 24 (3), 742–747. https://doi.org/10.1021/bp070481r.

Azizullah, A., Rehman, Z.U., Ali, I., Murad, W., Muhammad, N., Ullah, W., Häder, D.P., 2014. Chlorophyll derivatives can be an efficient weapon in the fight against dengue. Parasitol. Res. 113 (12), 4321–4326. https://doi.org/10.1007/s00436-014-4175-3.

Banerjee, D., Mondal, A., Gupta, M., Guha, A.K., Ray, L., 2014. Optimization of fermentation conditions for green pigment production from *Bacillus cereus* M 116 (MTCC 5521) and its pharmacological application. Lett. Appl. Microbiol. 58 (1), 25–30.

Bhosale, P., Gadre, R.V., 2001. Optimization of carotenoid production from hyper-producing *Rhodotorula glutinis* mutant 32 by a factorial approach. Lett. Appl. Microbiol. 33 (1), 12–16. https://doi.org/10.1046/j.1472-765X.2001.00940.x.

Blummenberg, M., 2017. Introductory chapter: melanin, a versatile guardian. In: Blumberg, M. (Ed.), Melanin. IntechOpen Limited, London, pp. 3–4.

Buey, R.M., Ledesma-Amaro, R., Balsera, M., de Pereda, J.M., Revuelta, J.L., 2015. Increased riboflavin production by manipulation of inosine 5′-monophosphate dehydrogenase in *Ashbya gossypii*. Appl. Microbiol. Biotechnol. 99 (22), 9577–9589. https://doi.org/10.1007/s00253-015-6710-2.

Buzzini, P., Innocenti, M., Turchetti, B., Libkind, D., van Broock, M., Mulinacci, N., 2007. Carotenoid profiles of yeasts belonging to the genera *Rhodotorula, Rhodosporidium, Sporobolomyces*, and *Sporidiobolus*. Can. J. Microbiol. 53 (8), 1024–1031. https://doi.org/10.1139/W07-068.

Carvalho, J.C., Oishi, B.O., Woiciechowski, A.L., Pandey, A., Babitha, S., Soccol, C.R., 2007. Effect of substrates on the production of *Monascus* biopigments by solid-state fermentation and pigment extraction using different solvents. Indian J. Biotechnol. 6, 194–199.

Carvalho, J.C., Cardoso, L.C., Ghiggi, A.L.W., Vandenberghe, L.P.S., Soccol, C.R., 2014. Microbial pigments. In: Brar, S.K., Dhillon, G.S., Fernandes, M. (Eds.), Biotransformation of Waste Biomass into High Value Biochemicals. Springer, New York, pp. 73–97.

Campas, C., Dalmau, M., Montaner, B., Barragán, M., Bellosillo, B., Colomer, D., Gil, J., 2003. Prodigiosin induces apoptosis of B and T cells from B-cell chronic lymphocytic leukemia. Leukemia 17 (4), 746.

Cecchi, T., Pezzella, A., Di Mauro, E., Cestola, S., Ginsburg, D., Luzi, M., Rigucci, A., santato, C., 2019. On the antioxidant activity of eumelanin biopigments: a quantitative comparison between free radical scavenging and redox properties. Nat. Prod. Res. 2, 1–9.

Celestino, J.D.R., de Carvalho, L.E., Lima, M.D.P., Lima, A.M., Ogusku, M.M., De Souza, J.V.B., 2014. Bioprospecting of Amazon soil fungi with the potential for pigment production. Process Biochem. 49 (4), 569–575. https://doi.org/10.1016/j.procbio.2014.01.018.

Ceresana, 2018. Market Study: Pigments, fourth ed. https://www.ceresana.com/en/market-studies/chemicals/pigments/.

Charkoudian, L.K., Fitzgerald, J.T., Khosla, C., Champlin, A., 2010. In living color: bacterial pigments as an untapped resource in the classroom and beyond. PLoS Biol. 8 (10), 1–6. https://doi.org/10.1371/journal.pbio.1000510.

Chen, M., Li, Y., Birch, D., Willows, R.D., 2012. cyanobacterium that contains chlorophyll f—a red-absorbing photopigment. FEBS Lett. 586 (19), 3249–3254. https://doi.org/10.1016/j.febslet.2012.06.045.

Cheng, X., Zhou, J., Huang, L., Li, K.T., 2011. Improved riboflavin production by Eremothecium ashbyii using glucose and yeast extract. Afr. J. Biotechnol. 10 (70), 15777–15782.

Chiba, S., Tsuyoshi, N., Fudou, R., Ojika, M., Murakami, Y., Ogoma, Y., et al., 2006. Magenta pigment produced by fungus. J. Gen. Appl. Microbiol. 52 (4), 201–207.

Cho, Y.J., Park, J.P., Hwang, H.J., Kim, S.W., Choi, J.W., Yun, J.W., 2002. Production of red pigment by submerged culture of *Paecilomyces sinclairii*. Lett. Appl. Microbiol. 35 (3), 195–202.

Choi, S.Y., Yoon, K.H., Lee, J.I., Mitchell, R.J., 2015a. Violacein: properties and production of a versatile bacterial pigment. BioMed Res. Int. 2015, 1–8. https://doi.org/10.1155/2015/465056.

Choi, S.Y., Kim, S., Lyuck, S., Kim, S.B., Mitchell, R.J., 2015b. High-level production of violacein by the newly isolated *Duganella violaceinigra* str. NI28 and its impact on *Staphylococcus aureus*. Sci. Rep. 5, 15598. https://doi.org/10.1038/srep15598.

Cordero, R.J.B., Casadevall, A., 2017. Functions of fungal melanin beyond virulence. Fungal Biol. Rev. 31 (2), 99–112. https://doi.org/10.1016/j.fbr.2016.12.003.

Delgado-Vargas, F., Jiménez, A.R., Paredes-López, O., 2000. Natural pigments: carotenoids, anthocyanins, and betalains—characteristics, biosynthesis, processing, and stability. Crit. Rev. Food Sci. Nutr. 40 (3), 173–289. https://doi.org/10.1080/10408690091189257.

Demain, A.L., Martens, E., 2017. Production of valuable compounds by molds and yeasts. J. Antibiot. 70 (4), 347–360. https://doi.org/10.1038/ja.2016.121.

Dewapriya, P., Kim, S.K., 2014. Marine microorganisms: an emerging avenue in modern nutraceuticals and functional foods. Food Res. Int. 56, 115–125. https://doi.org/10.1016/j.foodres.2013.12.022.

Díaz-Ruiz, C., Montaner, B., Pérez-Tomás, R., 2001. Prodigiosin induces cell death and morphological changes indicative of apoptosis in gastric cancer cell line HGT-1. Histol. Histopathol. 16 (2), 415–421. https://doi.org/10.14670/HH-16.415.

Dmytruk, K.V., Yatsyshyn, V.Y., Sybirna, N.O., Fedorovych, D.V., Sibirny, A.A., 2011. Metabolic engineering and classic selection of the yeast *Candida famata* (*Candida flareri*) for construction of strains with enhanced riboflavin production. Metab. Eng. 13 (1), 82–88. https://doi.org/10.1016/j.ymben.2010.10.005.

Dufossé, L., 2006. Microbial production of food grade pigments. Food Technol. Biotechnol. 44 (3), 313–321.

Dufossé, L., 2016. Current and potential natural pigments from microorganisms (bacteria, yeasts, fungi, microalgae). Handbook on Natural Pigments in Food and Beverages, pp. 337–354.

Dufossé, L., Galaup, P., Yaron, A., Arad, S.M., Blanc, P., Murthy, K.N.C., Ravishankar, G.A., 2005. Microorganisms and microalgae as sources of pigments for food use: a scientific oddity or an industrial reality? Trends Food Sci. Technol. 16 (9), 389–406.

Dufosse, L., Fouillaud, M., Caro, Y., Mapari, S.A., Sutthiwong, N., 2014. Filamentous fungi are large-scale producers of pigments and colorants for the food industry. Curr. Opin. Biotechnol. 26, 56–61.

Dufossé, L., 2018. Red colourants from filamentous fungi: Are they ready for the food industry? J. Food Compost. Anal. 69, 156–161.

El-Bialy, H.A., Abou El-Nour, S.A., 2015. Physical and chemical stress on *Serratia marcescens* and studies on prodigiosin pigment production. Ann. Microbiol. 65 (1), 59–68. https://doi.org/10.1007/s13213-014-0837-8.

El-Gamal, M.S., El-Bialy, H.A., Elsayed, M.A., Khalifa, M.A., 2017. Isolation and Characterization of melanized yeast form of *Aureobasidium pullulans* and physiological studies on the melanization process. J. Nucl. Technol. Appl. Sci. 5 (1), 57–72.

El-Naggar, N.E.A., El-Ewasy, S.M., 2017. Bioproduction, characterization, anticancer and antioxidant activities of extracellular melanin pigment produced by newly isolated microbial cell factories *Streptomyces glaucescens* NEAE-H. Sci. Rep. 7 (February) https://doi.org/10.1038/srep42129.

Embaby, A.M., Hussein, M.N., Hussein, A., 2018. Monascus orange and red pigments production by *Monascus purpureus* ATCC16436 through co-solid-state fermentation of corn cob and glycerol: an eco-friendly environmental low cost approach. PLoS One 10 (12), 13.

Ferdes, M., Ungureanu, C., Radu, N., Chirvase, A.A., 2009. Antimicrobial effect of *Monascus purpureus* red rice against some bacterial and fungal strains. N. Biotech. 25, 1.

Ferreira, V.,S., Sant'Anna, C., 2017. Impact of culture conditions on the chlorophyll content of microalgae for biotechnological applications. World J. Microbiol. Biotechnol. 33 (1), 20. https://doi.org/10.1007/s11274-016-2181-6.

Fouillaud, M., et al., 2018. Anthraquinones. In: Gutierrez-Uribe, J.A., Nollet, L.M. (Eds.), Phenolic Compounds in Food: Characterization and Analysis. CRC Press, Boca Raton, pp. 130−170.

Fujimoto, M., Tsuneyama, K., Chen, S.Y., Nishida, T., Chen, J.L., Chen, Y.C., et al., 2012. Study of the effects of monacolin k and other constituents of red yeast rice on obesity, insulin-resistance, hyperlipidemia, and nonalcoholic steatohepatitis using a mouse model of metabolic syndrome. Evid. Based Complement Altern. Med. *2012*, 1−12. https://doi.org/10.1155/2012/892697.

Gao, Q., Garcia-Pichel, F., 2011. Microbial ultraviolet sunscreens. Nat. Rev. Microbiol. 9 (11), 791−802. https://doi.org/10.1038/nrmicro2649.

General, T., Prasad, B., Kim, H.J., Vadakedath, N., Cho, M.G., 2014. *Saccharina japonica*, a potential feedstock for pigment production using submerged fermentation. Biotechnol. Bioproc. Eng. 19 (4), 711−719.

Gu, Z., Deming, C., Yongbin, H., Zhigang, C., Feirong, G., 2008. Optimization of carotenoids extraction from Rhodobacter sphaeroides. LWT-Food Sci. Technol. 41 (6), 1082−1088.

Gunasekaran, S., Poorniammal, R., 2008. Optimization of fermentation conditions for red pigment production from *Penicillium* sp. under submerged cultivation. Afr. J. Biotechnol. 7 (12).

Hardjito, L., Huq, A., Colwell, R.R., 2002. The influence of environmental conditions on the production of pigment by *Serratia marcescens*. Biotechnol. Bioproc. Eng. 7 (2), 100−104.

Harper, M., Lee, C.J., 2012. Genome-wide analysis of mutagenesis bias and context sensitivity of N-methyl-N′-nitro-N-nitrosoguanidine (NTG). Mutat. Res. Fund Mol. Mech. Mutagen 731 (1−2), 64−67. https://doi.org/10.1016/j.mrfmmm.2011.10.011.

Hasan, M., Azhar, M., Nangia, H., Bhatt, P.C., Panda, B.P., 2016. Influence of high-pressure homogenization, ultrasonication, and supercritical fluid on free astaxanthin extraction from β-glucanase-treated *Phaffia rhodozyma* cells. Prep. Biochem. Biotechnol. 46 (2), 116−122. https://doi.org/10.1080/10826068.2014.995807.

Heer, K., Sharma, S., 2017. Microbial pigments as a natural color: a review. Int. J. Pharm. Sci. Res. 8 (5), 1913−1922. https://doi.org/10.13040/IJPSR.0975-8232.8(5).1913-22.

Hernandez-Almanza, A., Montanez, J.C., Aguilar-Gonzalez, M.A., Martmez-Avila, C., Rodriguez-Herrera, R., Aguilar, C.N., 2014. *Rhodotorula glutinis* as source of pigments and metabolites for food industry. Food Biosci. 5, 64−72. https://doi.org/10.1016/j.fbio.2013.11.007.

Jalmi, P., Bodke, P., Wahidullah, S., Raghukumar, S., 2012. The fungus *Gliocephalotrichum simplex* as a source of abundant, extracellular melanin for biotechnological applications. World J. Microbiol. Biotechnol. 28 (2), 505−512. https://doi.org/10.1007/s11274-011-0841-0.

Jeong, B.Y., Wittmann, C., Kato, T., Park, E.Y., 2015. Comparative metabolic flux analysis of an *Ashbya gossypii* wild type strain and a high riboflavin-producing mutant strain. J. Biosci. Bioeng. 119 (1), 101−106. https://doi.org/10.1016/j.jbiosc.2014.06.014.

Jia, X.Q., Xu, Z.N., Zhou, L.P., Sung, C.K., 2010. Elimination of the mycotoxin citrinin production in the industrial important strain *Monascus purpureus* SM001. Metab. Eng. 12 (1), 1–7. https://doi.org/10.1016/j.ymben.2009.08.003.

Jiang, H., Liu, N.N., Liu, G.L., Chi, Z., Wang, J.M., Zhang, L.L., Chi, Z.M., 2016. Melanin production by a yeast strain XJ5-1 of *Aureobasidium melanogenum* isolated from the Taklimakan desert and its role in the yeast survival in stress environments. Extremophiles 20 (4), 567–577. https://doi.org/10.1007/s00792-016-0843-9.

Jin, J., Wang, Y., Yao, M., Gu, X., Li, B., Liu, H., et al., 2018. Astaxanthin overproduction in yeast by strain engineering and new gene target uncovering. Biotechnol. Biofuels 11 (1), 1–15. https://doi.org/10.1186/s13068-018-1227-4.

Jing, K., He, S., Chen, T., Lu, Y., Ng, I.S., 2016. Enhancing beta-carotene biosynthesis and gene transcriptional regulation in *Blakeslea trispora* with sodium acetate. Biochem. Eng. J. 114, 10–17. https://doi.org/10.1016/j.bej.2016.06.015.

Kang, B., Zhang, X., Wu, Z., Wang, Z., Park, S., 2014. Production of citrinin-free *Monascus* pigments by submerged culture at low pH. Enzym. Microb. Technol. 55, 50–57. https://doi.org/10.1016/j.enzmictec.2013.12.007.

Krairak, S., Yamamura, K., Irie, R., Nakajima, M., Shimizu, H., Chim-Anage, P., et al., 2000. Maximizing yellow pigment production in fed-batch culture of *Monascus* sp. J. Biosci. Bioeng. 90 (4), 363–367.

Kumar, A., Vishwakarma, H.S., Singh, J., Dwivedi, S., Kumar, M., 2015. Microbial pigments: production and their applications in various industries. Int. J. Pharm. Chem. Biol. Sci. 5 (1), 203–212.

Kurbanoglu, E.B., Ozdal, M., Ozdal, O.G., Algur, O.F., 2015. Enhanced production of prodigiosin by *Serratia marcescens* MO-1 using ram horn peptone. Braz. J. Microbiol. 46 (2), 631–637.

Langfelder, K., Streibel, M., Jahn, B., Haase, G., Brakhage, A.A., 2003. Biosynthesis of fungal melanins and their importance for human pathogenic fungi. Fungal Genet. Biol. 38 (2), 143–158. https://doi.org/10.1016/S1087-1845(02)00526-1.

Ledesma-Amaro, R., Kerkhoven, E.J., Revuelta, J.L., Nielsen, J., 2014. Genome scale metabolic modeling of the riboflavin overproducer *Ashbya gossypii*. Biotechnol. Bioeng. 111 (6), 1191–1199. https://doi.org/10.1002/bit.25167.

Lee, S.S., Lee, J.H., Lee, I., 2013. Strain improvement by overexpression of the laeA gene in *Monascus pilosus* for the production of monascus-fermented rice. J. Microbiol. Biotechnol. 23 (7), 959–965. https://doi.org/10.4014/jmb.1303.03026.

Li, Y., Li, X., Lee, U., Jung, S.K., Hong, D.C., Byeng, W.S., 2006. A new radical scavenging anthracene glycoside, asperflavin ribofuranoside, and polyketides from a marine isolate of the fungus *Microsporum*. Chem. Pharm. Bull. 54 (6), 882–883. https://doi.org/10.1248/cpb.54.882.

Łopusiewicz, Ł., Jedra, F., Bartkowiak, A., 2018. The application of melanin modified gelatin coatings for packaging and the oxidative stability of pork lard. World Sci. News 101, 108–119.

Liu, X., Wang, Y., Sun, S., Zhu, C., Xu, W., Park, Y., Zhou, H., 2013. Mutant breeding of *Serratia marcescens* strain for enhancing prodigiosin production and application to textiles. Prep. Biochem. Biotechnol. 43 (3), 271–284. https://doi.org/10.1080/10826068.2012.721850.

Manimala, M.R.A., Murugesan, R., 2014. In vitro antioxidant and antimicrobial activity of carotenoid pigment extracted from *Sporobolomyces* sp. isolated from natural source. J. Appl. Nat. Sci. 6 (2), 649–653.

Mantzouridou, F., Tsimidou, M.Z., 2008. Lycopene formation in *Blakeslea trispora*. Chemical aspects of a bioprocess. Trends Food Sci. Technol. 19 (7), 363–371. https://doi.org/10.1016/j.tifs.2008.01.003.

Mapari, S.A.S., Nielsen, K.F., Larsen, T.O., Frisvad, J.C., Meyer, A.S., Thrane, U., 2005. Exploring fungal biodiversity for the production of water-soluble pigments as potential natural food colorants. Curr. Opin. Biotechnol. 16 (2), 231–238. https://doi.org/10.1016/j.copbio.2005.03.004.

Mapari, S.A., Meyer, A.S., Thrane, U., Frisvad, J.C., 2009. Identification of potentially safe promising fungal cell factories for the production of polyketide natural food colorants using chemotaxonomic rationale. Microb. Cell Factories 8 (1), 24. https://doi.org/10.1186/1475-2859-8-24.

Mapari, S.A., Thrane, U., Meyer, A.S., 2010. Fungal polyketide azaphilone pigments as future natural food colorants? Trends Biotechnol. 28 (6), 300–307.

Martins, N., Roriz, C.L., Morales, P., Barros, L., Ferreira, I.C., 2017. Coloring attributes of betalains: a key emphasis on stability and future applications. Food Funct. 8 (4), 1357–1372. https://doi.org/10.1039/c7fo00144d.

Masi, C., Duraipandi, V., Yuvaraj, D., Vivek, P., Parthasarathy, N., 2014. Production and extraction of bacterial pigments from novel strains and their applications. R*es. J. Pharmaceut. Biol. Chem. Sci.* 5 (6), 584–593.

Mata-Gómez, L., Montañez, J., Méndez-Zavala, A., Aguilar, C., 2014. Biotechnological production of carotenoids by yeasts: an overview. Microb. Cell Factories 13 (12), 1–11. https://doi.org/10.1186/1475-2859-13-12.

Medina, K., Boido, E., Dellacassa, E., Carrau, F., 2018. Effects of non-*Saccharomyces* yeasts on color, anthocyanin, and anthocyanin-derived pigments of Tannat grapes during fermentation. Am. J. Enol. Vitic. 69 (2), 148–156. https://doi.org/10.5344/ajev.2017.17055.

Mehta, M., Shah, G., 2015. Extraction of pigment from *Serratia marcescens* and its application in candle industry. Adv. Appl. Res. 7 (2), 144–146. https://doi.org/10.5958/2349-2104.2015.00027.3.

Michelon, M., de Matos de Borba, T., da Silva Rafael, R., Burkert, C.A.V., de Medeiros Burkert, J.F., 2012. Extraction of carotenoids from *Phaffia rhodozyma*: a comparison between different techniques of cell disruption. Food Sci. Biotechnol. 21 (1), 1–8. https://doi.org/10.1007/s10068-012-0001-9.

Najafi, N., Ahmadi, A.-R., Hosseini, R., Golkhoo, S., 2011. Gamma irradiation as a useful tool for the isolation of astaxanthin-overproducing mutant strains of *Phaffia rhodozyma*. Can. J. Microbiol. 57 (9), 730–734. https://doi.org/10.1139/w11-060.

National Center for Biotechnology Information, 2018a. PubChem Database. Chlorophyll, CID=12085802. https://pubchem.ncbi.nlm.nih.gov/compound/12085802. (Accessed 15 April 2019).

National Center for Biotechnology Information, 2018c. PubChem Database. Lycopene, CID=446925. https://pubchem.ncbi.nlm.nih.gov/compound/446925. (Accessed 15 April 2019).

National Center for Biotechnology Information, 2018d. PubChem Database. beta-Carotene, CID=5280489. https://pubchem.ncbi.nlm.nih.gov/compound/5280489. (Accessed 15 April 2019).

National Center for Biotechnology Information, 2018e. PubChem Database. rivoflavin, CID=493570. https://pubchem.ncbi.nlm.nih.gov/compound/5280489. (Accessed 15 April 2019).

National Center for Biotechnology Information, 2018f. PubChem Database. Anthraquinone, CID=6780. https://pubchem.ncbi.nlm.nih.gov/compound/6780. (Accessed 15 April 2019).

Ni, H., Chen, Q.H., Ruan, H., Yang, Y.F., Li, L.J., Wu, G.B., et al., 2007. Studies on optimization of nitrogen sources for astaxanthin production by *Phaffia rhodozyma*. J. Zhejiang Univ. Sci. B 8 (5), 365–370.

Nigam, P.S., Luke, J.S., 2016. Food additives: production of microbial pigments and their antioxidant properties. Curr. Opin. Food Sci. 7, 93–100. https://doi.org/10.1016/j.cofs.2016.02.004.

Nimnoi, P., Lumyong, S., 2011. Improving solid-state fermentation of *Monascus purpureus* on agricultural products for pigment production. Food Bioprocess Technol. 4 (8), 1384–1390. https://doi.org/10.1007/s11947-009-0233-8.

Pagano, M.C., Dhar, P.P., 2015. Fungal pigments: an overview. Fungal biomolecules: sources, applications and recent developments. In: Gupta, V.K., Mach, R.L., Sreenivasaprasad, S. (Eds.), Fungal Biomolecules: Sources, Applications and Recent Developments. Wiley, New Jersey, pp. 173–182.

Panesar, R., 2014. Bioutilization of kinnow waste for the production of biopigments using submerged fermentation. Int. J. Food Sci. Nutr. 3, 9–13.

Panesar, R., Kaur, S., Panesar, P.S., 2015. Production of microbial pigments utilizing agro-industrial waste: a review. Curr. Opin. Food Sci. 1 (1), 70–76. https://doi.org/10.1016/j.cofs.2014.12.002.

Panis, G., Carreon, J.R., 2016. Commercial astaxanthin production derived by green alga *Haematococcus pluvialis*: a microalgae process model and a techno-economic assessment all through production line. Algal Res. 18, 175–190. https://doi.org/10.1016/j.algal.2016.06.007.

Park, P.K., Kim, E.Y., Chu, K.H., 2007. Chemical disruption of yeast cells for the isolation of carotenoid pigments. Separ. Purif. Technol. 53 (2), 148–152. https://doi.org/10.1016/j.seppur.2006.06.026.

Palencia, E.R., Hinton, D.M., Bacon, C.W., 2010. The black Aspergillus species of maize and peanuts and their potential for mycotoxin production. Toxins 2 (4), 399–416.

Pisareva, E.I., Kujumdzieva, A.V., 2010. Influence of carbon and nitrogen sources on growth and pigment production by *Monascus pilosus* C1 strain. Biotechnol. Biotechnol. Equip. 24 (Suppl. 1), 501–506.

Pombeiro-Sponchiado, S.R., Sousa, G.S., Andrade, J.C., Lisboa, H.F., Gonçalves, R.C., 2017. Production of melanin pigment by fungi and its biotechnological applications. In: Blumberg, M. (Ed.), Melanin. InTech, London, pp. 47–76.

Qiang, W., Ling-ran, F., Luo, W., Han-guang, L., Lin, W., Ya, Z., Xiao-bin, Y., 2014. Mutation breeding of lycopene-producing strain *Blakeslea trispora* by a novel atmospheric and room temperature plasma (ARTP). Appl. Biochem. Biotechnol. 174 (1), 452–460. https://doi.org/10.1007/s12010-014-0998-8.

Rao, M.P.N., Xiao, M., Li, W.J., 2017. Fungal and bacterial pigments: secondary metabolites with wide applications. Front. Microbiol. 8 (June), 1–13. https://doi.org/10.3389/fmicb.2017.01113.

Rastogi, R.P., Richa, Kumar, A., Tyagi, M.B., Sinha, R.P., 2010. Molecular mechanisms of ultraviolet radiation-induced DNA damage and repair. J. Nucleic Acids. https://doi.org/10.4061/2010/592980.

Reyes, L.H., Gomez, J.M., Kao, K.C., 2014. Improving carotenoids production in yeast via adaptive laboratory evolution. Metab. Eng. 21, 26–33. https://doi.org/10.1016/j.ymben.2013.11.002.

Ribeiro, B.D., Barreto, D.W., Coelho, M.A.Z., 2011. Technological aspects of β-carotene production. Food Bioprocess Technol. 4 (5), 693–701. https://doi.org/10.1007/s11947-011-0545-3.

Rosas, Á.L., Nosanchuk, J.D., Gómez, B.L., Edens, W.A., Henson, J.M., Casadevall, A., 2000. Isolation and serological analyses of fungal melanins. J. Immunol. Methods 244 (1–2), 69–80. https://doi.org/10.1016/S0022-1759(00)00255-6.

Sabarinathan, K.G., Muthukrishnan, G., Gopalswamy, G., 2010. Effect of carbon, nitrogen source and other conditions for enhancing phycocyanin production. Int. J. Agric. Sci. 6 (1), 315–318.

Said, F.M., Brooks, J., Chisti, Y., 2014. Optimal C:N ratio for the production of red pigments by *Monascus ruber.* World J. Microbiol. Biotechnol. 30 (9), 2471–2479.

Saini, D.K., chakdar, H., Pabbi, S., Shukla, P., 2019. Enhancing production of microalgal biopigments through metabolic and genetic engineering. Crit. Rev. Food Sci. Nutr. 1, 1–15.

Santos De Souza, A.C., Ferreira, C.V., Jucá, M.B., Aoyama, H., Martins Cavagis, A.D., Peppelenbosch, M.P., 2005. Riboflavina: Uma vitamina multifuncional. Quim. Nova 28 (5), 887–891. https://doi.org/10.1590/S0100-40422005000500028.

Schwechheimer, S.K., Becker, J., Wittmann, C., 2018. Towards better understanding of industrial cell factories: novel approaches for13C metabolic flux analysis in complex nutrient environments. Curr. Opin. Biotechnol. 54, 128–137. https://doi.org/10.1016/j.copbio.2018.07.001.

Sehrawat, R., Panesar, P.S., Panesar, R., Kumar, A., 2017. Biopigment produced by *Monascus purpureus* MTCC 369 in submerged and solid state fermentation: a comparative study. Pigment Resin Technol. 46 (6), 425–432.

Shanuja, S.K., Iswarya, S., Sridevi, J., Gnanamani, A., 2018. Exploring the UVB-protective efficacy of melanin precursor extracted from marine imperfect fungus: featuring characterization and application studies under in vitro conditions. Int. Microbiol. 1–13. https://doi.org/10.1007/s10123-018-0005-2.

Silva, M.P., Uhle, E.S.B., Graeff, C.F.O., 2014. Processo de produção de melanina sintética e produto obtido. Instituto Nacional de Propriedade Industrial (INPI), Brazil.

Soliev, A.B., Hosokawa, K., Enomoto, K., 2011. Bioactive pigments from marine bacteria: applications and physiological roles. Evid. Based Complement Altern. Med. 2011 https://doi.org/10.1155/2011/670349.

Sopandi, T., Wardah, A., Surtiningsih, T., Suwandi, A., Smith, J.J., 2013. Utilization and optimization of a waste stream cellulose culture medium for pigment production by *Penicillium* spp. J. Appl. Microbiol. 114 (3), 733–745.

Subhasree, R.S., Babu, P.D., Vidyalakshmi, R., Mohan, V.C., 2011. Effect of carbon and nitrogen sources on stimulation of pigment production by *Monascus purpureus* on jackfruit seeds. Int. J. Microbiol. Res. 2 (2), 184–187.

Subramaniam, S., Ravi, V., Sivasubramanian, A., 2014. Synergistic antimicrobial profiling of violacein with commercial antibiotics against pathogenic micro-organisms. Pharmaceut. Biol. 52 (1), 86–90. https://doi.org/10.3109/13880209.2013.815634.

Sun, S.Q., Wang, Y.J., Xu, W., Zhu, C.J., Liu, X.X., 2015. Optimizing ultrasound-assisted extraction of prodigiosin by response surface methodology. Prep. Biochem. Biotechnol. 45 (2), 101–108. https://doi.org/10.1080/10826068.2013.877029.

Tanaka, T., Shnimizu, M., Moriwaki, H., 2012. Cancer chemoprevention by carotenoids. Molecules 17 (3), 3202–3242. https://doi.org/10.3390/molecules17033202.

Taniguchi, H., Henke, N.A., Heider, S.A.E., Wendisch, V.F., 2017. Overexpression of the primary sigma factor gene sigA improved carotenoid production by *Corynebacterium glutamicum*: application to production of β-carotene and the non-native linear C50 carotenoid bisanhydrobacterioruberin. Metab. Eng. Commun. 4 (December), 1–11. https://doi.org/10.1016/j.meteno.2017.01.001.

Terán-Hilares, R., de Souza, R.A., Marcelino, P.F., da Silva, S.S., Dragone, G., Mussatto, S.I., Santos, J.C., 2018. Sugarcane bagasse hydrolysate as a potential feedstock for red pigment production by Monascus ruber. Food Chem. 245, 786–791.

Tuli, H.S., Chaudhary, P., Beniwal, V., Sharma, A.K., 2015. Microbial pigments as natural color sources: current trends and future perspectives. J. Food Sci. Technol. 52 (8), 4669–4678.

Uenojo, M., Maróstica, M.R., Pastore, G.M., 2007. Carotenoides: propriedades, aplicações e biotransformação para formação de compostos de aroma. Quim. Nova 30 (3), 616–622. https://doi.org/10.1590/S0100-40422007000300022.

Valduga, E., Valério, A., Treichel, H., Júnior, A.F., Di, M.L., 2009. Optimization of the production of total carotenoids by Sporidiobolus salmonicolor (CBS 2636) using response surface technique. Food Bioprocess Technol. 2 (4), 415–421. https://doi.org/10.1007/s11947-008-0066-x.

Varmira, K., Habibi, A., Moradi, S., Bahramian, E., 2018. Experimental evaluation of Airlift photobioreactor for carotenoid pigments production by *Rhodotorula rubra*. Rom. Biotech. Lett. 23 (4), 13843.

Varzakakou, M., Roukas, T., 2010. Identification of carotenoids produced from cheese whey by *Blakeslea trispora* in submerged fermentation. Prep. Biochem. Biotechnol. 40 (1), 76–82. https://doi.org/10.1080/10826060903400666.

Venil, C.K., Zakaria, Z.A., Ahmad, W.A., 2013. Bacterial pigments and their applications. Process Biochem. 48 (7), 1065–1079. https://doi.org/10.1016/j.procbio.2013.06.006.

Venil, C.K., Aruldass, C.A., Dufossé, L., Zakaria, Z.A., Ahmad, W.A., 2014. Current perspective on bacterial pigments: emerging sustainable compounds with coloring and biological properties for the industry-an incisive evaluation. RSC Adv. 4 (74), 39523–39529. https://doi.org/10.1039/c4ra06162d.

Venil, C.K., Wahidin, M.A.B., Aruldass, C.A., Ahmad, W.A., 2017. Production of bacterial pigments in low cost medium and formulation of biodegradable ink. Indian J. Exp. Biol. 55, 441–447.

Vidyalakshmi, R., Paranthaman, R., Murugesh, S., Singaravadivel, K., 2009. Microbial bioconversion of rice broken to food grade pigments. Glob. J. Biotechnol. Biochem. 4 (2), 84–87.

Vishnu, T., Palaniswamy, M., 2017. Impact of various fermentation conditions on the production of violacein by the novel isolate *Chromobacterium vaccinii* cv5. Int. J. Pharm. Biol. Sci. 8 (3), 514–522.

Wang, Q., Liu, D., Yang, Q., Wang, P., 2017. Enhancing carotenoid production in Rhodotorula mucilaginosa KC8 by combining mutation and metabolic engineering. Ann. Microbiol. 67 (6), 425–431.

Xu, X., Tian, L., Xu, J., Xie, C., Jiang, L., Huang, H., 2018. Analysis and expression of the carotenoid biosynthesis genes from Deinococcus wulumuqiensis R12 in engineered *Escherichia coli*. Amb. Express 8 (1), 1–13. https://doi.org/10.1186/s13568-018-0624-1.

Yakovlev, P., Garmanchuk, L., Falalyeyeva, T., Beregova, T., Ostapchenko, L., Nehelia, A., 2018. Cytostatic effect of polyphenol compound melanin in combination with Cisplatin on bladder cancer cell line T24/83. Eur. Urol. Suppl. 17 (10), e2553–e2554. https://doi.org/10.1016/s1569-9056(18)33031-8.

Ye, M., Wang, Y., Qian, M., Chen, X., Hu, X., 2011. Preparation and properties of the melanin from *Lachnum singerianum*. Int. J. Basic Appl. Sci. 11 (30), 51–58.

Yolmeh, M., Khomeiri, M., Ghorbani, M., Ghaemi, E., Ramezanpour, S.S., 2017. High efficiency pigment production from *Micrococcus roseus* (PTCC 1411) under ultraviolet irradiation. Biocatal. Agric. Biotechnol. 9, 156–161. https://doi.org/10.1016/j.bcab.2016.12.010.

Zhang, L., Li, Z., Dai, B., Zhang, W., Yuan, Y., 2013a. Effect of submerged and solid-state fermentation on pigment and citrinin production by *Monascus purpureus*. Acta Biol. Hung. 64 (3), 385–394.

Zhang, X.W., Wang, J.H., Chen, M.H., Wang, C.L., 2013b. Effect of nitrogen sources on production and photostability of *Monascus* pigments in liquid fermentation. IERI Procedia.

Zhang, C., Shen, H., Zhang, X., Yu, X., Wang, H., Xiao, S., et al., 2016. Combined mutagenesis of Rhodosporidium toruloides for improved production of carotenoids and lipids. Biotechnol. Lett. 38 (10), 1733–1738. https://doi.org/10.1007/s10529-016-2148-6.

Zhao, B., Ren, B., Guo, R., Zhang, W., Ma, S., Yao, Y., et al., 2017. Supplementation of lycopene attenuates oxidative stress induced neuroinflammation and cognitive impairment via Nrf2/Nf-κB transcriptional pathway. Food Chem. Toxicol. 109, 505–516. https://doi.org/10.1016/j.fct.2017.09.050.

Zheng, W., Campbell, B.S., McDougall, B.M., Seviour, R.J., 2008. Effects of melanin on the accumulation of exopolysaccharides by *Aureobasidium pullulans* grown on nitrate. Bioresour. Technol. 99 (16), 7480–7486. https://doi.org/10.1016/j.biortech.2008.02.016.

Zou, Y., Ma, K., 2018. Screening of *Auricularia auricula* strains for strong production ability of melanin pigments. Food Sci. Technol. 38 (1), 41–44. https://doi.org/10.1590/1678-457x.00117.

Zoz, L., Carvalho, J.C., Soccol, V.T., Casagrande, T.C., Cardoso, L., 2015. Torularhodin and torulene: bioproduction, properties and prospective applications in food and cosmetics – a review. Braz. Arch. Biol. Technol. 58 (2), 278–288. https://doi.org/10.1590/S1516-8913201400152.

CHAPTER

Biotechnological production and applications of ribonucleases

12

Rakesh Kumar, Shamsher Singh Kanwar

Department of Biotechnology, Himachal Pradesh University, Summer Hill, Shimla, India

Chapter outline

Biotechnological Production of Bioactive Compounds. https://doi.org/10.1016/B978-0-444-64323-0.00012-6

1. Introduction

All microorganisms contain different classes of RNases that are involved in RNA-degradation. RNA degradation is a crucial step in the cellular and biological processes of most cell types. Cellular RNA molecules must endure the specific cleavage reactions in order to accomplish their mature functional forms by ribonucleases (RNases). Microbial RNases are hydrolytic enzymes that help in degradation of phosphodiester bond of cellular RNA. Studies on RNases began when Walter Jones discovered a thermostable enzyme in the extracts of bovine pancreas that hydrolyzed transforming yeast RNA into soluble acid product without discharge of inorganic phosphate or pyrimidine and purine bases (Jones, 1920; Rajashree et al., 2002; Kumar and Kanwar, 2018). In 1924, nucleic acid degrading enzyme was found in commercial Takadiastase sourced from *Aspergillus oryzae* (Noguchi, 1924). Both RNase and DNase activities were observed in 36 strain of hemolytic group of Streptococci. RNase activity was also observed in actinomycetes cells and the presence of RNA-degrading enzyme(s) was also seen in growth media of many other fungi (Otani, 1935). Ribonuclease T1 was the first microbial RNase isolated from commercial enzyme mixture from *A. oryaze* (Sato and Egami, 1957) and other fungal species (Kumar and Kanwar, 2018). Their substrate specificity revealed that presence of two cyclic RNases from Takadiastase, which split specifically phosphodiester bond of 3′-gaunylic acid in RNA and smaller component have specificity for phosphodiester bonds of 3′-adenylic acid (Sato and Egami, 1957). The primary structure of RNase T1 was determined and successfully characterized by X-ray crystallography, NMR spectroscopy and chemical modification. A guanylic acid-specific RNase was referred as RNase T1 and an adenylic acid specific RNase was named as RNase T2 (Sato and Egami, 1957). RNases with varying specificities have been reported from microbes, plants and animals (Rajashree et al., 2002). Microbial RNase has ultimate capability to degrade cellular RNA into oligonucleotides and mononucleotides (Fig. 12.1).

The RNases are broadly designated as exoribonucleases and endoribonucleases, and these nucleases are further divided into several sub-classes. The major type of exoribonucleases include PNPase, RNase II and oligo-ribonuclease for RNA degradation, while others such as RNase PH, RNase BN, RNase D and RNase T are mainly responsible for the 3′ end maturation of RNA. On other hand, the major microbial endoribonucleases have been classified into RNase I, RNase III, RNase P, RNase E, RNase HI, HII and HIII (Nicholson, 1999; Aravind and Koonin, 2001; Condon and Putzer, 2002). A RNase uses different types of RNAs as substrates. Single stranded RNA, DNA/RNA hybrid, mRNA, rRNA and tRNA are the substrates commonly used by different type of ribonucleases. Moreover, each ribonuclease has its own specific substrate (Kumari et al., 2016; Kumar and Kanwar, 2017). Interestingly, quite often, the microbial RNases are highly cytotoxic toward *in vitro* cultured malignant cells. The RNases also often trigger apoptotic response; due to this property, they are thus also considered as alternative chemotherapeutic drugs (Ilinskaya et al., 2018).

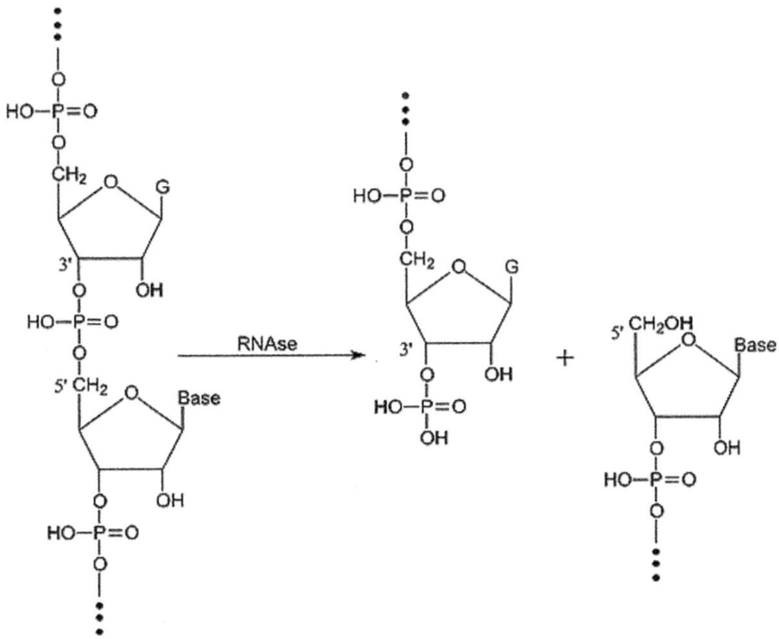

FIG. 12.1 A general mode of action of RNase.

2. Major sources of RNases

RNases are nucleases produced by various organisms, which include fungi, bacteria, plants, sub-mammalian vertebrates and the mammalian organs such as liver, kidney, brain, placenta, pancreas, milk and testes (Nicholson, 1999; Guan et al., 2007). RNases possess potent biological therapeutic activities like antitumor, antiproliferative, antiviral, immunosuppressive, antifungal, antiangiogenic and induction of apoptosis. These multitudes of diverse activities have drawn the attention of many scientists to exploit RNases as therapeutic tools to efficiently deal with malignancies (Cobaleda and Sanchez-Garcia, 2000; Eller et al., 2015). *Escherichia coli* has been used as a model organism for most of RNA decay studies. RNA decay has not been well characterized in Gram-positive bacteria with the exception of some preliminary studies conducted in *Bacillus subtilis* (Even et al., 2005; Mader et al., 2005), *Streptococcus pyogenes* (Barnett et al., 2007), *Streptococcus pneumoniae* (Domingues et al., 2009) and *Staphylococcus aureus* (Huntzinger et al., 2005; Boisset et al., 2007). Generally, RNases have been isolated from different archaea, bovine pancreas, plants, humans, bacteria like *E. coli*, *Bacillus amyloliquifaciens*, *Pseudomonas patiuda*, fungus such as *Aspergillus niger*, *A. oryzae*, and yeast like *Saccharomyces cerivisiae*, *S. pombe* and *Flammulina velutipes*.

2.1 Bacterial and archaeal ribonucleases

Bacterial RNases represent large group of nucleases having various functional mechanisms containing, wide spectrum of biological activities, particularly during the process of maturation and turnover of RNA. The RNA decay is well-organized process and it has half-life of less than a few minutes that allows the cell to rapidly degrade unwanted proteins in bacteria. Extracellular RNase production from *B. subtilis* and its stimulation by actinomycin D (antitumor antibiotic) have already been demonstrated (Coleman and Elliott, 1965). *E. coli* and *B. subtilis* are the two-model organisms, which have been extensively used to study the bacterial mRNA degradation. It has been revealed that surprising diversity in enzymes and pathways exists between Gram-negative and Gram-positive bacteria. RNases from *Bacillus intermedius* and *Bacillus amyloliquefaciens* often referred as Binase and Barnase, respectively have come in to existence due to their cytotoxic effect against different cancer cell lines. First bacterial RNase possessing antitumor activity was reported in year 1964 from *Serratia marcescens* (Fang and Ng, 2011; Ilinskaya et al., 2018). *Streptomyces* species, particularly *Streptomyces aureofaciens* is a promising bacterial RNase producer. It produces three small extracellular RNases namely Sa, Sa2 and Sa3. RNases from marine bacteria, *Alteromonas* and *Pseudomonas* possess properties such as salt tolerance, thermostability, barophilicity and cold adaptivity (Ghosh et al., 2005).

RNase production from archaeal species has mainly been aimed to study the characteristic, structure and evolutionary role of the RNases. RNase J from *Methanocaldococcus jannaschii* (Levy et al., 2011) and RNase R from *Haloferax volcanii* (Portnoy and Schuster, 2006) are considered as examples of archaeal RNases. Extremophile members of archaea can provide novel RNases with various characteristics contributing promising potential for their application in different fields.

2.2 Fungal ribonucleases

Fungal ribonucleases are divided into three broad families: (i) RNase T1 family *e.g.*, RNase T, (*A. oryzae*), RNase U2 (*Ustilago sphaerogena*), Bamase (*B. amyloliquefaciens* strain H); (ii) RNase T2 family *e.g.*, RNase T2 (*Aspergillus* oryzae), RNase Rh (*Rhizopus niveus*) and (iii) Ribotoxins, that specifically cleave a single phosphodiester bond from the eukaryotic 28S rRNA *e.g.*, α-Sarcin (*Aspergillus giganteus*) and Restrictocin (*Aspergillus restrictus*) (Lacadena et al., 2007; Kumar and Kanwar, 2017, 2018). *Aspergillus* is one of the most prominent genera than any other member of fungi in terms of both production and medical applications of RNases. One of the earlier studies showed that α-sarcin RNase produced from *A. giganteus* MDH 18894 showed anti-tumor activity toward different animal tumors. The molecular structure and characterization of RNases from different fungal producers were also investigated including those containing guanyl-specific RNase from *Penicillium chrysogenum* (Yakovlev et al., 1980) and *Penicillium brevicompactum* (Shlyapnikov et al., 1984), guanyl-specific RNase FI1 from *Fusarium moniliforme* (Yoshida et al., 1980), base-specific RNase Lu from *A. niger* (Horitsu et al., 1982), RNase C2 from *A. clavatus* (Bezborodova et al., 1983), minor RNase from *Aspergillus saito* (Flogel et al., 1988) and extracellular poly(A) specific RNase from *A. niger* (Gundampati et al., 2011; Kumar and Kanwar, 2017, 2018).

2.3 Plant ribonucleases

Many higher order plants possess RNases. These RNases are present as soluble enzymes in most plant tissues (Farkas, 1982). Although many plant RNases participate in classical activities such as processing and turnover of cellular RNA, but others are located either outside the cell or within the vacuole and probably do not take part in normal RNA metabolism. In a few instances, these "nonclassical" RNases were first described as effectors of some biological processes and later they were considered as RNases. In other cases, RNase activity or RNase-related sequences were discovered that accumulated in the cells in response to physiological or developmental signals or as a result of exterior signals, such as nutrient deficiency. These RNases might attack the pathogens. Most of the reports on RNases of plant origin facilitated the categorization of plant RNases into four main groups, which include RNase I, RNase II, nuclease I and exonuclease I, with representatives present in wheat, barley, rye, zinnia, lentil and *Arabidopsis thaliana* (Farkas, 1982; Wilson, 1982; Kumar and Kanwar, 2017, 2018). The Arabidopsis RNase E was found to be located in the chloroplast and it exhibited endoribonuclease activity present in the soluble fraction of the chloroplast as a high molecular weight complex (Schein et al., 2008). In plants, T2 RNases are represented by S-like RNases and these were the first secretory RNases. S-RNases were also studied as their functions are related to the mechanisms of gametophytic self-incompatibility. S-like RNases are not involved in generation of self-incompatibility and they were found both in self-compatible and self-incompatible plant species. Gametophytic self-incompatibility in which S-RNases are involved is a bright example of cytotoxic action of ribonucleases. S-like RNases belong to nonspecific endoribonucleases produced in pistils and secreted into the extracellular matrix. Pollen tube elongation is halted due to RNase activity because S-RNase can enter inside the cytoplasm of growing pollen tubes causing inhibition of process of translation (Sangaev et al., 2011). A previous study showed that the ribonucleolytic activity of pathogenesis-related (PR) proteins complex was important in plant to confer defense response against pathogens (Fan et al., 2015). The antifungal activity of PR-ribonucleases was associated both with their direct cytotoxic impact on pathogen cells and with their possible involvement in the induction of plant cell apoptosis and development of hypersensitive reactions (Filipenko et al., 2013; Kumar and Kanwar, 2017, 2018).

2.4 Animal ribonucleases

RNases of mammals and vertebrates constitute an extensive superfamily of homologous enzyme proteins that has been the subject of intense biochemical, structural and evolutionary studies for over half a century (Dyer and Rosenberg, 1996). The prototype of this superfamily, bovine pancreatic ribonuclease A, also known as RNase A successfully extracted from the bovine pancreas, was perhaps the best characterized of all known mammalian enzyme proteins (Raines, 1998). This vertebrate (RNase A) superfamily of proteins is distinct from two other well-investigated super families of T1 and T2 ribonucleases, which are not homologous with RNase A. In ruminants and other mammals with ruminant-like digestion, the major function of pancreatic RNases is to degrade dietary RNA and to recover RNA-nitrogen from the intestinal flora (Beintema and Kleinedam, 1998; Sorrentino, 2010; Shigematsu et al., 2018). However, many members of the vertebrate RNase superfamily have evolved over time to perform important nondigestive functions including neurotoxicity, angiogenesis, immunosuppressivity and antibacterial as well as antiviral actions (Rosenberg, 2008); indeed, the dynamic evolutionary history of this superfamily is one of its most intriguing features (Cho et al., 2005; Dyer and Rosenberg, 1996; Sorrentino, 2010; Lu et al., 2018). RNase 1, also known as RNase A or pancreatic-type RNase has been found in various organs, so its expression is not restricted to the exocrine pancreas (Morita et al., 1986; Koczera et al., 2016). This has been found to undergo different post-translational modifications, and purified samples from the urine, seminal plasma, kidney and brain showed different patterns of its glycosylation (Yasuda et al., 1993). There is further information on the ancestral origin and role of this superfamily, while results regarding the function of RNases in zebrafish (*Danio rerio*) suggested an ancestral origin consisting of angiogenesis-related ribonucleases, however, a host defense-associated role was proposed in studies on birds and mammals (Cho et al., 2005).

3. Potential applications of ribonucleases

The RNases are a heterogeneous group of enzymes mainly assigned in two categories according to their RNA-processing mechanisms: exoribonucleases, which cleave RNA from the terminal end and endoribonucleases that attack RNA at the internal sites (Nicholson, 1998; Mishra et al., 2014). Endoribonucleases include RNase A, RNase P, RNase H, RNase T1, RNase T2, RNase III, RNase U2, RNase V1, RNase I, RNase PhyM and RNase V. On other hand the exoribonucleases include polynucleotide phosphorylase (PNPase), oligoribonuclease, exoribonuclease I and exoribonucleases II, RNase PH, RNase II, RNase R, RNase D, RNase T, oligoribonuclease, exoribonuclease I and exoribonucleases II (Suri et al., 2007; Mohanty and Kushner, 2018). Many endoribonucleases exhibit substrate specificity but their target site is usually limited to one or few specific nucleotides in single-stranded RNA and usually in a context of a particular 3D-structure of the substrate (Table 12.1).

Table 12.1 Potential application of RNase against transformed/cancer lines and other processes.

Source	RNase	Effect on cell lines	Reference(s)
Bacillus amyloliquefaciens	RNase	Cytotoxicity and toxic for mouse, murine fibroblast cell line (L929), epidermal carcinoma, breast carcinoma, hepatoma and glioblastoma.	Prior et al. (1996)
Streptomyces aureofaciens	RNase Sa and variants	**i.** Human acute myeloid leukemia cells kasumi-1 and cytotoxicity by apoptosis-inducing effect. **ii.** Human erythroleukemia cells K-562.	Sevcik et al. (2002)
Bacillus intermedius	Binase	**i.** Leukemic Kasumi-1 cells, human ovarian cancer cells (SKOV3 and OVCAR5) and normal ovarian epithelial cells. **ii.** Chicken embryo fibroblasts (normal or Rous sarcoma virus-transformed) mouse fibroblasts or non-transformed NIH3T3 cells. **iii.** Human myelogenous leukemia cells R562, human peripheral blood lymphocytes, normal myeloid progenitor cells FDCP1 and transgenic myeloid progenitor cells expressing activated kit-oncogene.	Ilinskaya et al. (2007), Mitkevich et al. (2013), and Garipov et al. (2014)
Aspergillus fumigatus	Asp-fl (allergen)	Macrophage cells and helps to ribosomal RNA degradation and protein biosynthesis inhibition	Madan et al. (1997)
Aspergillus giganteus	α-Sarcin	**i.** Specific cleavage of 28S rRNA, protein biosynthesis inhibition and apoptosis observed in human rhabdomyosarcoma cells. **ii.** Cytotoxic effect on human liver, KB, Detroit 6, human heart, Maben and sarcoma 180.	Olson et al. (1965) and Olmo et al. (2001)
Aspergillus restrictus	Restrictocin Immune-conjugates	Specific cleavage of 28S rRNA and protein biosynthesis inhibition by blocking the ribosome elongation cycle and low toxicity for BALB/c mice and human breast carcinoma MCF-7; human colon carcinoma HT-29; human ovarian carcinoma MeWo and OVCA-432.	Heisterkamp et al. (1990)
Bovine pancreas	RNase A	Hydrolyzes RNA (but not DNA) phosphodiester bonds which covalently link ribonucleotides particularly those linked to pyrimidine bases such as uracil. $(NH_2)_2CO + H_2O \rightarrow CO_2 + 2NH_3$	Mitkevich et al. (2014), Kleinert et al. (2016), and Lu et al. (2018)
RNase H	Bacteria, archaea and eukaryotes	Cleaves the 3′—O—P bond of RNA in a DNA/RNA duplex to produce 3′-hydroxyl and 5′-phosphate terminated products. RNase H mediated viral RNA degradation.	Berkower et al. (1973) and Scott and Hu, 2013
Plants and humans	RNase and PNase	Nucleotidyl transferase activity.	Yehudai-Resheff et al. (2001)

Continued

Table 12.1 Potential application of RNase against transformed/cancer lines and other processes. *Continued*

Source	RNase	Effect on cell lines	Reference(s)
Archaea and eubacteria	RNase PH	Involved in tRNA processing.	Ishii et al. (2003)
E.coli	RNase II and RNase D	3′ to 5′ degradation of ssRNA, They help in the 3′-to-5′ maturation of many stable RNA molecules.	Coburn and Mackie (1996) and Zu et al. (2005)
Bacteria	RNase R	Close homolog of RNase II, but it can, unlike RNase II degrade RNA with secondary structures without the help of accessory factors.	Cheng and Deutscher (2005)
Aspergillus oryzae	RNAse T1	Sequence specific for ssRNA.	Pace et al. (1998)
Humans	RNase HI	Cleavage of RNA *via* a hydrolytic mechanism.	Ten Asbroek et al., 2002 and Lu et al. (2018)
Humans	RNase L	Interferon-induced ribonuclease which, upon activation, destroys all RNA within the cell (both cellular and viral).	Suhadolnik, 1997 and Lu et al. (2018)
Bovine seminal fluid	BS-RNase	*In vitro*, bind to cell membrane and disturb bilayer membrane and enter the cytosol to degrade the natural RNA in cell.	Sokurenko et al. (2016), Kumar and Kanwar (2017), Ilinskaya et al. (2018), Mironova and Zenkova (2018), and Lu et al. (2018)
Oocytes and early embryo of leopard frog	Onconase	Degradation of tRNA and inhibition of the protein synthesis which leads to activation of caspases and induced apoptosis.	
Aspergillus oryzae (Fungal)	RNase T1	Inhibition of protein synthesis and induced apoptosis.	
Aspergillus giganeus	α-Sacrin	Cleavage of phosphodiester bond of rRNA, block the protein synthesis.	
Human cell culture	RNase P	Removal of 5′- leader sequence of tRNA, blockage of protein synthesis and apoptosis.	
Human cell culture	RNase T2	Blocks blood supply to tumors cell by stopping angiogenesis.	
Aspergillus niger	Actibind	Also blocks blood supply to tumors cell by stopping process of angiogenesis.	
Streptomyces aureofaciens	RNase Sa	Internalized into cell by increasing net positive charge on cell surface.	

4. Ribonuclease A (RNase A)

RNase A was one of the first enzymes to be isolated by Dubos in the year 1938. It was purified in crystalline form by Kunitz in year 1939 (Kleinert et al., 2016). It has served as a model peptide in work related to protein folding, mechanism of kinetic reaction, disulfide bond formation, protein crystallography, spectroscopy and protein dynamics. RNase A is an enzyme that catalyses the degradation of RNA by cleaving the 3′-end of pyrimidine nucleosides. The mammalian RNase A superfamily consists of several peptides that develop in a range of mammals. The secretory members of mammalian RNase A superfamily have not only been found in the pancreas but also in other tissues and body fluids. They have been characterized by displaying sequence as well as structural and catalytic properties similar to those of bovine or human pancreatic ribonucleases. These secretory peptides contain a classic hydrophobic signal peptide followed by a mature peptide of 12–16 kDa in size. They also contain 6–8 cysteine residues that form 3–4 disulfide bonds, which play essential role(s) in maintaining the tertiary structure of these enzymes (Beintema and Kleinedam, 1998; Gupta et al., 2013; Lu et al., 2018). The non-secretory members of this family expressed in tissues other than pancreas are characterized by sequence, structural and catalytic properties similar to those of bovine kidney RNase K2 or human EDN/liver RNase. Frog ribonucleases include mainly Onconase from the leopard frog *Rana pipiens*. The ribonucleases from the Bullfrog *Rana catesbeiana* and the rice paddy frog *Rana japonica* are also the members of this family. In humans, the RNase A superfamily was originally thought to contain eight peptides on chromosome 14q11.2. These canonical RNases, which contain the catalytic motif (CKXXNTF), include pancreatic RNase (also known as RNase 1), eosinophil-derived neurotoxin (EDN/RNase 2), eosinophil cationic protein (ECP/RNase 3), RNase 4, angiogenin (RNase 5), RNase 6 (k6), RNase 7 and RNase 8 (Sorrentino, 2010). Analysis of the human genome identified five additional non-canonical ribonucleases, which were named as RNase 9–13 (Cho et al., 2005; Gho and Cho, 2009; Lu et al., 2018). The expansion and functional diversification of the mammalian ribonucleases, a superfamily epitomizes the efficiency of multigene families at emerging biological novelty. The structure and enzymatic activity of many RNase A superfamily members have been well characterized. X-ray diffraction experiments at high pressure showed that the secondary structure of RNase A is well preserved beyond 0.67 GPa at room temperature (Kurpiewska et al., 2016). Detailed structural analysis of RNase A conformation observed under high pressure revealed that pressure influences cavity size, packing of molecule and hydrogen bonds pattern (Prats-Ejargue et al., 2019).

5. Ribonuclease P

RNase P is a ribonucleoprotein complex consisting of a catalytic RNA component and one or more protein subunits, depending on the organism (Evans et al., 2006; Esakova and Krasilnikov, 2010). The RNA component is the catalytic component of the holoenzyme that contains the active site involved in the phosphodiester bond cleavage to generate mature tRNA product from the pre-tRNA molecules (Reiter et al., 2010; Mao, 2018). RNase P enzyme occurs in all three domains of microbes *i.e.,* bacteria, archaea and eukarya. Bacterial RNAse P differs from other two domains of microbes due to its protein composition and RNA structure. *In vivo*, bacterial RNase P has been associated with single protein (Garcia and Grutz, 1995). The bacterial RNase P holoenzyme is composed of an RNA subunit of 330–420 nucleotides and a protein subunit of around 120 amino acid residues (Kazantsev and Pace, 2006). The archaeal and eukaryotic RNase P holoenzyme is composed of one RNA subunit, but the number of protein subunits may vary. RNase P catalyses the removal of extra sequences in the pre-tRNA to form a mature tRNA product that can attain correct conformational shape, which thence takes part in the process of protein synthesis (Mao, 2018). The protein component of *Mycobacterium tuberculosis* RNase P, made under certain conditions, was able to process pre-tRNA independently, which is unusual for a bacterial RNase P (Singh et al., 2016; Singh and Batra, 2018). RNase P is a divalent cation-dependent endoribonuclease, which acts as a ribozyme and is encoded by protein subunit *rnpA* gene and RNA subunit rnpB. RNase P functions to breakdown the precursor sequence of RNA on RNA molecules. The ongoing cancer therapy approaches have a major problem to differentiate between the cancer cells and the normal cells. This problem may be overcome by specific chimeric molecules, which are specific to the cancer cells, and thus efficiently act on the specific targets. Another finding observed by scientists was about the ribonuclease inhibitor binding and the change of cytotoxic nature by producing mutants or mutation in base pair residues or change in structure of enzyme. In some cases, the combination of catalytic active region and ribonuclease inhibitor evasion, the active segment combines to produce new active molecules with more advantageous forms of RNases known as chimers (Cobaleda and Sanchez-Garcia, 2000; Mao, 2018; Singh and Batra, 2018). These techniques may allow the production of active microbial RNases with effective and efficient anticancer outcomes (Mao, 2018).

6. Ribonuclease H

RNase H enzymes belong to the nucleotidyl transferase superfamily whose members share a similar protein fold as well as the catalytic mechanism. There are two main types of RNases H, and at least one of them is present in most organisms. Eukaryotic RNases H are larger and more complex than their prokaryotic counterparts (Hostomsky et al., 1993). Eukaryotic RNase H1 has acquired a “Hybrid Binding Domain” that confers processivity and affinity for the substrate, whereas eukaryotic RNase H2 is composed of three different proteins: the catalytic subunit (2A), homogenous to the monomeric prokaryotic RNase H2 and two other subunits (2B and 2C) that have no prokaryotic counterparts but are necessary for catalysis (Cerritelli and Crouch, 2009). Mutations in anyone of the three subunits of human

RNase H2 cause Aicardi-Goutieres syndrome, a human neurological disorder with devastating consequences (Cerritelli and Crouch, 2009). This RNase family also includes the retroviral RNase H and integrase, including the human immunodeficiency virus enzymes. The RNA is subsequently degraded in a 3′-5′ direction during elongation of the minus end polarity in DNA strand (Summers and Mason, 1982). The RNase H had no obvious sequence specificity or positional dependence within the RNA, so it cleaves the RNA at multiple positions even within the minimal 14 nucleotides duplex. The endonucleolytic RNase H activity requiring a DNA:RNA duplex having fourteen nucleotides or more, could not digest a stem-loop in either the RNA or DNA strands and could tolerate a nick in the DNA strand but not a gap.

A number of prokaryotic and eukaryotic ribonucleases have been successfully purified from eukaryotes and prokaryotes with the use of chromatographic techniques. The main chromatographic techniques exploited to purify RNases are ion exchange chromatography and size exclusion chromatography.

7. Role of ribonucleases in cancerous diseases

In contrast to many chemotherapeutic drugs currently used in cancer therapy, which may target DNA synthesis and transcription, cytotoxic RNases primarily target RNA functions, such as protein formation or gene regulation (Tafech et al., 2006) and therefore, they are non-mutagenic antitumor drugs. The cytotoxic effects of RNase include the RNA cleavage leading to the inhibition of the protein synthesis and induction of apoptosis (Benito et al., 2005). In the endocytic pathway, cytotoxic RNases are translocated to the cytoplasm where they often digest cellular RNA(s). Triterpenoids and polysaccharides are the major constituents contributing to anticancer functions of RNases. The cytotoxic action of some RNases may be hampered by the action of a potent inhibitor, the ribonuclease inhibitor, which is found in the cytoplasm of mammalian cells (Fig. 12.2). Ribonuclease inhibitor is a 50 kDa protein that tightly binds to some RNases thus inhibiting their activity. The cytotoxic potential of a specific RNase also depends on its ability to bypass the action of ribonuclease inhibitor (Benito et al., 2008; Canestrari and Paroo, 2018). A sequence-specific RNase of 19 kDa was isolated from human colon carcinoma T84 cells. It has been recommended that RNases constitute markers for human cancer cells. The RNase L Arg462Gln and Asp541Glu polymorphisms are associated with prostate cancer risk and could be low-penetrance prostate cancer susceptibility biomarkers (Shapiro et al., 1986; Shlyakhovenko, 2016).

In the human HT-29 adenocarcinoma cell line, two enzymes exhibiting RNase activities have been purified from the cell culture medium. The angiogenin, one of these enzymes (Fett et al., 1985) catalyzes a limited number of endonucleolytic cleavages of 28S and 18S ribosomal RNA (Fig. 12.3). The other one is RNase 4, a secretory enzyme, which shares with the non-secretory RNases the ability to degrade poly (U) faster than poly (C), but its precise function is not known (Shapiro et al., 1986; Shlyakhovenko, 2016). The N1/T1 RNase superfamily comprises

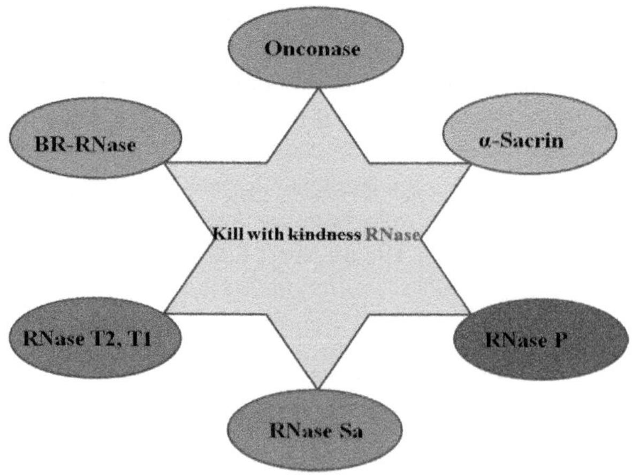

FIG. 12.2 The broader types of RNases.

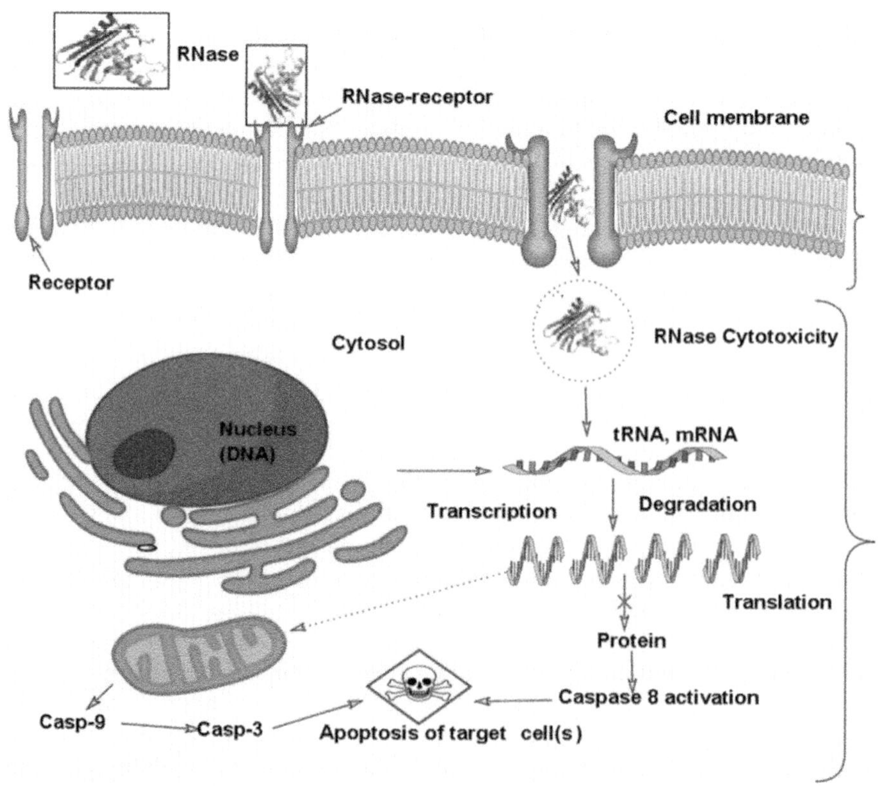

FIG. 12.3 RNase cytotoxic effect on the transformed mammalian cells.

enzymes with well-known antitumor effects, such as ribotoxins secreted by fungi, primarily by *Aspergillus* and *Penicillium* species as well as the bacterial RNases secreted by *Bacillus pumilus* (Binase) and *B. amyloliquefaciens* (Barnase). A new RNase with a high degree of structural similarity with Binase (73%) and Barnase (74%) was isolated and purified from *Bacillus licheniformis* (Balifase, calculated molecular weight 12.4 kDa). The unique features of Balifase include the synthesis of an operon together with a gene of its intracellular inhibitor, as well as high stability and inability to form natural dimmers (Sokurenko et al., 2016; Ilinskaya et al., 2018; Mironova and Zenkova, 2018).

8. Bovine seminal ribonuclease (BR-RNase)

The BS-RNase is only one enzyme, which occurs in quaternary structure and was discovered by Hosokawa and Irie, as well as Dostal and Matousek in year 1972. The BS-RNase is a natural dimer with two identical subunits held together by two intramolecular disulfide bonds and non-covalent interactions that prevent the binding of ribonuclease inhibitor due to steric reasons as well as mediated the cytotoxic effect (Makarova and Ilinskayab, 2003; Shlyakhovenko, 2016). The BS-RNase exists as dimmer forms and maintains the allosteric regulation by both substrate and product at nucleotide hydrolysis process. Single monomeric subunit of BS-RNase consists of 124 amino acid residues. BS-RNase is highly cytotoxic in its dimmer structure and shows high activity. If, BS-RNase is reduced to monomeric structure, its cytotoxicity greatly decreases without affecting it's the ribonuclease activity. BS-RNase amino acid sequence and crystallographic structure are similar to the members of the RNase A superfamily. Lys41, His12 and His119 are catalytic residues of BS-RNase and in these enzymes the four disulfide bridges of RNase A are strictly conserved. The two consecutive Cys residues at position 31 and 32 are present in BS-RNase and these form two disulfide bonds, which cause dimerization. However, BS-RNase monomeric unit is highly catalytically active than the dimeric form of BS-RNase, but it has little cytotoxicity (Makarova and Ilinskayab, 2003; Benito et al., 2005; Suri, 2007; Ilinskaya et al., 2018; Attery et al., 2018). Mammalian ribonuclease inhibitor energetically inhibits the BS-RNase monomeric unit but it does not interact with the natural dimeric enzyme.

The analyses of single subunit variants of BS-RNase that evaded ribonuclease inhibitor, ribonuclease activity and cytotoxic activity were conducted. These monomeric variants had cytotoxic activity, which was 30-fold higher than wild-type BS-RNase. The ribonuclease inhibitor is a 50 kDa protein present in cytosol and it forms a strong complex with mammalian RNases that eventually abolishes the RNase catalytic activity. Microbial and other RNases are unwilling to bind to mammalian RNase inhibitor. The RNase found in bull semen fluid shows antitumor, antispermatogenic and immunosuppressive activities due to its unique structure and enzymatic properties. It belongs to RNases group of enzymes endowed with "Special Biological Actions" (RISBASES) (Makarova and Ilinskayab, 2003; Benito et al., 2005; Arnold and Ulbrich, 2006; Suri, 2007). The cytotoxic effect of BS-RNase on tumor cells is accompanied by the induction of apoptosis. This has been proved by ultrastructure and flow cytometry evidence of apoptotic death following BS-RNase treatment in phytohemagglutinin-stimulated lymphocytes and normal cells. Further, BS-RNase monomers are highly toxic to human tumor cells than other variant or homologue of RNase A. The amphibian homologues are already in phase III clinical trials for the treatment of unrespectable malignant mesothelioma. A BS-RNase cytotoxic property renders it with potent antitumor, embryo toxic, immunosuppressive and aspermatogenic activities thus putting it in the category of therapeutic molecule. BS-RNase has been found to be a promising tool for the treatment of thyroid cancer, too (Kim et al., 1995; Sinatra et al., 2000; Ilinskaya et al., 2018; Attery et al., 2018; Prats-Ejargue et al., 2019).

9. Onconase

Onconase belongs to the family of RNases A. It was isolated from early embryos and oocytes of Northern leopard frog (*Rana pipiens*). Onconase shows strong cytotoxic and cytostatic activity and it possesses 30% sequence similarity with superfamily RNase A (Shlyakhovenko, 2016; Prats-Ejargue et al., 2019). Onconase enzyme has the potential to degrade RNA and their substrates, too. They also possess highly cytotoxic effect on several cancer cell lines. Onconase also named as Ranpiranse is the first therapeutic product studied by Alfacell Corporation. Its single chain consists of 104 amino acid residues, which are 20 residues less than RNase A. Ranpiranse shows cytotoxic effect *in vivo* and *in vitro* in tumor cells. Ranpiranse mediated cytotoxic activity has been observed by targeting the cellular tRNA. It also breaks down the double-stranded RNA at 40 °C under same conditions. Currently, a new mechanism followed by Ranpiranse activity for inducing cell death/apoptosis in malignant phenotypes *via* the RNA interference mechanism involving miRNA and siRNA has been postulated (Saxena et al., 2009; Shlyakhovenko, 2016). Onconase is the smallest enzyme in the RNase A superfamily and it has also been cited as the first RNase tested in clinical trial in US and Europe. This has been involved in recent phase III trial for treatment of malignant mesothelioma. Due to its cytotoxic and cytostatic properties, it has been used externally against the skin cancer.

In Onconase, Thr and Ser residues are present in 70:30 ratio of its molecule and both forms are equivalent with respect to their catalytic activity and cytotoxicity. RNase A catalytic residues are conserved in all variants of Onconase (His10, Lys31 and His97). According to amino acid sequence and crystallographic studies, it has been revealed that three disulfide bonds of RNase A are conserved in all variant of Onconase at position 19–68, 30–75 and 48–90. These catalytic residues and disulfide bond position lead to the prerequisite for cytotoxicity of RNases. Onconase is an exceptionally stable enzyme at temperature 90 °C or at 4.4 M, as respective values for RNase A have been found to be 62.4 °C and 2.8 M. Onconase because of its conformation is very much stable molecule. However, its unwilling leads to proteolysis. The unpredictable thermodynamic stability of Onconase appears to be due to the presence of the C-terminal disulfide bond and N-terminal network of hydrogen bonds, which interact with hydrophobic cluster. Disulfide bonds substantially contribute to enzyme stability and make Onconase more efficient than RNase A at recovering its catalytically and biologically active structure. The cytotoxicity of Onconase is same as that of BS-RNase because of catalysis of phosphodiester cleavage in mammalian cells. The Onconase practically does not interact with mammalian ribonuclease inhibitor and its catalytic and antitumour activities are also terminated by alkylation of His residue. Onconase evades ribonuclease inhibitor in cell cytosol because it lacks an amino acid sequence interaction with the ribonuclease inhibitor. Onconase may possibly binds to cell surface receptor like a polypeptide or hormone or moves inside the target tumor cells by endocytosis (Saxena et al., 2009; Shlyakhovenko, 2016; Jordaan et al., 2018; Ilinskaya et al., 2018; Prats-Ejargue et al., 2019).

10. RNase T1

RNase T1 is type of an endonuclease that degrades phosphodiester bonds of cellular RNA between 3′-guanylic residues and 5′-OH residues of adjoining nucleotides with the development of consequent intermediate 2′,3′- cyclic phosphates. RNase T1 has been isolated mainly from *Aspergillus* and *Penicillium* species. RNase T1 has cytotoxic nature toward malignant cells without metal ions and because of this character, it is also known as Ribotoxin. RNase T1 has been used to analyze RNA structure, mapping, RNA protection and removal of RNA form extracted DNA samples. RNase T1 from *A. oryzae* and RNases U1 from *U. sphaerogena* were inactivated by ribonuclease inhibitor. RNase T1 shows toxic effects by targeting the conserved sequence known as Sarcin-ricin loop of larger rRNA and thus it leads to inhibition of protein biosynthesis in the target cell followed by eventual cellular death by apoptosis. RNase T1 internalizes into cell-cytosol by Columbic interaction or Clathrin-independent protein. The Ribotoxin protein that targets cellular RNA eventually leads to cytotoxic effect toward the tumor cells. Ribotoxin preferentially kills the tumor cells by altered membrane permeability if no specific protein receptor has been present on the cell membrane, such as the cells infected with virus or transformed cells (Shlyakhovenko, 2016). RNase T1 consists of a small protein $\alpha + \beta$ dimer containing 104 amino acid residues with Cys2-Cys10 and Cys6-Cys103 disulfide bonds. The disulfide bonds help in folding and unfolding of RNase T1. RNase T1 specifically targets the guanine bases of cellular RNA for denaturing. Researchers have found that Hemagglutinating virus of Japan as a vector can be targeted to tumor cells to make it to act as a unique anticancer drug. RNase T1 has been internalized by cells *via* a novel gene transfer reagent; Hemagglutinating virus of Japan enveloped vector targeted to human tumor resulted in their death. Researchers observed that pre-treatment of Hemagglutinating virus of Japan envelope vector with protamine sulfate leads RNase T1 to show a tremendously improved cytotoxic activity (Yuki et al., 2004; Shlyakhovenko, 2016). RNase T1 is in phase III human clinical trials as a non-mutagenic cancer chemotherapeutic agent. Furthermore, internalized RNase T1 also induces the apoptotic cell death program in the tumor cells. However, previous studies have indicated that the RNase T1 cytotoxicity is unfortunately not specific to tumor cells but it could otherwise be overcome by Hemagglutinating virus of Japan envelope vector thus making the RNase a highly specific antitumor tool (Yuki et al., 2004; Shlyakhovenko, 2016; Kumar and Kanwar, 2017, 2018).

11. Binase

Binase, a low-molecular weight (12.2 kDa) protein containing 109 amino acids has been isolated from *B. intermedius*. Binase has been well studied in many aspects including its purification, sequencing, crystal structure to *in vitro* and *in vivo* antitumor activities. Binase unit contains two molecules forming a homodimer whose one of the subunits binds the nucleotide in the 3′GMP-Binase complex in which the guanyl base is located in the recognition loop of the enzyme (Ilinskaya et al., 2018). Binase revealed a broad spectrum and persuasive activity against malignant cells. In previous studies, it has been reported that Binase has antiproliferative and selective apoptosis-inducing activities against human myelogenic erythroleukemia K562 cells, peripheral blood mononuclear cells and human lung carcinoma A549 cells. However, Binase-mediated antiproliferative and apoptotic effects were not detected in normal human peripheral blood. In transformed myeloid cells, apoptosis was induced by Binase but it however, didn't induce the T-cell immune response. Extracellular cytotoxic RNases of bacterial origin, that trigger apoptosis/cytotoxicity in tumor cells, may have a great potential as anticancer drugs. A Binase sourced from *B. pumilus* was selectively cytotoxic toward lung carcinoma, leukemic and ovarian cancer cells. Interestingly, it was found that the cell sensitivity toward xenogeneic RNases is related to expression of oncogenes *ras*, *kit* and *AML*1. Binase selective apoptotic properties found toward malignant cells confirmed that it could be considered as an alternative antitumor chemotherapeutic drug. Binase possesses remarkable biological activities, which exert positive mutagenic properties at high concentrations and at low concentration, it stimulates the cell growth. The most fascinating feature of Binase is its antitumor activity but the specific mechanism(s) of antitumor activity remains unknown. These encouraging observations unveil the therapeutic potentials of bacterial RNases, especially Binase for the treatment of tumors mononuclear cells (Shlyakhovenko, 2016; Ilinskaya et al., 2018).

12. Barnase

Barnase is also a promising contender found as an anticancer therapeutic molecule, which possesses 84% sequence homology with Binase and it is also the closest natural homologue of Binase. Barnase was initially isolated from *Bacillus amiloliquefaciens* as an active proenzyme. It was processed by the removal of the amino-terminal signal peptide. It was secreted into the extracellular space. It showed promising results against target cancer cells. Importantly, it had low susceptibility against cytoplasmic ribonuclease inhibitor, was highly catalytic and stable (Arnold et al., 2006; Shlyakhovenko, 2016; Kumar and Kanwar, 2018). Currently, chemical modification and genetic engineering to target Barnase to enhance its productivity and activity are the hot areas of research work. Barnase has no disulfide bonds and thus it does not require any post-translational modifications, divalent cations or other non-peptide components. Due to these favorable features, Barnase is active in any cell type in which it is expressed. Barnase efficiently cleaves target RNA, and this property has been exploited in a wide variety of biological applications since introduction of this enzyme into cells causes' cell death. Specific ablation of particular cells is feasible by directing Barnase gene expression *via* the use of cell-specific promoters. Alternatively, proteins that target Barnase to specific cells endow specificity to Barnase action. Major problems in human pancreatic-type RNases is that these are sensitive to inhibition by cytoplasmic ribonuclease inhibitor found in every mammalian cell studied so far. While examining the susceptibility of Barnase to ribonuclease inhibitor, it was found that Barnase is fortunately insensitive to inhibition by ribonuclease inhibitor (Shlyakhovenko, 2016; Ilinskaya et al., 2018; Prats-Ejargue et al., 2019).

13. RNase Sa

Another promising microbial T1 family belonging to *Streptomyces* has representative members such as RNases Sa (strain BMK), Sa2 (strain R8/26) and Sa3 (*S. aureofaciens*, strain CCM 3239), which structurally resembled with each other and have identical amino acids in 48 out of 96 positions. Out of these three RNases, RNase Sa3 is potentially active and highly catalytic against cancer cells rather than RNase Sa2 and RNase Sa1. Although, RNase Sa3 tertiary structure is quite similar to RNase Sa2 but still RNase Sa3 is more cytotoxic toward many mammalian malignant cells (Ilinskaya et al., 2002, 2018; Shlyakhovenko, 2016). The cytotoxicity of RNase Sa3 depends on some structural elements and sequence motifs. The replacement of Asp and Glu residues with Lys on the surface of RNase Sa3 and variation in their charge form acidic to basic protein increases toxic effect against malignant cells (Shlyakhovenko, 2016). Due to reversing charges, it could generate sufficient cytotoxic effects in RNase Sa3. Cytotoxic activity of RNase Sa correlates with the change in its net charge from negative to positive. Site-directed mutation is one of approaches, which allows the creation of charges on enzymes and there by produces toxic effect with bare minimum side effects on the cells. These malignant cell-selective characteristics revealed that RNase Sa3 could be used as a potential anticancer agent for treating leukemia in human patients (Pace et al., 1998; Sevcik et al., 2002; Kumar and Kanwar, 2018).

14. α-Sarcin

The α-Sarcin is composed of 150-residues polypeptide toxin, while Mitogillin and Restrictocin are quite small (approximately 17 kDa) ribosome-inactivating proteins produced by the *A. giganteus*. α-Sarcin belongs to type 1 group of RIPs that are predominant members of the fungal Ribotoxins displaying 3D structure. His50, Glu96 and His137 residues are responsible for α-Sarcin catalytic activity toward malignant cells. These three fungal ribotoxins act on specific RNA to accomplish the cleavage of phosphodiester bonds in the universally conserved α-Sarcin domain of 28S rRNA and thus inhibit the protein synthesis. The α-Sarcin interacts with lipid bilayer of malignant cell membrane and fuses with the cell leading to inhibition of protein synthesis and subsequent induction of apoptosis. The α-Sarcin has been previously reported as a potent cytotoxin that promotes apoptosis in human rhabdomyosarcoma cells. Due to this application, the ribotoxins are used as tools in research, therapeutics and as diagnostic agents (Schwartz et al., 2007; Suri et al., 2007; Xu et al., 2011; Shlyakhovenko, 2016).

15. Actibind and RNase T2

Actibind is a protein isolated from black mold *A. niger*, which is a well-known microorganism used in bio- and food-industry. In previous finding, Actibind was found to control the cancer development by preventing the growth of malignant cells in mammalian cells. RNase T2 was also found to bind actin in migrating cells in both animals and humans. RNase T2 enzyme prevents the blood supply to tumor cells while Actibind prevents the malignant cells to move through the blood stream to form new metastases (Schwartz et al., 2007; Suri et al., 2007; Xu et al., 2011). The encouraging outcome was that the Actibind did not show any toxic effect on normal cells thus minimizing the possibility of its side effects. In future, fungal Actibind and RNase T2 could be used as frontline therapies in the fight against cancers (Kumar and Kanwar, 2017, 2018).

16. Antiviral activity of ribonucleases

To control viral diseases, biological treatment is superior to physicochemical methods, because the latter has higher treatment costs and the possibility of causing secondary pollution. The antiviral activity of RNases can result from degradation of viral RNA and viral infection in certain plants, causing RNase induction that may block viral replication (Saxena et al., 2009). A *Bacillus cereus* strain ZH14 with antiviral activity against TMV infection from the Chinese Anxi oolong tea (Zhou et al., 2008). The studies also demonstrated the ability of artificial ribonucleases (aRNases, small organic RNA cleaving compounds) to inactivate RNA-viruses *via* the synergetic effect of viral RNA cleavage and disruption of viral envelope synthesis

(Fedorova et al., 2011). Furthermore, another RNase isolated from *B. pumilus*, possesses antiviral activity against plant RNA-viruses *e.g.*, Red clover mottle virus, Potato virus X and Alfalfa Mosaic virus. The extensive inhibitory effect against actively replicating viruses has been observed when plants are treated with the enzyme concentration of 100 μg/mL prior to infection (Sharipova et al., 2015).

Human eosinophil-derived neurotoxin and eosinophil cationic protein are two RNase toxins secreted by the large cytoplasmic granules of human eosinophilic leukocytes (Torrent et al., 2009; Rosenberg, 2015). The human eosinophil derived neurotoxin (RNase 2), a rapidly evolving secretory protein resulting from eosinophilic leukocytes, mediates the ribonucleolytic degradation of extracellular virions of the single-stranded RNA virus *e.g.*, Respiratory syncytial virus. The interaction among EDN and extracellular virions of Respiratory syncytial virus is both saturable and specific. Eosinophils promote the destruction of extracellular virions of Respiratory syncytial virus *in vitro*, an antiviral effect that was reversed in the presence of soluble RNase inhibitor.

Human immunodeficiency virus and Hepatitis B virus reverse transcription require coordinated function of the reverse transcriptase and RNase H activities of the viral polymerase protein. The reverse transcriptase has been biochemically characterized but technical complications have prevented both assessment of the RNase H and development of high throughput inhibitor screens against the RNase H. The Hepatitis B virus RNase H and the more extensively studied Human immunodeficiency virus enzyme are less related, sharing just 23% identity (Zhang et al., 2019). Despite the large genetic distance between the Hepatitis B virus and Human immunodeficiency virus RNase H, both degrade the RNA template throughout the viral replication to enable plus-polarity DNA synthesis (Tang et al., 2011). Furthermore, three novel subtypes of 3-hydroxypyrimidine-2,4-dione (HPD) derivative were carefully designed to attain selective RNase H inhibition (Tang et al., 2016) for Human immunodeficiency virus infection. Biochemical studies displayed the two subtypes with an N-1 methyl group inhibited RNase H in low micromolar range without significantly inhibiting reverse transcription polymerase, but the N-1 non-substituted subtype inhibited RNase H at sub-micromolar range and reverse transcription polymerase in low micro-molar range. A new active recombinant hexahistidine-tagged Hepatitis B virus RNase H was produced that was appropriate for low-throughput drug screening. It was used it to discover all of the currently known inhibitors of the enzyme, many of which act against viral replication in cells (Tavis and Lomonosova, 2015).

17. Conclusion and future directions

RNases acting as therapeutic agents have the aptitude to abolish the growth transformed cells. RNase cytotoxic effects are based on the selective hydrolysis of RNA, intracellular routing and membrane-specific recognition processes occur based upon the cytotoxic nature of RNases. Onconase and microbial RNase have been evaluated in clinical trials, but have some obvious limitations. The cytotoxic pathways of RNases serve as working platform(s) for the creation of some potent new anticancer drugs in future by chemical modification in ligand/receptor and engineering process by incorporating an increase in the RNase cytotoxicity against cancer cell line(s). RNases also act as antiviral proteins as they inhibit the reproduction of viruses in infected cells and laboratory animals, but the molecular mechanisms of their antiviral activity remain unclear. There is a strong belief that further development in the mechanistic and structural elucidation of RNase will be the prerequisites for development of effective drug(s) against malignant cells and establishment of potent anticancer therapeutics.

In future, RNases may be assigned many important roles in gene regulation as well as treatment of cancers and viral diseases. Many recent studies on RNases have shown promise in their potential application to treat various types of cancers and how they can be manipulated to benefit and optimize anticancer drug development.

Acknowledgments

The authors are thankful University Grants Commission, Government of India, New Delhi for the financial support to one of the authors Rakesh Kumar in the form of a RGNF-SRF sanctioned vide a Letter No. F1-17.1/2013-14/RGNF-2013-14-SC-HIM-51265. Further, the authors declare that no conflict of interest exists among them or with their parent institution.

References

Aravind, L., Koonin, E.V., 2001. A natural classification of ribonucleases. Methods Enzymol. 341, 3–28.

Arnold, U., Ulbrich, H.R., 2006. Natural and engineered ribonucleases as potential cancer therapeutics. Biotechnol. Lett. 28, 1615–1622.

Arnold, U., Schulenburg, C., Schmidt, D., Ulbrich, H.R., 2006. Contribution of structural peculiarities of Onconase to its high stability and folding kinetics. Biochemistry 45, 3580–3587.

Attery, A., Dey, P., Tripathi, P., Batra, J.K., 2018. A ribonuclease inhibitor resistant dimer of human pancreatic ribonuclease displays specific antitumor activity. Int. J. Biol. Macromol. 107, 1965–1970.

Barnett, T.C., Bugrysheva, J.V., Scott, J.R., 2007. Role of mRNA stability in growth phase regulation of gene expression in the group A *streptococcus*. J. Bacteriol. 189, 1866–1873. https://doi.org/10.1128/JB.01658-06.

Beintema, J.J., Kleinedam, R.G., 1998. The ribonuclease A superfamily: general discussion. Cell. Mol. Life Sci. 54, 825–832.

Benito, A., Ribo, M., Vilanova, M., 2005. On the track of antitumour ribonucleases. Mol. Biosyst. 1, 294–302.

Benito, A., Vilanova, M., Ribo, M., 2008. Intracellular routing of cytotoxic pancreatic-type ribonucleases. Curr. Pharmaceut. Biotechnol. 9, 169–179.

Berkower, I., Leis, J., Hurwitz, J., 1973. Isolation and characterization of an endonuclease from *Escherichia coli* specific for ribonucleic acid in ribonucleic acid-deoxyribonucleic acid hybrid structures. J. Biol. Chem. 248, 5914–5921.

Bezborodova, S.I., Khodiva, O.M., Stepanov, V.M., 1983. The complete amino acid sequence of ribonuclease C_2 from *Aspergillus clavatus*. FEBS Letters 159, 256–258.

Boisset, S., Geissmann, T., Huntzinger, E., Fechter, P., Bendridi, N., Possedko, M., 2007. *Staphylococcus aureus* RNAIII coordinately represses the synthesis of virulence factors and the transcription regulator Rot by an anti sense mechanism. Genes Dev. 21, 1353–1366. https://doi.org/10.1101/gad.423507.

Canestrari, E., Paroo, Z., 2018. Ribonucleases as drug targets. Trends Pharmacol. Sci. 39, 855–866.

Cerritelli, S.M., Crouch, R.J., 2009. Ribonuclease H: the enzymes in eukaryotes. FEBS J. 276, 1494–1505.

Cheng, M., Deutscher, M.P., 2005. An important role for RNase R in mRNA decay. Mol. Cell 17, 313–318. https://doi.org/10.1016/j.molcel.2004.11.048.

Cho, S., Beintema, J.J., Zhang, J., 2005. The ribonuclease A superfamily of mammals and birds: identifying new members and tracing evolutionary histories. Genomics 85, 208–220.

Cobaleda, C., Sanchez-Garcia, I., 2000. *In vivo* inhibition by a site-specific catalytic RNA subunit of RNase P designated against the BCR-ABL oncogenic products: a novel approach for cancer treatment. Blood 95, 731–737.

Coburn, G.A., Mackie, G.A., 1996. Overexpression, purification and properties of *Escherichia coli* ribonuclease II. J. Biol. Chem. 271, 1048–1053. https://doi.org/10.1074/jbc.271.2.1048.

Coleman, G., Elliott, W.H., 1965. Extracellular ribonuclease formation in *Bacillus subtilis* and its stimulation by actinomycin D. J. Biochem. 95, 699–706.

Condon, C., Putzer, H., 2002. The phylogenetic distribution of bacterial ribonucleases. Nucleic Acids Res. 30, 5339–5346.

Domingues, S., Matos, R.G., Reis, F.P., Fialho, A.M., Barbas, A., Arraiano, C.M., 2009. Biochemical characterization of the RNase II family of exoribonucleases from the human pathogens *Salmonella typhimurium* and *Streptococcus pneumoniae*. Biochemistry 48, 11848–11857.

Dyer, K.D., Rosenberg, H.F., 1996. Molecular cloning and characterization of a novel human ribonuclease (RNase k6): increasing diversity in the enlarging ribonuclease gene family. Nucleic Acids Res. 24, 3507–3513.

Eller, C.H., Chao, T.-Y., Singarapu, K.K., Ouerfelli, O., Yang, G., Markley, J.L., Danishefsky, S.J., Raines, R.T., 2015. Human cancer antigen Globo H is a cell-surface ligand for human Ribonuclease 1. ACS Cent. Sci. 1, 181–190. https://doi.org/10.1021/acscentsci.5b00164.

Esakova, O., Krasilnikov, A.S., 2010. Of proteins and RNA: the RNase P/MRP family. RNA 16, 1725–1747.

Evans, D., Marquez, S.M., Pace, N.R., 2006. RNase P: interface of the RNA and protein worlds. Trends Biochem. Sci. 31, 333–341.

Even, S., Pellegrini, O., Zig, L., Labas, V., Vinh, J., Bréchemmier- Baey, D., Putzer, H., 2005. Ribonucleases J1 and J2: two novel endoribonucleases in *Bacillus subtilis* with functional homology to *E. coli* RNase E. Nucleic Acids Res. 33, 2141–2152. https://doi.org/10.1093/nar/gki505.

Fan, S., Jiang, L., Wu, J., Dong, L., Cheng, Q., Xu, P., Zhang, S., 2015. A Novel pathogenesis-related class 10 protein gly m 4l, increases resistance upon *Phytophthora sojae* infection in Soybean (*Glycine max* [L.] Merr.). PLoS One 10. https://doi.org/10.1371/journal.pone.0140364.

Fang, E.F., Ng, T.B., 2011. Ribonucleases of different origins with a wide spectrum of medicinal applications. Biochim. Biophys. Acta 1815, 65–67.

Farkas, G.L., 1982. RNA-spliting enzymes. In: Nucleic Acids and Proteins in Plants: Structure, Biochemistry and Physiology of Nucleic Acid. Encyclopedia Plant Physiol, vol. 14, pp. 224–262.

Fedorova, A.A., Azzami, K., Ryabchikova, E.I., Spitsyna, Y.E., Silnikov, V.N., Ritter, W., Gross, H.J., Tautz, J., Vlassov, V.V., Beier, H., Zenkova, M.A., 2011. Inactivation of a non-enveloped RNA virus by artificial ribonucleases: honey bees and acute bee paralysis virus as a new experimental model for *in vivo* antiviral activity assessment. Antivir. Res. 91, 267–277.

Fett, J.W., Strydom, D.J., Lobb, R.R., Alderman, E.M., Bethune, J.L., Riordan, J.F., Vallee, B.L., 1985. Isolation and characterization of angiogenin, an angiogenic protein from human carcinoma cells. Biochemistry 24, 5480–5486.

Filipenko, E.A., Kochetov, A.V., Kanayama, Y., Malinovsky, V.I., Shumny, V.K., 2013. PR-proteins with ribonuclease activity and plant resistance against pathogenic fungi. Russ. J. Genet. Appl Res. 3, 474–480.

Flogel, R., Zielenkiewicz, P., Saenger, W., 1988. The tertiary structure of *Aspergillus saitoi* minor ribonuclease (Ms) predicted from the structure of RNase T1. FEBS Lett. 240, 29–32.

Garcia, I.S., Grutz, G., 1995. Tumorigenic activity of the BCR-ABL oncogenes is mediated by BCL2. Proc. Natl. Acad. Sci. U.S.A. 92, 5287–5291.

Garipov, A.R., Nesmelov, A.A., Cabrera-Fuentes, H.A., Ilinskaya, O.N., 2014. *Bacillus intermedius* ribonuclease (BINASE) induces apoptosis in human ovarian cancer cells. Toxicon 92, 54–59.

Gho, S.M., Cho, S., 2009. The expansion and functional diversification of the mammalian ribonuclease a superfamily epitomizes the efficiency of multigene families at generating biological novelty. Genome Biol. Evol. 5, 2124–2140.

Ghosh, D., Saha, M., Sana, B., Mukherjee, J., 2005. Marine enzymes. Adv. Biochem. Eng. Biotechnol. 96, 189–218.

Guan, G.P., Wang, H.X., Ng, T.B., 2007. A novel ribonuclease with antiproliferative activity from fresh fruiting bodies of the edible mushroom *Hypsizigus marmoreus*. Biochim. Biophys. Acta 1770, 1593–1597.

Gundampati, R.K., Sharma, A., Kumari, M., Debnath, M., 2011. Extracellular poly (A) specific ribonuclease from *Aspergillus niger* ATCC 26550: purification, biochemical and spectroscopic studies. Process Biochem. 46, 135–141.

Gupta, S.K., Haigh, B.J., Griffin, F.J., Wheeler, T.T., 2013. The mammalian secreted RNases: mechanisms of action in host defence. J Innate Immunol 19, 86–97.

Heisterkamp, N., Jenster, G., Hoeve, J.T., Zovich, D., Pattengale, P.K., Groffen, 1990. Acute leukemia in BCR/ABL transgenic mice. Nature 344, 251–253.

Horitsu, H., Hara, N., Tohgane, M., Kawai, K., 1982. Purification of a base-specific ribonuclease Lu from *Aspergillus niger*. Agric. Biol. Chem. 46, 1145–1151.

Hostomsky, Z., Hostomska, Z., Matthews, D.A., 1993. Ribonuclease H. Cold Spring Harbor Laboratory Press, Plainview, NY, pp. 341–376. https://doi.org/10.1021/bi901105n.

Huntzinger, E., Boisset, S., Saveanu, C., Benito, Y., Geissmann, T., Namane, A., 2005. *Staphylococcus aureus* RNAIII and the endo ribonuclease III co-ordinately regulate spa gene expression. EMBO J. 24, 824–835. https://doi.org/10.1038/sj.emboj.7600572.

Ilinskaya, O., Ulyanova, V., Lisevich, I., Dudkina, E., Zakharchenko, N., Kusova, A., Faizullin, D., Zuev, Y., 2018. The native monomer of *Bacillus pumilus* ribonuclease does not exist extracellularly. BioMed Res. Int. 2018, 1–8.

Ilinskaya, O.N., Dreyer, F., Mitkevich, V.A., 2002. Changing the net charge from negative to positive makes ribonuclease Sa cytotoxic. Protein Sci. 11, 2522–2525.

Ilinskaya, O.N., Zelenikhin, P.V., Petrushanko, I.Y., Mitkevich, V.A., Prassolov, V.S., Makarov, A.A., 2007. Binase induces apoptosis of transformed myeloid cells and does not induce T-cell immune response. Biochem. Biophys. Res. Commun. 361, 1000–1005.

Ishii, I., Nureki, O., Yokoyama, S., 2003. Crystal structure of the tRNA processing enzyme RNase PH from *Aquifex aeolicus*. J. Biol. Chem. 278, 32397–32404. https://doi.org/10.1074/jbc.M300639200.

Jones, W., 1920. The action of boiled pancreas extract on yeast nucleic acid. Am. J. Physiol. 52, 203–207.

Jordaan, S., Akinrinmade, O.A., Nachreiner, T., Cremer, C., Naran, K., Chetty, S., Barth, S., 2018. Updates in the development of ImmunoRNases for the selective killing of tumor cells. Biomedicines 6, 28.

Kazantsev, A.V., Pace, N.R., 2006. Bacterial RNase P: a new view of an ancient enzyme. Nat. Rev. Microbiol. 4, 729–740.

Kim, J.S., Soucek, J., Matousek, J., Raines, R.T., 1995. Mechanism of ribonuclease cytotoxicity. J. Biol. Chem. 270, 31097–31102.

Kleinert, E., Langenmayer, M.C., Reichart, B., Kindermann, J., Griemert, B., Blutke, A., Troidl, K., Mayr, T., Grantzow, T., Noyan, F., Abicht, J.M., Fischer, S., Preissner, K.T., Wanke, R., Deindl, E., Guethoff, S., 2016. Ribonuclease (RNase) prolongs survival of grafts in experimental heart transplantation. J. Am. Heart Assoc. 5 https://doi.org/10.1161/JAHA.116.003429.

Koczera, P., Martin, L., Marx, G., Schuerholz, T., 2016. The ribonuclease A superfamily in humans: canonical RNases as the buttress of innate immunity. Int. J. Mol. Sci. 17, 1278–1294.

Kumar, R., Kanwar, 2018. New contenders for anticancer therapeutics: microbial ribonucleases. Adv. Biotech. Micro. 8. MS.ID.555739.

Kumar, R., Kanwar, S.S., 2017. Ribonuclease as anticancer therapeutics. Enzyme Eng. 6, 1–9. https://doi.org/10.4172/2329-6674.1000162.

Kumari, S., Pardeshi, S., Deshmukh, S.S., 2016. Extracellular RNase production and purification from *Streptomyces aureofaciens*. Int. J. Innov. Res. Dev. 17, 308–317.

Kurpiewska, K., Dziubek, K., Katrusiak, A., Font, J., Ribo, M., Vilanova, M., Lewin, K., 2016. Structural investigation of ribonuclease A conformational preferences using high-pressure protein crystallography. Chem. Phys. 468, 53–62.

Lacadena, J., Garcia, E.A., Sangra, N.C., Galan, E.H., Cebollada, J.A., Ortega, L.G., Onaderra, M., Gavilanes, J.G., Pozo, A.M.D., 2007. Fungal ribotoxins: molecular dissection of a family of natural killers. FEMS Microbiol. Rev. 31, 212–237.

Levy, S., Portnoy, V., Admon, J., Schuster, G., 2011. Distinct activities of several RNase J proteins in methanogenic archaea. RNA Biol. 8, 1073–1083.

Lu, L., Li, J., Moussaoui, M., Boix, E., 2018. Immune modulation by human secreted RNases at the extracellular space. Front. Immunol. 9, 1012. https://doi.org/10.3389/fimmu.2018.01012.

Madan, T., Arora, N., Sarma, P.U., 1997. Ribonuclease activity dependent cytotoxicity of Asp fl, a major allergen of A. fumigatus. Mol. Cell. Biochem. 175, 21–27.

Mader, U., Zig, L., Kretschmer, J., Homuth, G., Putzer, H., 2005. mRNA processing by RNases J1 and J2 affects *Bacillus subtilis* gene expression on a global scale. Mol. Microbiol. 70, 183–196. https://doi.org/10.1111/j.1365-2958.2008.06400.x.

Makarova, A.A., Ilinskayab, O.N., 2003. Cytotoxic ribonucleases: molecular weapons and their targets. FEBS Lett. 54, 15–20.

Mao, S., 2018. Structures of eukaryotic ribonuclease P. Struct. Biol. 362, 651–654.

Mironova, N.L., Zenkova, M.A., 2018. Vertebrate RNase 1 homologs: potential regulators of extracellular RNAs. Biotarget 2, 3.

Mishra, P., Singh, L., Wahid, Z.A., 2014. RNase purification from isolate Rns. Int. J. Eng. Technol. Sci. 1, 25–27.

Mitkevich, V.A., Kretova, O.V., Petrushanko, I.Y., Burnysheva, K.M., Sosin, D.V., Simonenko, O.V., Ilinskaya, O.N., Tchurikov, N.A., Makarov, A.A., 2013. Ribonuclease binase apoptotic signature in leukemic Kasumi-1 cells. Biochimie 95, 1344–1349.

Mitkevich, V.A., Makarov, A.A., Ilinskaya, O.N., 2014. Cell targets of antitumor ribonucleases. Mol. Biol. 48, 181–188.

Mohanty, B.K., Kushner, S.R., 2018. Enzymes involved in posttranscriptional RNA metabolism in Gram-negative bacteria. Microbiol. Spectr. 6, 10–36.

Morita, T., Niwata, Y., Ohgi, K., Ogawa, M., Irie, M., 1986. Distribution of two urinary ribonuclease-like enzymes in human organs and body fluids. J. Biochem. 99, 17–25.

Nicholson, A.W., 1998. Function, mechanism and regulation of bacterial ribonucleases. FEMS Microbiol. Rev. 23, 371–390.

Nicholson, A.W., 1999. Function, mechanism and regulation of bacterial ribonucleases. FEMS Microbiol. Rev. 23, 371–390.

Noguchi, J., 1924. On the decomposition of nucleic acids through takadiastase. Biochemistry 147–255.

Olmo, N., Turnay, J., Buitrago, G.G., et al., 2001. Cytotoxic mechanism of the ribotoxin-sarcin induction of cell death via apoptosis. Eur. J. Biochem. 268, 2113–2123.

Olson, B.H., Jennings, J.C., Roga, V., Junek, A.J., Schuurmans, D.M., 1965. Alpha-sarcin, a new antitumor agent. Appl. Microbiol. 13, 322–326.

Otani, H., 1935. The mold enzymes splitting nucleic acid. Acta. Schol. Med. Univ. Imp. Kioto. 17, 323.

Pace, C.N., Grimsley, G.R., Thomson, J.A., Barnett, B.J., 1998. Conformational stability and activity of ribonuclease T1 with zero, one, and two intact disulfide bonds. J. Biol. Chem. 263, 11820–11825.

Portnoy, V., Schuster, G., 2006. RNA polyadenylation and degradation in different Archaea; roles of the exosome and RNase R. Nucleic Acids Res. 34, 5923–5931.

Prats-Ejargue, G., Blanco, J.A., Salazar, V.A., Nogues, V.M., Moussaoui, M., Boix, E., 2019. Characterization of an RNase with two catalytic centers. Human RNase6 catalytic and phosphate-binding site arrangement favors the endonuclease cleavage of polymeric substrates. BBA-Gen. Subj. 1863, 105–117.

Prior, T.I., Kunwar, S., Pastan, I., 1996. Studies on the activity of barnase toxins in vitro and in vivo. Bioconjug. Chem. 7, 23–29.

Raines, R.T., 1998. Ribonuclease A. Chem. Rev. 98, 1045–1066.

Rajashree, A., Deshpande, Shankar, V., 2002. Ribonucleases from T2 family. Crit. Rev. Microbiol. 28, 79–122.

Reiter, N.J., Osterman, A., Torres-Larios, A., Swinger, K.K., Pan, T., Mondragon, A., 2010. Structure of a bacterial ribonuclease P holoenzyme in complex with tRNA. Nature 468, 784–789.

Rosenberg, H.F., 2008. RNase A ribonucleases and host defense: an evolving story. J. Leukoc. Biol. 83, 1079–1087.

Rosenberg, H.F., 2015. Eosinophil-Derived Neurotoxin (EDN/RNase 2) and the mouse Eosinophil-Associated RNases (mEars): expanding roles in promoting host defense. Int. J. Mol. Sci. 16, 15442–15455.

Sangaev, S.S., Kochetov, A.V., Ibragimova, S.S., Levenko, B.A., Shumny, V.K., 2011. The role of the extracellular ribonuclease in the physiological processes in higher plants. Russ. J. Genet. Appl. Res. 1, 44–50.

Sato, K., Egami, F., 1957. Studies on ribonucleases in takadiastase. J. Biochem. 44, 753.

Saxena, A., Saxena, S.K., Shogen, K., 2009. Effect of Onconase on double-stranded RNA *in vitro*. Anticancer Res. 29, 1067–1071.

Schein, A., Levin, S.S., Glaser, F., Schuster, G., 2008. The RNase E/G-type endoribonuclease of higher plants is located in the chloroplast and cleaves RNA similarly to the *E. coli* enzyme. RNA 14, 1057–1068.

Schwartz, B., Shoseyov, O., Vladislaya, O., 2007. ACTIBIND, a T2 RNase, competes with angiogenin and inhibits human melanoma growth, angiogenesis, and metastasis. Cancer Res. 67, 5258–5266.

Scott, A.J., Hu, J., 2013. Hepatitis B virus reverse transcriptase: diverse functions as classical and emerging targets for antiviral intervention. Emerg. Microb. Infect. 2, e56.

Sevcik, J., Urbanikova, L., Leland, P.A., Raines, R.T., 2002. X-ray structure of two crystalline forms of a streptomycete ribonuclease with cytotoxic activity. J. Biol. Chem. 277, 47325–47330.

Shapiro, R., Fett, J.W., Strydom, D.J., Vallee, B.L., 1986. Characteristic ribonucleolytic activity of human angiogenin. Biochemistry 25, 7255–7264.

Sharipova, M., Rockstroh, A., Balaban, N., Mardanova, A., Toymentseva, A., Tikhonova, A., Vologin, S., Stashevsky, Z., 2015. Antiviral effect of ribonuclease from *Bacillus pumilus* against phytopathogenic RNA-viruses. Agric. Sci. 6, 1357–1366.

Shigematsu, M., Kawamura, T., Kirino, Y., 2018. Generation of 2′,3′-cyclic phosphate-containing RNAs as a hidden layer of the transcriptome. Front. Genet. 9, 562. https://doi.org/10.3389/fgene.2018.00562.

Shlyakhovenko, V.O., 2016. Ribonucleases: possible new approach in cancer therapy. Exp. Oncol. 38, 2–8.

Shlyapnikov, S.V., Kulikov, V.A., Yakovlev, G.I., 1984. Amino acid sequence and S-S bonds of *Penicillium brevicompactum* guanyl specific ribonuclease. FEBS (Fed. Eur. Biochem. Soc.) Lett. 177, 246–248.

Sinatra, F., Callari, D., Viola, M., 2000. Bovine seminal RNase induces apoptosis in normal proliferating lymphocytes. Int. J. Clin. Lab. Res. 30, 191–196.

Singh, A., Batra, J.K., 2018. Insight into the functional role of unique determinants in RNA component of RNase P of *Mycobacterium tuberculosis*. Int. J. Biol. Macromol. 119, 937–944.

Singh, A., Ubaid-ullah, S., Ramteke, A.K., Batra, J.K., 2016. Influence of conformation of *M. tuberculosis* RNase P protein subunit on its function. PLoS One 11. https://doi.org/10.1371/journal.pone.0153798.

Sokurenko, Y., Nadyrova, A., Ulyanova, V., Ilinskaya, O., 2016. Extracellular Ribonuclease from *Bacillus licheniformis* (Balifase), a new member of the N1/T1 RNase superfamily. BioMed Res. Int. https://doi.org/10.1155/2016/4239375.

Sorrentino, S., 2010. The eight human "canonical" ribonucleases: molecular diversity, catalytic properties, and special biological actions of the enzyme proteins. FEBS Lett. 584, 2194–2200.

Suhadolnik, R.J., 1997. Biochemical evidence for a novel low molecular weight 2-5A-dependent RNase L in chronic fatigue syndrome. J. Interferon Cytokine Res. 17, 377–385. https://doi.org/10.1089/jir.1997.17.377.

Summers, J., Mason, W.S., 1982. Replication of the genome of a hepatitis B-like virus by reverse transcription of an RNA intermediate. Cell 29, 403–415.

Suri, S., Panda, B., Javed, S., Mohd, A., 2007. RNase: a novel enzyme for treatment of cancers. Internet J. Oncol. 5, 1–5.

Tafech, A., Bassett, T., Sparanese, D., Lee, C.H., 2006. Destroying RNA as a therapeutic approach. Curr. Med. Chem. 13, 863–881.

Tang, J., Liu, F., Nagy, E., Miller, L., Kirby, K.A., Wilson, D.J., Wu, B., Sarafianos, S.G., Parniak, M.A., Wang, Z., 2016. 3-Hydroxypyrimidine-2,4-diones as selective active site inhibitors of HIV Reverse transcriptase-associated RNase H: design, synthesis and biochemical evaluations. J. Med. Chem. 59, 2648–2659.

Tang, J., Maddali, K., Dreis, C.D., Sham, Y.Y., Vince, R., Pommier, Y., Wang, Z., 2011. N-3 hydroxylation of pyrimidine-2,4-diones yields dual inhibitors of HIV reverse transcriptase and integrase. ACS Med. Chem. Lett. 2, 63–67.

Tavis, J.E., Lomonosova, E., 2015. The hepatitis B virus ribonuclease H as a drug target. Antivir. Res. 118, 132–138.

Ten Asbroek, A.L., van Groenigen, M., Jakobs, M.E., Koevoets, C., Janssen, B., Baas, F., 2002. Ribonuclease H1 maps to chromosome 2 and has at least three pseudogene loci in the human genome. Genomics 79, 818–823. https://doi.org/10.1006/geno.2002.6776.

Torrent, M., Sanchez, D., Buzon, V., Nogues, M.V., Cladera, J., Boix, E., 2009. Comparison of the membrane interaction mechanism of two antimicrobial RNases: RNase 3/ECP and RNase 7. Biochim. Biophys. Acta 1788, 1116–1125.

Wilson, C.M., 1982. Plant nucleases: biochemistry and development of multiple molecular forms. In: Isoenzymes: Current Topics in Biological and Medical Research, vol. 5. Alan R. Liss, NY, pp. 33–54.

Xu, X., Yan, H., Chen, J., 2011. Bioactive proteins from mushrooms. Biotechnol. Adv. 29, 667–674.

Yakovlev, G.I., Karpeisky, M.Y., Bezborodova, S.I., Beletskaja, O.P., Sakharovsky, V.G., 1980. Guanyl-specific ribonuclease from the fungus *Penicillium chrysogenum* strain 152 and its complex with guanosine 30 -phosphate studied by nuclear magnetic resonance. Eur. J. Biochem. 109, 75–85.

Yasuda, T., Nadano, D., Takeshita, H., Kishi, K., 1993. Two distinct secretory ribonucleases from human cerebrum: purification, characterization and relationships to other ribonucleases. Biochem. J. 296, 617–625.

Yehudai-Resheff, S., Hirsh, M., Schuster, G., 2001. Polynucleotide phosphorylase functions as both an exonuclease and a poly (A) polymerase in spinach chloroplasts. Mol. Cell Biol. 21, 5408–5416. https://doi.org/10.1128/MCB.21.16.5408-5416.2001.

Yoshida, H., Fukuda, I., Hashiguchi, M., 1980. Purification by affinity chromatography and physicochemical properties of the guanine specific ribonuclease of *Fusarium moniliforme*. J. Biochem. 88, 1813–1818.

Yuki, S., Kondo, Y., Kato, F., Kato, M., Matsuo, N., 2004. Noncytotoxic ribonuclease, RNase T1, induces tumor cell death via hemagglutinating virus of Japan envelope vector. Eur. J. Biochem. 271, 3567–3572.

Zhang, K., Yang, O., Huang, W., Wang, K., Zhu, X., Xie, M., 2019. Detection of HIV-1 ribonuclease H activity in single-cell by using RNA mimics green fluorescent protein based biosensor. Sensor. Actuators B Chem. 281, 439–444.

Zhou, W.W., Zhang, L., Zhang, B., Wang, F., Liang, Z.H., Niu, T.G., 2008. Isolation and characterization of ZH14 with antiviral activity against *Tobacco mosaic virus*. Canadian J. Microbiol. 54, 441–449.

Zu, I., Wang, Y., Malhotra, A., 2005. Crystal structure of *Escherichia coli* RNase D, an exoribonucleases involved in structured RNA processing. Structure 13, 973–984. https://doi.org/10.1016/j.str.2005.04.015.

CHAPTER

Green nanotechnology for bioactive compounds delivery

13

Priya Sharma[a], Praveen Guleria[a], Vineet Kumar[b]

Department of Biotechnology, Faculty of Life Sciences, DAV University, Jalandhar, Punjab, India[a]; Department of Biotechnology, Faculty of Technology and Sciences, Lovely Professional University (LPU), Phagwara, Punjab, India[b]

Chapter outline

1. Introduction

Green chemistry is also known as sustainable chemistry. It is very well defined as the area of engineered chemistry and chemicals. Green chemistry is basically focused on designing the product with manipulating processes to reduce the use and emission of hazardous substances. Since 2010 till 2035, the total energy requirement is supposed to increase by more than 30% globally (International Energy Agency, 2011). The worldwide demand for safe drinking water is rising, with more than 800 million people having no access to water (WHO, 2010). These problems have raised the concern of various corporations, researchers and bureaucrats toward development of new technologies and their applications in an efficient and eco-friendly way. Among such effective new technologies, green innovation considerably possesses the potential of addressing the raising global challenges (OECD, 2012). Nanotechnology is also attracting the attention as it is expected to

Biotechnological Production of Bioactive Compounds. https://doi.org/10.1016/B978-0-444-64323-0.00013-8

contribute toward the transition. Hence, green innovation with nanotechnology termed as green nanotechnology can significantly regulate the environmental sustainability via manipulating processes producing negative emissions. Green nanotechnology can potentially find application in interdisciplinary technological fields giving rise to novel applications with promising improvement of life (Vartholomeos et al., 2011; Chahine et al., 2014; Kah, 2015; Prasad et al., 2016; Maksimovic and Omanovic-Miklicanin, 2017).

Green nanotechnology can and is expected to potentially help in development of clean studies minimizing the considerable environmental and human health associated risks. Encouraged replacement of existing nanotechnology products with environmentally safe nano-products offer reduction of risks arising from their chemical based manufacturing and applications (Iavicoli et al., 2014; Maksimovic and Omanovic-Miklicanin, 2017). Miniaturized nanoparticles have found potent application in imaging and targeted drug deliver for the detection and treatment of early human diseases (Koo et al., 2005). To support the biological applications of nanotechnology, green chemistry-based approaches for nanomaterial synthesis is highly demanded. The poor targeting of biological compounds to target tissues reduces their medicinal value (Koo et al., 2005; Terracciano et al., 2018). Hence to employ the bioactive compounds for their highest medicinal value, use of greener nanomaterials as target vehicles is expected to be efficient for slow and sustained release of bioactive compounds.

2. Green chemistry and its principles

Green chemistry is defined as the synthesis of chemical products with significant reduction in the production and elimination of harmful substances. It is also termed as sustainable chemistry. Green chemistry is an interdisciplinary approach to science instead of a separate scientific discipline. It maintains sustainability between natural resources, environment and economic growth (Riđanović et al., 2013). Twelve principles of sustainable chemistry efficiently define the green chemical synthesis making the processes environment friendly and safer for the human health (Taksande et al., 2017). These principles demonstrate green chemistry to include processes that offers the use of chemicals and manufacturing processes posing least/no harmful impact the ecology and environment (Jukić et al., 2005; Margetic, 2005; Cue and Zhang, 2009). The twelve principles defining green chemistry were given by Paul Anastas and John Warner of EPA. It is recommended to meet the principles of green chemistry during synthesis; however, it is impossible to meet all the twelve principles in a single reaction. It is based on twelve principles, given by Paul Anastas and John Warner of EPA. The twelve principle of green chemistry include "prevention of waste, atom economy, less hazardous chemical synthesis, safer chemical designing, use of safer solvents and auxiliaries, design for energy efficiency, use of renewable feed backs, reduction of derivatives, use of catalysis, design for degradation, real-time" analysis for pollution prevention, inherently safer chemistry for accident prevention (Anastas and Warner, 1998; Anastas et al., 2001; Collins, 1997; Li and Trost, 2008; Anastas and Eghbali, 2010).

The first principle of green chemistry recommends prevention of waste generation. It states the significance of waste prevention over waste clean-up/treatment. This principle is more prominent to the environment and to the humans. The United States has been reported to annually eliminate 12 billion tonnes of hazardous waste. 70% of the total hazardous industrial waste includes the organic toxic waste supplemented with methanol and xylene along with the toxic reactants used for organic synthesis (Ivankovic et al., 2017a,b).

Prevention of waste is also a challenge for pharmaceutical industry. Synthesis of eco-friendly process for medicine development poses a challenge to pharmaceutical industries. Sertraline production is an efficient example of green chemistry or prevention of waste. Using only one solvent ethanol eliminated the extensive processing and recovery of solvents during steraline production. Additionally, the solvent consumption for the process was reduced from 250 to 25 L/kg of steraline (Sheldon, 2011).

Principle of waste prevention is significantly related to the principle of Atom economy. Waste prevention designates the conversion of total raw material into final product with maximum utilization with minimum amount of waste. The principle was given by Barry Trost, Stanford University in 1991. The best example justifying atom economy is of ibuprofen biosynthesis (Murphy, 2018). The old synthesis method has lowest economic cost, with only 40% utilization of the input raw material. However, the new synthesis process showed high atom efficiency with least waste production. Further, the process contributed to pollution prevention (Anastas, 1998; Anastas et al., 2001).

Substances should be generated or used that produce less or no effect on human health and environment. Synthetic methods are being designed with same aim. Industrial processes can be cheaper and clearer if the harmful chemicals are replaced by biological enzymes (Li et al., 2012). Polycarbonate synthesis suitably justifies this principle. Substitution of carbonyl dichloride ($COCl_2$) with CO_2 has eliminated the use of dichloromethane as process solvent (Welton, 2015).

The green chemistry principle recommends designing chemical products of significant effectiveness with reduced toxicity. The greener chemical synthesis ensures reduced toxicity via balancing the optimal performance and function of chemical product. This principle manages the synthesis of target organism specific insecticides and pesticides with non-toxicity to surrounding organisms and eco-friendly degradation. Another example is successful replacement of tin-based highly toxic organic compound by Sea-Nine. Sea-Nine is biodegradable, non-toxic organic substitute for tin toxic compound (Riđanović et al., 2013). Production of polyphenylene sulfone polymer is another example. It is widely used for indoor airplanes and underground trains. Polyphenylene sulfone has flame resistant properties and is environmentally safe (Hill and Brady, 1976).

Solvents used for the chromatographic separations are problematic due to environmental pollution. Most of the organic solvents used in chromatographic separations are toxic, flammable and corrosive. Hence, it is recommended to avoid the use of auxiliary solvents during various separation processes as their recycling lead to huge energy losses. This principle thus proposes a process of synthesis, which reduced and avoids the use of auxiliary chemical substances. Hence during the choice of organic solvent, it should always be in mind that the process is safe for workers, safe for environment and possess sustainability. Further, the solvent is recommended to be physically and chemically stable, should possess low volatility and ease of recycling (Karkkainen, 2007; Guardia and Armenta, 2010).

The green chemistry proposes the designing of economical processes with recognition and minimization of environmental impact of energy. The chemical synthesis at ambient pressure and temperature is highly recommended. The 1973 oil crisis demonstrates the application of green chemistry principles by developing processes for complete energy consumption during oil production (Hoffert et al., 2002). A suitable example for the principle is Tomato. Tomatoes grown in green houses can be heated by using a waste steam from nearby chemical plants producing ammonia. Hence the CO_2 requirement of the green houses falls down by 50% (Garnett, 2011).

The seventh principle advocates the use of renewable energy sources. This principle says that the feedstock should be renewable instead of depleting. For example, irrespective of plastic that generates waste, use of biodegradable plastic is recommended. Food industries are efficiently using the biodegradable plastic for food packaging (Petersen et al., 1999; Stevens, 2002; Ivankovic et al., 2017a,b). The principle further proposes the application of renewable energy resources like hydropower, biomass, wind and sunlight rather than non-renewable energy sources (Owusu and Asumadu-Sarkodie, 2016).

The principle of derivatives reduction recommends the avoidance of elimination of chemical derivatives during target molecule synthesis. Production of antibiotic was modified by replacing the chemical penicillin synthesis by an enzymatic reaction. This substitution eradicated the requirement of ultra-low temperature, organic solvents and other complexities associated with chemical penicillin synthesis (Volpato et al., 2010).

The principle of catalysis promotes the use of biodegradable catalysts than organochlorine-based compounds during synthesis. Biodegradable catalysts are energy efficient and reduce water usage with least water wastage. The catalyst potentially reduces the activation energy of a reaction but accelerates the speed of reaction by several times. Also, the catalyst remains unchanged during the chemical reaction. The chemical synthesis of catechol from benzene is a multi-step process and generates undesirable byproducts. The biocatalytic synthesis of catechol from D-glucose in the presence of *E. coli* was summed up in one step without any production of byproducts (Anastas et al., 2001).

3. Medicinal importance of bioactive molecules

Molecules that are synthesized through biochemical process and derived from natural sources are known as bioactive molecules or bio molecules (Kang et al., 2013). Bioactive molecules are mainly known for their medicinal properties and can be obtained from plants, animals, microbes and their derived products. Table 13.1 tabulates various bioactive molecules along with their medical applications (Yew et al., 2013). Bioactive molecules act as potential lead molecules for new drug development. Isolation of penicillin from *Penicillium notatum* marked the focus of scientists for drug discovery from living sources. Identification of lead bioactive molecule and their mechanistic property can help in designing and designing new molecules with potent biological activities (Cragg and Newman, 2013).

Presence of multiple phytochemicals in several plant systems has been the basis for the development of therapeutics from biomolecules (Basu, 2013). The use of plant-derived biomolecules as herbal medicines traces its history back since the establishment of Ayurveda. But with the development of genetic and immunological

Table 13.1 Medicinal properties of distinct bioactive molecules.

Bio-active molecule	Medicinal property	References
Tea phenolics	Antioxidant	Rietveld and Wiseman (2003)
Quercetin	Antioxidant and anticancer potential., Tyrosine kinase mediated inhibition of cellular proliferation	Ferry et al. (1996) and Janssen et al. (1998)
Beta-carotene and lycopene	Anticancer	Kucuk et al. (2001)
Rutin	Antioxidant	Gill et al. (2007)
Isoflavones	Anticancer	Kumar et al. (2007)
Ascorbic acid	Anti-malignant	Hoffer et al. (2008)
Epigallocatechin gallate	Anticancer	Mukhtar and Ahmad (2000)
Curcumin	Anticancer and anti-inflammatory	Deodhar et al. (1980)
Ellagic acid (Pomegranate juice)	Anticancer	Pantuck et al. (2006)
Proanthocyanidin (Cranberry juice)	Antimicrobial	Raz et al. (2004)
Apigenin	Cardiovascular risk inhibition	Janssen et al. (1998)
Omega-3 and omega-6 fatty acids	Anticancer	Marwick (2001) and Wang (1998)
Podophyllotoxin (Mayapple)	Anticancer	Cragg and Newman (2005)

background behind the medicinal application of herbal compounds, the scope of bioactive molecules in therapeutics was assessed and implemented. Hence, naturally derived bioactive molecules were the extensively used for the development of new drugs (Rogerio et al., 2016).

Nanoparticles synthesized from various bulk counterparts of metals, non-metals, polymers can efficiently be internalized by the cells and can potentially induce biological responses. To further stabilize their biological applications, synthesis of nanoparticles from biological agents and their coating with biomolecules is experimentally validated (Ishihara et al., 2016). The chapter further discusses the synthesis of nanoparticles by greener approaches and their validated therapeutic applications.

4. Biological synthesis of nanoparticles

Nanoparticles have been reported to possess potential biological applications. They possess efficient biomedical and therapeutic abilities. However, certain physiochemical properties of nanoparticles significantly hinder their biological functionalities (Mou et al., 2015; Ishihara et al., 2016). Hence to enhance their biocompatibility, nanoparticle synthesis and engineering with molecules of natural origin have been recommendation. This has led to the development of strategies and procedures for the development of nanoparticles using biological molecules (Kumar and Yadav, 2009) (Fig. 13.1).

Various reports have documented the biological synthesis of nanoparticles using plant extracts (Kumar et al., 2010; Kumar and Yadav, 2011a, 2011b, 2012a, 2012b; Vijayaraghavan and Ashokkumar, 2017). The plant phytochemicals act as natural stabilizers and reducers supporting the nanoparticle synthesis. The plant-mediated synthesis of nanomaterials has been reported as economical and eco-friendly (Kumar et al., 2010; Vijayaraghavan and Ashokkumar, 2017). Likewise, microbial nanoparticle synthesis has also been documented (Li et al., 2011; Fariq et al., 2017; Hulkoti and Taranath, 2014). It is also a green and environmentally safe approach like plant derived nanoparticle preparation. Microbial enzymes and proteins are documented to be responsible for the synthesis and stabilization of nanoparticles (Nasreen et al., 2014; Fariq et al., 2017).

5. Nanomaterials based bioactive compound delivery

The various physicochemical properties of nanoparticles contribute for their medicinal potential (Bar et al., 2009; Tuutijarvi et al., 2009; Rassaei et al., 2008). The stability of nanoparticles is dependent on their way of synthesis. The aggregation of nanoparticles during non-biological synthesis can lead to nanoparticle aggregates having harmful and non-ecofriendly side effects. Eco-friendly and non-toxic nanoparticles can be synthesized by improvement of natural methodologies that potentially blend nanoparticles and their biological properties (Kowshik et al., 2003;

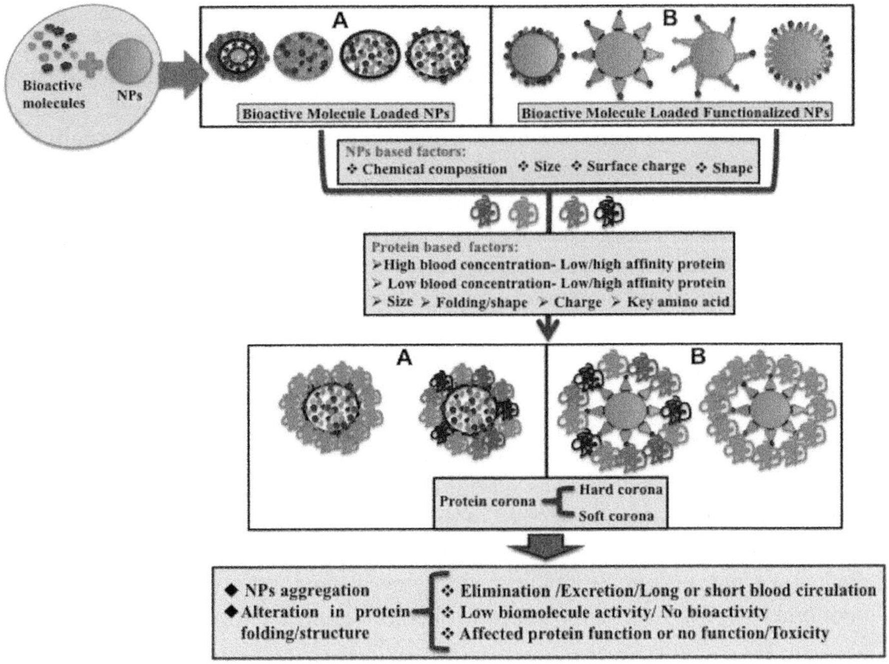

FIG. 13.1

Diagram showing formation of nanoparticles (non-functionalized nanoparticles) and their functionalization for drug delivery application. Functionalized as well as non-functionalized nanoparticles undergo different fate once inside body.

Kalimuthu et al., 2008; Shahverdi et al., 2007; Sangi and Verma, 2009). Among the natural biological methodologies for nanoparticles synthesis, plant mediated synthesis is the most significant process. Plant based nanoparticles synthesis is the eco-friendliest technique than several physical and chemical processes (Willner et al., 2006; Sreeram et al., 2008).

5.1 Nanomaterials for delivery of anti-oxidant bioactive molecules

Due to the reliability and simplicity of chitosan synthesis route, chitosan based nano carriers have been widely explored. Tripolyphosphate mediated ionic gelation method was used for the synthesis of naringenin (NAR) nanoparticles encapsulated with chitosan designated as CS-NPs/NAR. DLS, SEM, zeta potential, FTIR and Energy-dispersive X-ray spectroscopy analysis were used for nanoparticle characterization. The synthesized nanoparticles possessed 53.2 nm size. However, encapsulation of nanoparticles increased the size to 407.47 nm. The HPLC reported the efficiency of encapsulation of particles with chitosan to be 80%. The CS-NPs/NAR showed efficient free radical scavenging efficiency as indicated by 2,2-diphenyl-2-picryl hydrazyl hydrate assay, hydroxyl radical and nitrate-scavenging assay. Hence chitosan loaded nanoparticles were documented as efficient naringenin delivery vehicle with considerable antioxidant activity (Kumar et al., 2015). Likewise, nanoceria, the cerium oxide nanoparticles have been reported as significant inorganic cancer preventive agents. Biodegradable as well as non-biodegradable polyelectrolyte microcapsules can be efficiently used for the intercellular delivery of therapeutic nanoceria (Popov et al., 2018).

PLGA polymeric nanoparticles have already been reported for the encapsulation and controlled release of quercetin and catechin. The solvent displacement method was used for nanoparticles synthesis. The nanoparticles physicochemical characterization was done by using methodologies like SEM, light scattering, X-ray diffraction, zeta potential and infrared spectroscopy. The characterized nanoparticles were non-crystalline and of size approximately 400 nm. The formation of hydrogen bonds was responsible for the interaction of nanoparticles with loaded flavonoids catechin and quercetin. The encapsulation efficiency, *in vitro* release assay and antioxidant potential of polymeric nanoparticles was also assayed. The *in vitro* release profile was found to be pH dependent. The least the pH, the rapid was the release of flavonoids from the nanoparticles. Loading of quercetin and catechin onto PLGA nanoparticles inhibited the development of free radicals and chelating properties. Hence, systematized designing and fabrication of polymeric nanoparticles offers their use as efficient oral delivery vehicle for the controlled release of biomolecules (Pool et al., 2012).

5.2 Nanomaterials for delivery of anti-cancer bioactive molecules

The green nanoparticles has also been efficiently reported for the loading and release of anticancerous compounds (Kumar et al., 2014; Kumari et al., 2012). Anticancerous effect of chitosan nanoparticles thiolated individually with curcumin and 5-fluorouracil (CRC-TCS-NPs/5-FU-TCS-NPs) has been evaluated using colon cancer cells and swiss albino mice. The CRC-TCS-NPs and 5-FU-TCS-NPs used for study were 150 nm in size and were documented to be blood compatible. These nanoparticles showed sustained release of anti-cancerous compound at pH 4.5 and 7.4 over a period of 4 days. Anticancerous effect of curcumin and 5-fluorouracil was enhanced by 2.5- to 3-folds by loading them on chitosan nanoparticles. Nano-formulation of 5-FU and CRC improved their release and accumulation in the plasma in comparison to bare biomolecules (Anitha et al., 2014). Hence, nano-formulations have the potential of significant and consistent release of anticancerous biomolecules and drugs that enhances their bioavailability *in vivo* as well as *in vitro*.

The biological phytochemicals have been known to possess significant anticancerous potential. However, reduced bioavailability and retention of biomolecules constrain their clinical and pharmaceutical usage. Hence, the combined application of anti-tumorous phytochemicals has been shown to improve their anti-cancerous potential. Liposome loaded with curcumin co-administered with liposomal resveratrol showed significant *in vivo* anti-tumorous activity against prostatic adenocarcinoma. Co-administration of curcumin with resveratrol inhibited the cancerous cell growth by inducing cellular apoptosis (Narayanan et al., 2009).

Similarly, another study had reported the consistent and stable release of anticancerous molecules tamoxifen (Tmx) and quercetin (QT) from orally administered polymeric PLGA nanoparticles. The PLGA nanoparticles were co-encapsulated with Tmx and QT. The nanoparticles were 185.3 nm in size and showed 67.16% entrapment efficiency toward Tmx and 68.60% for QT. Further, the biomolecules were documented to retain their functional structure even in the freeze-dried form of synthesized nanoparticles. The oral bioavailability of Tmx and QT was three- to fivefold higher in nano-formulation rather than their free forms, respectively. Hence, the oral administrations of Tmx/QT loaded PLGA NPs showed significant tumor angiogenesis. Further unlike drugs and drug combinations, the release of Tmx-QT-NPs did not show any symptomatic hepatotoxicity and oxidative stress (Jain et al., 2013). Therefore, loading of multiple biomolecules on polymeric nanoparticles can be an efficient strategy for oral delivery of anti-cancerous molecules with significant stability.

5.3 Nanomaterials for delivery of anti-malarial bioactive molecules

Nanosystem has been shown to deliver resveratrol in brain without allowing its degradation in the blood stream to target malaria. The spherical lipid nanoparticles loaded with resveratrol were targeting to LDL receptors of blood-brain barrier by functionalization with apolipoprotein E. The efficiency of resveratrol release was documented to be 90%. Incubation of resveratrol loaded lipid nanoparticles were non-toxic up to concentration 50 μM as indicated by MTT and LDH assays (Neves et al., 2016).

The intra-erythrocytic growth and progression of malarial parasite depends on the nutrient uptake by the parasitized red blood cells like uptake of human serum albumin (HSA). Hence, the possibility of HSA nanoparticles as potential delivery vehicle for drug anti-malarial molecules has been evaluated. HAS nanoparticles loaded with anti-malarial drug Artemether were evaluated for their *in vitro* as well as *in vivo* activity against *Plasmodium falciparum* and *P. berghei*. The confocal imaging indicated selective internalization of drug by the plasmodium-infected red blood cells over non-infected healthy red blood cells. The drug loaded onto nanoparticles achieved IC_{50} levels of inhibition at a dose lower than 50% of the only drug control. Further, twofold higher drug concentration was documented in the red blood cells via HAS nanoparticles with higher anti-malarial activity at 50% lesser dosage on the only drug control. Hence this study reports a novel strategy of using HSA-based nanoparticles for secure and effective delivery of anti-malarial drugs (Sidhaye et al., 2016).

The silver nanoparticles synthesized from aqueous leaf extract of *Artemisia nilagirica* has also been documented as an efficient anti-malarial agent. The synthesized nanoparticles were spherical in shape and 6.723 in size. The nanoparticles showed inhibitory effect against all the four developmental stages of malarial parasite

Anopheles stephensi and dengue parasite *Aede aegypti*. Hence, synthesized nanoparticles from *A. nilagirica* could be a greener anti-malarial and anti-dengue approach that is larvicidal as well as eco-friendly (Nalini et al., 2017).

5.4 Nanomaterials for delivery of anti-bacterial bioactive molecules

Delivery of anti-microbial drug molecules using metallic nanoparticles has been suggested to enhance their activity against microbes. The synthesis of silver nanoparticles via curcumin showed interaction of curcumin with silver, leading to aggregate formation and a dark brown/black silver oxide. XRD confirmed the crystallization of curcumin covered Ag NPs and the size was estimated to be 13.27 nm. However, the individual nanoparticle was spherical with size ranging from 10 to 50 nm. The Ag-NPs synthesis showed linearly enhancement with increase in curcumin concentration suggesting it to have first order kinetics (Khoury et al., 2015).

Ubiquicidin 29-41 (UBI) is an antimicrobial peptide cation and is documented to possess sensitivity for bacterial and fungal contamination. The UBI encapsulated silver and gold nanoparticles has shown significant antimicrobial potential against bacterial strains *E. coli* and *Pseudomonas aeruginosawa*. The UBI coated silver nanoparticles showed higher antibacterial activity than non-functionalized Ag NPs. The spectroscopic analysis showed NH_2 groups interaction of nanoparticles with UBI on functionalization (Morales-Avila et al., 2017).

The silver nanoparticles synthesized from leaf extract of *Cestrum nocturnum* have shown significant bacteriostatic and bactericidal activity against bacterial strains including *E. coli*, *Vibrio cholera* and *Salmonella typhii*. The crystalline nanoparticles were spherical with 20 nm in size. The synthesized nanoparticles showed antioxidant activity higher than vitamin C. Further, bioactive molecules loaded onto nanoparticles from the leaf extract of *C. nocturnum* were responsible for their antioxidant and antimicrobial activity (Keshari et al., 2019).

5.5 Nanomaterials for delivery of wound healing bioactive molecules

Biologically synthesized nanoparticles have also been documented to show significant wound healing abilities. The silver nanoparticles synthesized from *Naringi cernulata* leaf extracts has documented significant wound repairing potential. The plant derived bioactive compounds, like alkaloids, phenols, saponins and quinines induced the reduction of cationic silver into Ag^0. The synthesized nanoparticles were spherical to cubic with their size ranging from 72 to 98 nm. These nanoparticles showed 100% wound healing potential when topically applied to wounded male Wistar rats showed in a percent concentration of 5%. Hence, the study supported and encouraged the possibilities of using plant derived biomolecule loaded nanoparticles as efficient treatment of topical infections including wounds (Bhuveneshwari et al., 2014).

The silver nanoparticles synthesized using callus and leaf extracts of *Cumunic sativus* has also been comparatively evaluated their wound healing potential. The nanoparticles synthesized from callus were polygonal and 21–23 nm in size. On the contrary, nanoparticles synthesized from *C. sativus* leaf extract were spherical and 11–19 nm in size. Both the synthesized particles showed significant wound healing abilities. The 5% concentration of nanoparticles constituting an ointment base showed 100% wound healing potential on topical application than used control drugs. Hence, the biomolecules present in *C. sativus* extract implicated effective wound healing potential to the synthesized nanoparticles (Venkatachalam et al., 2015). Similarly, the excision wounds of male Wistar rats have been reported to be efficiently healed on topical exposure of silver and gold nanoparticles synthesized from root extracts of *Coleus forskohlii*. The antioxidant and antimicrobial potential of synthesized nanoparticles was responsible for their potent wound healing capability (Naraginti et al., 2017). Likewise, several other reports have documented the significant wound healing potential of green nanoparticles (Archana et al., 2013; Sandri et al., 2013; Han et al., 2012).

6. Conclusion and future directions

The essential objective of every industry is to produce cash from accessible crude materials and fundamental input capital. However, every feasible modern exercise must address the present issues without imperiling the necessities of who and what is to come, implying that synthetic procedures need to utilize crude materials, water and vitality in a way that should be monetarily suitable and eco-friendly. The use of the idea of green science presents compound well-being that suggests satisfactory lawful control of specific systems and exercises that are unavoidable for the execution of such an idea. The objectives of green science in ecological insurance furthermore, correspond to financial gain. Use of biocatalyst, catalyst, utilization of elective sustainable crude materials (biomass), elective response media (water, ionic fluids, and supercritical liquids), and elective response conditions (microwave actuation) demonstrate new photocatalytic responses. Further, the tremendous capabilities of biological agents including plant biomolecules and microbes are predominantly going to replace chemical agents thus, transforming the chemistry into a more sustainable and eco-friendlier "green" program.

Acknowledgments

Priya Sharma and Praveen Guleria would like to thank vice-chancellor and chancellor DAV University Jalandhar for encouragement and support. Vineet Kumar would like to thank vice-chancellor and chancellor LPU Jalandhar for encouragement and support.

References

Anastas, P., Eghbali, N., 2010. Green chemistry: principles and practice. Chem. Soc. Rev. 39, 301–312.

Anastas, P.T., Warner, J.C., 1998. Green Chemistry: Theory and Practice. Oxford University Press, Oxford.

Anastas, P.T., Warner, J.C., 2000. Green Chemistry Theory and Practice. Oxford University Press, New York, ISBN 9780198506980, pp. 10–55.

Anastas, P.T., Kirchhoff, M.M., Williamson, T.C., 2001. Catalysis as foundational pillar of green chemistry. Appl. Catal. A: General. 221, 3–13.

Anitha, A., Deepa, N., Chennazhi, K.P., Lakshmanan, V.K., Jayakumar, R., 2014. Combinatorial anticancer effects of curcumin and 5-fluorouracil loaded thiolated chitosan nanoparticles towards colon cancer treatment. Biochim. Biophys. Acta 1840, 2730–2743.

Archana, D., Dutta, J., Dutta, P.K., 2013. Evaluation of chitosan nano dressing for wound healing: characterization, in vitro and in vivo studies. Int. J. Biol. Macromol. 57, 193–203.

Bar, H., Bhui, D.K., Sahoo, G.P., Sarkar, P., De, S.P., Misra, A., 2009. Green synthesis of silver nanoparticles using latex of *Jatrobha curcas*. Colloids Surf. B 339, 134–139.

Basu, K.S., 2013. New bioactive molecules from Indian medicinal plants for pharmaceutical and insecticidal applications. Biochem. Pharmacol. 2, 4.

Bhuveneshwari, T., Thiyagarajan, M., Geetha, N., Venkatachalam, P., 2014. Bioactive compound loaded stable silver nanoparticle synthesis from microwave irradiated aqueous extracellular leaf extracts of *Naringi crenulata* and its wound healing activity in experimental rat model. Acta Trop. 135, 55–61.

Chahine, N.O., Collette, N.M., Thomas, C.B., Genetos, D.C., Loots, G.G., 2014. Nanocomposite scaffold for chondrocyte growth and cartilage tissue engineering: effects of carbon nanotube surface functionalization. Tissue Eng. Part A 20, 2305–2315.

Collins, T.J., 1997. Green chemistry, Macmillan Encyclopedia of Chemistry. Simon and Schuster, New York, pp. 691–697.

Cragg, G.M., Newman, D.J., 2005. Plants as a source of anti-cancer agents. J. Ethnopharmacol. 100, 72–79.

Cragg, G.M., Newman, D.J., 2013. Natural products: a continuing source of novel drug leads. Biochim. Biophys. Acta 1830, 3670–3695.

Cue, B.W., Zhang, J., 2009. Green process chemistry in the pharmaceutical industry. Green Chem. Lett. Rev. 2, 193–211.

Deodhar, S.D., Sethi, R., Srimal, R.C., 1980. Preliminary study on antirheumatic activity of curcumin (diferuloyl methane). Indian J. Med. Res. 71, 632–634.

Fariq, A., Khan, T., Yasmin, A., 2017. Microbial synthesis of nanoparticles and their potential applications in biomedicine. J. Appl. Biomed. 15, 241–248.

Ferry, D.R., Smith, A., Malkhandi, J., Fyfe, D.W., deTakats, P.G.A., Anderson, D., Baker, J., Kerr, D.J., 1996. Phase I clinical trial of the flavonoid quercetin: pharmacokinetics and evidence for in vivo tyrosine kinase inhibition. Clin. Cancer Res. 2, 659–668.

Garnett, T., 2011. Where are the best opportunities for reducing greenhouse gas emissions in the food system (including the food chain)? Food Policy 36, S23–S32.

Gill, C.I., Haldar, S., Boyd, L.A., Bennett, R., Whiteford, J., Butler, M., Pearson, J.R., Bradbury, I., Rowland, I.R., 2007. Watercress supplementation in diet reduces lymphocyte DNA damage and alters blood antioxidant status in healthy adults. Am. J. Clin. Nutr. 85, 504–510.

Guardia, M., Armenta, S., 2010. Green solvents for analytical separation and analyses. In: Encyclopedia of Analytical Chemistry: Applications, Theory and Instrumentation, pp. 1–15. https://doi.org/10.1002/9780470027318.a917.

Han, G., Nguyen, L.N., Macherla, C., Chi, Y., Friedman, J.M., Nosanchuk, J.D., Matrinez, L.R., 2012. Nitric oxide-releasing nanoparticles accelerate wound healing by promoting fibroblast migration and collagen deposition. Am. J. Pathol. 180, 1465–1473.

Hill, W.H., Brady, D.G., 1976. Properties, environmental stability, and molding characteristics of polyphenylene sulfide. Polm. Eng. Sci. 16, 831–835.

Hoffer, L.J., Levine, M., Assouline, S., Melnychuk, D., Padayatty, S.J., Rosadiuk, K., Rousseau, C., Robitaille, L., Miller, W.H., 2008. Phase I clinical trial of i.v. ascorbic acid in advanced malignancy. Ann. Oncol. 377, 1969–1974.

Hoffert, M.I., Caldeira, K., Benford, G., David, R., Criswell, D.R., Christopher, G.C., Herzog, H., Jain, A.K., Kheshgi, H.S., Lackner, K.S., Lewis, J.S., Lightfoot, H.D., Manheimer, W., Mankins, J.C., Mauel, M.E., Perkins, L.J., Schlesinger, M.E., Volk, T., Wigley, T., 2002. Advanced technology paths to global climate stability: energy for a greenhouse planet. Science 298, 981–987.

Hulkoti, N.I., Taranath, T.C., 2014. Biosynthesis of nanoparticles using microbes – a review. Colloids Surf. B 121, 474–483.

Iavicoli, I., Leso, V., Ricciardi, W., Hodson, L.L., Hoover, M.D., 2014. Opportunities and challenges of nanotechnology in the green economy. Environ. Health 13, 78.

International Energy Agency, World energy outlook, 2011. Energy projections to 2035. Re-energising the global economy? France, ISBN 9789264124134.

Ishihara, K., Chen, W., Liu, Y., Tsukamoto, Y., Inoue, Y., 2016. Cytocompatible and multi-functional polymeric nanoparticles for transportation of bioactive molecules into and within cells. Sci. Technol. Adv. Mater. 17, 300–312.

Ivankovic, A., Dronjic, A., Bevanda, A.M., Talic, S., 2017a. Review of 12 principles of green chemistry in practice. Int. J. Sustain. Green Energy 6, 39–48.

Ivankovic, A., Zeljko, K., Talic, S., Martinovic, B.A., Lasic, M., 2017b. Biodegradable packaging in the food industry. Arch. Leb. 68, 23–52.

Jain, A.K., Thanki, K., Jain, S., 2013. Co-encapsulation of tamoxifen and quercetin in polymeric nanoparticles: implications on oral bioavailability, antitumor efficacy, and drug-induced toxicity. Mol. Pharm. 10, 3459–3474.

Janssen, K., Mensink, R.P., Cox, F.J., Harryvan, J.L., Hovenier, R., Hollman, P.C., Katan, M.B., 1998. Effects of the flavonoids quercetin and apigenin on hemostasis in healthy volunteers: results from an in vitro and a dietary supplement study. Am. J. Clin. Nutr. 67, 255–262.

Jukić, M., Djaković, S., Filipović-Kovačević, Ž., Kovač, V., Vorkapić-Furač, J., 2005. Dominant trends of green chemistry. Kem Ind 54, 255–272.

Kah, M., 2015. Nanopesticides and nanofertilizers: emerging contaminants or opportunities for risk mitigation? Front Chem. 3, 64.

Kalimuthu, K., Babu, R.S., Venkataraman, D., Bilal, M., Gurunathan, S., 2008. Biosynthesis of silver nanocrystals by *Bacillus licheniformis*. Colloids Surf. 65, 150–153.

Kang, Y.B., Mallikarjuna, P.R., Fabian, D.A., Gorajana, A., Lim, C.L., Tan, E.L., 2013. Bioactive molecules: current trends in discovery, synthesis, delivery and testing. Int. J. Educ. Soc. 7, 32–46.

Karkkainen, J., 2007. Preparation and Characterization of Some Ionic Liquids and Their Use in the Dimerization Reaction of 2-methylpropene (Dissertation). Oulu University Press.

Keshari, A.K., Srivastava, A., Singh, A., Yadav, V.B., Nath, G., 2019. Antioxidant and antibacterial activity of silver nanoparticles synthesized by *Cestrum nocturnum*. J. Ayurveda Integr. Med. https://doi.org/10.1016/j.jaim.2017.11.003.

Khoury, E.E., Abiad, M., Kassaify, Z.G., Patra, D., 2015. Green synthesis of curcumin conjugated nanosilver for the applications in nucleic acid sensing and anti-bacterial activity. Colloids Surf. B 127, 274–280.

Koo, O.M., Rubinstein, I., Onyuksel, H., 2005. Role of nanotechnology in targeted drug delivery and imaging: a concise review. Nanomedicine 1, 193–212.

Kowshik, M., Ashtaputure, S., Kharazi, S., Vogel, N., Urban, J., Kullkarni, S.K., Panknikar, K.M., 2003. Extracellular synthesis of silver nanoparticles by a silver-tolerant yeast strain MKY3. Nanotechnology 14, 95–100.

Kucuk, O., Sarkar, F.H., Sakr, W., Djuric, Z., Pollak, M.N., Khachik, F., Li, Y.W., Banerjee, M., Grignon, D., Bertram, J.S., Crissman, J.D., Pontes, E.J., Wood, D.P., 2001. Phase II randomized clinical trial of lycopene supplementation before radical prostatectomy. Cancer Epidemiol. Biomark. Prev. 10, 861–868.

Kumar, N.B., Krischer, J.P., Allen, K., Riccardi, D., Besterman-Dahan, K., Salup, R., Kang, L., Xu, P., Pow-Sang, J., 2007. A phase II randomized, placebo-controlled clinical trial of purified isoflavones in modulating steroid hormones in men diagnosed with localized prostate cancer. Nutr. Canc. 59, 163–168.

Kumar, S.P., Birundha, K., Kaveri, K., Devi, R., 2015. Antioxidant studies of chitosan nanoparticles containing naringenin and their cytotoxicity effects in lung cancer cells. Int. J. Biol. Macromol. 78, 87–95.

Kumar, V., Kumari, A., Kumar, D., Yadav, S.K., 2014. Biosurfactant stabilized anticancer biomolecule-loaded poly (d,l-lactide) nanoparticles. Colloids Surf. B 117, 505.

Kumar, V., Yadav, S.C., Yadav, S.K., 2010. *Syzygium cumini* leaf and seed extract mediated biosynthesis of silver nanoparticles and their characterization. J. Appl. Chem. Biotechnol. 85, 1301–1309.

Kumar, V., Yadav, S.K., 2009. Plant-mediated synthesis of silver and gold nanoparticles and their applications. J. Appl. Chem. Biotechnol. 84, 151–157.

Kumar, V., Yadav, S.K., 2011a. Synthesis of stable, polyshaped silver and gold nanoparticles using leaf extract of *Lonicera japonica* L. Int. J. Green Nanotechnol. 3, 281–291.

Kumar, V., Yadav, S.K., 2011b. Synthesis of variable shaped gold nanoparticles in one solution using leaf extract of *Bauhinia variegata* L. Dig. J. Nanomater. Bios. 6, 1685–1693.

Kumar, V., Yadav, S.K., 2012a. Synthesis of different-sized silver nanoparticles by simply varying reaction conditions with leaf extracts of *Bauhinia variegata* L. IET Nanobiotechnol. 6, 1–8.

Kumar, V., Yadav, S.K., 2012b. Characterisation of gold nanoparticles synthesised by leaf and seed extract of *Syzygium cumini* L. J. Exp. Nanosci. 7, 440–451.

Kumari, A., Kumar, V., Yadav, S.K., 2012. Plant extract synthesized PLA nanoparticles for controlled and sustained release of quercetin: a green approach. PLoS One 7, e41230.

Li, S., Yang, X., Yang, S., Zhu, M., Wang, X., 2012. Technology prospecting on enzymes: application, marketing and engineering. Comput. Struct. Biotechnol. J. 2, e201209017.

Li, C.J., Trost, B.M., 2008. Green chemistry for chemical synthesis. Proc. Natl. Acad. Sci. U.S.A. 105, 13197–13202.

Li, X., Xu, H., Chen, Z.S., Chen, G., 2011. Biosynthesis of nanoparticles by microorganisms and their applications. J. Nanomater. 2011, 270974.

Maksimovic, M., Omanovic-Miklicanin, E., 2017. Towards green nanotechnology: maximizing benefits and minimizing harm. In: Badnjevic, A. (Ed.), CMBEBIH 2017. IFMBE Proceedings, vol. 62. Springer, Singapore.

Margetic, D., 2005. Mechanic-chemical organic reactions without the use of solvents. Kem. Ind. 54, 351–358.

Marwick, C., 2001. Natural compounds show antiangiogenic activity. J. Natl. Cancer Inst. 93, 1685–1690.

Morales-Avila, E., Ferro-Flores, G., Ocampo-García, B.E., López-Téllez, G., López-Ortega, J., Rogel-Ayala, D.G., Sánchez-Padilla, D., 2017. Antibacterial efficacy of gold and silver nanoparticles functionalized with the ubiquicidin (29-41) antimicrobial peptide. J. Nanomater. 2017, 5831959.

Mou, X., Ali, Z., Li, S., He, N., 2015. Applications of magnetic nanoparticles in targeted drug delivery system. J. Nanosci. Nanotechnol. 15, 54–62.

Mukhtar, H., Ahmad, N., 2000. Tea polyphenols: prevention of cancer and optimizing health. Am. J. Clin. Nutr. 71, 1698–1702.

Murphy, M.A., 2018. Early industrial roots of green chemistry and the history of the BHC ibuprofen process invention and its quality connection. Found. Chem. 20, 121–165.

Nalini, M., Lena, M., Sumathi, P., Sundaravadivelan, C., 2017. Effect of phyto-synthesized silver nanoparticles on developmental stages of malaria vector, *Anopheles stephensi* and dengue vector, *Aedes aegypti*, Egypt. J. Basic Appl. Sci. 4, 212–218.

Naraginti, S., Kumari, P.L., Das, R.K., Sivakumar, A., Patil, S.H., Andhalkar, V.V., 2017. Amelioration of excision wounds by topical application of green synthesized, formulated silver and gold nanoparticles in albino Wistar rats. Mater Sci Eng C Mater Biol Appl 62, 293–300.

Narayanan, N.K., Nargi, D., Randolph, C., Narayanan, B.A., 2009. Liposome encapsulation of curcumin and resveratrol in combination reduces prostate cancer incidence in PTEN knockout mice. Int. J. Cancer 125, 1–8.

Neves, A.R., Queiroz, J.F., Reis, S., 2016. Brain-targeted delivery of resveratrol using solid lipid nanoparticles functionalized with apolipoprotein E. J. Nanobiotechnol. 14, 27.

OECD, 2012. Transitioning to green innovation and technology. In: OECD Science, Technology and Industry Outlook 2012. OECD Publishing, pp. 61–95. https://doi.org/10.1787/sti_outlook-2012-5-en.

Owusu, P.A., Asumadu-Sarkodie, S., 2016. A review of renewable energy sources, sustainability issues and climate change mitigation. Cogent Eng. 3, 1167990.

Pantuck, A.J., Leppert, J.T., Zomorodian, N., Aronson, W., Hong, J., Barnard, R.J., Seeram, N., Liker, H., Wang, H., Elashoff, R., Heber, D., Aviram, M., Ignarro, L., Belldegrun, A., 2006. Phase II study of pomegranate juice for men with rising prostate-specific antigen following surgery or radiation for prostate cancer. Clin. Cancer Res. 12, 4018–4026.

Petersen, K., Nielsen, P.V., Bertelsen, G., Lawther, M., Olsen, M.B., Nilson, N.H., Mortensen, G., 1999. Potential of biobased materials for food packaging. Trends Food Sci. Technol. 10, 52–68.

Pool, H., Quintanar, D., Figueroa, J.D., Mano, C.M., Bechara, J.E.H., Godınez, L.A., Mendoza, S., 2012. Antioxidant effects of quercetin and catechin encapsulated into PLGA nanoparticles. J. Nanomater. 2012, 145380.

Popov, A.L., Popova, N.R., Tarakina, N.V., Ivanova, O.S., Ermakov, A.M., Ivanov, V.K., Sukhorukov, G.V., 2018. Intracellular delivery of antioxidant CeO_2 nanoparticles via polyelectrolyte microcapsules. ACS Biomater. Sci. Eng. 4, 2453–2462.

Prasad, R., Pandey, R., Barman, I., 2016. Engineering tailored nanoparticles with microbes: quo vadis? Wiley Interdiscip. Rev. Nanomed. Nanobiotechnol. 8, 316–330.

Rassaei, L., Sillanpaa, M., French, R.W., Compton, R.G., Marken, F., 2008. Arsenite determination in the presence of phosphate at electro-aggregated gold nanoparticle deposits. Electroanalysis 20, 1286–1292.

Raz, R., Chazan, B., Dan, M., 2004. Cranberry juice and urinary tract infection. Clin. Infect. Dis. 38, 1413–1419.

Riđanović, L., Ćatović, F., Riđanović, S., 2013. The Green Chemistry-Ecological Revolution in the Classroom, 8th Research/Expert Conference with International Participations "QUALITY 2013", Neum. B&H, pp. 447–452.

Rietveld, A., Wiseman, S., 2003. Antioxidant effects of tea: evidence from human clinical trials. J. Nutr. 133, 3285–3292.

Rogerio, A.P., Carlo, T., Ambrosio, S.R., 2016. Bioactive natural molecules and traditional herbal medicine in the treatment of airways diseases. Evid. Based Complement. Alternat. Med. 2016, 9872302.

Sandri, G., Bonferoni, M.C., D'Autilia, F., Rossi, S., Ferrari, F., Grisoli, P., Sorrenti, M., Catenacci, L., Del Fante, C., Perotti, C., Caramella, C., 2013. Wound dressings based on silver sulfadiazine solid lipid nanoparticles for tissue repairing. Eur. J. Pharm. Biopharm. 84, 84–90.

Sangi, R., Verma, P., 2009. Biometric synthesis and characterization of protein capped silver nanoparticles. Bioresour. Technol. 100, 501–504.

Shahverdi, A.R., Minacian, S., Shahverdi, H.R., Jamalifar, H., Nohi, A.A., 2007. Rapid synthesis of silver nanoparticles using culture supernatants of enterobacteria: a novel biological approach. Process Biochem. 42, 919–923.

Sheldon, R.A., 2011. Utilisation of biomass for sustainable fuels and chemicals: molecules, methods and metrics. Catal. Today 167, 3–13.

Sidhaye, A.A., Bhuran, K.C., Zambare, S., Abubaker, M., Nirmalan, N., Singh, K.K., 2016. Bio-inspired artemether-loaded human serum albumin nanoparticles for effective control of malaria-infected erythrocytes. Nanomedicine 11, 2809–2828.

Sreeram, K.J., Nidhin, M., Nair, B.U., 2008. Microwave assisted template synthesis of silver nanoparticles. Bull. Mater. Sci. 31, 937–942.

Stevens, E.S., 2002. How green are green plastic? Biocycle 42–45.

Taksande, A.K., Raut, R.D., Choudhary, M.D., Dahake, K.R., 2017. Green chemistry: challenges and its application in daily life. International Conference on Recent Trends in Engineering Science and Technology 5, 504–506.

Terracciano, M., De Stefano, L., Rea, I., 2018. Diatoms green nanotechnology for biosilica-based drug delivery systems. Pharmaceutics 10, 242.

Tuutijarvi, T., Lu, J., Sillanpaa, M., Chen, G., 2009. As (V) adsorption in maghemite nanoparticles. J. Hazard Mater. 166, 1415–1420.

Vartholomeos, P., Fruchard, M., Ferreira, A., Mavroidis, C., 2011. MRI-guided nanorobotic systems for therapeutic and diagnostic applications. Annu. Rev. Biomed. Eng. 13, 157–184.

Venkatachalam, P., Sangeetha, P., Geetha, N., Sahi, S.K., 2015. Phytofabrication of bioactive molecules encapsulated metallic silver nanoparticles from *Cucumis sativus* L. and its enhanced wound healing potential in rat model. J. Nanomater. 2015, 753193. https://doi.org/10.1155/2015/753193.

Vijayaraghavan, K., Ashokkumar, T., 2017. Plant-mediated biosynthesis of metallic nanoparticles: a review of literature, factors affecting synthesis, characterization techniques and applications. J. Environ. Chem. Eng. 5, 4866–4883.

Volpato, G., Rodrigues, R.C., Fernandez-Lafuente, R., 2010. Use of enzymes in the production of semi-synthetic penicillins and cephalosporins: drawbacks and perspectives. Curr. Med. Chem. 17, 3855–3873.

Wang, H.K., 1998. Plant-derived anticancer agents currently in clinical use or in clinical trials. Idrugs 1, 92–102.

Welton, T., 2015. Solvents and sustainable chemistry. Proc. Math. Phys. Eng. Sci. 471, 20150502.

Willner, I., Baron, R., Willner, B., 2006. Growing metal nanoparticles by enzymes. Adv. Mater. 18, 1109–1120.

World Health Organization, United Nations Children's Fund, 2010. Progress on sanitation and drinking-water: Joint Monitoring Programme 2010 update, ISBN 9789241563956.

CHAPTER

14

Recent trends in the development of nano-bioactive compounds and delivery systems

Avinash P. Ingle[a], Sudhir Shende[b], Indarchand Gupta[c], Mahendra Rai[b]

Department of Biotechnology, Engineering School of Lorena, University of Sao Paulo, Lorena, SP, Brazil[a]; Department of Biotechnology, Sant Gadge Baba Amravati University, Amravati, Maharashtra, India[b]; Department of Biotechnology, Institute of Science, Aurangabad, Maharashtra, India[c]

Chapter outline

1. Introduction

Bioactive compounds are considered as a special type of chemicals which are usually found in plants (fruits, vegetables, whole grains, etc.) microorganisms (fungi, bacteria, algae, etc.), marine organisms, animals, etc. in varying amounts. These compounds are distinct from nutrients because nutrients are the essential components of daily intake values for proper animal growth and body functions, however, currently bioactive compounds are not recommended as essential compound but they are required for the modulation of metabolic processes and some other important body functions which promote good health (Lanham-New et al., 2010; Correia

Biotechnological Production of Bioactive Compounds. https://doi.org/10.1016/B978-0-444-64323-0.00014-X

et al., 2012). In general, bioactive compounds are among the heterogeneous compounds which mainly include various polyphenolic compounds like carotenoids, tocopherols, phytosterols, and organosulfur compounds having different chemical structures (i.e. hydrophilic or lipophilic). In addition, these compounds are diverse in their distribution, concentrations in both foods and in the human body, their possible site of action, effectiveness against oxidative species, specificity and biological action (Porrini and Riso, 2008; Carbonell-Capella et al., 2014). As far as the specific examples of bioactive compounds are concerns it mainly include antioxidants, probiotics, polyunsaturated fatty acids, proteins, lycopene, resveratrol, lignan, tannins, indoles, flavonoids, carnitine, choline, coenzyme Q, creatine, dithiolthiones, polysaccharides, phytoestrogens, glucosinolates, anthocyanins, prebiotics, taurine, etc. (Fig. 14.1). All the above-mentioned bioactive compounds have different sources and function.

It is proposed that almost all the bioactive compounds irrespective of their source have positive impacts on human health (Carson and Clarke, 2018). Among these bioactive compounds like antioxidants, polyunsaturated fatty, probiotics, acids, and proteins are commonly used as food supplement which helps to improve the nutritional value of food. Moreover, lycopene, resveratrol, lignan, tannins, indoles some other bioactive compounds recovered from marine algae have been studied for their important role in the prevention of various diseases including cancer, heart disease, central nervous system diseases, antioxidant and anti-inflammatory activity (Raposo et al., 2013). Fig. 14.2 illustrated the schematic presentation of some important biological function of bioactive compounds. The bioaccessibility and

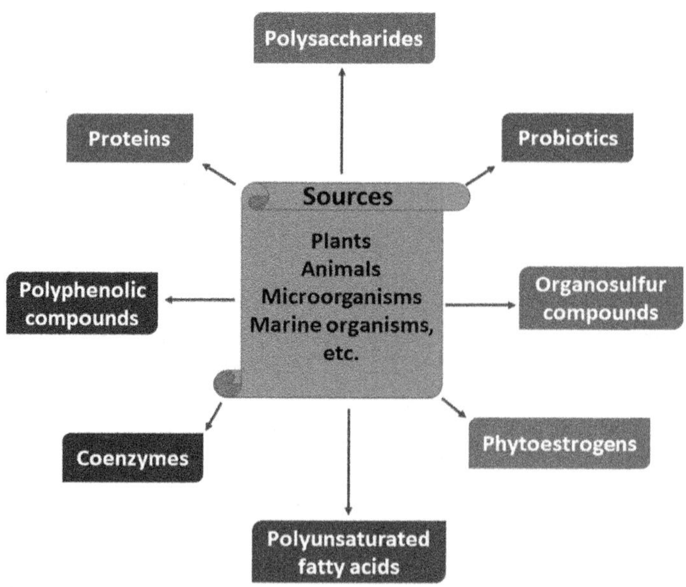

FIG. 14.1

Important bioactive compounds and their sources.

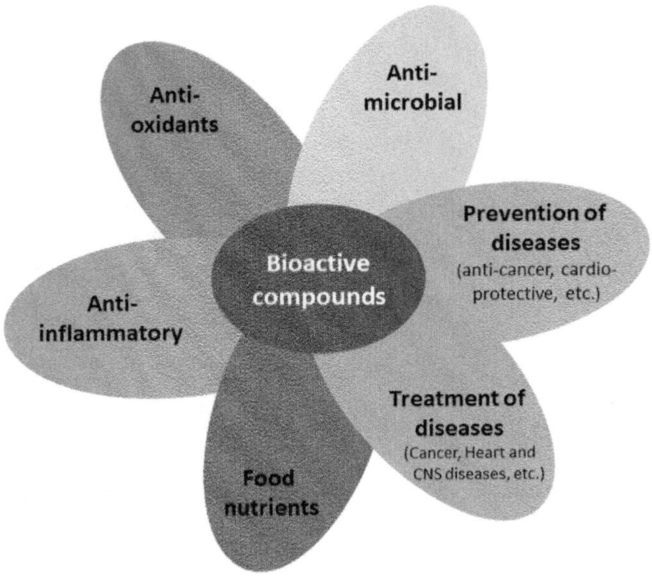

FIG. 14.2

Schematic presentation of some important biological functions of bioactive compounds.

bioavailability of various bioactive compounds greatly differ from each other, however, there are several factors which interfere with it (Parada and Aguilera, 2007). Moreover, bioaccessibility, uptake, bioavailability, solubility, degradation, etc. are major concerns associated with the use of different bioactive compound in various applications including food and health promotion (Rein et al., 2012). These limitations collectively restrict the delivery of bioactive compounds for human benefits.

In this context, it is proposed that the effective delivery of such bioactive compounds their efficiency can be improved by reducing their size to nano level (in between 1 and 100 nm) and making them nano-bioactive compound from bioactive compounds. In addition, it can be achieved by using many novel nano-based delivery systems. Development of nano-bioactive compounds and application of various nano-delivery systems (nanoparticles, solid lipid nanoparticles, polymeric nanoparticles, nanoliposomes, nanostructured lipid carrier, nanoniosomes, etc.) have attracted a great deal of attention all over world due to their multiple benefits, including higher bioavailability, permeability, fast delivery, improve circulation time and lower toxicity (Ganesan et al., 2015). Each of the above nano-based techniques has its own advantages and disadvantages in the delivery of bioactive compounds to their target site. The devolvement of nano-bioactive compounds and nano-delivery systems mostly overcome the problems associated with bioavailability of bioactive compounds and it also help to (i) increase the solubility of bioactive compounds which are used in both food and health sectors (ii) protect the degradation of various sensitive ingredients such as aroma, antimicrobial, antioxidant or nutritionally relevant compounds (iii) increase the bioavailability of bioactive compounds including the controlled release of encapsulated compounds or (iv) modify physical properties of food materials, e.g. rheological properties (when used in food) (Borel and Sabliov, 2014; Oehlke et al., 2014).

Considering the emerging applications of nanotechnology in various industrial fields, in the present chapter, we have emphasized the role of nanotechnology in the development of nano-bioactive compounds and novel nano-based delivery systems. In addition, a brief overview of bioactive compounds and their biological functions has been discussed. Moreover, the focus has been also given on bioavailability compounds and various factors influencing their bioavailability.

2. Bioaccessibility and bioavailability of bioactive compounds

Bioaccessibility is considered as the first step for making the bioactive compounds bioavailable. Generally, in this step, the bioactive compounds liberated from the food matrix transformed into a suitable chemical form which can bind to gut cells and enter in it or pass between them. The process of chewing, enzymatic digestion of the food in the mouth, mixing of food with acid and enzymes in the gastric juice and release into the small intestine, are the unit operations of the process which increase the bioaccessibility of bioactive compounds in food (Galanakis, 2017).

Among the various sources discussed earlier, it is demonstrated that most of the bioactive compounds responsible for the positive effects on human health are predominantly derived from the plant kingdom (Rein et al., 2012). The bioavailability any of dietary (foods, supplement, etc.) phytochemicals is generally defined as a measurement of their distribution in body fluids and exposure to tissues. Therefore, for proper positive health promotion, it is important that the required bioactive compounds should withstand food processing, should be released from the food matrix post-ingestion and should be bioaccessible in the gastrointestinal tract, undergo metabolism and reach the target tissue of action (Espìn et al., 2007). Ultimately, bioavailability is critical for bioactivity, because exposure to target tissues determines the potential of such bioactive compounds to deliver beneficial activity (Manach et al., 2005; Neilson et al., 2017). As far as pharmacological applications (as pharmaceutical agents) of bioactive compounds is concern bioavailability can be defined as "the rate and extent to which the active ingredient or active moiety is absorbed from a drug product and becomes available at the site of action". Therefore, bioavailability is considered as one of the most important aspects during their use in various applications because it addresses several processes such as liberation from a food matrix, absorption, distribution, metabolism and elimination phases. Hence, ideal bioactive compound must have properties like, (i) it should be present at the site of action to impart its biological activity, and (ii) the concentration and the length of exposure of the bioactive compound at the site of action determine the magnitude of activity (Neilson et al., 2017).

The evaluation of bioavailability and bioaccessibility of bioactive compounds responsible for health promotion is necessary to understand the relationship between food and nutrition. The rate and extent of absorption can vary widely between individuals. However, the variation in bioavailability between each individual mainly depends on different important factors like diet, genetic background, gut microbiota composition and activity. The bioavailability of bioactive compounds like carbohydrates, proteins and fats is usually very high, e.g. more than 90% of the amount ingested. On the other hand, it was proposed that the food bioactive compounds e.g. polyphenols have relatively poor absorption rate (i.e. from 0.3% to 43%), and therefore, circulating plasma concentrations of their metabolites can be low (Manach et al., 2005). Such lower bioavailability of particular compounds hinders their use as functional ingredients (Scalbert and Williamson, 2000; Manach et al., 2004). Therefore, understanding the mechanisms of absorption of various bioactive compounds is the only way to enhance their bioavailability and thus the potential for greater health benefits be realized.

Several approaches have been proposed for the improvement of bioaccessibility and bioavailability of bioactive compound because it is the fundamental way to improve their bioefficacy. These approaches mainly include technological and chemical modifications of the bioactive compounds to improve their solubility or the site of absorption, use of colloidal systems like micelles and vesicles, and most recent and effective is the development of nano-based delivery systems using different nanomaterials (Yang et al., 2008). As mentioned earlier, nanotechnological approaches can be effectively used as alternative strategies to mitigate bioavailability challenges by modifying different bioactive compounds to nano-bioactive compounds (to nano-size) which enhances their availability to cells, leading to enhanced bioactivity.

3. Nano-bioactive compounds

As discussed earlier, reduction in the size of normal bioactive compounds to nanometer range or their association and delivery with any nanomaterials makes them nano-bioactive compounds which results in improved bioavailability and ultimately enhances their bioactivities for several applications. Ganesan et al. (2015) reviewed some important nanobioactive from plants origin and also commented on their biomedical applications particularly in the treatment of Parkinson's disease. However, in the present chapter, we have also briefly discussed some important nano-bioactive compounds.

The curcumin is globally known bioactive compound for its biomedical applications. It has unique medicinal properties which mainly include antioxidant, antimicrobial and anti-inflammatory activities. In addition, many reported claimed its promising role in the management and treatment of many diseases including Alzheimer, Parkinson, cancer, etc. (Saranya et al., 2018). However, poor solubility of curcumin in water, its fast degradation and low bioavailability restrict its medicinal use up to certain extents (Khezri et al., 2018). Certain previous studies performed proved that these problems can be overcome by using nanocurcumin, which is water soluble and has significantly improved biological potential. Development of nano-curcumin will help to increase its bioavailability and also possesses the ability to cross the blood-brain barrier (Zhao et al., 2014). Luz et al. (2012) developed nano-curcumin by loading it on PLGA (poly-lactic-*co*-glycolic acid) nanospheres by using nanoprecipitation technique and demonstrated its potential antibacterial efficacy. Similarly, Anand et al. (2010) synthesized curcumin-loaded PLGA nanoparticles showing enhanced bioavailability compared with other nanodelivery systems and claimed that it is effective in the treatment of Parkinson's disease. Recently, it was reported that preparation nano-curcumin by loading it on polymeric nanoparticles having different polymers like Eudragit®, PLGA and polycaprolactone (PCL) significantly improved the bioavailability of curcumin. Moreover, it was also observed that curcumin-loaded Eudragit® RLPO (ERL) showed smaller particle diameters (245 ± 2 nm) than either PLGA or PCL nanoparticles (Umerska et al., 2018).

Apart from curcumin, ginsenosides is another plant-derived bioactive compound generally found in ginseng and are known to reduce oxidative stress in the liver, brain, and other organs by scavenging hydrogen peroxide radicals. It was reported that nanosizing of its formulation can serve as the effective alternative to increase its bioactivity and bioavailability. Li et al. (2015a) developed two different nanoginsenosides (Rg1 and Rb1) using a nanoemulsion technique and observed that resulting nanoginsenosides particles showed the size of 19.9 nm. Similarly, resveratrol, a polyphenolic compound usually found in grapes, peanut, peanut sprouts, blueberry, cranberry, and mulberry and showed several health benefits, including anti-aging, anticancer, cardioprotective, etc. (Ganesan et al., 2015). The reports clearly indicated that the bioefficacy and bioavailability of this compound can improve by transforming normal resveratrol into nanoresveratrol. Various attempts have been made which mainly include the encapsulation of resveratrol and development of PLGA-loaded resveratrol nanoparticles with enhanced bioavailability (Singh and Pai, 2014).

In addition, there are some other examples of bioactive compounds which get converted to their nano-form to achieve higher bioefficacy. "Catechins", a group of polyphenols from plants like tea, fruits, and beverages having multiple health benefits viz. anti-aging, anticancer, antimicrobial, antiviral, and antioxidative effects (Goh et al., 2015; Jaworska et al., 2015). Various nanotechnological approaches like nanoliposome, nanoemulsion, or nanoencapsulation have been used to make the nanosized catechins (nanocatechins) (Zou et al., 2104). "Lycopene", a carotenoid naturally occurring in tomato, watermelon and pink guava highly effective against various neurological disorders (Altas et al., 2011; Gerszberg et al., 2015). Preparation of nanolycopene using a self-emulsifying nanodelivery system or nanoemulsion promisingly improved the bioavailability of the lycopene (Ross et al., 2011; Kim et al., 2014; Ha et al., 2015). In addition, other utilization of other nano-based delivery systems like nanoemulsion delivery and nanostructured lipid carrier delivery was also found efficient in the development of nanolycopene (Ha et al., 2015; Okonogi and Riangjanapatee, 2015). Similarly, some of the studies anticipated that the potential biomedical applications of phytobioactive compounds like quercetin, kaempferol, silibinin, etc. can be much better improved by enhancing their bioavailability through the development of their respective nano-form i.e. nanoquercetin, nanokaempferol, nanosilibinin, etc. using different nanotechnological approaches (Dhawan et al., 2011; Tzeng et al., 2011; Elmowafy et al., 2013). The more details about the all the above-mentioned plant-derived nanobioactive compounds have been critically reviewed by Ganesan et al. (2015). Table 14.1 showed the list some important bioactive compounds which are improved to their nano-form using different approaches (Oehlke et al., 2014).

4. Nano-based systems for the delivery of nanobioactive compounds

It was proposed that delivery of nano-bioactive components using nano-based systems significantly improve various properties associated with it which mainly includes stability (Trombino et al., 2009), solubility, functionality (Swarnakar et al., 2011), cellular uptake (Wang et al., 2012), and bioavailability (Anand et al., 2010). In addition, it is anticipated that this approach also helps in the controlled release bioactive compounds which increases their efficacy (Kumari et al., 2010). However, development of ideal delivery systems requires utmost care because it plays a crucial role in the delivery of bioactive components at the target site without any adverse effects (Kumar et al., 2013; Wu et al., 2014). Till date many nano-based delivery systems like polymeric nanoparticles, solid lipid nanoparticles, nanostructured lipid carriers, liposomes, nanoemulsions, nanoniosomes, etc. (Fig. 14.3) have been designed for the delivery of bio- and nano-bioactive compounds (Wu et al., 2014; Yao et al., 2014). Moreover, all these have distinct advantages and disadvantages based on the activity of the bioactive molecule.

4.1 Polymeric nanoparticles

Among the various nano-based delivery systems, polymeric nanoparticles have attracted a great deal of attention over the last few years because of their distinguishing properties and behaviors resulted due to their small size (Farokhzad and Langer, 2009; Crucho and Barros, 2017). Many reports suggested the extensive range of applications polymeric nanoparticles in various sectors related diagnostics and drug delivery (Patel et al., 2012; Crucho, 2015; Lin et al., 2015). It possesses many advantages as carriers, however, most important comprises their ability to combine both, therapy and imaging, protection of bioactive compounds and drug which are loaded or incorporated in it. In addition, their ability of specific targeting and controlled release makes them one of the ideal nano-based delivery systems (Kamaly et al., 2012; Zhong et al., 2014). The uptake of nanomaterials into cells

Table 14.1 Important bioactive compounds which are improved to their nano-form using different approaches (Oehlke et al., 2014).

Bioactive compounds	Nano-based delivery system	Resulting size (nm)	Increase in bioavailability (in folds)
Curcumin	Liposomes	263	5
	Nanocrystal dispersion	250	16
	Solid lipid nanoparticles	135	39
	Nanoemulsion	196	9
	Organogel-based nanoemulsion	128	9.3
Quercetin	Solid lipid nanoparticles	155	5.7
	Microemulsion	39	1.6
Capsaicin	Nanoemulsion (Alginate-1)	50	132
	Nanoemulsion (Alginate-2)	100	81
	Nanoemulsion (Alginate-2)	150	72
	Nanoemulsion (Chitosan)	100	77
	Nanoemulsion (Chitosan + Alginate)	100	8
Hoodia gordonii steroid glycoside	Colloidal particles	110	No difference
	Nanoemulsion	200	
Co-enzyme Q10	Nanoemulsion	60	1.7
	Nanomicelles (Nanosolve)	30–60	5
Vitamin E	Nanomicelles (Nanosolve)	30–60	10
Tocopherol acetate	Nanomicelles (Aquanova)	50	1.5
Lignan glycosides	Nanoparticles	200	1.7
Sitosterol	Crystalline nanoscales	550	No difference
	Emulsified nanoscales	100–130	
Iron	Amorphous $FePO_4$ particles	11	0.96
	Amorphous $FePO_4$ particles	31	0.7
	Amorphous $FePO_4$ particles	64	0.6
	Emulsifier coated $FePO_4$ particles	300	0.83
Calcium	Nanoparticles (Pearl powder)	84	1.4
	Nanoparticles (Pearl powder)	40–80	2
	Nanometer sized carbonate	398	1.38
	Nanometer sized citrate	151	1.53
Chromium	Nanometer sized chromium picolinate	70	2

occurs through the endocytosis processes and it is strictly depending on the size and surface chemistry of nanomaterials (Owens and Peppas, 2006; Iversen et al., 2011). However, these properties can be managed through the optimization of process of nanomaterials synthesis (Reis et al., 2006; Rao and Geckeler, 2011).

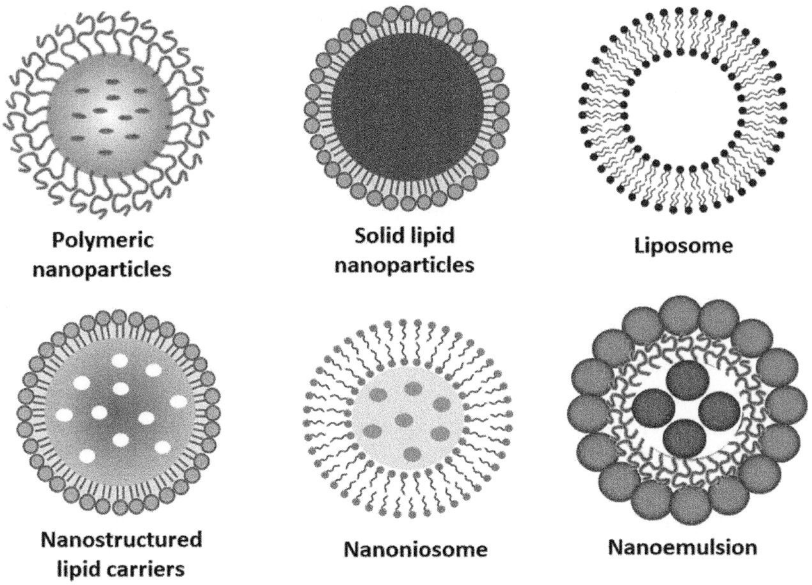

FIG. 14.3

Different nano-based delivery systems for bioactive compounds.

Depending upon the preparation technique and the organic phase composition, nanospheres or nanocapsules can be obtained (Crucho, 2015). In nanospheres, the matrix has even distribution of active compounds and the polymer, whereas in case of nanocapsules, the particle has core-shell morphology with an oily or aqueous cavity in which the active compounds are incorporated and enclosed by a polymer shell. The common mechanism of controlled release of molecules incorporated in polymeric nanoparticles is accomplished by regulation of the rates of polymer biodegradation and the dispersion of compounds out of the polymer matrix (Kamaly et al., 2016; Crucho and Barros, 2017).

Importantly the basis of delivery of bioactive compounds or any drug molecule via polymeric nanoparticles is greatly relied on the biocompatibility and biodegradability of the polymers used in its preparation. Because application of biodegradable polymers has an added advantage that they can be completely removed from the body by normal metabolic pathways (Marin et al., 2013). It is well fact that the natural polymers are usually biocompatible as well as biodegradable. But the continuous deviations in their properties from batch to batch limits their use, in addition, some of the natural polymers also showed slightly immunogenic properties. On the contrary, synthetic polymers are well known for their precise chemical composition. Considering the novel properties, various natural and synthetic polymers have been successfully exploited in the preparation of polymeric nanoparticles viz. proteins, sugars or other natural macromolecules, biodegradable polymers and non-biodegradable, pharmaceutically active polymers (Kumari et al., 2010; Mansour et al., 2010; Joshi and Patel, 2012).

As mentioned, above different approaches have been proposed for the preparation of polymeric nanoparticles, however, selection of appropriate approach is basically depending on some parameters which significantly include nature of polymer used in the preparation of polymeric nanoparticles, physicochemical properties of the compounds to be encapsulated and finally the desired characteristics of the polymeric nanoparticles (Allouche, 2013; Nasir et al., 2015). Above-mentioned are the advantages why polymeric nanoparticles are extensively used as a carrier for many bioactive compounds obtained from different sources with almost no adverse effects or toxicity (Mahapatro and Singh, 2011). Danhier et al. (2012) demonstrate the absorption and incorporation of bioactive compounds like curcumin and resveratrol in polymeric nanoparticles using various approaches such as nano-precipitation or emulsion-diffusion techniques respectively. Similarly, Luz et al. (2012) demonstrated the use of PLGA nanospheres for the encapsulation of curcumin using nanoprecipitation technique for its effective antibacterial efficacy and in the treatment of Parkinson's disease (Anand et al., 2010). Moreover, it was claimed polymeric nanoparticles with different polymers like Eudragit®, PLGA and PCL can be significantly used for successful loading the curcumin which enhances its bioavailability (Umerska et al., 2018).

4.2 Solid lipid nanoparticles

Solid lipid nanoparticles are composed of solid lipids or lipid blends which is analogous to an oil-in-water emulsion where the liquid lipid (oil) of the emulsion has been substituted by solid lipid in solid lipid nanoparticles (Kokardekar and Mody, 2011). Generally, solid lipid nanoparticles include numerous highly pure biocompatible/biodegradable lipids, which are solid at room temperature, and recognized as safe and economical. In general, it includes solid lipids such as triglycerides, carnauba wax, beeswax, cetyl alcohol, emulsifying wax, complex glycerides, cholesterol, cholesterol butyrate, etc. (Loxley, 2009). Solid lipid nanoparticles can be fabricated using a templated synthesis from a microemulsion of the molten lipid in aqueous surfactants, or by precipitation of the wax from a solution in a non-ionic surfactant on addition of water, or by emulsifying the molten lipid into a hot aqueous surfactant solution with high-shear mixing to acquire the desired submicron particle size (Mumper, 2006; Loxley, 2009).

In addition, a variety of techniques like high-pressure homogenization method, through microemulsions, double emulsion method, multiple emulsion, solvent emulsification-evaporation technique, and ultrasonication have been exploited for the synthesis of solid lipid nanoparticles (Mehnert and Mader, 2001; Mukherjee et al., 2009; Scalia et al., 2015; Ganesan et al., 2015). Out of these techniques, high-pressure homogenization approach is commonly used, which is performed at high temperatures or at or below room temperature (Kokardekar and Mody,

2011). The approach which is carried high pressure is called as hot high-pressure homogenization, however, which performed at low pressure is referred as cold high-pressure homogenization. The second approach is through micro-emulsions, in which a hot microemulsion containing the lipids and surfactants is prepared and dispersed into cold water. The excess water is removed by lyophilization to produce solid lipid nanoparticles (Cavalli et al., 1996). In solvent emulsification-evaporation method, a lipid-containing solution is emulsified with an aqueous phase followed by solvent evaporation, which leads to the precipitation of lipids and solid lipid nanoparticles (Wissing et al., 2004). Even double emulsification and multiple emulsion methods can be used for the production of solid lipid nanoparticles. Solid lipid nanoparticles have also been amalgamated with the help of high-speed stirring in combination with ultra-sonication. This method is usually performed at higher temperatures (Kokardekar and Mody, 2011).

The important advantages of solid lipid nanoparticles include its cost-effectiveness and easy large-scale production. The development of solid lipid nanoparticles involved the use of not toxic components; therefore, they are safe. It provides delivery systems which have control on the release of any bioactive compounds which is ultimately effective in the treatment of disease. Additionally, encapsulation of bioactive ingredients in solid lipid nanoparticles offers protection from degradation. It also has advantages of site-specific targeting and of being stable over a period of time (Wissing et al., 2004; Kokardekar and Mody, 2011).

Generally, the solid lipid present in solid lipid nanoparticles integrate the bioactive compounds in a lipophilic and hydrophilic shell which is bounded by a phospholipid layer for controlled release of the bioactive compound to the target site (Sachdeva et al., 2015; Zhao et al., 2015). According to Ganesan et al. (2015) the mobility of the bioactive compound is significantly reduced due to a solid lipid core; consequently, it remains in the gut, which increases the controlled release of the bioactive compound for a prolonged period of time. The lipophilic bioactive compounds profoundly present in the lipid matrix, on the other hand, the hydrophilic bioactive compounds found outside to the lipid matrix. It was also suggested that the dispersion of bioactive compounds in lipid, requires an appropriate solvent or mechanical force (Ganesan et al., 2015). In addition, stabilization of lipid integrated in solid lipid nanoparticles is very necessary which commonly performed by the coating of a stabilizing agent like polyethylene glycol (PEG). Such coating with stabilizing agents enhances the stability of the bioactive compound in blood plasma with minimizing phagocytic uptake (Nance et al., 2014). Many synthetic drugs and bioactive compounds have been delivered using SNLs to prevent different diseases including Parkinson's diseases (Bicker et al., 2014; Bose et al., 2014; Rostami et al., 2014). Esposito et al. (2008) demonstrated the development of bromocriptine-loaded SLNs for control of Parkinson's diseases and the results obtained showed that such nanomaterials are greatly effective in tumbling dyskenesia. Apart from these, it was reported that the plant-derived nano-bioactive compounds developed using a solid lipid nanoparticles delivery system showed higher bioavailability (Neves et al., 2013). Similarly, Vandita et al. (2012) demonstrated that curcumin-loaded solid lipid nanoparticles showed significantly higher (155 times) curcumin delivery than that of natural curcumin in cancer cells. It was reviewed that the incorporation of various bioactive food components e.g. carotenoids, omega-3 fatty acids, or phytosterols in food system can be successfully performed by encapsulation these compounds in solid lipid nanoparticles (Weiss et al., 2008). Moreover, the release of curcumin in brain is also highly efficient when administered in the form of curcumin-loaded solid lipid nanoparticles. In case of oral administration in rat it was found to be 16.5 times more and about 30 times more when given it intravenously (Kakkar et al., 2013). Some other reports claimed that bio-availabilities bioactive compounds like quercetin (Dora et al., 2011) and resveratrol (Pandita et al., 2014) can be increased significantly when used in the form of formulation with solid lipid nanoparticles.

Apart from these, solid lipid nanoparticles can be also used to incorporate and deliver both hydrophobic and hydrophilic proteins. It is well known that encapsulation of bioactive compounds in solid lipid nanoparticles avoid their degradation and protect them, similarly, the protein integrated in solid lipid nanoparticles does not expose to external environment and thus it gets protection from various proteolytic degradation which resulted enhanced bioavailability. A variety of different proteins such as insulin, cyclosporins, and somatostatin have been successfully incorporated within solid lipid nanoparticles matrix and deliberated. Similarly, it is possible to administered vaccine effectively with the help of solid lipid nanoparticles system (Almeida and Souto, 2007; Kokardekar and Mody, 2011). Solid lipid nanoparticles system can also be applied to deliver some bioactive compounds and drugs through oral and pulmonary routes. For instance, camptothecin-loaded solid lipid nanoparticles were successfully produced and stabilized by stearic acid and poloxamer 188 (Müller et al., 2000; Kokardekar and Mody, 2011).

4.3 Liposomes

Liposomes are generally used as a carrier system for the delivery of a variety of bioactive compounds, nutrients and pharmaceutical drugs due to their small and size and unique hydrophobic and hydrophilic properties (Akbarzadeh et al., 2013). It was anticipated that phospholipids of liposomes spontaneously form closed structures when they are hydrated in aqueous solutions. Hence, liposomes having one or more phospholipid bilayer membranes can easily transport various molecules depending on the nature of those molecules/compounds (Chrai et al., 2001). Liposomes composed of the hydrophilic head along with two hydrophobic tails. The size of liposomes in generally ranges in between 30 nm to a few microns (Hadian et al., 2014; Li et al., 2015b). Various bioactive compounds encapsulated in liposomes have shown improved and prolong bioactivities like antioxidant activity due increase in surface area of exposure (Wu et al., 2015). In addition, to encapsulation of bioactive compounds liposomes have promising biomedical applications in nano-therapy (e.g. cancer therapy, diagnosis, gene delivery), food technology, cosmetics and agriculture (Mozafari, 2010; Fakhravar et al., 2016). Liposomes are able to increase the performance of bioactive compounds thereby improving their solubility, stability and bioavailability, and loading of bioactive compounds in liposome also avoid their undesirable interactions with other molecules. Moreover, it facilitates cell-specific targeting which is an important aspect as far as the bioefficacy of any bioactive compound is concerned (Mozafari, 2010). Till date, liposomes have been used for the encapsulation of a variety of bioactive compounds which are used for various health benefits. Aisha et al. (2014) demonstrated the successful encapsulation of phytobioactive compounds present in the extract of *Orthosiphon stamineus*, the resulting extract showed greater bioavailability and *in vitro* antioxidant activity. Similarly, the synthesis of curcumin-encapsulated liposomes found to be effective for oral treatment in rats due to improved bioavailability and antioxidant activity (Takahashi et al., 2009). Moreover, according to Mourtas et al. (2014), curcumin loaded-liposomes has the ability to cross the blood-brain barrier and were effective in the management of Alzheimer's disease.

4.4 Nanostructured lipid carriers (NLCs)

Nanostructured lipid carriers are generally prepared by the fusion of lipids in solid and liquid phases in which the desired bioactive compound can be encapsulated (Neves et al., 2013; Zhang et al., 2014). Nanostructured lipid carriers are considered as the next generation of the discovered at the ending of the 1990s to overcome the possible disadvantages of solid lipid nanoparticles. Nanostructured lipid carriers can significantly enhance the stability, loading capacity of desired biomolecules or bioactive compounds and also prevent it from degradation from external factors and increase storage capacity (Naseri et al., 2015). Among two different lipid phases used, solid phase lipids commonly used include glycerol, acetyl alcohol, stearic acid and monostearate. However, the liquid phase lipids such as used for the oleic acid, caprylic triglycerides and cupric triglycerides usually preferred as a carrier for nanostructured lipid carriers (Nanjwade et al., 2013). The type and concentration of components used in both solid and liquid phase lipids are crucially responsible for the stability of bioactive compound incorporated in the matrix (Iqbal et al., 2012; Neves et al., 2013). Baicalin (flavone glycoside), a phytobioactive compound naturally found in several species in the genus *Scutellaria*, like *Scutellaria baicalensis* with many medicinal properties have been encapsulated in nanostructured lipid carriers to developed baicalin-loaded nanostructured lipid carriers. Thus, prepared baicalin-loaded nanostructured lipid carriers showed improved bioavailability and sustained release of baicalin (Luan et al., 2015). In another study, Aditya et al. (2013, 2014) reported a significant increase in the solubility and bioavailability of poorly soluble and showing lower bioavailability of curcumin and genistein which ultimately demonstrated a stronger effect in inhibiting the prostate cancer. Omega-3 fatty acids are considered as one of most essential fatty acids and these are susceptible to oxidative deterioration. The resulting products after oxidation of omega-3 fatty acids exert undesirable effect on sensory acceptance of enriched foodstuffs. In this situation, stabilization is among the safe alternative to protect such fatty acids from deteriorating factors (Zimet et al., 2011). Therefore, encapsulation of omega-3 fatty acids in nanostructured lipid carriers significantly decreased the rate of its oxidation (in enriched foods) (Tamjidi et al., 2013). Recently, Nitthikan et al. (2018) reported the significant improvement in the stability and transdermal delivery of some bioactive compounds from green Robusta coffee beans when loaded on nanostructured lipid carriers. All these reports confirmed that nanostructured lipid carriers can be effectively used as one of the suitable nano-based delivery systems for stabilization of variety of bioactive compounds.

4.5 Nanoniosomes

Niosomes are the type of liposome but these vesicles mainly composed of the hydrated non-ionic surfactants with cholesterol (CHOL) or its derivatives, in many cases. The specialized structures of niosomes allow the encapsulation of both hydrophilic as well as lipophilic compounds. The encapsulation is generally achieved by adsorbing the desired compound on the bilayer surfaces or by entrapment of hydrophilic compounds in vesicular aqueous core whereas the lipophilic substances are encapsulated by their partitioning into the lipophilic area of the bilayers. The niosomes having size in the range of 10–1000 nm commonly referred to as "nanoniosomes" (Mukherjee et al., 2007; Bragagni et al., 2012).

In recent years, among the all vesicular systems, nanoniosomes have been attracted a great of attention around the globe as far as the nano-based delivery systems are concerned for efficient delivery of a variety of drugs and compounds through different routes of administration. Till date, nanoniosomes have been exploited for the delivery of various drugs, genes, proteins and vaccines due to their promising capability of entrapping molecules. Nanoniosomes have many advantages over other liposomes which mainly include their higher chemical stability, lower toxicity due to their non-ionic nature, enhanced protection of bioactive compounds, non-immunogenicity, as well as enhanced oral bioavailability (Moghassemi and Hadjizadeh, 2014). Through both *in vitro* and *in vivo* studies, it was confirmed that smaller size of nanoniosomes facilitates better retention of bioactive compounds at the target site (Junyaprasert et al., 2012; Hamishehkar et al., 2013). Some bioactive compounds encapsulated in nanoniosomes showed improved antioxidant, antifungal, antimalarial, and anti-Alzheimer's disease activities (Gaafar et al., 2014). Different bioactive compounds e.g. ellagic acid, doxorubicin and ganciclovir have been successfully encapsulated and loaded on nanoniosomes for best possible delivery at specific target sites (Akhter et al., 2012; Ganesan et al., 2015).

4.6 Nanoemulsions

Nanoemulsions are called as clear stable emulsion made up of two immiscible liquids having particles size less 100 nm. Encapsulation of bioactive compound in such nanoemulsions provide better bioavailability (McClements and Rao, 2011). The application of nanoemulsions, for the controlled delivery of bioactive compounds, is a most recent advancement in the field of food science and technology and drug delivery system (Sarker, 2005; Solans et al., 2005). In nanoemulsions, the functional compounds are restricted from reacting with the food matrix, which facilitates in the preservation of organoleptic properties, through enhancing the bioavailability of the bioactive compounds of interest, due to favoring passive transport through biological membranes (Kumar and Sarkar, 2018). Nanoencapsulation enhances the solubility of bioactive nutrients and non-nutrients, and their presence suiting absorption at the gastric and intestinal levels, during the digestion. Furthermore, the application of these compounds using nanoemulsions affords improved protection, lowering the activity of efflux transporters, and reducing their metabolism (Yu and Huang, 2013).

Up till now, the encapsulation of various functional compounds, for example, curcumin and resveratrol have been performed (Sessa et al., 2014; Sari et al., 2015), by nanoemulsions to protect these compounds from digestion. It is mostly used to the design of novel functional foods, as this technology allows regulating the release of bioactive nutrients and non-nutrients which are poorly soluble in water (Donsìa et al., 2011; Ahmed et al., 2012). Young et al. (2014) reported that bioavailability of curcumin can be increase using nanoemulsified curcumin in mice which showed reduced inflammation through the oral administration. Similarly, other compounds like seed oil of pomegranate and other plant-derived bioactive compounds (betlunic acid) have been deliberated by nanoemulsion delivery for better-enhanced bioavailability and bioefficacy (Mizrahi et al., 2014; Cavazos-Garduno et al., 2015).

5. Conclusion and future directions

The bioactive compounds from various sources play a crucial role in human health care, they can act as food supplements, nutrients, etc. All such bioactive compounds possess potential bioactivities including antimicrobial, antioxidant, anti-inflammatory, activities. In addition, many of bioactive compounds like curcumin showed promising biomedical applications in the management of dreadful diseases like Parkinson's and Alzheimer's diseases. However, poor solubility and lower bioavailability are the major concerns associated with bioactive compounds and such problems are mainly responsible for the limited use of these bioactive compounds. Considering the recent developments in the field of nanotechnology is believed that nanotechnology has the potential to solve these disadvantages through the preparation of nano-bioactive compounds and using different nano-based delivery systems. The reports available confirmed that both the above mentioned nanotechnological approaches are most efficient for enhancement of bioavailability, protection and improved bioefficacy of bioactive compounds. The small size of nano-bioactive compound helps in the delivery of specific bioactive component to target organs and cross the blood-brain barrier.

Considering the great attraction of scientific community toward utilization of natural bioactive compounds in biomedical science, it is expected that in next coming five years there will be chances of path breaking research in this field. Moreover, nanotechnology is rapidly emerging field of science having enormous application in various sectors including biomedicine. Various nanomaterials mentioned-above reported to have potential role in the detection and management of various diseases. However, some of these nanomaterials can be effectively used as delivery system for different bioactive compounds. In this context, it is believed that nanotechnological investigations will be a promising approaches in the next five years which will be the preferable choice for the management of various diseases which ultimately beneficial for all mankind's.

References

Aditya, N.P., Shim, M., Lee, I., Lee, Y., Im, M.H., Ko, S., 2013. Curcumin and genistein coloaded nanostructured lipid carriers: in vitro digestion and antiprostate cancer activity. J. Agric. Food Chem. 61 (8), 1878–1883.

Aditya, N.P., Shim, M., Yang, H., Lee, Y., Ko, S., 2014. Antiangiogenic effect of combined treatment with curcumin and genistein on human prostate cancer cell line. J. Funct. Foods 8, 204–213.

Ahmed, K., Li, Y., McClements, D.J., Xiao, H., 2012. Nanoemulsion- and emulsion-based delivery systems for curcumin: encapsulation and release properties. Food Chem. 132 (2), 799–807.

Aisha, A.F.A., Majid, A.M.S.A., Ismail, Z., 2014. Preparation and characterization of nano liposomes of *Orthosiphon stamineus* ethanolic extract in soybean phospholipids. BMC Biotechnol. 14 (23), 1–11.

Akbarzadeh, A., Rezaei-Sadabady, R., Davaran, S., Joo, S.W., Zarghami, N., Hanifehpour, Y., Samiei, M., Kouhi, M., Nejati-Koshki, K., 2013. Liposome: classification, preparation, and applications. Nanoscale Res. Lett. 8 (1), 102. https://doi.org/10.1186/1556-276X-8-102.

Akhter, S., Kushwaha, S., Warsi, M.H., Anwar, M., Ahmad, M.Z., Ahmad, I., Talegaonkar, S., Khan, Z.I., Khar, R.K., Ahmad, F.J., 2012. Development and evaluation of nanosized niosomal dispersion for oral delivery of Ganciclovir. Drug Dev. Ind. Pharm. 38 (1), 84–92.

Allouche, J., 2013. Synthesis of organic and bioorganic nanoparticles: an overview of the preparation methods. In: Brayner, R., Fiévet, F., Coradin, T. (Eds.), Nanomaterials: A Danger or a Promise? A Chemical and Biological Perspective. Springer, London, pp. 27–74.

Almeida, A.J., Souto, E., 2007. Solid lipid nanoparticles as a drug delivery system for peptides and proteins. Adv. Drug Deliv. Rev. 59, 478–490.

Altas, S., Kizil, G., Kizil, M., Ketani, A., Haris, P.I., 2011. Protective effect of Diyarbakir watermelon juice on carbon tetrachloride-induced toxicity in rats. Food Chem. Toxicol. 49 (9), 2433–2438.

Anand, P., Nair, H.B., Sung, B.K., Kunnumakkara, A.B., Yadav, V.R., Tekmal, R.R., Aggarwal, B.B., 2010. Design of curcumin-loaded PLGA nanoparticles formulation with enhanced cellular uptake, and increased bioactivity in vitro and superior bioavailability *in vivo*. Biochem. Pharmacol. 79 (3), 330–338.

Bicker, J., Alves, G., Fortuna, A., Falcao, A., 2014. Blood-brain barrier models and their relevance for a successful development of CNS drug delivery systems: a review. Eur. J. Pharm. Biopharm. 87 (3), 409–432.

Borel, T., Sabliov, C.M., 2014. Nano-delivery of bioactive components for food applications: types of delivery systems, properties, and their effect on ADME profiles and toxicity of nanoparticles. Annu. Rev. Food Sci. Technol. 5, 197–213.

Bose, T., Latawiec, D., Mondal, P.P., Mandal, S., 2014. Overview of nano-drugs characteristics for clinical application: the journey from the entry to the exit point. J. Nanoparticle Res. 16 (8), 1–25.

Bragagni, M., Mennini, N., Ghelardini, C., Mura, P., 2012. Development and characterization of niosomal formulations of doxorubicin aimed at brain targeting. J. Pharm. Pharm. Sci. 15 (1), 184–196.

Carbonell-Capella, J.M., Buniowska, M., Barba, F.J., Esteve, M.J., Frígola, A., 2014. Analytical methods for determining bioavailability and bioaccessibility of bioactive compounds from fruits and vegetables: a review. Compr. Rev. Food Sci. Food Saf. 13, 155–171.

Carson, M.A., Clarke, S.A., 2018. Bioactive compounds from marine organisms: potential for bone growth and healing. Mar. Drugs 16 (9), 340.

Cavalli, R., Marengo, E., Rodriguez, L., Gasco, M.R., 1996. Effects of some experimental factors on the production process of solid lipid nanoparticles. Eur. J. Pharm. Biopharm. 43, 110–115.

Cavazos-Garduno, A., Flores, A.A.O., Serrano-Nino, J.C., Martinez-Sanchez, C.E., Beristain, C.I., Garcia, H.S., 2015. Preparation of betulinic acid nanoemulsions stabilized by omega-3 enriched phosphatidylcholine. Ultrason. Sonochem. 24, 204–213.

Chrai, S.S., Murari, R., Imran, A., 2001. Liposomes: a review. Bio Pharm 14 (11), 10–14.

Correia, R.T., Borges, K.C., Medeiros, M.F., Genovese, M.I., 2012. Bioactive compounds and phenolic-linked functionality of powdered tropical fruit residues. Food Sci. Technol. Int. 18 (6), 539–547.

Crucho, C.I.C., 2015. Stimuli-responsive polymeric nanoparticles for nanomedicine. ChemMedChem 10, 24–38.

Crucho, C.I.C., Barros, M.T., 2017. Polymeric nanoparticles: a study on the preparation variables and characterization methods. Mater. Sci. Eng. C Mater. Biol. Appl. 80, 771–784.

Danhier, F., Ansorena, E., Silva, J.M., Coco, R., Le Breton, A., Preat, V., 2012. PLGA-based nanoparticles: an overview of biomedical applications. J. Control. Release 161 (2), 505–522.

Dhawan, S., Kapil, R., Singh, B., 2011. Formulation development and systematic optimization of solid lipid nanoparticles of quercetin for improved brain delivery. J. Pharm. Pharmacol. 63 (3), 342–351.

Donsìa, F., Sessaa, M., Mediounic, H., Mgaidic, A., Ferrari, G., 2011. Encapsulation of bioactive compounds in nanoemulsion based delivery systems. Procedia Food Sci. 1, 1666–1671.

Dora, C.L., Silva, L.F.C., Tagliari, M.P., Silva, M.A.S., Lemos-Senna, E., 2011. Formulation study of quercetin-loaded lipid-based nanocarriers obtained by hot solvent diffusion method. Lat. Am. J. Pharm. 30 (2), 289–296.

Elmowafy, M., Viitala, T., Ibrahim, H.M., Abu-Elyazid, S.K., Samy, A., Kassem, A., Yliperttula, M., 2013. Silymarin loaded liposomes for hepatic targeting: in vitro evaluation and HepG2 drug uptake. Eur. J. Pharm. Sci. 50 (2), 161–171.

Espìn, J.C., Garcìa-Conesa, M.T., Tomàs-Barberàn, F.A., 2007. Nutraceuticals: facts and fiction. Phytochemistry 68, 2986–3008.

Esposito, E., Fantin, M., Marti, M., Drechsler, M., Paccamiccio, L., Mariani, P., Sivieri, E., Lain, F., Menegatti, E., Morari, M., Cortesi, R., 2008. Solid lipid nanoparticles as delivery systems for bromocriptine. Pharm. Res. (N. Y.) 25 (7), 1521–1530.

Fakhravar, Z., Ebrahimnejad, P., Daraee, H., Akbarzadeh, A., 2016. Nanoliposomes: synthesis methods and applications in cosmetics. J. Cos. Laser Ther. 18 (3), 174–181.

Farokhzad, O.C., Langer, R., 2009. Impact of nanotechnology on drug delivery. ACS Nano 3, 16–20.

Gaafar, P.M.E., Abdallah, O.Y., Farid, R.M., Abdelkader, H., 2014. Preparation, characterization and evaluation of novel elastic nano-sized niosomes (ethoniosomes) for ocular delivery of prednisolone. J. Liposome Res. 24 (3), 204–215.

Galanakis, C., 2017. Nutraceutical and Functional Food Components: Effects of Innovative Processing Techniques, first ed. Academic Press, Elsevier, p. 382.

Ganesan, P., Ko, H.M., Kim, I.S., Choi, D.K., 2015. Recent trends of nano bioactive compounds from ginseng for its possible preventive role in chronic disease models. RSC Adv. 5, 98634–98642.

Gerszberg, A., Hnatuszko-Konka, K., Kowalczyk, T., Kononowicz, A.K., 2015. Tomato (*Solanum lycopersicum* L.) in the service of biotechnology. Plant Cell. Tissue Org. 120 (3), 881–902.

Goh, R., Gao, J., Ananingsih, V.K., Ranawana, V., Henry, C.J., Zhou, W.B., 2015. Green tea catechins reduced the glycaemic potential of bread: an in vitro digestibility study. Food Chem. 180, 203–210.

Ha, T.V.A., Kim, S., Choi, Y., Kwak, H.S., Lee, S.J., Wen, J., Oey, I., Ko, S., 2015. Antioxidant activity and bioaccessibility of size-different nanoemulsions for lycopene-enriched tomato extract. Food Chem. 178, 115–121.

Hadian, Z., Sahari, M.A., Moghimi, H.R., Barzegar, M., 2014. Formulation, characterization and optimization of liposomes containing eicosapentaenoic and docosahexaenoic acids; a methodology approach. Iran. J. Pharm. Res. 13 (2), 393–404.

Hamishehkar, H., Rahimpour, Y., Kouhsoltani, M., 2013. Niosomes as a propitious carrier for topical drug delivery. Expert Opin. Drug Deliv. 10 (2), 261–272.

Iqbal, M.A., Md, S., Sahni, J.K., Baboota, S., Dang, S., Ali, J., 2012. Nanostructured lipid carriers system: recent advances in drug delivery. J. Drug Target. 20 (10), 813–830.

Iversen, T.-G., Skotland, T., Sandvig, K., 2011. Endocytosis and intracellular transport of nanoparticles: present knowledge and need for future studies. Nano Today 6, 176–185.

Jaworska, G., Pogon, K., Bernas, E., Duda-Chodak, A., 2015. Nutraceuticals and antioxidant activity of prepared for consumption commercial mushrooms *Agaricus bisporus* and *Pleurotus ostreatus*. J. Food Qual. 38 (2), 111–122.

Joshi, J.R., Patel, R.P., 2012. Role of biodegradable polymers in drug delivery. Int. J. Curr. Pharm. Res. 4, 74–81.

Junyaprasert, V.B., Singhsa, P., Suksiriworapong, J., Chantasart, D., 2012. Physicochemical properties and skin permeation of Span 60/Tween 60 niosomes of ellagic acid. Int. J. Pharm. 423 (2), 303–311.

Kakkar, V., Muppu, S.K., Chopra, K., Kaur, I.P., 2013. Curcumin loaded solid lipid nanoparticles: an efficient formulation approach for cerebral ischemic reperfusion injury in rats. Eur. J. Pharm. Biopharm. 85 (3), 339–345.

Kamaly, N., Xiao, Z., Valencia, P.M., Radovic-Moreno, A.F., Farokhzad, O.C., 2012. Targeted polymeric therapeutic nanoparticles: design, development and clinical translation. Chem. Soc. Rev. 41, 2971–3010.

Kamaly, N., Yameen, B., Wu, J., Farokhzad, O.C., 2016. Degradable controlled-release polymers and polymeric nanoparticles: mechanisms of controlling drug release. Chem. Rev. 116, 2602–2663.

Khezri, A., Karimi, A., Yazdian, F., Jokar, M., Mofradnia, S.R., Rashedi, H., Tavakoli, Z., 2018. Molecular dynamic of curcumin/chitosan interaction using a computational molecular approach: emphasis on biofilm reduction. Int. J. Biol. Macromol. 114, 972–978.

Kim, S.O., Ha, T.V.A., Choi, Y.J., Ko, S., 2014. Optimization of homogenization-evaporation process for lycopene nanoemulsion production and its beverage applications. J. Food Sci. 79 (8), N1604–N1610.

Kokardekar, R.R., Mody, H.R., 2011. Solid lipid nanoparticles: a drug carrier system. Chron. Young Sci. 2 (1), 26–28.

Kumar, A., Chen, F., Mozhi, A., Zhang, X., Zhao, Y., Xue, X., Hao, Y., Zhang, X., Wang, P.C., Liang, X.J., 2013. Innovative pharmaceutical development based on unique properties of nanoscale delivery formulation. Nanoscale 5 (18), 8307–8325.

Kumar, K., Sarkar, P., 2018. Encapsulation of bioactive compounds using nanoemulsions. Environ. Chem. Lett. 16, 59–70.

Kumari, A., Yadav, S.K., Pakade, Y.B., Singh, B., Yadav, S.C., 2010. Development of biodegradable nanoparticles for delivery of quercetin. Colloids Surf. B: Biointerfaces 80 (2), 184–192.

Lanham-New, S.A., Macdonald, I.A., Roche, H.M. (Eds.), 2010. Nutrition and Metabolism (The Nutrition Society Textbook), second ed. Wiley-Blackwell.

Li, T., Shu, Y.J., Cheng, J.Y., Liang, R.C., Dian, S.N., Lv, X.X., Yang, M.Q., Huang, S.L., Chen, G., Yang, F., 2015a. Pharmacokinetics and efficiency of brain targeting of ginsenosides Rg1 and Rb1 given as Nao-Qing microemulsion. Drug Dev. Ind. Pharm. 41 (2), 224–231.

Li, T., Yang, S.B., Liu, W., Liu, C., Liu, W., Zheng, H., Zhou, W., Tong, G., 2015b. Preparation and characterization of nanoscale complex liposomes containing medium-chain fatty acids and vitamin C. Int. J. Food Prop. 18 (1), 113–124.

Lin, G., Zhang, H., Huang, L., 2015. Smart polymeric nanoparticles for cancer gene delivery. Mol. Pharm. 12, 314–321.

Loxley, A., 2009. Solid lipid nanoparticles for the delivery of pharmaceutical actives. Drug Deliv. Technol. 9 (8), 1–5.

Luan, J.J., Zheng, F., Yang, X.Y., Yu, A.H., Zhai, G.X., 2015. Nanostructured lipid carriers for oral delivery of baicalin: in vitro and in vivo evaluation. Colloids Surf., A 466, 154–159.

Luz, P., Magalhães, L., Pereira, A., Cunha, W., Rodrigues, V., Andrade, E.S.M.L., 2012. Curcumin-loaded into PLGA Nanoparticles Preparation and *in vitro* schistosomicidal activity. Parasitol. Res. 110 (2), 593–598.

Mahapatro, A., Singh, D.K., 2011. Biodegradable nanoparticles are excellent vehicle for site directed in-vivo delivery of drugs and vaccines. J. Nanobiotechnol. 9, 55.

Manach, C., Scalbert, A., Morand, C., Rémésy, C., Jiménez, L., 2004. Polyphenols: food sources and bioavailability. Am. J. Clin. Nutr. 79, 727–747.

Manach, C., Williamson, G., Morand, C., Scalbert, A., Remesy, C., 2005. Bioavailability and bioefficacy of polyphenols in humans. I. Review of 97 bioavailability studies. Am. J. Clin. Nutr. 81 (1 Suppl. 1), 230S–242S.

Mansour, H.M., Sohn, M., Al-Ghananeem, A., Deluca, P.P., 2010. Materials for pharmaceutical dosage forms: molecular pharmaceutics and controlled release drug delivery aspects. Int. J. Mol. Sci. 11, 3298–3322.

Marin, E., Briceño, M.I., Caballero-George, C., 2013. Critical evaluation of biodegradable polymers used in nanodrugs. Int. J. Nanomed. 8, 3071–3091.

McClements, D.J., Rao, J., 2011. Food-grade nanoemulsions: formulation, fabrication, properties, performance, biological fate, and potential toxicity. Crit. Rev. Food Sci. 51 (4), 285–330.

Mehnert, W., Mader, K., 2001. Solid lipid nanoparticles- Production, characterization and applications. Adv. Drug Deliv. Rev. 47 (2–3), 165–196.

Mizrahi, M., Friedman-Levi, Y., Larush, L., Frid, K., Binyamin, O., Dori, D., Fainstein, N., Ovadia, H., Ben-Hur, T., Magdassi, S., Gabizon, R., 2014. Pomegranate seed oil nanoemulsions for the prevention and treatment of neurodegenerative diseases: the case of genetic CJD. Nanomed. Nanotechnol. 10 (6), 1353–1363.

Moghassemi, S., Hadjizadeh, A., 2014. Nano-niosomes as nanoscale drug delivery systems: an illustrated review. J. Control. Release 185, 22–36.

Mourtas, S., Lazar, A.N., Markoutsa, E., Duyckaerts, C., Antimisiaris, S.G., 2014. Multifunctional nanoliposomes with curcumin lipid derivative and brain targeting functionality with potential applications for Alzheimer disease. Eur. J. Med. Chem. 80, 175–183.

Mozafari, M.R., 2010. Nanoliposomes: preparation and analysis. Methods Mol. Biol. 605, 29–50.

Mukherjee, B., Patra, B., Layek, B., Mukherjee, A., 2007. Sustained release of acyclovir from nano-liposomes and nano-niosomes: an in vitro study. Int. J. Nanomed. 2 (2), 213–225.

Mukherjee, S., Ray, S., Thakur, R.S., 2009. Solid lipid nanoparticles: a modern formulation approach in drug delivery system. Indian J. Pharm. Sci. 71 (4), 349–358.

Müller, R.H., Mäder, K., Gohla, S., 2000. Solid lipid nanoparticles (SLN) for controlled drug delivery: a review of the state of the art. Eur. J. Pharm. Biopharm. 50, 161–177.

Mumper, R.J., 2006. Microemulsions as Precursors to Solid Nanoparticles. US Patent No. 7153525.

Nance, E., Timbie, K., Miller, G.W., Song, J., Louttit, C., Klibanov, A.L., Shih, T.-Y., Swaminathan, G., Tamargo, R.J., Woodworth, G.F., Hanes, J., Price, R.J., 2014. Non-invasive delivery of stealth, brain-penetrating nanoparticles across the blood-brain barrier using MRI-guided focused ultrasound. J. Control. Release 189, 123–132.

Nanjwade, B.K., Kadam, V.T., Manvi, F.V., 2013. Formulation and characterization of nanostructured lipid carrier of ubiquinone (coenzyme Q10). J. Biomed. Nanotechnol. 9 (3), 450–460.

Naseri, N., Valizadeh, H., Zakeri-Milani, P., 2015. Solid lipid nanoparticles and nanostructured lipid carriers: structure, preparation and application. Adv. Pharmaceut. Bull. 5 (3), 305–313.

Nasir, A., Kausar, A., Younus, A., 2015. A review on preparation, properties and applications of polymeric nanoparticle-based materials. Polym. Plast. Technol. Eng. 54, 325–341.

Neilson, A.P., Goodrich, K.M., Ferruzzi, M.G., 2017. Bioavailability and metabolism of bioactive compounds from foods. In: Coulston, A., Boushey, C., Ferruzzi, M., Delahanty, L. (Eds.), Nutrition in the Prevention and Treatment of Disease. Elsevier, pp. 301–319.

Neves, A.R., Lucio, M., Martins, S., Lima, J.L.C., Reis, S., 2013. Novel resveratrol nanodelivery systems based on lipid nanoparticles to enhance its oral bioavailability. Int. J. Nanomed. 8, 177–187.

Nitthikan, N., Leelapornpisid, P., Natakankitkul, S., Chaiyana, W., Mueller, M., Viernstein, H., Kiattisin, K., 2018. Improvement of stability and transdermal delivery of bioactive compounds in green Robusta coffee beans extract loaded nanostructured lipid carriers. J. Nanotechnol. 2018, 7865024. https://doi.org/10.1155/2018/7865024.

Oehlke, K., Adamiuk, M., Behsnilian, D., Graf, V., Mayer-Miebach, E., Walz, E., Greiner, R., 2014. Potential bioavailability enhancement of bioactive compounds using food-grade engineered nanomaterials: a review of the existing evidence. Food Funct. 5, 1341–1359.

Okonogi, S., Riangjanapatee, P., 2015. Physicochemical characterization of lycopene-loaded nanostructured lipid carrier formulations for topical administration. Int. J. Pharm. 478 (2), 726–735.

Owens, D.E., Peppas, N.A., 2006. Opsonization, biodistribution, and pharmacokinetics of polymeric nanoparticles. Int. J. Pharm. 307, 93–102.

Pandita, D., Kumar, S., Poonia, N., Lather, V., 2014. Solid lipid nanoparticles enhance oral bioavailability of resveratrol, a natural polyphenol. Food Res. Int. 62, 1165–1174.

Parada, J., Aguilera, J.M., 2007. Food microstructure affects the bioavailability of several nutrients. J. Food Sci. 72 (2), R21–R32.

Patel, T., Zhou, J., Piepmeier, J.M., Saltzman, W.M., 2012. Polymeric nanoparticles for drug delivery to the central nervous system. Adv. Drug Deliv. Rev. 64, 701–705.

Porrini, M., Riso, P., 2008. Factors influencing the bioavailability of antioxidants in foods: a critical appraisal. Nutr. Metabol. Cardiovasc. Dis. 18 (10), 647–650.

Rao, J.P., Geckeler, K.E., 2011. Polymer nanoparticles: preparation techniques and size control parameters. Prog. Polym. Sci. 36, 887–913.

Raposo, M.F.J., de Morais, R.M.S.T., de Morais, A.M.M.B., 2013. Health applications of bioactive compounds from marine microalgae. Life Sci. 93 (15), 479–486.

Rein, M.J., Renouf, M., Cruz-Hernandez, C., Actis-Goretta, L., Thakkar, S.K., Pinto, M.S., 2012. Bioavailability of bioactive food compounds: a challenging journey to bioefficacy. Br. J. Clin. Pharmacol. 75 (3), 588–602.

Reis, C.P., Neufeld, R.J., Ribeiro, A.J., Veiga, F., 2006. Nanoencapsulation I. Methods for preparation of drug-loaded polymeric nanoparticles. Nanomedicine 2 (1), 8–21.

Ross, A.B., Vuong, L.T., Ruckle, J., Synal, H.A., Schulze-König, T., Wertz, K., Rümbeli, R., Liberman, R.G., Skipper, P.L., Tannenbaum, S.R., Bourgeois, A., Guy, P.A., Enslen, M., Nielsen, I.L., Kochhar, S., Richelle, M., Fay, L.B., Williamson, G., 2011. Lycopene bioavailability and metabolism in humans: an accelerator mass spectrometry study. Am. J. Clin. Nutr. 93 (6), 1263–1273.

Rostami, E., Kashanian, S., Azandaryani, A.H., Faramarzi, H., Dolatabadi, J.E.N., Omidfar, K., 2014. Drug targeting using solid lipid nanoparticles. Chem. Phys. Lipids 181, 56–61.

Sachdeva, A.K., Misra, S., Kaur, I.P., Chopra, K., 2015. Neuroprotective potential of sesamol and its loaded solid lipid nanoparticles in ICV-STZinduced cognitive deficits: behavioral and biochemical evidence. Eur. J. Pharmacol. 747, 132–140.

Saranya, T.S., Rajan, V.K., Biswas, R., Jayakumar, R., Sathianarayanan, S., 2018. Synthesis, characterization and biomedical applications of curcumin conjugated chitosan microspheres. Int. J. Biol. Macromol. 110, 227–233.

Sari, T.P., Mann, B., Kumar, R., Singh, R.R.B., Sharma, R., Bhardwaj, M., Athira, S., 2015. Preparation and characterization of nanoemulsion encapsulating curcumin. Food Hydrocoll. 43, 540–546.

Sarker, D.K., 2005. Engineering of nanoemulsions for drug delivery. Curr. Drug Deliv. 2, 297–310.

Scalbert, A., Williamson, G., 2000. Dietary intake and bioavailability of polyphenols. J. Nutr. 130 (8S Suppl. l), 2073S–2085S.

Scalia, S., Young, P.M., Traini, D., 2015. Solid lipid microparticles as an approach to drug delivery. Expert Opin. Drug Deliv. 12 (4), 583–599.

Sessa, M., Balestrieri, M.L., Ferrari, G., Servillo, L., Castaldo, D., D'Onofrio, N., Donsì, F., Tsao, R., 2014. Bioavailability of encapsulated resveratrol into nanoemulsion-based delivery systems. Food Chem. 147, 42–50.

Singh, G., Pai, R.S., 2014. Optimized PLGA nanoparticle platform for orally dosed trans-resveratrol with enhanced bioavailability potential. Expert Opin. Drug Deliv. 11 (5), 647–659.

Solans, C., Izquierdo, P., Nolla, J., Azemar, N., Garcia-Celma, M.J., 2005. Nano-emulsions. Curr. Opin. Colloid Interface Sci. 10, 102–110.

Swarnakar, N.K., Jain, A.K., Singh, R.P., Godugu, C., Das, M., Jain, S., 2011. Oral bioavailability, therapeutic efficacy and reactive oxygen species scavenging properties of coenzyme Q10-loaded polymeric nanoparticles. Biomaterials 32 (28), 6860–6874.

Takahashi, M., Uechi, S., Takara, K., Asikin, Y., Wada, K., 2009. Evaluation of an oral carrier system in rats: bioavailability and antioxidant properties of liposome-encapsulated curcumin. J. Agric. Food Chem. 57 (19), 9141–9146.

Tamjidi, F., Shahedi, M., Varshosaz, J., Nasirpour, A., 2013. Nanostructured lipid carriers (NLC): a potential delivery system for bioactive food molecules. Innov. Food Sci. Emerg. Technol. 19, 29–43.

Trombino, S., Cassano, R., Muzzalupo, R., Pingitore, A., Cione, E., Picci, N., 2009. Stearyl ferulate-based solid lipid nanoparticles for the encapsulation and stabilization of beta-carotene and alpha-tocopherol. Colloids Surf. B: Biointerfaces 72 (2), 181–187.

Tzeng, C.W., Yen, F.L., Wu, T.H., Ko, H.H., Lee, C.W., Tzeng, W.S., Lin, C.C., 2011. Enhancement of dissolution and antioxidant activity of kaempferol using a nanoparticle engineering process. J. Agric. Food Chem. 59 (9), 5073–5080.

Umerska, A., Gaucher, C., Oyarzun-Ampuero, F., Fries-Raeth, I., Colin, F., Villamizar-Sarmiento, M.G., Maincent, P., Sapin-Minet, A.N., 2018. Polymeric nanoparticles for increasing oral bioavailability of curcumin. Antioxidants 7 (46). https://doi.org/10.3390/antiox7040046.

Vandita, K., Shashi, B., Santosh, K.G., Pal, K.I., 2012. Enhanced apoptotic effect of curcumin loaded solid lipid nanoparticles. Mol. Pharm. 9 (12), 3411–3421.

Wang, J., Wang, Y., Liang, W., 2012. Delivery of drugs to cell membranes by encapsulation in PEG–PE micelles. J. Control. Release 160 (3), 637–651.

Weiss, J., Decker, E.A., McClements, D.J., Kristbergsson, K., Helgason, T., Awad, T., 2008. Solid lipid nanoparticles as delivery systems for bioactive food components. Food Biophys. 3 (2), 146–154.

Wissing, S.A., Kayser, O., Muller, R.H., 2004. Solid lipid nanoparticles for parenteral drug delivery. Adv. Drug Deliv. Rev. 56, 1257–1272.

Wu, J.L., Liu, H., Ge, S.Y., Wang, S., Qin, Z., Chen, L., Zheng, Q., Liu, Q., Zhang, Q., 2015. The preparation, characterization, antimicrobial stability and in vitro release evaluation of fish gelatin films incorporated with cinnamon essential oil nanoliposomes. Food Hydrocoll. 43, 427–435.

Wu, W., Lee, S.Y., Wu, X.B., Tyler, J.Y., Wang, H., Ouyang, Z., Park, K., Xu, X.M., Cheng, J.X., 2014. Neuroprotective ferulic acid (FA)-glycol chitosan (GC) nanoparticles for functional restoration of traumatically injured spinal cord. Biomaterials 35 (7), 2355–2364.

Yang, C.S., Sang, S., Lambert, J.D., Lee, M.J., 2008. Bioavailability issues in studying the health effects of plant polyphenolic compounds. Mol. Nutr. Food Res. 52 (Suppl. 1), S139–S151.

Yao, M.F., Xiao, H., McClements, D.J., 2014. Delivery of lipophilic bioactives: assembly, disassembly, and reassembly of lipid nanoparticles. Annu. Rev. Food Sci. Technol. 5, 53–81.

Young, N.A., Bruss, M.S., Gardner, M., Willis, W.L., Mo, X., Valiente, G.R., Cao, Y., Liu, Z., Jarjour, W.N., Wu, L.-C., 2014. Oral administration of nano-emulsion curcumin in mice suppresses inflammatory-induced NF kappa B signaling and macrophage migration. PLoS One 9 (11), e111559.

Yu, H., Huang, Q., 2013. Investigation of the cytotoxicity of food-grade nanoemulsions in caco-2 cell monolayers and hepg2 cells. Food Chem. 141, 29–33.

Zhang, W., Li, X., Ye, T., Chen, F., Yu, S., Chen, J., Yang, X., Yang, N., Zhang, J., Liu, J., Pan, W., Kong, J., 2014. Nanostructured lipid carrier surface modified with Eudragit RS 100 and its potential ophthalmic functions. Int. J. Nanomed. 9, 4305–4315.

Zhao, J.W., Dyson, S.C., Kriegel, C., Tyers, P., He, X., Fahmy, T.M., Metcalfe, S.M., Barker, R.A., 2014. Modelling of a targeted nanotherapeutic 'stroma' to deliver the cytokine LIF, or XAV939, a potent inhibitor of Wnt-β-catenin signaling, for use in human fetal dopaminergic grafts in Parkinson's disease. Dis. Model Mech. 7 (10), 1193–1203.

Zhao, S., Zhang, Y.L., Han, Y.Z., Wang, J., Yang, J., 2015. Preparation and characterization of cisplatin magnetic solid lipid nanoparticles (MSLNs): effects of loading procedures of Fe_3O_4 nanoparticles. Pharm. Res. (N. Y.) 32 (2), 482–491.

Zhong, Y., Meng, F., Deng, C., Zhong, Z., 2014. Ligand-directed active tumor-targeting polymeric nanoparticles for cancer chemotherapy. Biomacromolecules 15, 1955–1969.

Zimet, P., Rosenberg, D., Livney, Y.D., 2011. Re-assembled casein micelles and casein nanoparticles as nano-vehicles for ω-3 polyunsaturated fatty acids. Food Hydrocoll. 25, 1270–1276.

Zou, L.Q., Peng, S.F., Liu, W., Gan, L., Liu, W.L., Liang, R.H., Liu, C.M., Niu, J., Cao, W.L., Liu, Z., Chen, X., 2014. Improved *in vitro* digestion stability of (-)-epigallocatechin gallate through nanoliposome encapsulation. Food Res. Int. 64, 492–499.

CHAPTER

Nanotechnology for enhanced bioactivity of bioactive compounds

15

Moumita Majumdar[a, 1], Saurabh Shivalkar[b, 1], Ayantika Pal[c], Madan L. Verma[d], Amaresh Kumar Sahoo[b], Dijendra Nath Roy[e]

Department of Chemistry, National Institute of Technology, Agartala, Tripura, India[a]; Department of Applied Sciences, Indian Institute of Information Technology, Allahabad, Allahabad, India[b]; Department of Human Physiology, Tripura University, Agartala, Tripura, India[c]; Department of Biotechnology, Dr Y. S. Parmar University of Horticulture and Forestry, Neri Campus, Himachal Pradesh, India[d]; Department of Bio Engineering, National Institute of Technology, Agartala, Tripura, India[e]

Chapter outline

[1]Equal contribution.

Biotechnological Production of Bioactive Compounds. https://doi.org/10.1016/B978-0-444-64323-0.00015-1

1. Introduction

There are immense interests and prospects of bioactive compounds-components of food that possess the biological activity-owing to its multifarious functions and effects in various biological pathways including metabolism, cell signaling, cell proliferation and growth and development (Skrovankova et al., 2015). Sources of bioactive compounds are foods; however, these are not imperative to sustain the life, still have huge influence on the cellular functions. Basically, it is a class of active food molecules provide better health in addition to their caloric values. For example, caffeine is a bioactive compound of coffee. There are various kinds of bioactive compounds which could be derived from numerous sources like plant, animals or synthetically. Some of the commonly use bioactive compounds are flavonoids, carotenoids, Collin, coenzyme Q, phytosterols, polysaccharides, glycosylates, polyphenols, and Vitamins. These all are having vital biological functions. For example, several flavonoids serve as free-radical scavenger, help in prevention of coronary heart disease, and shows anticancer activity (Vanacker et al., 1995). Likewise, polyphenolic compounds also act as antioxidant and anti-inflammatory agents (Soobrattee et al., 2005). Similarly, other bioactive compounds such as the coenzyme Q are essential for the electron transport chain that aids in biochemical generation of ATP by oxidative phosphorylation (Littarru and Tiano, 2007). Therefore, any incongruity in the cellular level of bioactive compounds may lead to serious impairment in downstream pathways.

It would be mentioned here that sellers of bioactive substances offer health benefit however; it may show adverse effects also. Additionally, a number of bioactive substances elicit immune responses too. However, insufficient number of uptakes or malfunctions of these agents lead to serious health impact which has been a matter of concern for long time (Bernhoft et al., 2010). Continuous supplies of the agents along with the foods are thus essential for normal biological functions. Otherwise, intra cellular deliveries of these classes of compounds possess huge challenges due to its physicochemical properties which are not easily amenable for biological delivery. Predominantly, the bioactive compounds are having the polarity which is not favourable for cellular uptakes. Thus, for last few decades, various research groups have been proposed numerous delivery strategies to increase the efficacy of delivery of the bioactive compounds (Li et al., 2012).

One of the effective ways of tuning of effectiveness of the delivery of bioactive compounds is incorporation nanotechnology. Nanotechnology deals with the phenomena and materials which occur in the length scale of nanometer (1 nm = 10^{-9} m). In this size realm, the materials exhibit unique physiochemical behaviors which are distinctly different from the bulk counterparts. Several studies revealed that using the principles and techniques of nanotechnology, activity and functioning of bioactive compounds may be monitored and improved (Huang et al., 2010).

Moreover, nanotechnology offers new horizons of scope and opportunities to employ these bioactive compounds in various other applications such as it can be used to make nanosized materials for controlled and targeted delivery system, development of new therapeutic agents, design of biosensors, synthesis of scaffold for tissue and cellular engineering. Recent development in this line of interest also helps in synthesis of new derivatives and analogous molecules and polymers based on the available bioactive molecules (Goddard and Hotchkiss, 2007). There are plenty of derivatives of bioactive molecules which are also found to be effective in various biological applications too. Some example derivatives of bioactive molecules are chitosan, poly (lactic-*co*-glycolic acid) (PLGA), poly (lactic acid), poly (glycolic acid). These are easily amenable for various biological applications and overcome some of the inherent limitations of the naturally available bioactive molecules such as hydrophobicity, water solubility and availability of suitable functional groups (Fontana et al., 2013).

Therefore, it's conceivable that prospects of bioactive compounds have been enlarging day-by-day due to advent of the nanotechnology. In the present chapter, the various nanotechnological aspects for delivering and enhancing the activity of bioactive compounds is critically discussed. The use of nanotechnology for design scaffolds suitable for tissue engineering based on bioactive molecules is discussed.

2. Role of "nano" to enhance the activity of bioactive compounds

2.1 Enzyme catalysis

Enzymes are the biocatalysts that accelerate the rate of reaction with undergoing any changes itself. Undoubtedly, enzymes exhibit remarkable specificity and efficiency in favourable as well as unfavourable varied reaction conditions of temperature, pH, and solvents (Kanwar et al., 2005, 2006). Enzymes, as catalysts, have applications in industrial, biomedical, and analytical processes. The applications of enzymes at industrial level are limited due to high cost and low reusability. In addition, many enzymes are active in mild condition, unstable at high temperature, and require an aqueous medium for their function, which diminishes their applications in various fields. However, in the immobilized state, activity, specificity, and stability of the enzymes is an alternative to such impressing issues (Verma and Kanwar, 2008; Verma et al., 2008a, 2008b).

Immobilization of enzymes is a well-established methodology that offers multiple advantages over free enzymes including improved stability against solvents and temperature, increased functional efficiency, enhanced reproducibility, continuous use, reusability, lower reaction time, fewer chances of contamination in a product, a higher enzyme to substrate ratio, and also use of the multi-enzyme system to obtain a desired product is possible. In addition, this process is cheap, easy to use and environment-friendly (Verma and Kanwar, 2012; Verma et al., 2011). On the other side, it has disadvantages including high cost for isolation and purification of active enzyme, lower or loss of activity of some enzymes due to conformational changes after binding to support, low stability, and alteration in their properties and reduced efficiency against insoluble substrates such as cellulose (Verma and Kanwar, 2010; Verma et al., 2009).

The enzyme immobilization methods including adsorption, entrapment, covalent binding and cross-linking have been used successfully for reusability and other applications (Verma and Kanwar, 2008; Kanwar and Verma, 2010). Almost all enzymes are proteins, and proteins are composed of amino acids. The enzyme immobilization methods take advantage of functional groups present in the side chain of amino acid. These amino acids bind with a carrier through various type of interactions including the weak bonds (ionic interaction, hydrogen bonds, and Van der Waal forces), and covalent bonding. In covalent bonding method, the functional group of enzymes that provide chemical groups to form covalent bonds with functional groups of protein (Kanwar et al., 2007a, 2007b, 2008). The choice of enzyme immobilization method depends upon desired enzyme property, existing structural and mechanistic information of protein, and high throughput screening method (Verma and Barrow, 2015). The characteristics of materials used for immobilization of enzymes have utmost importance. The ideal support or carrier used for enzyme immobilization should be cost-effective, eco-friendly, inert, highly stable, thermally and mechanically resistant, able to be regenerated, resistant to microorganisms, and enzyme specific (Puri et al., 2013). The support or matrix used for enzyme immobilization allows the exchange of medium containing substrate or cofactor or product molecules.

Recent advancement in nanotechnology has led to the development of various nanomaterials of industrial importance. The nanostructured materials including nanoparticles, nanofibers, nanocomposites, and nanotubes are favored as excellent support for enzyme immobilization (Verma et al., 2012, 2013a). They are expensive but they offer ideal characteristics that determine the efficiency of enzymes. They have inherently large surface area and high mechanical properties, which allows high enzyme loading with minimum diffusion limitation (Verma, 2017a; 2017b, 2017c; Verma et al., 2016). The nanomaterials-based immobilization has multiple advantages as compared to conventional bulk material immobilization techniques such as easy synthesis in high enzyme loading, quick separation, ease of surface modification, precise control over synthesis of particle size, and homogeneity (Verma et al., 2013a, 2013b). In addition, the nanoparticles have optical, electrical, thermal, chemical, mechanical, and catalytic properties.

The synthesis of suitable nanocarrier for enzyme immobilization with potential industrial biotechnology applications has attracted significant attention. Recent developments in nanocarrier, the combination of protein immobilization methods for improved enzyme function, for biofuels production.

2.2 Detection of pathogen and biomarkers

Detection of pathogen and biomarkers by conventional bacterial detection methods such as PCR, fluorescent immunoassays and ELISA exhibit high sensitivity and specificity, they are relatively laborious and require sophisticated instruments. In this situation, simpler and more sensitive diagnostic technique is required for human benefits. Some of the recent major advancement in pathogen detection has been described in the following sections.

Reduced graphene oxide has drawn a great attention as a potential nanomaterial for reliable detection of pathogenic microorganism such as bacteria due to its outstanding properties like ultrahigh electron transfer ability, great surface to volume ratio, biocompatibility, and its distinctive interactions with DNA bases of the aptamer. Reduced graphene oxide-azophloxine nanocomposite aptasensor was invented for a sensitive, rapid, and robust detection of food-borne pathogens along with it is acting as an electroactive indicator for redox reactions. Under optimum conditions, this aptasensor showed a linear range of detection with good linearity ($R^2 = 0.98$) and a detection limit of 10^1 cfu/mL (Muniandy et al., 2017). Another one study reported that multiwalled carbon nanotube based zinc nanocomposite demonstrated the capability to detect of an agricultural pathogen i.e. Chili leaf curl betasatellite (ChLCB) (Tahir et al., 2017). The multiwalled carbon nanotubes (MWCNTs) having a diameter of 50—100 nm with zinc nanoparticles (Zn-NPs) of 25—500 nm was used for the immobilization of probe DNA strands. The electrochemical performance of DNA biosensor was measured in presence and absence of the complementary DNA during cyclic and differential pulse voltammetry scans. This kind of designed material can be used to recognition the plant viruses and biomolecules.

Tristeza is one of the critical diseases of citrus caused by citrus tristeza virus (CTV). For detection of CTV, a label-free impedimetric biosensor was developed for the detection of nucleic acid of CTV (Khater et al., 2019). The sensing platform is based on a screen-printed carbon electrode (SPCE), which was modified by electrodeposited gold nanoparticles (AuNPs). This process helps to optimize efficiently immobilizing thiolate dss DNA probes as well to increase the electrode conductivity. In view of selective binding affinity between viral pathogens and normal live cells, nanoparticles covered cellular membranes are attributed for virus targeting. One group used influenza virus along with erythrocyte's membrane bound with magnetic functionalities nanoparticles (RBC-mNPs) (Chen et al., 2017). The manufactured nanoparticles carry a distinctive membrane corona with sialylated surface and form colloidally stable clusters with influenza viruses. The manifestation of pathogen targeting by RBC-mNPs highlights a biologically inspired approach toward better treatment and analysis against infectious disease threats. The quantum dots (QDs) is well known for their photoluminescence properties, which may be used for detection of bacteria. In this regard, one study was carried out with *Escherichia coli* labeled with CdSe-QDs and that were examined both on an Agar nutrient and ground fish substrates to observe the growth rate in different environments in the Lab (Mohamadi et al., 2017). The bacterial growth was examined under the irradiation of UV light. The growth pattern was investigated, which indicates *Escherichia coli* can stay alive and it can be dispersed on food products to monitor the growth easily. This finding helps to detect bacterial growth on food products, which may be used as an easy detection of food borne pathogens grown on a food sample. An immunological biosensor was proposed by a group based on peroxidase mimetic activity and photothermal effect of nickel oxide nanoparticle (NiOGs) to find out and destroy the food borne pathogen like *Salmonella typhimurium* (Pandian et al., 2017). NiOGs demonstrated peroxidase activity and it was coated with citric acid followed by conjugated with primary antibody (anti-*S. typhimurium*) that effectively bound *S. typhimurium*. This detection method has advantage of no cross reactivity against other pathogens. NiOGs exposed to laser irradiation confirmed dose dependent increase in temperature and singlet oxygen that induced membrane damage and reduced bacterial survival rate. The bi-functional peroxidase-mimetic activity and photothermal effect of NiOGs can be used in selective sensing and killing of target pathogens respectively. Furthermore, one research team used multi-parametric magneto-fluorescent nanosensors (MFnS) to detect bacterial contamination with concentrations as little as 1 colony-forming unit (CFU) (Banerjee et al., 2016). The magnetic relaxation property of this MFnS allowed for sensitive screening at lowest value of CFU, which is justified by fluorescence measurements of higher CFU samples. Together, these qualities allowed for detection and quantification of broad-spectrum contaminations in various samples ranging from aquatic reservoirs to commercially produced food.

2.3 Hormonal balance

Inside body each and every biological procedure is under jurisdiction of specific biomolecules and their interactions termed as signaling. The principal signaling molecules that control the structural and behavioral regulation of multicellular organism is designated as hormones. Circulatory system is the primary transportation system of hormone inside body. Hormonal activity mainly depends on their chemical composition such as steroids, eicosanoids and hormones composed of amino acid derivatives and pH niche. Hormones communicate within organs as chemical messengers and thereby help to accomplish the physiological processes including metabolism, respiration, reproduction, growth and development (Folkman, 2013; Neave, 2007). These chemical messengers often initiate the expression of target proteins of a metabolic pathway as secondary messenger. Some of the authoritative hormones are insulin, progesterone, estrogen, testosterone, follicle-stimulating hormone, thyroid stimulating hormone etc. Minute change in Homeostasis in hormonal level leads to severe life threating disease conditions.

In the present era, nanotechnology is invincibly involved in field of medicine because of its capabilities to act on molecular levels inside cell to regulate biochemical pathways. It's possible to target the disease instigating molecules and signaling pathways often simultaneously. Application of nanotechnology in medical field includes direct and indirect exposure to disease site.

In the treatment practice of hormonal imbalance there is a tremendous success achieved through application of nanodrug as medicine and nanotechnology in indirect targeting (Bartneck et al., 2014; Gonzaga et al., 2012). The role of nanotechnology in enhancing the activity of bioactive compounds is now a days has become the principal attention of research in the field of modern medicine as because of its biocompatibility and less side effect on health. Generally chemicals obtained from plant source; those proficient in controlling metabolic pathways and growth promotion are designated as bioactive compounds. The constructive role of these compounds in several biochemical pathways such as inhibition of receptor proteins, antioxidant activity, inhibition or stimulation of enzyme action and reserve the gene expression (Correia et al., 2012). These promising activities lead to the elevation of their application with nanotechnology in targeting and to get more accurate result in comparatively shorter time span.

Very common and life threatening health problem associated with hormonal dysfunction is diabetes in which blood sugar level of individual is above normal threshold. Most common type diabetes is diabetes mellitus type 2 caused by insulin resistance or lack of insulin in body, thereby elevated level of blood glucose. Polysaccharides and its derivatives of Ginseng saponin are in use to treat diabetes (Kim and Kim, 2008; Yuan et al., 2011, 2012; Chen et al., 2012). Ginseng saponin or Ginsenosides is extracted from plant genus *Panax* (Qi et al., 2011). In a recent study nanoencapsulation of ginseng compounds have showed a greater result in diabetes patients with potential addition health effects (Chen et al., 2012). This is because pituitary has an indirect effect on secretion of insulin, secreted from beta cell of pancreatic Islets. In modern medicine it has become a recent practice to use nanosized compounds extracted from plant derivatives as it easily cross the blood brain barrier to control the insulin signaling (Niu et al., 2018). Experimental result suggest the enhancing role of nano ginseng in secretion of testicular hormones and improved sperm production along with lesser damage of DNA (Kim et al., 1998; Melnyk and Marcone, 2011; Achike and Kwan, 2003; Tambi et al., 2012). This outcome has a very positive impact on developing nanoparticles of bioactive compounds to facilitate alternative and more effective therapeutic approach for management of hormonal disorders. In a research work, administration of ecofriendly synthesized gold nanoparticles ranging 50 nm with phytochemicals; extracted from *Gymnema sylvestre R. Br,* has showed significant inhibitory effects against diabetes in experimental albino rats along with superior inhibitory activity against inflammation (Karthick et al., 2014). Application of bioactive ginseng stabilized silver nanoparticle ranging of 40 and 30 nm showed more enhanced antidiabetic effects (Han et al., 2014). Proliposomes are developed artificially for oral delivery of drugs using saponins compounds from ginseng fruit with sodium deoxycholate, exhibited superior effects on bioavailability (Hao et al., 2016). Targeting biomolecules of signaling pathway of specific hormone with quantum dots has become evolutionary in medicinal field to treat hormonal dysfunction in molecular level. Quantum dots emit light of specific wavelengths which clues to assist the probable pathways of targeted biomolecules inside cell (Shrivastava and Dash, 2009).

Melanocyte stimulating hormone (MSH) is the key hormone that controls the pigmentation in layers of skin and hair. Its activity is regulated via G protein coupled receptor, melanocortin receptor 1. MSH stimulate the production of melanin in melanocyte cells of skin and hair through melanogenesis (D'Mello et al., 2016). There is a disease phenomenon where melanoma cells show overexpression of a melanocortin receptor 1 (Salazar-Onfray et al., 2002). To combat this problem an innovative approach has been taken where peptides derivatives of MSH are treated with nanosized particle to target the receptor protein to regulate the specific signaling (Wong and Minchin, 1996; Singh and Lillard, 2009; Wen and Chase, 1999; Tao et al., 2010). In a study, depending on the anti-inflammatory nature of melanocortin peptides, application of nanoparticle-peptide conjugate in treatment of inflammation caused by combined periodontic-endodontic lesions and ulcerative colitis (Fioretti et al., 2010; Yan et al., 2009) in mouse model was demonstrated to have promising healing effect.

Hormonal imbalance in menopause is a very common problem among women. This often leads to several health problems such as increased rick of UTI, lesser bone density with risk of osteoporosis, stress, anxiety which leads to depression, cardiovascular issues. To withstand this health issue conventional hormone therapy was formerly used to treat menopausal symptoms via oral drug delivery system. A new concept of transdermal hormone replacement therapy (HRT) (Bakour and Williamson, 2015) has become one of the alternative medication because controlled release of drugs has a significance positive impact with less side effect (Correia et al., 2012; Gonzaga et al., 2012). Drug penetration efficacy with drug concentration in every distinct layer of skin was evaluated by confocal Raman spectroscopy in androgen deficient groups of people. Nano emulsified testosterone therapy significantly increase the serum testosterone levels in treated group than untreated control groups (Gonzaga et al., 2012; Botelho et al., 2013). One characteristic feature of body during menopause is elevated level of follicle-stimulating hormone (FSH) with estrogen and progesterone deficiency (Wellons et al., 2012). Patients get rid of these uncomfortable physical condition with administration of exogenous hormones through HRT (Rosano et al., 2003). However, the long term use of HRT has lees significant effect than shot term treatments (Gottsäter et al., 2001; Daly et al., 1996; Canonico et al., 2007). HRT also have therapeutic effects on menopause related heart problems and recovery from osteoporosis (Teede et al., 2000; Szekacs et al., 2000).

5 year long term study of HRT on a group of women showed a promising result for delivery of estriol and estradiol with help of nanotechnology. A combination of estriol, estradiol and progesterone was used to design a nanostructure for exogenous hormone delivery. The effect was monitored clinically through biochemical tests, Raman spectroscopic study for real time drug concentration (Gonzaga et al., 2012; Botelho et al., 2013, 2014). Nanostructures were absorbed in to layer of skin and abled to deliver estriol and estradiol more effectively within short time of administration. There was no such skin irritation reported. This research data has a very positive impact on evaluation of estriol and estradiol nanoemulsification technique as a very steady system of drug release in treatment of menopausal complications.

2.4 Application in healing from wounds

With expansion of its wings in various filed of medicine, nanotechnology has grasped the threshold where it's invincibility for almost every aspect of medical practice. One of the very important and primary applications of nanocrystals is in the arena of wound healing. The biochemical and cellular progression by which body reinstate the integrity of previously disrupted skin by regeneration of cells is termed as wound healing. Three distinctive steps are involved in wound healing process namely inflammation or swelling, proliferation or regeneration via neo-angiogenesis, granule formation and construction of epithelium and growth or maturation through modulation of extracellular matrix (Groeber et al., 2011; Pereira and Bartolo, 2016; Reinke and Sorg, 2012). When body cannot able to repair the damage by itself; there is a need for some aid to accelerate the healing process. Natural products extracted from plants, have become one of the alternative therapeutics for their antimicrobial, anti-inflammatory, antidiabetic properties due to presence of bioactive compounds such as alkaloids, phenolic compounds, essential oils, flavonoids, terpenoids etc. (Upadhyay, 2016). These natural products are considered as more superior than that of conventional antibiotics and modern synthetic drug as these traditional medicines are cost effective with low cytotoxicity (Gamit et al., 2017; Farzaei et al., 2014). Having such promising bioactive compounds; different plant extract are subjected to convert into nano form for more bioavailability as in nano form they have high surface area to volume ratio with physical and chemical modifications (Alvarez-Román et al., 2004; Mordorski and Prow, 2016). In recent years plant extract mediated metal nanoparticle synthesis reported to exhibit wound healing activity through surface treatment to quicken blood clotting with inhibition on bacterial infection spread (Vatankhah et al., 2014; Liao et al., 2015). Silver is a well-known antiseptic as it possess anti-inflammatory and antibacterial properties (Prabhu and Poulose, 2012). By the use of these characteristic of silver nanoparticles are synthesized using plant extracts composed of bioactive molecule which is more effective in drug release and less toxic than that of chemically synthesized silver nanoparticle (Bindhani and Panigrahi, 2015).

Growth factors or nucleic acids loaded nanoparticles are very effective implements for treatment of membrane injury as well as in wound healing (Hamdan et al., 2017). Lipid-based short-interfering RNA (siRNA) nanoparticle is designed to treat internal wounds in genetic levels. Lipid or gold based siRNA NPs are very effective against tumor necrosis factor coding alpha gene which is excessively produced in chronic wounds by macrophages (Kasiewicz and Whitehead, 2016). Wound healing in diabetic animal was achieved by diminution of the ganglioside-monosialic acid 3 synthase level through activation of IGF-1andinsulin receptor via siRNA NPs (Randeria et al., 2015).

Another aspect to design nanoparticle in dressing and wound healing purpose is to recruit natural polymers; derived from extracellular matrix of plant mainly derivatives of cellulose such as collagen, gelatin, silk fibroin polysaccharides and also from animals namely chitosan, hyaluronan (Mano et al., 2007; Gasperini et al., 2014). Proteins and polysaccharides are not always applied as nanoparticles but rather functional in the form of scaffolds (a nanosized network of polymers) in treatment of bone damage by regeneration of stem cells. Drugs and often stem cell entrapped scaffolds, are injected to injured parts of body and stem cells have readily accustomed in the niche and nerve ends starts to adjoin with scaffolds ultimately leads to healing the wound site (Fathi and Zaminy, 2017).

Chitosan is a carbohydrate polymer composed of β-(1−4) linked D-glucosamine and *N*-acetyl-D-glucosamine which possess antimicrobial activity. It is also biocompatible as well as biodegradable in nature. In a very recent research, chitosan NPs were designed in combination with cathelicidin; a family of antimicrobial peptides (AMPs) derived from King kobra (OH30). OH30 encapsulated carboxymethyl chitosan nanoparticles (CMCS-OH30 NP) accelerated wound healing by advancement of granule formation via neovascularization through functional microvascular networks and reconstruction of epithelial tissues in animal models (Sun et al., 2018).

To reduce the effect of reactive oxygen species on individual cells more specifically in membrane proteins and DNA, natural antioxidants such as aloe vera, curcumin, melatonin vitamin E are used with chitosan in form of hydrogels (Lü et al., 2010; Rahman et al., 2017). This hydrogel technique is useful as it helps to restore the hydrated state of mortiferous skin tissue and facilitates the auto elimination of dead tissues debris. This is mainly constructed with polysaccharide polymers for their mucoadhesive and high-water affinity behaviors (Das et al., 2015).

2.5 Application in immune responses

The chemical change in host immune system to salvage homeostasis inside cell after exposure to any antigens is termed as immune responses. Immune response is of different types depending on the nature of antigen. For example, host activates its immune system against invaded or adhered microorganisms through innate immune response and autoimmune response is triggered by production of antigen produced by own-self along with graft rejections (Ulvestad, 2007; Kagan, 2006).

Natural bioactive compounds are very effective against combating theses type of health issues (Balunas and Kinghorn, 2005). Formation of nanocrystals of natural products increases their bioavailability due to certain accrued characteristics. When these compounds are converted in to nano form; with increasing surface area the active functional groups are further accessible to become involved in chemical reactions (Khan et al., 2017). Biologically active compounds when administrated with nanocarriers, due to high surface volume ratio through appropriate chemical formulation led to formation of nanocapsule with additional adhesive properties. These type of nanocapsule helps to slowdown the transportation of active compound there by facilitate to enhance the bioavailability through supplementary absorption (Bengoechea et al., 2011; Berton-Carabin et al., 2013; Cerqueira et al., 2014). Liposome infusion with Polyethylene glycol (PEG) grafting reported to have promising effect on enhancing the hypersensitivity of non-IgE-mediated signaling (Moghimi et al., 2005). CpG DNA is a very new approach to treat immunological disease conditions. CpG DNA is single stranded DNA molecule with cytosine triphosphate deoxynucleotide ("C") followed by a guanine triphosphate deoxynucleotide ("G") which has phosphodiester bond (p) in between successive nucleotides. Unmethylated CpG motifs are effective to stimulate immune responses (Weiner et al., 1997). In innate immune system CpG DNA have a captivating role by being ligand of Toll Like Receptor-9 (TLR-9). Due to this promising outcome these nucleic acids can be used in treating different diseases like bacterial, viral infection, and allergies (Godfrey et al., 2000; Baron et al., 1991). Naturally occurring CpG DNA have some disadvantages to be used as therapeutics due to little in vivo circulation time and its phosphodiester bond is prone to enzymatic degradation along with very low cell membrane permeability (Emlen and Mannik, 1984; Sands et al., 1994). To circumvent this problem cationic liposome and scaffold are used as carrier to deliver CpG DNA to TLR9-positive cells (Latz et al., 2004; Smolarczyk et al., 2005).

With the benefit of conception of nanotechnology there is a gigantic uprising in the field of pharmacokinetics and dynamics of drug design. In a recent study chitosan was used as eco-friendly nanocarrier to deliver leaf extract of *Ziziphus mauritiana* containing various bioactive compounds. *Z. mauritiana* leaf reported to have modulatory effect on immune response. To evaluate the restorative action of this phytochemical chitosan nanocapsule loaded with leaf extract of *Z. mauritiana was* applied to previously hydrocortisone treated Swiss albino mice with suppressed immune system. The data showed significant uplifting of immune response in host after oral administration the designed drug (Bhatia et al., 2011). Crude extract and pure ginseng compounds have reported to exhibit immunomodulatory effect with varying concentration in animal models (Kim et al., 2015; Lee et al., 2015; Lim et al., 2015; Yu et al., 2015). From earlier reports on analytic and chemical studies of ginseng, presence of polysaccharide and saponin was confirmed (Jovanovski et al., 2014). These bioactive compounds are also demonstrating to involve in regulation of immune response (Ma et al., 2014; Kim et al., 2014; Lee et al., 2014; Mukherjee et al., 2014). Construction of nano forms ranging from 70 to 107 nm of ginseng metabolites and their superior effects in enhancement of immune response in immunosuppressed animal and cell models were recently reported (Song et al., 2009; Zhang et al., 2012). Co administration of antigen ovalbumin and ginsosome nanoparticles in mice shows generation of specific antibodies namely IgG3, IgG2a, IgG1 and IgG2b with a significant higher level of B and T lymphocytes proliferation (Song et al., 2009). Nano emulsion of ginseng saponin with a droplet size of 72 nm demonstrate to have positive effect on production of Immunoglobulin 2 (IgG2) in ovalbumin antigen treated mice along with improved Th1 and Th2 immune response (Zhu et al., 1998). Depending on these promising data nano ginseng is also employed on vaccine designing. The vaccine adjuvant activity of nanoginsosomes in combination with recombinant vaccine on coccidiosis in chickens against apicomplexan parasite *Eimeriatenella* was studied. It was found that nanoginsosomes was capable to induce vaccine subunits with Interlukin-1 secretion and proliferation of lymphocyte (Zhang et al., 2012). With help of all these spectacular applications of nanoencapsulated and nanoemulsified bioactive compounds in the field of medicine specifically in hormonal treatment, wound healing and generation of immune response nanotechnology has opened up an alternative gateway to amalgamate the nature with technology toward immense human benefit.

3. Increase the biological activity of bioactive molecules by using nanotechnology

Bioactive compounds have huge effect on the living organism and thus essential for normal growth, differentiation and prevention from many diseases by controlling biological mechanism. For example, collections of bioactive compounds from various sources have helps in growth, development and prevention of diseases (Table 15.1). Moreover, recent studies established that several bioactive compounds have been shown to influence the effect of the existing drugs such aspaclitaxel and cyclophosphamide are coupled with bioactive compounds like curcumin, baicalein and reseratol to enhance the therapeutic effect of breast cancer (Mangla and Kohli, 2018). Therefore, the combination of bioactive compounds along with drugs paves new avenue in the treatment of several diseases including cancers. Essentially, the combination of the bioactive molecules along with conventional drugs complements the drawback of the overdose of drug molecules also. However, there are cases where bioactive compounds are not essentially taken in desired amount by normal food uptake. Thus, unregulated uptake causes several unwanted health impacts. Thus, food supplements are provided to address these issues. Apart from this

Table 15.1 List of some of the bioactive compounds from different sources and their mode of biological actions.

Bioactive compounds	Source	Mode of action	References
β-carotene	Fresh fruits and vegetables	Antioxidant helps in recovery of heart disease and age-related diseases	Gloria et al. (2014)
Catechin	Green Tea	Antioxidant	Baker and Bauer (2015)
Flavonoids	Vegetables (Water Mint)	Anti-proliferative activity of MCF7	Conforti et al. (2008)
Diallyl disulfide	Garlic	Antioxidant, anti-bacterial, anti-fungal and Detoxification of carcinogens	Tsubura et al. (2011)
Capsaicin	Red Chilli	Notch pathway inhibition of breast cancer stem cell	Shim and Song (2015)

huge biological importance and relevance, bioactive compounds found to be offered a great deal of scope of using these in various other fields such as optoelectronics, biomedical, and healthcare with little modifications and/or functionalization. There have been great repertoires of researches going on development of nanoformulations based on bioactive molecules (McClements et al., 2007). Nanoformulations eventually reduce the inherent limitations of the bioactive molecules such as issues of solubility, bioavailability and specificity. That provides opportunities of employing these nanoformulations in development of advanced materials particularly suitable for various biomedical applications (Danhier et al., 2012). Thus, the various biomedical aspects and scope of nanoengineering by using these bioactive molecules have been discussed in the subsequent topics.

3.1 Design of nanocarriers for delivery of bioactive molecules

Delivery of any biologically relevant agents including drugs, proteins and nucleic acid is a conventional mechanism for transporting to achieve specific therapeutic effect. These therapeutic agents cross various level of biological barriers in our body to reach to the predetermined sites. First level of barrier includes cellular linings of epithelial and endothelial cells designed to prevent foreign molecules to enter inside the body. This is crucial for tropical applications of any therapeutics. Moreover, intra cellular uptakes provide another level of barriers owing its selectively permeable nature. This creates additional challenge for transportation of any agents, if the chemical properties are not suitable to overcome the chemical properties physiological barrier. Based on the chemical properties of the physiological barriers various approaches, technologies and methods have been employed in the delivery of therapeutics (Hans and Lowman, 2002).

Conventional route of administration of drugs are oral delivery, inhalation, intranasal, pulmonary, vaginal, rectal, transdermal (skin) and topical (injections) (Vélez et al., 2017). However, all these method of delivery systems have been having inherent demerits of distribution of the drugs throughout the body. That eventually increases the drugs dose and frequency. In addition, these methods suffer from the targeting and specificity issues. Whereas the primary objective of drug targeting is to bring the drug substances as close to the site of action as possible and avoid the side effects in other parts of the body.

Nano sized delivery system (<200 nm) enhances the efficacy of the drugs as it increases the chemical activity and/or reactivity, and ease of bioavailability (Ezhilarasi et al., 2013). As bioactive molecules are having less toxicity issues, hence, there are plenty of studies directed globally to explore the use of these bioactive molecules as integrated components of the nanocarriers suitable for delivery of drugs, proteins and nucleic acids. Besides, these are easily available economical materials suitable for large scale applications. This also provides the opportunities to tag specific ligands on the surface of the nanocarriers for targeted delivery of the bioactive compounds. The engineered nanocarriers offer great deals of scope of controlled releases at the specific sites by monitoring intra cellular stimuli such as pH or temperature, magnetic or electric filed.

Further, the delivery of bioactive compounds along with drug molecules is well known aspects as we have stated above. This increases the shelf life of drug in the blood, effective route of administration and drug targeting. An increase in the self-life of the drug inside body is required for increasing the time of action of the molecules. There are various types of nanoformulations available which are given below which increases the efficiency of the delivery of bioactive molecules.

3.1.1 Methods of synthesis of nanocarriers

Development of the engineered nanocarriers is plausible by employing the suitable bioactive molecules. These nanocarriers provide the opportunities of encapsulation and controlled release of the drugs/therapeutic agents to the targeted site. The nanocarriers are formed using various wet chemical methods such as emulsification, nanoprecipitation, coacervation, emulsification solvent evaporation, and freeze drying etc. (Singh et al., 2017b) (Table 15.2).

3.1.1.1 Nanoemulsions

Colloidal biphasic dispersion of two or more immiscible liquids results into spontaneous formation of emulsions with internal phase droplets of solid spheres. These emulsions are termed as nanoemulsions while the particle size is within nanometre range. Size of the sphere may vary from widely based on synthesis techniques and solvent properties. However, for drug delivery purpose 50 nm–200 nm sizes of the particles are found to be more effective. The major components of nanoemulsions are two phases i.e. oil/dispersed phase and aqueous/continuous phase formed from dispersion of immiscible liquids (Oil in Water or Water in Oil) and an

Table 15.2 Different encapsulation techniques and use of bioactive molecules.

Encapsulation techniques	Bioactive molecules	References
Spray drying	Polyphenols molecules (both water insoluble as well as water soluble form of polyphenols)	Ferreira et al. (2007)
Vesicle formation (e.g., liposomes)	Polyphenols molecules within phospholipid bilayer	De Kruif et al. (2004)
Phase separation (e.g., polymeric nanoparticles)	Water insoluble polyphenols as oil phase	Reis et al. (2006)
Freeze drying	Polyphenols molecules	Choi et al. (2004)
Emulsion	Polyphenols molecules using emulsifying agent	Jafari et al. (2007b)

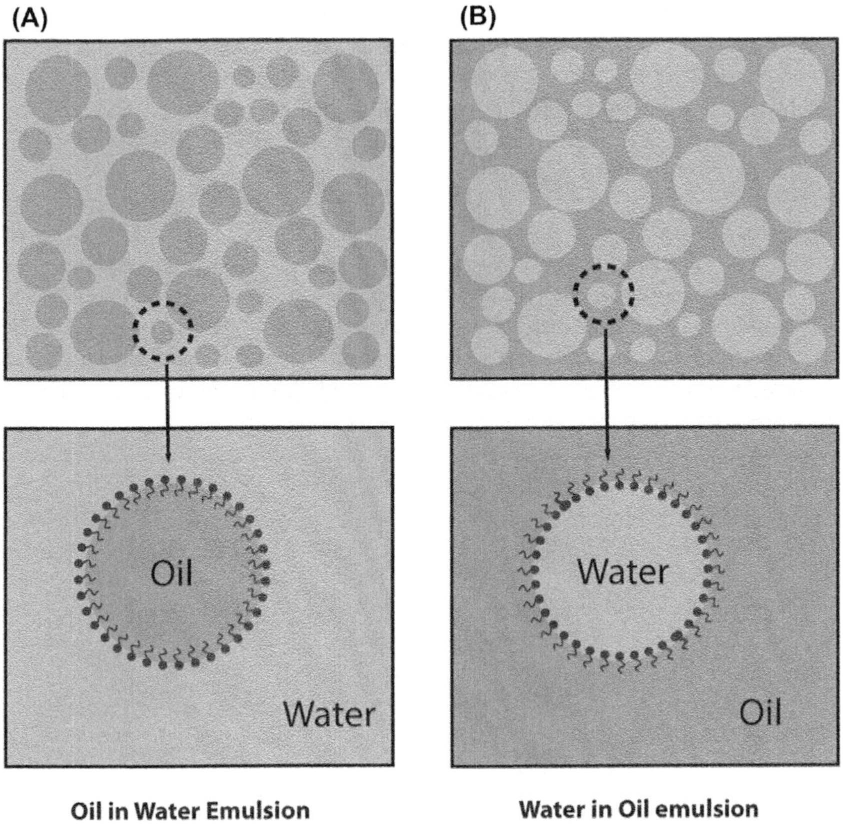

FIG. 15.1

Illustration of oil in water and water in oil emulsion methods for synthesis of nanoemulsions.

amphiphilic surfactant to stabilize the emulsion droplets. Oils mostly used for nanoemulsion formation are castor oil, corn oil, linseed oil, mineral oil, evening primrose oil. The number and overall stability of droplets are determined by comparative volume of external and internal phase, measured as phase volume ratio. The phase having higher volume is the external phase (Sharma et al., 2013) (Fig. 15.1).

Emulsion droplets size easily varies based on the oil to surfactant ratio. For example, higher the oil to surfactant ratio results greater the size of nanoemulsion. Therapeutics loaded within these smaller size nanoemulsions provides greater absorption due to higher surface area leading enhanced bioavailability. The unique properties of these nanoemulsion based carriers such as suitable size (50–200 nm), favourable viscoelasticity, tuneable surface properties, thermodynamic stability (Belhaj et al., 2010) and high drug loading efficiency cater to resolve various intricacies in present day drug delivery system. Nanoemulsions is very essential strategy for delivery of various poor water-soluble bioactive compounds like retinol, β-carotene, plant sterols and fatty acids are loaded and delivered using oil in water nanoemulsion (Wang et al., 2008; Yuan et al., 2008; Miladi et al., 2016). And water-soluble bioactive compounds like polyphenols and folic acids are delivered using water. This is essential while normal food uptakes unable to provide sufficient number of bioactive molecules inside the body.

3.1.1.2 Nanoprecipitation technique

Nanoprecipitation is a technique to develop nanocarriers for encapsulating and delivering the hydrophobic bioactive molecules such as retinoic acid and curcumin (Anand et al., 2010). Nanoprecipitation is a continuous emulsification of an organic solvent (hydrophobic bioactive molecules) in the organic phase (biodegradable polymer) with aqueous phase. It is also known as solvent displacement. In this technique the polymer from the organic phase is precipitated and organic solvent is diffused into the aqueous phase to form a vesicular nanostructure termed as nanospheres and nanocapsules (Fig. 15.2).

These nanospheres and nanocapsules having bioactive nanoformulations are then targeted to desired site for its delivery. Some of the bioactive compounds

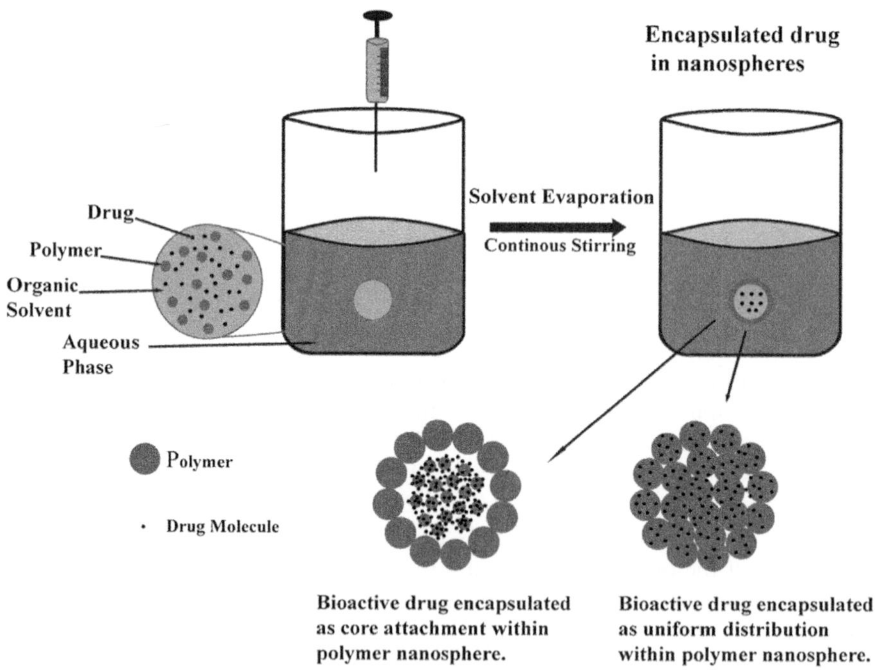

FIG. 15.2

Illustration of nanoprecipitation technique for the formation of drug encapsulated nanospheres.

encapsulated with this technique are heparin curcumin, β-carotene, astaxanthin. The biodegradable polymers which are constantly used for this purpose are polycaprolactone, poly (lactide-*co*-glicolide), poly (lactide), poly (alkylcyanoacrylate) (Ribeiro et al., 2008). However, there are some limitations such chemical stability and issues of irreversible aggregation. This leads to the leakage of encapsulated bioactive molecules/compounds (Tachaprutinun et al., 2009; Reis et al., 2017; Guterres et al., 2009; Jafari et al., 2007a).

3.1.1.3 Drying techniques for producing nanoparticles

The dried form of organic encapsulation suspension of bioactive compounds is obtained using a combination of nanoencapsulation method and drying technique. The dried nanostructures (nanospheres and nanocapsules) have core shell nanosuspension where core is encapsulated bioactive molecules and shell provides higher stability for better control over release of these bioactive compounds as compared to the original nanosuspension. Recent studies showed that the use of spray drying techniques for effective encapsulations of several bioactive molecules. For example, limonene and fish oil are the bioactive compounds which have been encapsulated and delivered by using this technique (Jafari et al., 2008; Fathi et al., 2014; Mundargi et al., 2008). But the problem with this technique is that in some cases it induces additional stress on the nanocapsules thus affecting its stability (Fig. 15.3).

3.2 Design of nanocarriers by using bioactive molecules

Another domain of the research which gains huge attention from the scientific communities for last few decades is the use of bioactive molecules and its derivatives for designing of various kinds of nanoformulations. These nanoformulations are particularly suitable for synthesis of nanocarriers for drug, protein and gene delivery.

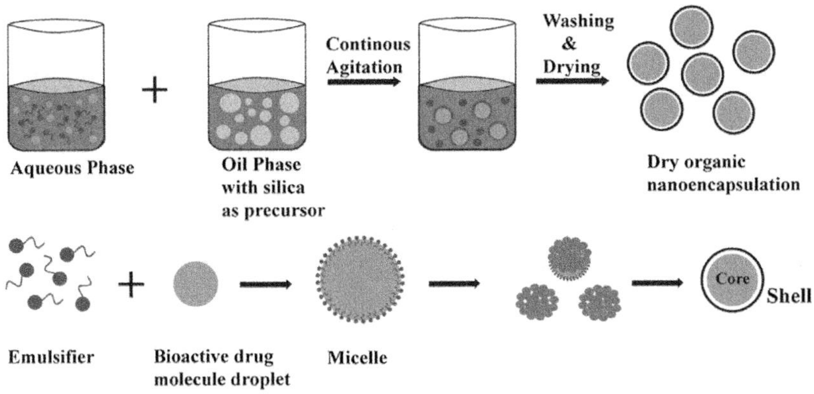

FIG. 15.3

Description of formation of organic encapsulation suspension of bioactive molecules using both nanoencapsulation and drying technique together.

The "smart" nanocarriers could be easily developed for targeted and controlled delivery of any therapeutic agents. Development of drug delivery cargos would be obtained by slight modification of the nanocarriers for the delivery of bioactive molecules as discussed earlier. Therefore, in the following write up we are mainly focus on the development of nanocarriers suitable for the protein and gene delivery.

3.2.1 Bioactive molecules based nanocarriers for protein delivery

Proteins are one of the vital components and found most abundantly in the biological cells. They are large, complex biomolecules and are required for the structure, function, and regulation of the body's tissues and organs. Thus, differential expression of protein leads to variety of human diseases. Most proteins itself falls in the category of bioactive compounds such as enzymes, growth factors and hormones. Efficient delivery of the proteins molecules will have significant impact on the treatment of several diseases. However, proteins are highly vulnerable to chemical degradation due to poor biological or physiochemical stability thus creating intricacies for its ultimate utility. Additionally, proteins do its function while it's in native 3D conformation. To cope up with these limitations nanocarriers are offer huge promised to deliver the proteins. Nanocarriers tend to minimize the change in the conformation of the proteins and/or proteins coated with derivatives of bioactive molecules while delivering to target site.

3.2.1.1 Current protein delivery methods and challenges

Presently, there are several protein delivery methods reported by various research groups, however, quite a few are showed promising *in-vivo* outcomes to understand the effectiveness of the methods. Hence, it's very much essential to evaluate all these methods based on standard performances to make it feasible for next level application. Some of the most prominent issues are given below.

- Maintaining the structural integrity (3D conformation) of proteins after delivery to the patient is a huge challenge as misfolded proteins lead to degradation that may cause several side-effects or biological malfunction.
- Maintaining the amount of proteins to be released to the target site with the intrinsic need of body at specific course of time is very crucial. Any discrepancy between these two events may lead to unavoidable side effects.
- There are several cases whereas the delivery methods may not be able to meet the essential clinical or commercial demands as due to its small-scale operation which is not feasible for pilot scale production.
- Frequently, it is also observed that the production of large is not favourable due to high manufacturing cost.

Nanoformulations can be used to manufacture novel systems for protein delivery using bioactive compounds from food using encapsulation technique. Nanocarriers provide several advantages like control release, high encapsulation efficiency and stability, highly soluble (Muller and Lucks, 1996) and the most important one is to improve the bioavailability of proteins (Fig. 15.4).

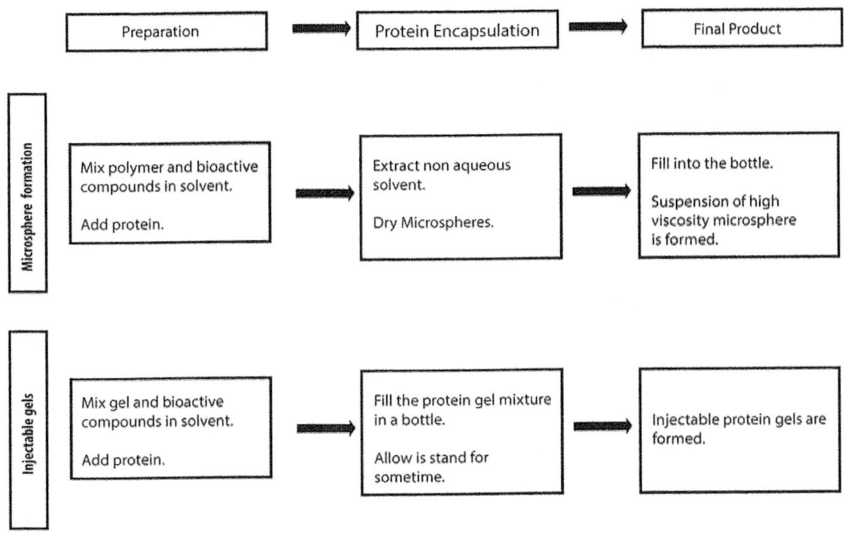

FIG. 15.4

The manufacturing processes for microspheres and injectable gel formulations for protein delivery.

Emulsification technique is also used to encapsulate proteins into the nanoparticles either by oil in water or water in oil techniques but they are prone to denaturation, oxidation and cleavage, thus reducing the biological activity of proteins. Adding bioactive compounds such as trehalose and mannitol during protein phase and some carrier proteins such as albumin during emulsion phase increases the integrity of the proteins. Other methods for increasing the integrity of proteins are encapsulating it as a solid rather the keeping in solution form (Eldridge et al., 1990). Some of the carrier nanostructures for efficient protein delivery are prepared using bioactive compounds either directly such as polysaccharides, lipids (micelles based), vitamin B12 or their derivatives like poly (lactic-*co*-glycolic acid) (PLGA), poly (lactic acid), poly (glycolic acid).

Some of the solid-lipid based bioactive compounds such as phytosterol, tocopherol, triacylglycerol and carotenoids are using developing nanostructured system for the delivery of proteins. Solid-lipid nanostructure is a novel delivery system to overcome the challenges faced by traditional carriers like emulsions, polymeric nanoparticles and micelles or liposomes (Zhang et al., 2000). Solvent emulsification evaporation method (water-oil-water double emulsion) was used to prepare solid lipid nanoparticles. The unique properties of these nanostructures are, it can be easily taken up by M-cells on Payer's patches of intestinal epithelium due to its hydrophobic surface. Thus, it helps proteins to diffuse from high concentration of solid lipid nanostructure matrix toward the gastrointestinal epithelium cells. Most commonly insulin proteins have been delivered using this lipid-based nanostructure (Bekerman et al., 2004; Komori, 2009).

3.3 Tissue engineering

Tissue engineering also known as regenerative medicine refers to create functional tissues from the cells in the laboratory. Its ultimate goal is to be a cure not merely a treatment by repairing and replacing the tissues or organs that fail due to disease, genetic errors, congenital abnormalities or traumatic injuries. Tissue engineering relies on four important factors such as

- **Specific cells:** Depending on the desired site of applications specific stem cell need to be seeded into the scaffold.
- **Scaffold material:** Once the seeded cells starts secreting their own extracellular matrix or performing natural function, the scaffolds material should be such that it is resorbed inside the body. So biodegradable material or materials derived from natural sources such as collagen, elastin etc. are preferred.
- **Suitable Scaffold and its architecture:** The physical and mechanical forces to ensure the proper growth and development of the cells. There several factors like porosity, relative density, and mechanical properties need to be optimized to support the cells and cell growth.
- **Proper microenvironment and growth factors:** To make those cells healthy and productive sufficient amount of growth factors has to be supplied.

The cells can be directly harvested from the target organs, precursors or stem cells or taken from the lines grown in the lab ideally from the patient, which limits the possibilities of the rejection. The supporting structures can be derived from the donor tissue or from some of the natural or synthetic polymers made to order their strength or endurance. The bioactive compounds are added directly or cokes from its cells to enhance the assembly of the scaffolds in the body. Sometimes scaffolds dissolve over time and others remains attached to the body to provide support to the organs. Some examples of tissues and organs that have been successfully tissue engineered and being implanted in humans that include bladders, skin graft, even a full trachea and cartilage.

3.3.1 Scaffold preparation or fabrication

The developed scaffolds should be bioactive for efficient cell adhesion, cell proliferation and differentiation. This tissue engineering scaffolds should substitute the extracellular matrix. The bioactive compounds in the scaffolds enhance the expression of cells or stem cells in their native environment and promote growth and development of tissues. On example in bone tissue engineering where bioactive compounds used for activating the signaling pathways and to stimulate the expression of Runx2, a transcription factor for bone regeneration by differentiation of mesenchymal stem cells (Komori, 2005; Choy et al., 2012).

3.3.2 Bioactive molecules for efficient tissue engineering application using nanotechnology

Nanotechnology in incorporated to engineer nanosized scaffolds using biomaterial and bioactive compounds to mimic the native tissues such as cardiac tissues, bone marrow etc. There are several nanofabrication techniques in tissue engineering employed to prepare controlled release nanoparticle-based scaffolds, nanofibers, and nanopatterns.

Collagen and glycosaminoglycans (GAGs) being abundant bioactive molecules inside the body, is used as biomaterials for developing natural function mimicking scaffolds. For better control over GAG binding collagens are modified by aminated, deaminated and methylated. This modified collagen is coprecipitated GAGs to form Collagen-GAG (CG) coprecipitate meshes of around 50 nm, which can be visualized under SEM after homogenization. The CG coprecipitates are photochemically cross linked using argon laser in presence of rose Bengal and then encapsulated as the core of collagen shell to from a biphasic scaffold (Xu et al., 2004) (Fig. 15.5).

Nanofabrication techniques for scaffold development are electrospun nanofibers, nanostructured substrate and self-assembled nanomaterials. Electrospunnan o fibers

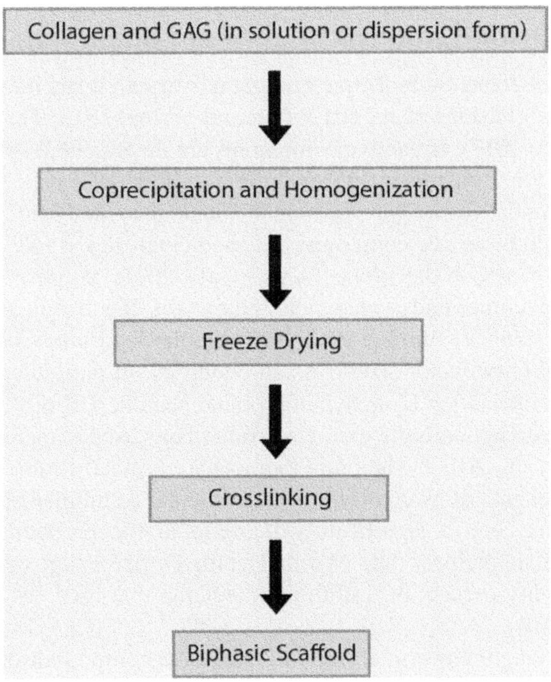

FIG. 15.5

Flow chart of collagen-GAG biphasic scaffold formation.

of poly (L-lactic-*co*-ε-caprolactone) are aligned to form blood vessel like structure which helps in guiding the cell orientation (Ma et al., 2005). Nanofibers of poly (lactic-*co*-glycolide) and poly (L-lactic acid) used to develop scaffold which can bio-mimic the cell adhesion function and its core-shell structure sometimes used for controlled release of bioactive molecules (Schindler et al., 2005).

Nanosurface of poly (lactic-*co*-glycolide) can also be treated with NaOH to increase cell density function. These nanostructures are formed using E-Beam lithography which helps in the improving cell adhesion and vascularisation. Improved adhesion of osteoclast cells using carbon nanotube based epithelial basement membrane is an example of nanostructure formed using E-Beam lithography for tissue engineering (Zhang et al., 2008). Another example of increased cell adhesion and higher rate of cell proliferation of mesenchymal stem cells in bone tissue engineering is achieved using scaffolds of hydroxyapatite/chitosan nanofibers (Lee et al., 2010). Due to the short half-life of bioactive compounds the scaffold from these compounds can help in efficient and organized bone regeneration. The only challenge needs to be overcome is the delivery of bioactive compounds without losing its stability and functionality.

4. Scope and limitation of bioactive compounds

Recently, the prime focus on bioactive compounds research has increased considerably all over the world (Puri et al., 2012; Puri and Verma, 2013; Thakur et al., 2017; Thakur and Verma, 2017). Bioactive compounds are documented from diverse sources including microorganisms, plants and animals (Banerjee et al., 2017). However, bioactive compounds from the medicinal plant provide unique platform for exploring the noble bioactive compounds of medicinal importance due to rich and varied biodiversity across the globe (Ares et al., 2018; Singh et al., 2017a). In view of bioactive compounds, medicinal plants are the chemical factories that capable to synthesize vast number of highly complex and unusual chemical compounds (Yuan and Grotewold, 2015). Asian countries, in particular to India, china and Japan, is repertories for bioactive compound search. Till now, more than two hundred bioactive compounds are extracted from plants used in medicine throughout the world (Luca et al., 2019). The plant extracts are typically subjected to pharmacological and biochemical evaluation to determine the suitability of bioactive compounds. There has been a significantly increase in the production of bioactive compounds-based medicines. The World Health Organization (WHO) estimates approximately eighty percent of traditional medicine that involves the use of plant extracts (Ekor, 2014).

There is a growing interest in correlating bioactive compounds of a plant with its pharmacological activity (Xi, 2017; Vardanega et al., 2014). Scientists have even started correlating the phytochemistry with their pharmacological activity (Wang et al., 2019). The multidisciplinary research aimed at correlating botanical and phyto-chemical properties to specific pharmacological activities is expected (Lele, 2010). Bioprospecting needs research and development of purified and biochemically characterized safe bioactive molecules from plants (Gu et al., 2014). The work done on rasayana group of plants is a good example of the above statement (Kuchewar et al., 2014).

However, there are very few studies involving evaluation of biochemically characterized immuno-modulating bioactive molecules. Any publication related to bioactive should ideally provide data on the authentication and standardization of the plant products. Not all the research areas selected by the scientists were relevant to needs of current research. Infectious diseases are some of the major problems that have not been extensively explored. Majority of the bioactive drugs have to still undergo clinical trials (Katiyar et al., 2012). Currently an active and strategic research has to be done to address the problem of adverse drug interactions (Hafner-Blumenstiel, 2011).

5. Conclusion and future directions

Nanotechnology is an effective tool for enhancing bioactivity of bioactive compounds like phytomedicine or food ingredients. To achieve the desired goal, the bioactivity of the product must be equipped with nanoformulation process, which is under the scope of nanotechnology. Bioactive compounds have served the foundation of the current strategy of nanoparticle researchers to increase the value addition of those compounds by means of their choice but bioactive compounds is always creating limitations due to their poor aqueous solubility, poor permeation, low systemic availability, and instability.

There has been a constant demand for novel natural bioactive compound for their improved quality of being bio-degradable, bio-compatible, readily available, renewable and low toxicity. Beyond categorizing such polysaccharides, DNA/RNA and proteins as natural bioactive compounds, research on other bioactive compounds like phytochemicals, food ingredients, is among the most advanced research area nowadays. One of the great awareness in the progress of nanomedicine in recent years related to the integration of therapy and diagnosis has drawn severe attention of the scientific community for nanotechnology-based products and clinical trials. Further newer techniques include synthetic polymer particles incorporation; liposome formulations; micellar nanoparticles creation; protein nanoparticles synthesis; nanocrystals and many others often in combination with drugs. Even though regulatory mechanisms for nanotechnology mediated improved bioactive compounds along with safety/toxicity assessments will be the subject of future development in the new directions. Nanotechnology has already revolutionized the way by which we have discovered and administered drugs in biological systems. This chapter have tried to discuss in detail about how nanotechnology can enhance the activity of the biological active molecules and how it can be further improved in coming years.

References

Achike, F.I., Kwan, C.Y., 2003. Nitric oxide, human diseases and the herbal products that affect the nitric oxide signalling pathway. Clin. Exp. Pharmacol. Physiol. 30 (9), 605–615.

Alvarez-Román, R., Naik, A., Kalia, Y., Guy, R.H., Fessi, H., 2004. Skin penetration and distribution of polymeric nanoparticles. J. Control. Release 99 (1), 53–62.

Anand, P., Nair, H.B., Sung, B., Kunnumakkara, A.B., Yadav, V.R., Tekmal, R.R., Aggarwal, B.B., 2010. Design of Curcumin-Loaded PLGA Nanoparticles Formulation with Enhanced Cellular Uptake, and Increased Bioactivity In Vitro and Superior Bioavailability In Vivo. Elsevier.

Ares, A.M., Valverde, S., Bernal, J.L., Nozal, M.J., Bernal, J., 2018. Extraction and determination of bioactive compounds from bee pollen. J. Pharm. Biomed. Anal. 147, 110–124.

Baker, K.M., Bauer, A.C., 2015. Green tea catechin, egcg, suppresses pcb 102-induced proliferation in estrogen-sensitive breast cancer cells. Int. J. Breast Cancer 2015, 7.

Bakour, S.H., Williamson, J., 2015. Latest evidence on using hormone replacement therapy in the menopause. Obstet. Gynaecol. 17 (1), 20–28.

Balunas, M.J., Kinghorn, A.D., 2005. Drug discovery from medicinal plants. Life Sci. 78 (5), 431–441.

Banerjee, J., Singh, R., Vijayaraghavan, R., MacFarlane, D., Patti, A.F., Arora, A., 2017. Bioactives from fruit processing wastes: green approaches to valuable chemicals. Food Chem. 225, 10–22.

Banerjee, T., Sulthana, S., Shelby, T., Heckert, B., Jewell, J., Woody, K., Karimnia, V., McAfee, J., Santra, S., 2016. Multiparametric magneto-fluorescent nanosensors for the ultrasensitive detection of *Escherichia coli* O157: H7. ACS Infect. Dis. 2 (10), 667–673.

Baron, S., Tyring, S.K., Fleischmann, W.R., Coppenhaver, D.H., Niesel, D.W., Klimpel, G.R., Stanton, G.J., Hughes, T.K., 1991. The interferons: mechanisms of action and clinical applications. JAMA 266 (10), 1375–1383.

Bartneck, M., Warzecha, K.T., Tacke, F., 2014. Therapeutic targeting of liver inflammation and fibrosis by nanomedicine. Hepatobiliary Surg. Nutr. 3 (6), 364.

Bekerman, T., Golenser, J., Domb, A., 2004. Cyclosporin nanoparticulate lipospheres for oral administration. J. Pharm. Sci. 93 (5), 1264–1270.

Belhaj, N., Arab-Tehrany, E., Linder, M., 2010. Oxidative kinetics of salmon oil in bulk and in nanoemulsion stabilized by marine lecithin. Process Biochem. 45 (2), 187–195.

Bengoechea, C., Jones, O.G., Guerrero, A., McClements, D.J., 2011. Formation and characterization of lactoferrin/pectin electrostatic complexes: impact of composition, pH and thermal treatment. Food Hydrocoll. 25 (5), 1227–1232.

Bernhoft, A., Siem, H., Bjertness, E., Meltzer, M., Flaten, T., Holmsen, E., 2010. Bioactive Compounds in Plants–Benefits and Risks for Man and Animals. The Norwegian Academy of Science and Letters, Oslo.

Berton-Carabin, C.C., Coupland, J.N., Elias, R.J., 2013. Effect of the lipophilicity of model ingredients on their location and reactivity in emulsions and solid lipid nanoparticles. Colloid. Surf. Physicochem. Eng. Asp. 431, 9–17.

Bhatia, A., Shard, P., Chopra, D., Mishra, T., 2011. Chitosan nanoparticles as carrier of immunorestoratory plant extract: synthesis, characterization and Immunorestoratory efficacy. Int. J. Drug Deliv. 3 (2), 381–385.

Bindhani, B.K., Panigrahi, A.K., 2015. Biosynthesis and characterization of silver nanoparticles (SNPs) by using leaf extracts of *Ocimum sanctum* L (Tulsi) and study of its antibacterial activities. J. Nanomed. Nanotechnol. (S6), 1.

Botelho, M.A., Queiroz, D.B., Barros, G., Guerreiro, S., Fechine, P., Umbelino, S., Lyra, A., Borges, B., Freitas, A., Queiroz, D.C.D., 2014. Nanostructured transdermal hormone replacement therapy for relieving menopausal symptoms: a confocal Raman spectroscopy study. Clinics 69 (2), 75−82.

Botelho, M., Queiroz, D., Freitas, A., Guerreiro, S., Umbelino, S., Barros, G., 2013. Effects of a new testosterone transdermal delivery system, Biolipid B2-testosterone in healthy middle aged men: a confocal Raman spectroscopy study. J. Pharm. Sci. Innov. 2 (2), 1−7.

Canonico, M., Oger, E., Plu-Bureau, G., Conard, J., Meyer, G., Lévesque, H., Trillot, N., Barrellier, M.-T., Wahl, D., Emmerich, J., 2007. Hormone therapy and venous thromboembolism among postmenopausal women: impact of the route of estrogen administration and progestogens: the ESTHER study. Circulation 115 (7), 840−845.

Cerqueira, M.A., Pinheiro, A.C., Silva, H.D., Ramos, P.E., Azevedo, M.A., Flores-López, M.L., Rivera, M.C., Bourbon, A.I., Ramos, O.L., Vicente, A.A., 2014. Design of bio-nanosystems for oral delivery of functional compounds. Food Eng. Rev. 6 (1−2), 1−19.

Chen, F., Chen, Y., Kang, X., Zhou, Z., Zhang, Z., Liu, D., 2012. Anti-apoptotic function and mechanism of ginseng saponins in Rattus pancreatic β-cells. Biol. Pharm. Bull. 35 (9), 1568−1573.

Chen, H.-W., Fang, Z.-S., Chen, Y.-T., Chen, Y.-I., Yao, B.-Y., Cheng, J.-Y., Chien, C.-Y., Chang, Y.-C., Hu, C.-M.J., 2017. Targeting and enrichment of viral pathogen by cell membrane cloaked magnetic nanoparticles for enhanced detection. ACS Appl. Mater. Interfaces 9 (46), 39953−39961.

Choi, M.J., Briancon, S., Andrieu, J., Min, S.G., Fessi, H., 2004. Effect of freeze-drying process conditions on the stability of nanoparticles. Dry. Technol. 22, 335−346.

Choy, A., Chan, B., Leong, K., 2012. Collagen-glycosaminoglycan coprecipitate and photochemically cross-linked collagen for disk replacement. Glob. Spine J. 2 (1_Suppl. 1) https://doi.org/10.1055/s-0032-1319996.

Conforti, F., Ioele, G., Statti, G.A., Marrelli, M., Ragno, G., et al., 2008. Antiproliferative activity against human tumor cell lines and toxicity test on Mediterranean dietary plants. Food Chem. Toxicol. 46, 3325−3332.

Correia, R.T., Borges, K.C., Medeiros, M.F., Genovese, M.I., 2012. Bioactive compounds and phenolic-linked functionality of powdered tropical fruit residues. Food Sci. Technol. Int. 18 (6), 539−547.

D'Mello, S., Finlay, G., Baguley, B., Askarian-Amiri, M., 2016. Signaling pathways in melanogenesis. Int. J. Mol. Sci. 17 (7), 1144.

Daly, E., Vessey, M.P., Hawkins, M.M., Carson, J.L., Gough, P., Marsh, S., 1996. Risk of venous thromboembolism in users of hormone replacement therapy. Lancet 348 (9033), 977−980.

Danhier, F., Ansorena, E., Silva, J.M., Coco, R., Le Breton, A., Préat, V., 2012. PLGA-based nanoparticles: an overview of biomedical applications. J. Control. Release 161 (2), 505−522.

Das, A., Kumar, A., Patil, N.B., Viswanathan, C., Ghosh, D., 2015. Preparation and characterization of silver nanoparticle loaded amorphous hydrogel of carboxymethylcellulose for infected wounds. Carbohydr. Polym. 130, 254−261.

De Kruif, C.G., Weinbreck, F., De Vries, R., 2004. Complex coacervation of proteins and anionic polysaccharides. Curr. Opin. Colloid Interface Sci. 9 (5), 340–349.
Ekor, M., 2014. The growing use of herbal medicines: issues relating to adverse reactions and challenges in monitoring safety. Front. Pharmacol. 4, 177.
Eldridge, J.H., Hammond, C.J., Meulbroek, J.A., Staas, J.K., Gilley, R.M., Tice, T.R., 1990. Controlled vaccine release in the gut-associated lymphoid tissues. I. Orally administered biodegradable microspheres target the Peyer's patches. J. Control. Release 11 (1–3), 205–214.
Emlen, W., Mannik, M., 1984. Effect of DNA size and strandedness on the in vivo clearance and organ localization of DNA. Clin. Exp. Immunol. 56 (1), 185.
Ezhilarasi, P., Karthik, P., Chhanwal, N., Anandharamakrishnan, C., 2013. Nanoencapsulation techniques for food bioactive components: a review. Food Bioprocess Technol. 6 (3), 628–647.
Farzaei, M.H., Abbasabadi, Z., Shams-Ardekani, M.R., Abdollahi, M., Rahimi, R., 2014. A comprehensive review of plants and their active constituents with wound healing activity in traditional Iranian medicine. Wounds 26 (7), 197–206.
Fathi, M., Martin, A., McClements, D.J., 2014. Nanoencapsulation of food ingredients using carbohydrate based delivery systems. Trends Food Sci. Technol. 39 (1), 18–39.
Fathi, S.S., Zaminy, A., 2017. Stem cell therapy for nerve injury. World J. Stem Cells 9 (9), 144.
Ferreira, I., Rocha, S., Coelho, M., 2007. Encapsulation of antioxidants by spray-drying. Chem. Eng. Trans. 11 (9), 713–717.
Fioretti, D., Iurescia, S., Fazio, V.M., Rinaldi, M., 2010. DNA vaccines: developing new strategies against cancer. BioMed Res. Int. 2010.
Folkman, S., 2013. 'Stress: appraisal and coping. In: Encyclopedia of Behavioral Medicine. Springer, pp. 1913–1915.
Fontana, A.R., Antoniolli, A., Bottini, R.N., 2013. Grape pomace as a sustainable source of bioactive compounds: extraction, characterization, and biotechnological applications of phenolics. J. Agric. Food Chem. 61 (38), 8987–9003.
Gamit, R., Nariya, M., Acharya, R., Shukla, V.J., 2017. Wound healing potential of some medicinal plants with their screening models: a review. Pharma Sci. Monit. 8 (1).
Gasperini, L., Mano, J.F., Reis, R.L., 2014. Natural polymers for the microencapsulation of cells. J. R. Soc. Interface 11 (100), 20140817.
Gloria, N.F., Soares, N., Brand, C., Oliveira, F.L., Borojevic, R., et al., 2014. Lycopene and beta-carotene induce cell-cycle arrest and apoptosis in human breast cancer cell lines. Anticancer Res. 34, 1377–1386.
Goddard, J.M., Hotchkiss, J., 2007. Polymer surface modification for the attachment of bioactive compounds. Prog. Polym. Sci. 32 (7), 698–725.
Godfrey, D.I., Hammond, K.J., Poulton, L.D., Smyth, M.J., Baxter, A.G., 2000. NKT cells: facts, functions and fallacies. Immunol. Today 21 (11), 573–583.
Gonzaga, L.W., Botelho, M.A., Queiroz, D.B., Fechine, P., Freire, R., Azevedo, E., Morais, A., Ruela, R., Lyra, A., Gomes, S., 2012. Nanotechnology in hormone replacement therapy: safe and efficacy of transdermal estriol and estradiol nanoparticles after 5 years follow-up study. Lat. Am. J. Pharm. 31 (3), 442–450.
Gottsäter, A., Rendell, M., Hulthen, U., Berntorp, E., Mattiasson, I., 2001. Hormone replacement therapy in healthy postmenopausal women: a randomized, placebo-controlled study of effects on coagulation and fibrinolytic factors. J. Intern. Med. 249 (3), 237–246.

Groeber, F., Holeiter, M., Hampel, M., Hinderer, S., Schenke-Layland, K., 2011. Skin tissue engineering—*in vivo* and *in vitro* applications. Adv. Drug Deliv. Rev. 63 (4–5), 352–366.

Gu, R., Wang, Y., Long, B., Kennelly, E., Wu, S., Liu, B., Li, P., Long, C., 2014. Prospecting for bioactive constituents from traditional medicinal plants through ethnobotanical approaches. Biol. Pharm. Bull. 37 (6), 903–915.

Guterres, S.S., Beck, R.C.R., Pohlmann, A.R., 2009. Spray-drying technique to prepare innovative nanoparticulated formulations for drug administration: a brief overview. Braz. J. Phys. 39 (1A), 205–209.

Hafner-Blumenstiel, V., 2011. Herbal drug-drug interaction and adverse drug reactions. Ther. Umsch. 68 (1), 54–57.

Hamdan, S., Pastar, I., Drakulich, S., Dikici, E., Tomic-Canic, M., Deo, S., Daunert, S., 2017. Nanotechnology-driven therapeutic interventions in wound healing: potential uses and applications. ACS Cent. Sci. 3 (3), 163–175.

Han, J.W., Gurunathan, S., Jeong, J.-K., Choi, Y.-J., Kwon, D.-N., Park, J.-K., Kim, J.-H., 2014. Oxidative stress mediated cytotoxicity of biologically synthesized silver nanoparticles in human lung epithelial adenocarcinoma cell line. Nanoscale Res. Lett. 9 (1), 459.

Hans, M., Lowman, A., 2002. Biodegradable nanoparticles for drug delivery and targeting. Curr. Opin. Solid State Mater. Sci. 6 (4), 319–327.

Hao, F., He, Y., Sun, Y., Zheng, B., Liu, Y., Wang, X., Zhang, Y., Lee, R.J., Teng, L., Xie, J., 2016. Improvement of oral availability of ginseng fruit saponins by a proliposome delivery system containing sodium deoxycholate. Saudi J. Biol. Sci. 23 (1), S113–S125.

Huang, Q., Yu, H., Ru, Q., 2010. Bioavailability and delivery of nutraceuticals using nanotechnology. J. Food Sci. 75 (1), R50–R57.

Jafari, S.M., Assadpoor, E., Bhandari, B., He, Y., 2008. Nano-particle encapsulation of fish oil by spray drying. Food Res. Int. 41 (2), 172–183.

Jafari, S.M., He, Y., Bhandari, B., 2007a. Encapsulation of nanoparticles of d-limonene by spray drying: role of emulsifiers and emulsifying techniques. Dry. Technol. 25 (6), 1069–1079.

Jafari, S.M., He, Y., Bhandari, B., 2007b. Production of submicron emulsions by ultrasound and microfluidization techniques. J. Food Eng. 82 (4), 478–488.

Jovanovski, E., Peeva, V., Sievenpiper, J.L., Jenkins, A.L., Desouza, L., Rahelic, D., Sung, M.K., Vuksan, V., 2014. Modulation of endothelial function by Korean red ginseng (Panax ginseng CA Meyer) and its components in healthy individuals: a randomized controlled trial. Cardiovasc. Ther. 32 (4), 163–169.

Kagan, H., 2006. The Psychological Immune System: A New Look at Protection and Survival. Author House.

Kanwar, S.S., Verma, M.L., 2010. Lipase. In: Encyclopedia of Industrial Biotechnology: Bioprocess, Bioseparation, and Cell Technology. John Wiley and Sons, Canada.

Kanwar, S.S., Gehlot, S., Verma, M.L., Gupta, R., Kumar, Y., Chauhan, G.S., 2008. Synthesis of geranyl butyrate with the poly (acrylic acid-co-hydroxy propyl methacrylate-cl-ethylene glycol dimethacrylate) hydrogel immobilized lipase of *Pseudomonas aeruginosa* MTCC-4713. J. Appl. Polym. Sci. 110 (5), 2681–2692.

Kanwar, S., Kaushal, R., Verma, M., Kumar, Y., Azmi, W., Gupta, R., Chimni, S., Chauhan, G., 2007a. Synthesis of ethyl oleate employing synthetic hydrogel-immobilized lipase of *Bacillus coagulans* MTCC-6375. Indian J. Biotechnol. 6, 68–73.

Kanwar, S., Kaushal, R., Verma, M., Kumar, Y., Chauhan, G., Gupta, R., Chimni, S., 2005. Synthesis of ethyl laurate by hydrogel immobilized lipase of *Bacillus coagulans* MTCC-6375. Indian J. Microbiol. 45 (3), 187–193.

Kanwar, S., Verma, H., Pathak, S., Kaushal, R., Kumar, Y., Verma, M., Chimni, S., Chauhan, G., 2006. Enhancement of ethyl propionate synthesis by poly (AAc-co-HPMA-cl-MBAm)-immobilized *Pseudomonas aeruginosa* MTCC-4713, exposed to Hg^{2+} and NH_4^+ ions. Acta Microbiol. Immunol. Hung. 53 (2), 195–207.

Kanwar, S.S., Verma, M.L., Maheshwari, C., Chauhan, S.S., Chimni, S.S., Chauhan, G.S., 2007b. Properties of poly (AAc-co-HPMA-cl-EGDMA) hydrogel-bound lipase of *Pseudomonas aeruginosa* MTCC-4713 and its use in synthesis of methyl acrylate. J. Appl. Polym. Sci. 104 (1), 183–191.

Karthick, V., Kumar, V.G., Dhas, T.S., Singaravelu, G., Sadiq, A.M., Govindaraju, K., 2014. Effect of biologically synthesized gold nanoparticles on alloxan-induced diabetic rats-an in vivo approach. Colloids Surf. B: Biointerfaces 122, 505–511.

Kasiewicz, L.N., Whitehead, K.A., 2016. Silencing TNFα with lipidoid nanoparticles downregulates both TNFα and MCP-1 in an *in vitro* co-culture model of diabetic foot ulcers. Acta Biomater. 32, 120–128.

Katiyar, C., Gupta, A., Kanjilal, S., Katiyar, S., 2012. Drug discovery from plant sources: an integrated approach. Ayu 33 (1), 10.

Khan, I., Saeed, K., Khan, I., 2017. Nanoparticles: properties, applications and toxicities. Arab. J. Chem. https://doi.org/10.1016/j.arabjc.2017.05.011.

Khater, M., de la Escosura-Muñiz, A., Quesada-González, D., Merkoçi, A., 2019. Electrochemical detection of plant virus using gold nanoparticle-modified electrodes. Anal. Chim. Acta 1046, 123–131.

Kim, H., Woo, D., Lee, G., Kim, J., 1998. The relaxation effects of ginseng saponin in rabbit corporal smooth muscle: is it a nitric oxide donor? Br. J. Urol. 82 (5), 744–748.

Kim, J., Ahn, H., Han, B.-C., Lee, S.-H., Cho, Y.-W., Kim, C.H., Hong, E.-J., An, B.-S., Jeung, E.-B., Lee, G.-S., 2014. Korean red ginseng extracts inhibit NLRP3 and AIM2 inflammasome activation. Immunol. Lett. 158 (1–2), 143–150.

Kim, K., Kim, H.Y., 2008. Korean red ginseng stimulates insulin release from isolated rat pancreatic islets. J. Ethnopharmacol. 120 (2), 190–195.

Kim, K.-H., Kim, S.-H., Yook, H.-S., 2015. Effect of flower-buds of *Panax ginseng* fermented by various microorganisms on the activation of T cell immune responses. Food Sci. Biotechnol. 24 (3), 1061–1067.

Komori, T., 2005. Regulation of skeletal development by the Runx family of transcription factors. J. Cell. Biochem. 95 (3), 445–453.

Komori, T., 2009. Regulation of osteoblast differentiation by Runx2. Osteoimmunology 43–49. Springer.

Kuchewar, V.V., Borkar, M.A., Nisargandha, M.A., 2014. Evaluation of antioxidant potential of Rasayana drugs in healthy human volunteers. Ayu 35 (1), 46.

Latz, E., Schoenemeyer, A., Visintin, A., Fitzgerald, K.A., Monks, B.G., Knetter, C.F., Lien, E., Nilsen, N.J., Espevik, T., Golenbock, D.T., 2004. TLR9 signals after translocating from the ER to CpG DNA in the lysosome. Nat. Immunol. 5 (2), 190.

Lee, J.S., Cho, M.K., Hwang, H.S., Ko, E.-J., Lee, Y.-N., Kwon, Y.-M., Kim, M.-C., Kim, K.-H., Lee, Y.-T., Jung, Y.-J., 2014. Ginseng diminishes lung disease in mice immunized with formalin-inactivated respiratory syncytial virus after challenge by modulating host immune responses. J. Interferon Cytokine Res. 34 (11), 902–914.

Lee, J.S., Lee, Y.-N., Lee, Y.-T., Hwang, H.S., Kim, K.-H., Ko, E.-J., Kim, M.-C., Kang, S.-M., 2015. Ginseng protects against respiratory syncytial virus by modulating multiple immune cells and inhibiting viral replication. Nutrients 7 (2), 1021–1036.

Lee, K., Silva, E.A., Mooney, D.J., 2010. Growth factor delivery-based tissue engineering: general approaches and a review of recent developments. J. R. Soc. Interface 8 (55), 153–170.

Lele, R., 2010. Beyond reverse pharmacology: mechanism-based screening of ayurvedic drugs. J. Ayurveda Integr. Med. 1 (4), 257.

Li, Y., Zheng, J., Xiao, H., McClements, D.J., 2012. Nanoemulsion-based delivery systems for poorly water-soluble bioactive compounds: influence of formulation parameters on polymethoxyflavone crystallization. Food Hydrocoll. 27 (2), 517–528.

Liao, N., Unnithan, A.R., Joshi, M.K., Tiwari, A.P., Hong, S.T., Park, C.-H., Kim, C.S., 2015. Electrospun bioactive poly (ε-caprolactone)–cellulose acetate–dextran antibacterial composite mats for wound dressing applications. Colloid. Surf. Physicochem. Eng. Asp. 469, 194–201.

Lim, C.-Y., Moon, J.-M., Kim, B.-Y., Lim, S.-H., Lee, G.-S., Yu, H.-S., Cho, S.-I., 2015. Comparative study of Korean White Ginseng and Korean Red Ginseng on efficacies of OVA-induced asthma model in mice. J. Ginseng Res. 39 (1), 38–45.

Littarru, G.P., Tiano, L., 2007. Bioenergetic and antioxidant properties of coenzyme Q 10: recent developments. Mol. Biotechnol. 37 (1), 31–37.

Lü, J.M., Lin, P.H., Yao, Q., Chen, C., 2010. Chemical and molecular mechanisms of antioxidants: experimental approaches and model systems. J. Cell Mol. Med. 14 (4), 840–860.

Luca, S.V., Macovei, I., Bujor, A., Miron, A., Skalicka-Woźniak, K., Aprotosoaie, A.C., Trifan, A., 2019. Bioactivity of dietary polyphenols: the role of metabolites. Crit. Rev. Food Sci. Nutr. 1–34.

Ma, J., Liu, H., Wang, X., 2014. Effect of ginseng polysaccharides and dendritic cells on the balance of Th1/Th2 T helper cells in patients with non-small cell lung cancer. J. Tradit. Chin. Med. 34 (6), 641–645.

Ma, Z., Kotaki, M., Inai, R., Ramakrishna, S., 2005. Potential of nanofiber matrix as tissue-engineering scaffolds. Tissue Eng. 11 (1–2), 101–109.

Mangla, B., Kohli, K., 2018. Combination of natural agent with synthetic drug for the breast cancer therapy. Int. J. Drug Dev. Res. 10 (1).

Mano, J., Silva, G., Azevedo, H.S., Malafaya, P., Sousa, R., Silva, S.S., Boesel, L., Oliveira, J.M., Santos, T., Marques, A., 2007. Natural origin biodegradable systems in tissue engineering and regenerative medicine: present status and some moving trends. J. R. Soc. Interface 4 (17), 999–1030.

McClements, D., Decker, E., Weiss, J., 2007. Emulsion-based delivery systems for lipophilic bioactive components. J. Food Sci. 72 (8), R109–R124.

Melnyk, J.P., Marcone, M.F., 2011. Aphrodisiacs from plant and animal sources—a review of current scientific literature. Food Res. Int. 44 (4), 840–850.

Miladi, K., Sfar, S., Fessi, H., Elaissari, A., 2016. Nanoprecipitation process: from particle preparation to in vivo applications. In: Polymer Nanoparticles for Nanomedicines. Springer, pp. 17–53.

Moghimi, S.M., Hunter, A.C., Murray, J.C., 2005. Nanomedicine: current status and future prospects. FASEB J. 19 (3), 311–330.

Mohamadi, E., Moghaddasi, M., Farahbakhsh, A., Kazemi, A., 2017. A quantum-dot-based fluoroassay for detection of food-borne pathogens. J. Photochem. Photobiol. B Biol. 174, 291–297.

Mordorski, B., Prow, T., 2016. Nanomaterials for wound healing. Curr. Dermatol. Rep. 5 (4), 278–286.

Mukherjee, P.K., Nema, N.K., Bhadra, S., Mukherjee, D., Braga, F.C., Matsabisa, M.G., 2014. Immunomodulatory leads from medicinal plants. Int. J. Tradit. Knowl. 13, 235–256.

Muller, R., Lucks, J., 1996. Azneistoffträger aus festen Lipidteilchen feste Lipid Nanosphären (SLN) European Patent 0605497. CIT0052.

Mundargi, R.C., Babu, V.R., Rangaswamy, V., Patel, P., Aminabhavi, T.M., 2008. Nano/micro technologies for delivering macromolecular therapeutics using poly (D,L-lactide-co-glycolide) and its derivatives. J. Control. Release 125 (3), 193–209.

Muniandy, S., Dinshaw, I.J., Teh, S.J., Lai, C.W., Ibrahim, F., Thong, K.L., Leo, B.F., 2017. Graphene-based label-free electrochemical aptasensor for rapid and sensitive detection of foodborne pathogen. Anal. Bioanal. Chem. 409 (29), 6893–6905.

Neave, N., 2007. Hormones and Behaviour: A Psychological Approach. Cambridge University Press.

Niu, X., Chen, J., Gao, J., 2018. Nanocarriers as a powerful vehicle to overcome blood-brain barrier in treating neurodegenerative diseases: focus on recent advances. Asian J. Pharm. Sci. https://doi.org/10.1016/j.ajps.2018.09.005.

Pandian, C.J., Palanivel, R., Balasundaram, U., 2017. Green synthesized nickel nanoparticles for targeted detection and killing of *S. typhimurium*. J. Photochem. Photobiol. B Biol. 174, 58–69.

Pereira, R.F., Bartolo, P.J., 2016. Traditional therapies for skin wound healing. Adv. Wound Care 5 (5), 208–229.

Prabhu, S., Poulose, E.K., 2012. Silver nanoparticles: mechanism of antimicrobial action, synthesis, medical applications, and toxicity effects. Int. Nano Lett. 2 (1), 32.

Puri, M., Verma, M.L., 2013. Innovation in technology development with reference to enzymatic extraction of flavonoids. In: Ghosh, D., Das, S., Bagchi, D., Smarta, R.B. (Eds.), Innovation in Healthy and Functional Foods. CRC Press, Boca Raton, Fl, pp. 331–342.

Puri, M., Barrow, C.J., Verma, M.L., 2013. Enzyme immobilization on nanomaterials for biofuel production. Trends Biotechnol. 31 (4), 215–216.

Puri, M., Verma, M.L., Mahale, K., 2012. Processing of citrus peel for the extraction of flavonoids for biotechnological applications. In: Handbook on Flavonoids: Dietary Sources, Properties and Health Benefits. Biochemistry Research Trends: Nutrition and Diet Research Progress. Nova Science Publishers, Inc., Hauppauge, New York, pp. 443–459.

Qi, L.-W., Wang, C.-Z., Yuan, C.-S., 2011. Ginsenosides from American ginseng: chemical and pharmacological diversity. Phytochemistry 72 (8), 689–699.

Rahman, S., Carter, P., Bhattarai, N., 2017. *Aloe vera* for tissue engineering applications. J. Funct. Biomater. 8 (1), 6.

Randeria, P.S., Seeger, M.A., Wang, X.-Q., Wilson, H., Shipp, D., Mirkin, C.A., Paller, A.S., 2015. siRNA-based spherical nucleic acids reverse impaired wound healing in diabetic mice by ganglioside GM3 synthase knockdown. Proc. Natl. Acad. Sci. USA 112 (18), 5573–5578.

Reinke, J., Sorg, H., 2012. Wound repair and regeneration. Eur. Surg. Res. 49 (1), 35–43.

Reis, C.P., Neufeld, R.J., Veiga, F., 2017. Preparation of Drug-Loaded Polymeric Nanoparticles. In: Nanomedicine in Cancer. Pan Stanford, pp. 197–240.

Reis, C.P., Neufeld, R.J., Ribeiro, A.J., Veiga, F., 2006. Nanoencapsulation I. Methods for preparation of drug-loaded polymeric nanoparticles. Nanomed. Nanotechnol. Biol. Med. 2, 8–21.

Ribeiro, H.S., Chu, B.-S., Ichikawa, S., Nakajima, M., 2008. Preparation of nanodispersions containing β-carotene by solvent displacement method. Food Hydrocoll. 22 (1), 12–17.

Rosano, G.M., Vitale, C., Silvestri, A., Fini, M., 2003. Hormone replacement therapy and cardioprotection: the end of the tale? Ann. N. Y. Acad. Sci. 997 (1), 351–357.
Salazar-Onfray, F., Lopez, M., Lundqvist, A., Aguirre, A., Escobar, A., Serrano, A., Korenblit, C., Petersson, M., Chhajlani, V., Larsson, O., 2002. Tissue distribution and differential expression of melanocortin 1 receptor, a malignant melanoma marker. Br. J. Cancer 87 (4), 414.
Sands, M.S., Vogler, C., Kyle, J.W., Grubb, J.H., Levy, B., Galvin, N., Sly, W.S., Birkenmeier, E.H., 1994. Enzyme replacement therapy for murine mucopolysaccharidosis type VII. J. Clin. Investig. 93 (6), 2324–2331.
Schindler, M., Ahmed, I., Kamal, J., Nur-E-Kamal, A., Grafe, T.H., Chung, H.Y., Meiners, S., 2005. A synthetic nanofibrillar matrix promotes in vivo-like organization and morphogenesis for cells in culture. Biomaterials 26 (28), 5624–5631.
Sharma, N., Mishra, S., Sharma, S., Deshpande, R.D., Sharma, R.K., 2013. Preparation and optimization of nanoemulsions for targeting drug delivery. Int. J. Drug Dev. Res. 5 (4), 37–48.
Shim, Y., Song, J.M., 2015. Quantum dot nanoprobe-based high-content monitoring of notch pathway inhibition of breast cancer stem cell by capsaicin. Mol. Cell Probes 29, 376–381.
Shrivastava, S., Dash, D., 2009. Applying nanotechnology to human health: revolution in biomedical sciences. J. Nanotechnol. 2009.
Singh, M., Kumar, A., Singh, R., Pandey, K.D., 2017a. Endophytic bacteria: a new source of bioactive compounds. 3 Biotech 7 (5), 315.
Singh, R., Lillard Jr., J.W., 2009. Nanoparticle-based targeted drug delivery. Exp. Mol. Pathol. 86 (3), 215–223.
Singh, Y., Meher, J.G., Raval, K., Khan, F.A., Chaurasia, M., Jain, N.K., Chourasia, M.K., 2017b. Nanoemulsion: concepts, development and applications in drug delivery. J. Control. Release 252, 28–49.
Skrovankova, S., Sumczynski, D., Mlcek, J., Jurikova, T., Sochor, J., 2015. Bioactive compounds and antioxidant activity in different types of berries. Int. J. Mol. Sci. 16 (10), 24673–24706.
Smolarczyk, R., Cichoń, T., Sochanik, A., Szala, S., 2005. Negligible induction of IFN-γ, IL-12 and TNF-α by DNA-PEI 750 kDa/albumin complexes. Cytokine 29 (6), 283–287.
Song, X., Zang, L., Hu, S., 2009. Amplified immune response by ginsenoside-based nanoparticles (ginsomes). Vaccine 27 (17), 2306–2311.
Soobrattee, M.A., Neergheen, V.S., Luximon-Ramma, A., Aruoma, O.I., Bahorun, T., 2005. Phenolics as potential antioxidant therapeutic agents: mechanism and actions. Mutat. Res. Fund Mol. Mech. Mutagen 579 (1), 200–213.
Sun, T., Zhan, B., Zhang, W., Qin, D., Xia, G., Zhang, H., Peng, M., Li, S.-A., Zhang, Y., Gao, Y., 2018. Carboxymethyl chitosan nanoparticles loaded with bioactive peptide OH-CATH30 benefit nonscar wound healing. Int. J. Nanomed. 13, 5771.
Szekacs, B., Vajo, Z., Acs, N., Hada, P., Csuzi, L., Bezeredi, J., Magyar, Z., Brinton, E.A., 2000. Hormone replacement therapy reduces mean 24-hour blood pressure and its variability in postmenopausal women with treated hypertension. Menopause 7 (1), 31–35.
Tachaprutinun, A., Udomsup, T., Luadthong, C., Wanichwecharungruang, S., 2009. Preventing the thermal degradation of astaxanthin through nanoencapsulation. Int. J. Pharm. 374 (1–2), 119–124.
Tahir, M.A., Hameed, S., Munawar, A., Amin, I., Mansoor, S., Khan, W.S., Bajwa, S.Z., 2017. Investigating the potential of multiwalled carbon nanotubes based zinc nanocomposite as a recognition interface towards plant pathogen detection. J. Virol Methods 249, 130–136.

Tambi, M., Imran, M., Henkel, R., 2012. Standardised water-soluble extract of Eurycoma longifolia, Tongkat ali, as testosterone booster for managing men with late-onset hypogonadism? Andrologia 44, 226–230.

Tao, R., Coleman, M.C., Pennington, J.D., Ozden, O., Park, S.-H., Jiang, H., Kim, H.-S., Flynn, C.R., Hill, S., McDonald, W.H., 2010. Sirt3-mediated deacetylation of evolutionarily conserved lysine 122 regulates MnSOD activity in response to stress. Mol. Cell 40 (6), 893–904.

Teede, H.J., McGrath, B.P., Smolich, J.J., Malan, E., Kotsopoulos, D., Liang, Y.-L., Peverill, R.E., 2000. Postmenopausal hormone replacement therapy increases coagulation activity and fibrinolysis. Arterioscler. Thromb. Vasc. Biol. 20 (5), 1404–1409.

Thakur, R., Kumar, S., Awasthi, C.P., Verma, M.L., 2017. Biochemical evaluation of tartary buckwheat (*Fagopyrum tataricum* Gaertn.) genotypes of cold desert of Himachal Pradesh. Biosci. Biotechnol. Res. Asia 14 (1), 1–16.

Thakur, R., Verma, M.L., 2017. Food Bioactives with Special Reference to Himalayan Tartary Buckwheat. LAMBERT Academic Publishing, Saarbrücken, Germany.

Tsubura, A., Lai, Y.C., Kuwata, M., Uehara, N., Yoshizawa, K., 2011. Anticancer effects of garlic and garlic-derived compounds for breast cancer control. Anti Cancer Agents Med. Chem. 11, 249–253.

Ulvestad, E., 2007. Defending Life: The Nature of Host-Parasite Relations. Springer Science & Business Media.

Upadhyay, R.K., 2016. Antidiabetic potential of plant natural products: a review. Int. J. Green Pharm. 10 (3).

Vanacker, S.A., Tromp, M.N., Haenen, G.R., Vandervijgh, W., Bast, A., 1995. Flavonoids as scavengers of nitric oxide radical. Biochem. Biophys. Res. Commun. 214 (3), 755–759.

Vardanega, R., Santos, D.T., Meireles, M.A.A., 2014. Intensification of bioactive compounds extraction from medicinal plants using ultrasonic irradiation. Phcog. Rev. 8 (16), 88.

Vatankhah, E., Prabhakaran, M.P., Jin, G., Mobarakeh, L.G., Ramakrishna, S., 2014. Development of nanofibrous cellulose acetate/gelatin skin substitutes for variety wound treatment applications. J. Biomater. Appl. 28 (6), 909–921.

Vélez, M.A., Perotti, M.C., Santiago, L., Gennaro, A.M., Hynes, E., 2017. Bioactive compounds delivery using nanotechnology: design and applications in dairy food. Nutr. Deliv. 221–250. Elsevier.

Verma, M.L., Kanwar, S.S., 2010. Purification and characterization of a low molecular mass alkaliphilic lipase of Bacillus cereus MTCC 8372. Acta Microbiol. Immunol. Hung. 57 (3), 191–207.

Verma, M., Kanwar, S., 2012. Harnessing the potential of thermophiles: the variants of extremophiles. Dyn. Biochem. Process Biotechnolol. Mol. Biol. 6 (1), 28–39.

Verma, M.L., 2017a. Enzymatic nanobiosensors in the agricultural and food industry. In: Nanoscience in Food and Agriculture 4. Springer, pp. 229–245.

Verma, M.L., 2017b. Nanobiotechnology advances in enzymatic biosensors for the agri-food industry. Environ. Chem. Lett. 15 (4), 555–560.

Verma, M.L., 2017c. Fungus-mediated bioleaching of metallic nanoparticles from agro-industrial by-products. In: Prasad, R. (Ed.), Fungal Nanotechnology. Fungal Biology, Springer, Cham. https://doi.org/10.1007/978-3-319-68424-6_5.

Verma, M.L., Barrow, C.J., 2015. 'Recent advances in feedstocks and enzyme-immobilised technology for effective transesterification of lipids into biodiesel. In: Microbial Factories. Springer, pp. 87–103.

Verma, M.L., Kanwar, S.S., 2008. Properties and application of poly (methacrylic acid-co-dodecyl methacrylate-cl-N, N-methylene bisacrylamide) hydrogel immobilized *Bacillus cereus* MTCC 8372 lipase for the synthesis of geranyl acetate. J. Appl. Polym. Sci. 110 (2), 837–846.

Verma, M.L., Azmi, W., Kanwar, S.S., 2011. Enzymatic synthesis of isopropyl acetate by immobilized *Bacillus cereus* lipase in organic medium. Enzym. Res. 2011.

Verma, M.L., Barrow, C.J., Kennedy, J., Puri, M., 2012. Immobilization of β-d-galactosidase from *Kluyveromyces lactis* on functionalized silicon dioxide nanoparticles: characterization and lactose hydrolysis. Int. J. Biol. Macromol. 50 (2), 432–437.

Verma, M.L., Chaudhary, R., Tsuzuki, T., Barrow, C.J., Puri, M., 2013a. Immobilization of β-glucosidase on a magnetic nanoparticle improves thermostability: application in cellobiose hydrolysis. Bioresour. Technol. 135, 2–6.

Verma, M.L., Naebe, M., Barrow, C.J., Puri, M., 2013b. Enzyme immobilisation on amino-functionalised multi-walled carbon nanotubes: structural and biocatalytic characterisation. PLoS One 8 (9), e73642.

Verma, M.L., Puri, M., Barrow, C.J., 2016. Recent trends in nanomaterials immobilised enzymes for biofuel production. Crit. Rev. Biotechnol. 36 (1), 108–119.

Verma, M., Azmi, W., Kanwar, S., 2008a. Microbial lipases: at the interface of aqueous and non-aqueous media: a review. Acta Microbiol. Immunol. Hung. 55 (3), 265–294.

Verma, M., Azmi, W., Kanwar, S., 2009. Synthesis of ethyl acetate employing celite-immobilized lipase of *Bacillus cereus* MTCC 8372. Acta Microbiol. Immunol. Hung. 56 (3), 229–242.

Verma, M., Chauhan, G., Kanwar, S., 2008b. Enzymatic synthesis of isopropyl myristate using immobilized lipase from *Bacillus cereus* MTCC 8372. Acta Microbiol. Immunol. Hung. 55 (3), 327–342.

Wang, J., Xu, J., Gong, X., Yang, M., Zhang, C., Li, M., 2019. Biosynthesis, chemistry, and pharmacology of polyphenols from Chinese salvia species: a review. Molecules 24 (1), 155.

Wang, X., Jiang, Y., Wang, Y.-W., Huang, M.-T., Ho, C.-T., Huang, Q., 2008. Enhancing anti-inflammation activity of curcumin through O/W nanoemulsions. Food Chem. 108 (2), 419–424.

Weiner, G.J., Liu, H.-M., Wooldridge, J.E., Dahle, C.E., Krieg, A.M., 1997. Immunostimulatory oligodeoxynucleotides containing the CpG motif are effective as immune adjuvants in tumor antigen immunization. Proc. Natl. Acad. Sci. USA 94 (20), 10833–10837.

Wellons, M., Ouyang, P., Schreiner, P.J., Herrington, D.M., Vaidya, D., 2012. Early menopause predicts future coronary heart disease and stroke: the multi-ethnic study of atherosclerosis (MESA). Menopause 19 (10), 1081.

Wen, L., Chase, C.D., 1999. Pleiotropic effects of a nuclear restorer-of-fertility locus on mitochondrial transcripts in male-fertile and S male-sterile maize. Curr. Genet. 35 (5), 521–526.

Wong, W., Minchin, R.F., 1996. Binding and internalization of the melanocyte stimulating hormone receptor ligand [Nle4, D-Phe7] α-MSH in B16 melanoma cells. Int. J. Biochem. Cell Biol. 28 (11), 1223–1232.

Xi, J., 2017. Ultrahigh pressure extraction of bioactive compounds from plants-a review. Crit. Rev. Food Sci. Nutr. 57 (6), 1097–1106.

Xu, C., Inai, R., Kotaki, M., Ramakrishna, S., 2004. Aligned biodegradable nanofibrous structure: a potential scaffold for blood vessel engineering. Biomaterials 25 (5), 877–886.

Yan, Y., Kolachala, V., Dalmasso, G., Nguyen, H., Laroui, H., Sitaraman, S.V., Merlin, D., 2009. Temporal and spatial analysis of clinical and molecular parameters in dextran sodium sulfate induced colitis. PLoS One 4 (6), e6073.

Yu, J., Chen, Y., Zhai, L., Zhang, L., Xu, Y., Wang, S., Hu, S., 2015. Antioxidative effect of ginseng stem-leaf saponins on oxidative stress induced by cyclophosphamide in chickens. Poultry Sci. 94 (5), 927–933.

Yuan, H.D., Quan, H.Y., Jung, M.S., Kim, S.J., Huang, B., Kim, D.Y., Chung, S.H., 2011. Anti-diabetic effect of pectinase-processed ginseng radix (GINST) in high fat diet-fed ICR mice. J. Ginseng Res. 35 (3), 308.

Yuan, H.-D., Kim, J.T., Kim, S.H., Chung, S.H., 2012. Ginseng and diabetes: the evidences from in vitro, animal and human studies. J. Ginseng Res. 36 (1), 27.

Yuan, L., Grotewold, E., 2015. Metabolic engineering to enhance the value of plants as green factories. Metab. Eng. 27, 83–91.

Yuan, Y., Gao, Y., Mao, L., Zhao, J., 2008. Optimisation of conditions for the preparation of β-carotene nanoemulsions using response surface methodology. Food Chem. 107 (3), 1300–1306.

Zhang, D.-F., Xu, H., Sun, B.-B., Li, J.-Q., Zhou, Q.-J., Zhang, H.-L., Du, A.-F., 2012. Adjuvant effect of ginsenoside-based nanoparticles (ginsomes) on the recombinant vaccine against *Eimeria tenella* in chickens. Parasitol. Res. 110 (6), 2445–2453.

Zhang, Q., Yie, G., Li, Y., Yang, Q., Nagai, T., 2000. Studies on the cyclosporin A loaded stearic acid nanoparticles. Int. J. Pharm. 200 (2), 153–159.

Zhang, Y., Venugopal, J.R., El-Turki, A., Ramakrishna, S., Su, B., Lim, C.T., 2008. Electrospun biomimetic nanocomposite nanofibers of hydroxyapatite/chitosan for bone tissue engineering. Biomaterials 29 (32), 4314–4322.

Zhu, J.-S., Halpern, G.M., Jones, K., 1998. The scientific rediscovery of an ancient Chinese herbal medicine: *Cordyceps sinensis* Part I. J. Altern. Complement. Med. 4 (3), 289–303.

Index

'*Note:* Page numbers followed by "f" indicate figures, "t" indicate tables.'

B

F

O

Q

R

S